D0936465

# A TALE OF
# TWO PLANTATIONS

# A TALE OF TWO PLANTATIONS

## *Slave Life and Labor in Jamaica and Virginia*

RICHARD S. DUNN

Harvard University Press

Cambridge, Massachusetts
London, England
2014

Printed in the United States of America

First Printing

Library of Congress Cataloging-in-Publication Data is available
from the Library of Congress
Library of Congress Cataloging-in-Publication Control Number: 2014014075
ISBN 978-0-674-73536-1

*To my family*

*Mary*

*Rebecca and Andrea* — *Cecilia and Lee*

*Cady B* — *Benjamin and Frederic*

# *Contents*



Biographical information about 430 members of seven multigenerational slave families at Mesopotamia and Mount Airy is digitally displayed on an accompanying website, *www.twoplantations.com.*

# *Appendixes*

# A TALE OF TWO PLANTATIONS

# *Prologue*

THIS BOOK HAS TWO GOALS. First, it reconstructs the individual lives and collective experiences of some 2,000 slaves on two large plantations—Mesopotamia sugar estate in western Jamaica and Mount Airy plantation in Tidewater Virginia—during the final three generations of slavery in both places. Second, it highlights the chief differences between Afro-Caribbean slave life in the British West Indies and African American slave life in the antebellum U.S. South by comparing Mesopotamia with Mount Airy. I have spent forty years working on this project, which is not a recommended *modus operandi* for historians. This long gestation period reflects the difficulty in bringing to life some very hidden people, in articulating the experiences of two quite different communities of slaves, and in situating Mesopotamia and Mount Airy within a meaningful larger framework. While I have by no means fully accomplished these aims, I do believe that I have found an effective approach to a central historical problem: the multifaceted character of racial slavery in America.

Anyone trying to study slave life faces the challenge that almost all of the surviving evidence was written down and preserved by the slaveholders, which is the case with Mesopotamia and Mount Airy. These slaveholders were compiling business records about their property, so they listed their slaves in plantation ledgers mixed in with accounts of their livestock, crop yields, and so on. But some ledgers are more informative than others. Back in the 1970s I encountered two sets of exceptionally detailed slave records. The Barham Papers in the Bodleian Library at the University of Oxford make it possible to recover the individual lives of 1,103 people on Mesopotamia estate between 1762 and 1833, and the Tayloe Papers at the Virginia Historical Society in Richmond similarly open up the lives of 973 people on Mount Airy plantation between 1808 and 1865. The Barham and Tayloe Papers also document the demographic pattern, the labor pattern, the role of motherhood, family formation, and racial mixing in both places. They illuminate the process by which replacement slaves were moved into Mesopotamia, and surplus slaves were moved out of Mount Airy. They show how the Mesopotamia people reacted to slave rebellions in western Jamaica and how the Mount Airy people responded to the U.S. Civil War. Most important, these plantation records portray two intergenerational communities in action over many years, tracing the numerous followers as well as the much smaller number of leaders. This, I believe, gives a more representative view of slave life than can be obtained by focusing (as many historians of slavery are inclined to do) on the most visible people, those who ran away, or wrote about themselves, or were in other ways remarkable.

Turning to my second goal, historians agree that there were big differences between slavery in the Caribbean islands and the North American mainland. There were differences in the number of slaves imported, in slave demographics, in the ratio of blacks to whites, in the distribution of blacks among white owners, and in slave labor conditions, slave living conditions, and slave resistance and rebellion. Indeed, the two slave systems seem to have so little in common that relatively few historians have tried to study them in tandem. This, I think, is unfortunate. To my mind, the best way to understand the most distinctive features of the slave experience in the British Caribbean as well as the most distinctive features of the slave experience in the antebellum U.S. South is by focusing on the major contrasts between the two systems. This enables us to see more clearly how slavery caused terrible suffering to the black people in both regions, but in strikingly dissimilar ways.

By choosing to compare slave life in one Jamaican slave community with slave life in one Virginia slave community, one faces the obvious hazard that Mesopotamia is not typical of the entire Caribbean slave system, nor is Mount Airy representative of the whole U.S. slave system. Of course they are not. Yet Mesopotamia and Mount Airy *are* entirely representative on the most fundamental point: the demographic difference between the two slave systems. Mesopotamia had a continually shrinking slave population, with many more deaths than births, which was standard on Jamaican sugar estates and elsewhere throughout the British Caribbean. To keep this plantation operable, the Barhams had to bring in many new workers, some from the African slave ships and some from local plantations in Jamaica. The slaves were thrown together in an environment in which ill health and death dictated the course of events. Mount Airy by contrast had a continually expanding slave population, with many more births than deaths, which was standard everywhere in Virginia and generally throughout the Old South. Here the Tayloes were constantly moving their surplus laborers to distant work sites or were selling them to new owners. The slaves lived in an environment made familiar to us by *Uncle Tom's Cabin,* in which normal social relations were impossible because families were routinely ripped apart.

Before examining these two slave plantations, we need to visualize the larger framework within which they operated—the dual slave systems established in the British West Indies from about 1640 to 1834, and on the North American mainland from about 1680 to 1865. The first point to notice is that the island colonists quickly received many slaves from Africa and plunged into slaveholding, while the mainland colonists entered the process more slowly. By 1680 there were nearly 40,000 black slaves in Barbados as against 3,000 in Virginia. By 1700 over 310,000 Africans had been imported to the British West Indies and only about 15,000 to British North America. During the early eighteenth century the southern mainland colonists fully embraced black slavery, but the Atlantic slave traders continued to focus primarily on the Caribbean. Some 850,000 Africans were landed in the British islands by 1750, and 2.3 million by the close of the British slave trade in 1807, compared with 150,000 Africans sold to the mainland colonists by 1750 and 388,000 by 1807, when the U.S. government also terminated the African trade. However, the slaves who were brought to the Caribbean died off and were replaced by new Africans, while those who came to North America expanded their numbers through natural increase at roughly the same rate as the white population. Thus, the 2.3 million Africans imported to the British sugar islands generated a slave population of about 775,000 by 1807, while the 388,000 Africans imported to North America during the slave trade generated a population double the Caribbean total—1.4 million U.S. slaves in 1807.

During the seventeenth and eighteenth centuries the slaves in both regions were mostly put to menial agricultural labor, producing sugar in the Caribbean, tobacco in the Chesapeake, and rice in lowcountry South Carolina. Sugar was the most profitable of these export crops, enabling the Caribbean slaveholders to establish a hierarchical pattern in which the big planters dominated the scene economically, politically, and socially. In Barbados, 175 sugar planters in 1680 with sixty or more slaves held more than half of the bondsmen on the island, as well as all the top political and military offices. On the much bigger island of Jamaica, about 650 sugar planters (including Joseph Foster Barham at Mesopotamia) held 60 percent of the total slave population in 1768, or 150 slaves each on average. In the southern mainland colonies, the distribution of slaves was far less concentrated. South Carolina was closest to the Caribbean model, with many large rice plantations, but in the Chesapeake there were only a few elite planters (such as the Tayloes of Mount Airy) with a hundred or more slaves, and the great majority of the blacks were held by smaller farmers in units of less than twenty-five people. And throughout 200 years of slavery in the southern colonies/states, more than half of the white householders never owned any slaves.

Over time, the black/white ratio became very different in the two regions. In 1750 Jamaica blacks outnumbered whites by ten to one, while Virginia whites outnumbered blacks by six to four. Partly because the number of whites in the islands was so small, they were not able to fully control the blacks, and there was much more slave unrest and rebelliousness than on the mainland. A further differentiating issue was absenteeism. By the mid-eighteenth century a large proportion of the Caribbean planters (such as the Barhams at Mesopotamia) retired to England and left the management of their slaves to salaried managers and overseers, whereas the North American slaveholders (such as the Tayloes at Mount Airy) lived among their slaves. Increasingly, the British islands were bases for nakedly exploitative profit making, while the slave-based southern mainland colonies/states were residential settlements.

The American Revolution dramatized the strengths of the mainland system and exposed the weaknesses of the island system. The Caribbean sugar planters, like their North American cousins, protested against British imperial policy in the 1760s, but they were far too dependent on British economic and military support to rebel against the mother country. The mainland rebels, on the other hand, were led to a large extent by big slaveholders like Washington and Jefferson. Though the American Revolutionary War damaged sugar output in the British islands, a boom period of large production and high profits followed from the mid-1780s through the 1810s. Yet the island planters were losing political influence at home, and were increasingly unable to counter the claims of British antislavery critics that they were horribly mistreating their black bondsmen. Parliament debated ending the British slave trade in the 1790s and legislated its close in 1807.

Meanwhile, as the victorious North American rebels set about establishing a national republican government, there were parallel protests from American abolitionists against slavery, but these protests had much less traction. Starting in 1780, all of the northern states (where slavery was marginal) gradually emancipated their slaves, but southern delegates to the Constitutional Convention of 1787 insisted on securing added representation in Congress via their slave populations. Four of the first five presidents were Virginia slaveholders who claimed to dislike slavery but did nothing to curtail it. The American Revolution opened up huge new territories to settlement, and slaveholders on the eastern seaboard began to move their bondsmen westward. The European market for slave-grown Chesapeake tobacco collapsed after the Revolution, but from the 1790s onward a new labor-intensive export crop—cotton—was seen as ideally suited to slave labor. And the Louisiana Purchase of 1803 triggered the expansion of cotton planting halfway across the continent, from Georgia to the Mississippi River valley.

The divergence between the two slave systems became much greater in the early nineteenth century. The close of the British slave trade in 1807 exposed the vulnerability of the West Indian slaveholders. With no fresh infusion from Africa, the slave population began to shrink, the slave workers grew ever more restive, sugar prices fell, and Caribbean output declined. In Britain, criticism of the West Indian slave system kept mounting, and in 1817 the home government obtained an amazingly comprehensive name-and-age census of all the 745,000 slaves who were then living in the twenty British Caribbean colonies, with follow-up registration every three years in order to track the condition of these people. A major slave rebellion in Jamaica in 1831 accelerated the demand by abolitionists at home that British slavery be terminated, and the passage of the Reform Bill in 1832 led to a new Parliament that passed an Abolition Act in 1833. The Caribbean slaveholders fumed, but when offered £20 million in compensation for the loss of their human property they submitted to Westminster, and every island assembly declared that slavery was abolished on August 1, 1834. The 665,000 ex-slaves now became "apprenticed laborers" in training for manumission; they still worked shortened hours for their old masters without pay, but after four years this limbo status was abandoned, and on August 1, 1838, the apprenticed laborers became fully free.

Quite the opposite trend was taking place in the antebellum U.S. South. As cotton production expanded across the Deep South, the American slaveholders grew wealthier, more powerful, and more self-assured. From the 1790s to 1860, they moved about one million slaves from the eastern seaboard to the Deep South in order to plant more cotton. Ironically, the cotton planters in the Gulf states and the Mississippi valley were establishing thousands of high-intensity production units that rather closely resembled eighteenth-century Caribbean sugar estates at the very time when slavery in the British West Indies was collapsing. From the 1830s onward the southern slaveholders were continually bombarded with criticism from outraged abolitionists, sparked by firsthand reports from slave runaways who described the terrible conditions in the South to northern audiences. But these attacks only toughened the southerners' belief in the superiority of their slave-based society. North/South sectionalism was now entrenched, with slavery the dividing issue in antebellum politics, and the slaveholders got most of what they wanted in a long series of political confrontations as the nation expanded west and cotton culture spread to Texas. By 1860, when Lincoln won the presidency with 39 percent of the national vote on a platform that rejected the expansion of slavery, the slaveholders felt strong enough to secede from the Union. The eleven Confederate States of America, inhabited by 5.4 million whites and 3.5 million slaves, had only one-third of the American population, but they battled for four extraordinarily bloody years before succumbing to the Union Army. Finally, after unprecedented violence, their slaves were forcibly freed in 1864–1865.

Clearly these two slave systems were on differing trajectories. The Caribbean slaves were trapped within a system that was at peak strength between 1760 and 1807 and then began to unravel. What was it like to live in an increasingly dysfunctional environment where Africans from many ethnicities were haphazardly flung together with locally born

Creoles, where slaves greatly outnumbered masters and many masters were absentees, where rebelliousness was common and rising? The U.S. slaves, on the other hand, were trapped within a system that gained steadily in strength and in geographical spread between 1800 and 1860. How did they cope with the frequent slave sales and family breakups, with the migrations that took them to distant new work sites, with aggressive white surveillance that made black protest difficult and dangerous? These are some of the questions that we will be grappling with as we examine slave life in Mesopotamia and Mount Airy.

## *Looking for Mesopotamia*

On a Saturday morning in February, I navigated a rental car through the crowded streets of Kingston on my way toward the western end of Jamaica, equipped with sunglasses, a camera, a guide book, a road map of the island, and a detailed ordinance survey map of the district in Westmoreland parish that was my destination.

I drove for 130 miles along Highway A2 through endlessly changing and continuously alluring scenery: the green plains of St. Catherine and Clarendon, the lush mountains of Manchester, the hills and valleys of St. Elizabeth, and the coastal road that brought me at last into Westmoreland. Suddenly the highway dipped down and a beautiful vista opened up: the wide sweep of the Westmoreland plain, a low-lying expanse of brilliant green fields, ringed by bluish-green mountains on three sides and the Caribbean Sea on the fourth. Here sugar has been the prime crop for 300 years. At Ferris Cross I left the highway, pulled out my ordinance survey map, and ventured into a maze of minor roads that crisscross the plain. Even in February the humid heat was oppressive. It was crop time, and although the field hands were not working in the cane pieces on Saturday, I passed several trucks loaded with cut cane, and spotted the Frome sugar factory in the distance—the factory that processes all of the cane grown in the region—spewing black smoke from its two tall chimneys.

Between the 1760s and the 1830s some sixty sugar estates were operating in Westmoreland, each with its own mill and boiling house.[1] Collectively, these production units were powered by a workforce that expanded from 10,000 slaves in the 1760s to 13,000 slaves in the early nineteenth century. Most of the sugar plantations bore pleasingly romantic names such as Belle Isle, Blackheath, Blue Castle, Canaan, Carawina, Cornwall, Fontabelle, Friendship, Grandvale, Jerusalem, King's Valley, Mount Tirzah, Nonpareil, Retreat, Roaring River, Shrewsbury, Spring Garden, Three Mile River, or Windsor Forest. And Mesopotamia, whose euphonious name evokes the fabled fruitfulness of the ancient Tigris and Euphrates river valleys. Mesopotamia sugar estate was situated on the northern edge of the Westmoreland plain along the winding Cabarita River, five and a half miles inland from the port town of Savanna la Mar and in the shadow of the rugged Hanover Mountains.

Mesopotamia was my destination, but today it is called Barham Farm, named for the family that owned this property from the 1720s to the 1850s. I searched on the map for the Cabarita River and tried to figure out how to get there. Every few miles I came to a little settlement with a jerk center (where barbecued meat was sold) and a church (whose signboard revealed the name of the place). After several false turns, I crossed a bridge over a river where people were picnicking on the bank and washing themselves and playing in the water, and came to another village with a clapboard church. A signboard announced the order of service of the Barham Go Between Ministry. I had come too far, and doubled back across the Cabarita River. My guidebook told me that the entrance to Barham Farm is marked by a pair of stone gate pillars at the driveway entrance. But only one stone gate pillar was still standing.

I entered the driveway past the gate pillar. Nothing looked like the paintings and engravings in my history books of Jamaican sugar estates in the slave era. There was no Great House for the white owner, no overseer's house and bookkeeper's barracks for the white supervisory staff, no village of little thatched huts for the black slaves, and no complex of sugar works buildings: water mill, boiling house, curing house, still house, and trash house.[2] What I did see were many acres of standing cane and many acres of cut cane, and a modern estate manager's bungalow. I walked up to the bungalow, where several workmen were gathered, and asked the black manager if I had come to Barham Farm. Yes, I had.

I asked the manager if he knew the location of the slave village, or the unmarked place where hundreds of Mesopotamia slaves were buried, or the old graveyard where the white people who operated the estate in slavery days have their tombstones. The manager did not, but he obligingly sent for a retired employee who lived on the farm and had worked here all his life. A cheerful grizzled person who called himself an old-age pensioner appeared, and told me that many years ago a white man came to Barham Farm, inspected the graveyard, and recorded the inscriptions on the tombstones. I know that this must have been Philip Wright, who published in 1966 a book entitled *Monumental Inscriptions of Jamaica,* which includes a list of tombstone inscriptions at Mesopotamia. So I asked to be taken to the graveyard.[3]

The old-age pensioner and his younger companion took me a short distance to a barbed wire enclosure, completely filled with trees, invasive plants, and tall grass. We passed through the barbed wire, and they warned me to watch out for the stinging nettles. Once inside the enclosure we were surrounded and entrapped by luxuriant vegetation. Twenty feet away the only tombstone I could see was jutting upward at a crazy angle, entangled in tree roots, and I realized that it would be very difficult to climb through the tropical growth to inspect it.

What to do? Suddenly I realized that the ground under my feet was as flat and firm as a tabletop. I must be standing on a tombstone. My companions hacked away the plants and grass enveloping us with their machetes, cleared the surface film of earth, and sure enough, they found a large stone slab under our feet. I wiped away the residual dirt and uncovered an incised date: 1735. Then I uncovered a name: BARHAM. Gradually the entire inscription came into view: HERE LIETH THE BODY OF SARAH ARCEDACKNE SISTER TO MARY BARHAM WHO DEPARTED THIS LIFE MAY 22 IN THE YEAR OF OUR LORD 1735 AGED 58. I took out my camera and photographed the tombstone, delighted to know that I had made direct contact with the eighteenth-century Barhams and that I had literally stumbled onto the grave of Sarah Arcedeckne, whose brother-in-law Dr. Henry Barham owned and operated Mesopotamia in 1735.[4] I was tempted to try to find the grave of Henry Barham's wife Mary Stephenson Barham, who also died in 1735 and was buried here, but decided to be satisfied with this one Barham monument. My two guides were relieved to find that I don't wish to look for any more tombstones, and the younger one fetched a long pole, knocked down a coconut, and cut it open with his machete. Together we celebrated our Barham discovery. My generous guides then pointed out the decayed remains of a water mill and an abandoned mid-nineteenth-century sugar works building half hidden in tropical vegetation. After inspecting several of the cane pieces and walking along the Cabarita River to get a feel for the topography of the place, I took my leave.

During my short visit to Barham Farm I saw no visible trace of past slave life. It would take a team of archaeologists many months to uncover the site of the slave village and the slave cemetery.[5] But if physical remains of slavery are lacking, this estate has one of the most comprehensive collections of written records yet discovered for any Jamaican plantation. Some 4,600 miles distant from Barham Farm, there is a treasure trove of Mesopotamia plantation records at the Bodleian Library at Oxford, as I discovered back in 1974. Wishing to make a comparative study of two slave plantations, one in the British Caribbean and the other in the United States, I had compiled a list of the largest collections of Caribbean sugar estate records in British depositories and was traveling from archive to archive checking out these collections when I came to Oxford to examine the Barham Papers. My Eureka moment came as I sat at a desk in the medieval splendor of the Duke Humfry Library at the Bodleian and opened five large boxes of Mesopotamia slave inventories. Most of these lists identified the hundreds of men, women, and children on the estate by name, age, occupation, and physical condition. I saw at once that if they could be connected chronologically, it might be possible to bring the Mesopotamia slaves to life.[6]

The Barhams owned Mesopotamia from the 1730s to the 1850s. Like many other Jamaican proprietors, they lived in England and operated their plantation from long distance. But both Joseph Foster Barham, who was Henry Barham's stepson and owned the estate from 1750 to 1789, and his son Joseph Foster Barham II, the owner from 1789 to 1832, required their bookkeepers to send them unusually detailed annual slave inventories (taken on December 31 or January 1) so that they could track changes in the population. Eighty-seven Mesopotamia slave inventories survive, taken over a span of more than a century, from 1727 to 1833. The most complete listings run from 1762 to 1833. By correlating the inventories for 1762–1833 I have reconstructed the skeletal biographies of 1,103 people—602 males and 501 females—who lived at Mesopotamia during these seventy-two years.[7] There are many defects in the Mesopotamia inventories, as we shall see, but they open a window onto slave life.[8]

Much additional information about the Mesopotamia slaves can be found in the Barhams' plantation records and in their correspondence with estate agents and overseers at Mesopotamia. The slaves' sugar and rum production is detailed in Mesopotamia crop accounts preserved in the Jamaica Archives.[9] And Mesopotamia also has exceptional religious records. The two Joseph Foster Barhams wanted to convert their slaves to Christianity, and they persuaded missionaries from the Moravian Church to come to Mesopotamia. The Moravians established a station at the estate in 1758, built a chapel for the slaves, gathered a congregation of baptized members, and maintained this mission for seventy-eight years, until 1836. The Mesopotamia missionaries wrote thirty-eight diaries that have survived, recording their day-to-day struggles to reach the slaves; these richly detailed diaries are now in the Unitätsarchiv in Herrnhut, Saxony.[10] And there are supplemental records concerning the Moravian mission at Mesopotamia in the Jamaica Archives.[11] In sum, the multifarious documentation for this sugar estate enables us to observe a large population in motion during the last two generations of slavery in Jamaica.[12]

## *Looking for Mount Airy*

On a Sunday morning in May I drove with my wife, Mary, from Williamsburg to the Northern Neck of Tidewater Virginia. We had been invited by Gwynne Tayloe, whom I met while doing research at the Library of Virginia in Richmond, to see his family home at Mount Airy—an eighteenth-century Palladian villa and one of the finest architectural showplaces of colonial Virginia. The Tayloes have been living in this grand house for over 250 years.

We crossed the mile-long bridge that spans the broad Rappahannock River and entered Richmond County (not to be confused with Virginia's capital city of Richmond), some thirty-five miles upstream from Chesapeake Bay. We followed Highway 360 for three miles and turned onto an unmarked road. Mount Airy is a private residence, not open to the public, and the Tayloes try to discourage uninvited sightseers. As we climbed up the driveway we passed a large and ominous signboard alerting us that we were entering a firing zone, a trophy that Gwynne's brother Bill Tayloe brought back from his tour of duty in the Vietnam

War and placed here to scare off casual tourists. Emerging at the top of a ridge, we found ourselves in an open park with a sweeping drive leading up to an imposing stone mansion with twin dependencies. Mount Airy was built for John Tayloe II between 1748 and 1758, and it is modeled on the villas designed in sixteenth-century Italy by Andrea Palladio. The central building, flanked by a pair of simpler two-story side buildings, is constructed of rough reddish-brown sandstone, with smooth honey-colored stone for the projecting pavilions, window frames, string coursing, and corner quoins. Immediately we felt the power of the design. Here, quite unlike my visit to Barham Farm, we were carried back two centuries and saw a gentleman's seat that unabashedly celebrated the hierarchical authority of the Virginia plantation master.

Mount Airy is not a large house by European aristocratic standards, but it *is* a grand house, partly because it stands elevated upon two raised terraces, partly because its entrance loggia, supported by four columns, is strongly assertive, and partly because its flanking satellite buildings—which stand well in advance, like guard posts—are substantial structures that clearly defer to the main edifice. These dependencies, originally a bedroom block and a kitchen block, are connected to the Great House by enclosed quadrant passageways. In the eighteenth century a second smaller pair of symmetrically placed outbuildings, one of which contained the plantation manager's office, stood further out on the periphery. The whole complex articulates the social dynamics of elite Virginians vis-à-vis plebeian Virginians in the eighteenth century. The office and the kitchen were the entry points for the common white people and the black slaves, while the Great House loggia was the entry point for the privileged few. Climbing a broad flight of stairs from the lower terrace to the second terrace, we ascended a higher flight of stairs to the loggia and entered a large central passage that leads to the stately public rooms on the first floor.[13]

We arrived to find that Gwynne's parents, Colonel and Mrs. H. Gwynne Tayloe Jr., were getting ready for a large luncheon party with fifty guests. They welcomed us cordially as they busily unpacked their heirloom china and silver for this event. Their friendly greeting pleased me, because the Tayloes knew that I was using their family papers to study the slave community on their estate—a touchy and uncomfortable issue for them and also for me.[14] Gwynne Tayloe took us on a tour of the house. Originally the interior had elaborate carvings and fittings executed by the English craftsman William Buckland between 1761 and 1764, but these were almost entirely destroyed when Mount Airy was gutted in 1844 by a fire set by a discontented slave.[15] The Tayloes' furniture and paintings were rescued from this disaster, the interior was rebuilt in Victorian style, and the house today displays a large collection of beautiful heirloom antiques. The splendid Tayloe portraits are especially notable, and I paused to view the two family members I am particularly interested in. John Tayloe III (1771–1828) confronted me with an air of suave mastery, while his son William Henry Tayloe (1799–1871) offered a kindly, quizzical gaze.

Exiting from the rear of the house, Gwynne Tayloe led us out into a very extensive garden. He didn't show us the family cemetery, but I have learned subsequently that John III and William Henry are buried here along with many other members of the Tayloe family, and that their graves are carefully maintained—in total contrast to the derelict cemetery at Barham Farm. We passed the ruins of an orangery and walked down a long pathway bordered by giant boxwood to the edge of the ridge where the land falls steeply downward toward the Rappahannock River valley. We saw the outline of a former racecourse below (the eighteenth- and nineteenth-century Tayloes had a passion for thoroughbred horses) and the spreading fields where they raised tobacco in the seventeenth and eighteenth centuries, corn and wheat in the nineteenth century, and soybeans today.[16] We didn't see Sabine Hall, built in 1733–1734 by the diarist Landon Carter, which stands closer to the Rappahannock on the same ridge, but we knew that this Carter mansion is only two miles away from Mount Airy as the crow flies and three miles away by road.

I saw no visible reminders of slavery at Mount Airy. The old slave quarters and workshops are long gone, and the slave graveyard has not been identified. But the magnificent collection of manuscripts that the Tayloe family presented to the Virginia Historical Society in Richmond contains abundant slave documentation.[17] Indeed, the slave records for Mount Airy in the Tayloe Family Papers make a very good match with the Mesopotamia slave records in the Barham Papers.

Back in the 1970s, after discovering the Mesopotamia inventories at the Bodleian, I set about trying to find a similar set of slave inventories for a North American plantation. I started looking in South Carolina, where I knew there were excellent records of planters who used their slaves to grow rice in the colonial era and then switched quickly to cotton production after the Revolution and the invention of the cotton gin. But none of the South Carolina plantation records that I saw contained long series of annual inventories in which the slaves are identified by name and age. So I moved north, only to find the same negative result when I searched in the North Carolina and Maryland archives. By the time I turned to Virginia, I feared that my comparative plan was unworkable. But when I came to the Virginia Historical Society and examined the Tayloe Family Papers, I found a large volume of Mount Airy slave inventories spanning the years 1808–1829 that was similar in character to the Mesopotamia inventories.[18] I soon discovered that John Tayloe III and his son William Henry Tayloe had listed their Mount Airy slaves almost every year (except for a five-year break in the 1850s) from 1808 to 1865. Altogether there are fifty-two inventories of the Mount Airy slaves.[19] As at Mesopotamia, the Tayloes identified the Mount Airy people by name, age, and occupation, so that the inventories can be correlated. This has enabled me to track 973 slaves—524 males and 449 females—over a fifty-eight-year period. As at Mesopotamia, the Mount Airy inventories contain demographic information, reveal individual lives, family connections, and occupational histories, and offer a skeletal view of slave life.

The Tayloe Papers, a larger collection of documents than the Barham Papers, contain much supplemental information about the slaves who lived at Mount Airy from 1808 to 1865.[20] John Tayloe III, the grandest planter and largest slaveholder in the history of the family, kept—in addition to his slave inventories—an unusually elaborate array of shop books and work logs that describe the yearly labor routine of his Mount Airy craft and agricultural workers.[21] There are nearly 14,000 documents in William Henry Tayloe's section of this family archive, including long files of correspondence with frequent reference to his slaves. Of particular importance for my study, William Henry Tayloe kept a detailed record of the process by which he transferred 218 of his Mount Airy people to new cotton plantations in Alabama.[22] Further data on the Mount Airy slaves can be found in another collection of Tayloe Papers at the Alderman Library at the University of Virginia.[23] All in all, the documentation about the Mount Airy people has the same multidimensional character as at Mesopotamia, and enables us to observe a large community of black Virginians in action during the final three generations of North American slavery.[24]

## *Methodology*

Having found two long series of slave inventories that suited my purpose, I began to examine slave life at Mesopotamia and Mount Airy, and to compare these communities in order to highlight the differences between British Caribbean and antebellum U.S. slavery. I felt sure that the plantations would show a strong demographic contrast, but was quite uncertain about what the other differences would add up to. Initially, I wished that the two sets of records were coterminous, and regretted that the Mount Airy records started so much later than the Mesopotamia records. I soon saw the advantages of focusing on the last seventy years of slavery in both places, but feared that the Mount Airy story would lose most of its interest with the death of John Tayloe III in 1828, because the Mount Airy slave force was then split up among his sons. It took me quite a while to realize how lucky I was that I could document William Henry Tayloe's movement of 218 slaves to Alabama from the 1830s to the 1860s. Only gradually did I gain a full understanding of my evidence and my argument.

Initially, my biggest challenge was how to record all of the biographical information in the two sets of slave inventories. There were no personal computers in the mid-1970s, and since I was determined to present the Mesopotamia and Mount Airy slaves as people rather than as digits, I was averse to using punch cards. This was the cliometric era, and after reading *Time on the Cross* by Robert Fogel and Stanley Engerman in 1974, I wanted to find a nonmechanical, nonstatistical way of handling my data. Adopting an old-fashioned handicraft method, I entered the correlated year-by-year information from the inventories into four large loose-leaf ledgers: two ledgers for the males and females at Mesopotamia, and two for the males and females at Mount Airy. I organized each ledger alphabetically, dating seventy-two columns (1762 to 1833) for each Mesopotamia alphabet letter, and fifty-eight columns (1808 to 1865) for each Mount Airy alphabet letter. Then, starting with the Mesopotamia inventory in 1762 and the Mount Airy inventory in 1808, I entered each new person who appeared on the record on a separate row under the first letter of his/her name. For each person, I initially listed place of birth and parentage if known, and (when relevant) information about purchase, and in the entry column I then noted age, occupation, health, value, or any other information to be found. As I slowly worked my way year by year and column by column through the two sets of inventories, I constructed lengthening annual reports on more than 2,000 people, usually ending with final illness and death. I was always adding new rows for new babies or newly acquired slaves, while terminating old rows. I was tracking two communities of enslaved men, women, and children in slow motion.[25]

I soon found that the biographies I was constructing for the Mesopotamia and Mount Airy men and women had differing characteristics. The men's biographies on both plantations displayed a more interesting variety of occupations, but the women's biographies disclosed motherhood, and identified newborn children by birth month at Mesopotamia and by birth year at Mount Airy.[26] The Tayloe inventories were less informative for demographic purposes than the Barham inventories, because birth and death dates are usually missing in the Virginia population. But the Tayloe inventories were better for family reconstruction, because they often identify both parents of a child, whereas the Barham inventories identify only the mothers.

Having completed my data entry, I assigned numbers to all the 2,076 people in the four ledgers. Then I began to construct multigenerational family networks on both plantations, two of which are presented in Chapters 2 and 3. Eventually I incorporated all of the most statistically relevant data from my ledgers on Excel spreadsheets. But as I drafted the chapters of this book I turned to my ledgers for insight and information far more frequently than to my Excel spreadsheets.

Soon after I started work on this project I published my first essay on Mesopotamia and Mount Airy in the *William and Mary Quarterly* in 1977, which was a mistake because my presentation was premature. Not having yet tracked many of the 2,000 people in my slave cohort, I attempted a twenty-year analysis of each plantation, a span of time too brief to bring out the chief characteristics of either place.[27] In retrospect the only thing I like about this initial essay is the title ("A Tale of Two Plantations"). Otherwise it contains many factual and interpretive errors, large and small. For example, I claimed that the Mount Airy slaves worked at a leisurely pace and that their benevolent owner, John Tayloe III, was a relaxed gentleman and "no profit-maximizing entrepreneur"—statements that I now see are completely untrue. In comparing the two communities I put far more emphasis on the harshness of slave life under Barham management than under Tayloe management, and ended up by declaring that "if one had to be a slave, Mount Airy was a better place than Mesopotamia."[28] A Virginia newspaper printed a review of my article under the apt headline "Early Slave Life in State Better Than in Jamaica."[29]

Sensing that I was off to a bad start, I published nothing further on Mesopotamia and Mount Airy during the next ten years. Turning to alternate projects, I revised and expanded *The Age of Religious Wars*, my interpretation of Early Modern European society in the sixteenth and seventeenth centuries.[30] I also laid plans for an Early American History program, housed at the University of Pennsylvania, that would offer dissertation fellowships to graduate students who wished to do research in the Philadelphia-area libraries and archives. Receiving a generous grant from the Andrew W. Mellon Foundation to fund this program, I opened the Philadelphia Center for Early American Studies in 1978. In the same year, my wife and I began a select edition of *The Papers of William Penn*. Working with a series of associate editors, we published our four volumes between 1981 and 1987.[31] And during the late 1980s and early 1990s I devoted much of my time to another complex editorial project, *The Journal of John Winthrop*, which I produced in collaboration with Laetitia Yeandle in 1996.[32] During these years Robert L. McNeil Jr. became a major benefactor to the Philadelphia Center, which was renamed the McNeil Center for Early American Studies in 1998. Thanks to Bob, the Center was securely established by the time I retired as the director in 2000.

During the 1980s and 1990s I continued to give talks about my two slave communities, trying new approaches. And as I tried to imagine what slave life was like at Mesopotamia and Mount Airy, I was challenged by the differing characteristics of the two sets of records. Most obviously, the Moravian records at Mesopotamia provide detailed information about worship and church membership, whereas I know nothing about the religious life of the Mount Airy people in Virginia or in Alabama. And the Mesopotamia inventories contain annual reports on each slave's physical condition, so that the impact of job assignments on health and life expectancy can be measured. Motherhood is open to analysis on both plantations—except that the children fathered by white men are always identified as mulatto or quadroon in the Mesopotamia inventories and never in Mount Airy. Mesopotamia has a complete listing of sugar and rum production and sales from 1751 to 1833, whereas the Tayloes' records of grain production and sales at Mount Airy and cotton production and sales in Alabama are scattered and fragmentary. On the other hand, the Mount Airy work logs and shop books kept by John Tayloe III show a seasonal pattern of work on the Virginia plantation that can only be surmised at Mesopotamia. Family and kinship linkages are much better documented at Mount Airy. William Henry Tayloe's detailed records help to illustrate the process by which a million African Americans were moved from the upper South to the lower South in the two generations before the Civil War. And the Tayloes (especially William Henry Tayloe) were always in close personal touch with their slaves, while the Barhams lived an ocean away.

Working on this book for so long, I have been inevitably influenced by the many changes during the past forty years in the interpretation of racial slavery in the Americas. Like many of my colleagues, I was drawn to the subject by the Black Freedom movement and by Winthrop Jordan's *White over Black,* published in 1968. Little was known in those days about any aspect of slave life in the Americas, so the tendency was to go for the big picture, to compare English slavery with Spanish slavery, or Protestant slavery with Catholic slavery, in order to find out which was "better" or "worse"—the trap that I fell into with my first article comparing Mesopotamia with Mount Airy in 1977.[33] There was also much initial discussion as to whether Europeans enslaved Africans primarily because they were racists or because this grossly exploitive labor system was cost-effective. In retrospect, this debate seems beside the point, since racism justified exploitation, and the two attitudes became so intertwined that scarcely anyone in Europe or white America ventured to question the enslavement of African people until the mid-eighteenth century. The liveliest dispute in the 1970s was between the cliometricians who discovered that many slave records could be subjected to quantitative analysis and the social historians who were trying to view the slaves as people, not digits. Though I was dealing with digits, I was very averse to mathematical modeling, and found common ground with historians like Herbert Gutman and Lawrence Levine who were beginning to explore slave family life and slave folk culture.[34]

The most ambitious statement about United States slavery published in the 1970s, indeed the most ambitious interpretation that has yet appeared, was Eugene Genovese's *Roll, Jordan, Roll: The World the Slaves Made*. This big book was much admired at the time for its boldly organic unity, but I thought then, as now, that Genovese was hugely mistaken in substituting class for race as the core feature of American slavery. His formulaic and time-frozen portrayal of the antebellum slave system has been roundly rejected by subsequent historians, who have uncovered huge regional variations in North American slavery, and very significant long-range changes over time. And Genovese's celebration of planter paternalism warned me that a scholar studying the correspondence of the slaveholders can be seduced into adopting their self-serving assumptions—a danger that I face when reading the letters of William Henry Tayloe, a slaveholder whom I admire in many ways. Genovese's highly restrictive view of the possibilities for black achievement during slavery has also been challenged. During the past forty years the chief driving force among students of United States slavery has been a search for slave agency, for evidences of slave activism and exceptionalism, to be found in particular among slave rebels and slave runaways.[35]

In the British Caribbean, a quite different style of historical inquiry took shape in the 1970s and 1980s, illustrated by the impressive work of David Eltis and Barry Higman. Eltis is the lead member of a team of scholars who have established a huge digitized database (www.slavevoyages.org) detailing more than 35,000 voyages in the transatlantic slave trade, a project of vital importance to understanding the evolution of the African slave system in the British Caribbean. Higman has meanwhile used slave registration data to analyze the composition of the entire slave population in the twenty British West Indian colonies from 1807 to 1834, and he has also written a series of deeply researched books on Jamaica during the final generation of slavery—examining the economy, the labor force, the plantation layout, and archaeological reconstitution of the slave village. Together, Eltis and Higman have established an admirable framework for investigating British Caribbean slave life. But there is scarcely any human dimension to their heavily quantitative accounts. And this has led a younger group of historians to portray the slave trade and Jamaican slave life very differently—to study (for example) the transformation of African persons into commodities via the Middle Passage, the tyrannical behavior of white Jamaican slave managers, and the centrality of death in Jamaican social and psychological experience.[36]

In recent years, historians of slavery in both the Caribbean and the United States have increasingly taken a cultural turn in order to evoke the interiority, thoughts, and emotions of the enslaved blacks—cultural attributes that I cannot possibly capture for the people of Mesopotamia and Mount Airy.[37] But there is also a strong current interest in slave biography, to illuminate through individual life stories the personal dimensions of the slave experience.[38] And here I think that my intergenerational study of the slaves at Mesopotamia and Mount Airy may have relevance. Two of the most striking recent black biographies are Jon Sensbach's life of Rebecca Protten, an evangelist for the Moravian Church on the island of St. Thomas, and Sydney Nathans's life of Mary Walker, an escaped slave from North Carolina who spent the next seventeen years trying to free her mother and children.[39] Sensbach and Nathans have combined informative sources with meticulous research in order to flesh out the lives of two obscure mulatto women and narrate their adventures in fascinating detail. What strikes me, however, is that Rebecca's biography bears no relation to the experiences of any Mesopotamia slave woman, whereas Mary's biography directly evokes the plight that many of the Mount Airy slave mothers faced. Rebecca Protten was born into slavery, learned to read and write as a child, converted to Christianity and was freed by her owners in her early teens, and after baptism by the Moravians became an evangelist to the local slaves, traveling around St. Thomas and instructing them in Christianity—a course of action that would have been impossible in Mesopotamia. Rebecca then married a white missionary, was rescued from jail by Count Zinzendorf, spent twenty years in Germany, and ended up in West Africa. She was a remarkable person, but her adventurous life story sheds little light on West Indian slavery. Mary Walker, by contrast, became a domestic servant to white abolitionists in Pennsylvania and Massachusetts after escaping from her master, grieved from 1848 to 1865 over the loss of her loved ones, and tried hard but failed to recover her mother and children. Only after the Civil War was she finally reunited with a son and a daughter. As we shall see in Chapters 7 through 9, Mary's story echoes the experiences of a great many Mount Airy people during slavery and after emancipation.[40]

While Jon Sensbach and Sydney Nathans were composing their black biographies, I was finally drafting my account of Mesopotamia and Mount Airy. Between 1987 and 2011, I published a series of essays on the two plantations that illuminate pieces of the puzzle but do not present a coherent or comprehensive overall interpretation. So in 2009–2013 I pulled the whole story together. And I now believe that I have wrapped my arms around this big subject.

The plan of my book is as follows. Chapter 1 establishes the framework by comparing population decrease at Mesopotamia in 1762–1833 with population growth at Mount Airy in 1808–1865; this chapter also situates the Mesopotamia people within a slave system that was weakening and the Mount Airy people within a slave system that was gaining strength. The next pair of chapters turn from the general to the particular and look at family life on the two plantations. Chapter 2 traces a Mesopotamia female field hand through every stage in her bondage, comparing her harsh experiences with those of her rebellious mulatto son and her clever quadroon granddaughter, to illustrate the workings of interracial sex in a Caribbean setting. Chapter 3 similarly tracks a Mount Airy female craft and domestic worker in a stormy tale of slave escape, family breakup, forced migration to Alabama, and interracial sex—this time in an Old South setting. The next pair of chapters explore the labor pattern on the two plantations. Chapter 4 examines the impact of sugar labor on the health and longevity of the Mesopotamia workers, the disciplinary system for slaves, and the role of motherhood among the female field hands. Chapter 5 looks at John Tayloe III's management of his Mount Airy field hands, domestics, and craft workers, noting especially his movement or sale of surplus slaves. Chapter 6 describes the Moravian mission at Mesopotamia, the establishment of a Christian congregation, and the interaction between the missionaries and the slave converts on this estate from 1758 to the 1830s. Chapter 7 describes the process by which William Henry Tayloe moved 218 Mount Airy people to Alabama between 1833 and 1862, and what happened to these migrants in the land of cotton. Chapter 8, the counterpart to Chapter 1, discusses the social contrast between the two plantations, slave leadership and slave rebelliousness in both places, the response of Mesopotamia people to the Jamaican slave rebellion of 1831–1832, and the response of the Mount Airy people to the U.S. Civil War. And Chapter 9 follows the Mesopotamia people as they become emancipated in 1834–1844 and the Mount Airy people in 1865–1870—a huge advance, yet still a very restricted kind of freedom.

# 1

## *Mesopotamia versus Mount Airy: The Demographic Contrast*

When Philip Curtin published *The Atlantic Slave Trade* in 1969, he boldly discarded previous estimates that fifteen million or more Africans had been shipped to the New World. Instead, he postulated that a total of 9,391,000 African slaves were landed in the Americas, with 4,683,000 coming to the Caribbean islands and 399,000 to North America.[1] Critics insisted that Curtin's total of less than ten million was far too low, but what struck me when I read his book was the revised shape that he gave to the distribution of the transatlantic slave trade. He was one of the first historians to emphasize that in the grand scheme of things the shipment of African captives to what is now the United States was relatively small, and that the importation to the Caribbean sugar islands was a great deal larger. In the forty-five years since Curtin published his book, investigators have vastly improved our understanding of the transatlantic slave trade, and Curtin's figures still look pretty good. In 2001, after analyzing the records of 27,000 slave voyages, David Eltis calculated that 9,468,000 Africans were landed in America, with 4,371,100 coming to the Caribbean and 361,100 to North America.[2] By 2013 the editors of the Trans-Atlantic Slave Trade Database, having analyzed 35,000 slave voyages, arrived at a larger total, estimating that 10,702,656 Africans landed in the Americas, with 5,065,117 coming to the Caribbean and 388,747 to North America.[3] With these revisions, the proportion of African captives landed in the Caribbean dropped from Curtin's 49.8 percent to 47.3 percent, and the North American proportion also dropped from Curtin's 4.2 percent to 3.6 percent, leaving the relation between the two destinations much the same. It is now indisputable that twelve or thirteen African slaves were brought to the Caribbean for every one African slave brought to North America.

The reason for this huge disparity is demographic. The Africans who were landed in North America expanded their numbers through natural increase, whereas the Africans brought to the Caribbean died off and were continually replaced with new slaves. Thus, the slaves in Jamaica and Virginia—the two largest American slave societies established by the British—had dramatically different population histories. Approximately 1,017,000 slaves were imported from Africa to Jamaica between the English conquest in 1655 and the close of the slave trade in 1807.[4] Some 210,000 of these people were reexported to the Spanish colonies, but the rest—over 800,000 African slaves—were sold to the Jamaican planters.[5] Despite this huge importation, the attrition rate in Jamaica was so high that in 1807 there were only about 385,000 people of African origin living on the island: 355,000 slaves and 30,000 free colored. In Virginia, by contrast, about 101,000 slaves were imported from Africa between 1607 and 1778, and then the Virginia state government prohibited further importation of slaves because they had as many African slaves as they wanted.[6] Indeed, the Virginia slave population quickly quadrupled through natural increase, and in 1807 the Old Dominion had a larger population of African origin than Jamaica: 380,000 slaves and 30,000 free blacks.

To get a firsthand sense of what population decrease meant for the Jamaica slaves and what population increase meant for the Virginia slaves, we can compare the demographic history of Mesopotamia in western Jamaica with the demographic history of Mount Airy in Tidewater Virginia. At Mesopotamia there were 331 more recorded slave deaths than births between 1762 and 1833, and the owners—Joseph Foster Barham I (1729–1789) and his son Joseph Foster Barham II (1759–1832)—continually brought in new slaves in order to keep the place going. At Mount Airy there were 293 more recorded births than deaths between 1809 and 1863, and the owners—John Tayloe III (1771–1828) and his son William Henry Tayloe (1799–1871)—took full advantage of this population growth. They moved their surplus laborers to new work sites or made money by selling them. The demographic contrast between these two plantations epitomizes the most fundamental difference between Caribbean and Old South slavery.

Compounding the demographic disparity was another very important underlying difference: the role of the slaveholding owners on these two plantations. The Barhams, along with a great many other British Caribbean planters, were absentee proprietors who lived in England. They delegated the management of Mesopotamia to attorneys in Jamaica who hired and fired the overseers and other white staff members and had very little direct knowledge of their slaves. The scene at Mount Airy was a world apart. Here the Tayloes were hands-on managers, like almost all southern slaveholders. They lived among their slaves and knew them personally.

## *Mesopotamia*

The detailed Mesopotamia inventories dating from 1762 to 1833 that the two Joseph Foster Barhams received in England record a total slave population of 1,103 males and females. Appendix 1 tabulates the population changes under the elder Barham during the years 1762–1789, followed by the changes under his son's management during the years 1790–1833. The birth and death figures drive all of the other numbers. During this seventy-two-year span there were 420 recorded births (5.8 per annum) as against 751 deaths (10.4 per annum). These totals are by no means complete. Though the Barhams' bookkeepers kept slave birth and death registers, they never reported abortions or miscarriages and only occasionally reported stillbirths, and they seem to have omitted a large number of infants who died within a few hours or days after birth. The true birth total was undoubtedly much higher than 420, which would make the true death total equally higher. The bottom line was the same—331 more deaths than births.

To sustain their Mesopotamia workforce, the Barhams bought 415 new slaves between 1762 and 1833—137 directly from the African slave ships and 278 from other estates in Jamaica. The first Joseph Foster Barham had a slave population of 268 in 1762 that dipped to a low of 238 in 1769, stabilized at around 260 throughout the 1770s, then climbed to 303 in 1786. The second Joseph Foster Barham made major purchases that expanded the population to a peak of 383 in 1792 and to a higher peak of 421 in 1820, after which he stopped buying and the total shrank year by year down to 329 in 1833. Though a great many slaves ran away temporarily, there was very little permanent movement out of the estate. Only seven slaves escaped for good, another four chronic runaways were sold for transportation off the island, and twelve mulattoes (all fathered by the white men who managed the estate) were manumitted. A particularly interesting feature of the Mesopotamia population is the gender balance, which kept shifting. A female majority in the earliest Mesopotamia 1727 inventory morphed into parity in 1736 and into strong male dominance in the inventories from 1751 to 1762. The males continued firmly in the majority through the 1790s, then gender parity was briefly reestablished in 1807–1810, after which (except for the year 1825) the females held the majority during the final twenty-four years of slavery.

To put the population figures of 1762–1833 in perspective, we need to look back to the early days of this estate. In the 1670s Edmund Stephenson staked out 540 acres on the Cabarita River and named his plantation Mesopotamia ("land of rivers"), in evocation of the ancient region watered by the Tigris and Euphrates.[7] Stephenson quickly entered into sugar production. A detailed map of Jamaica, published in 1684, identifies his sugar works on the Cabarita River, one of only seven sugar estates at this date in the Westmoreland plain.[8] We know nothing about Edmund Stephenson's sugar making, but he must have been working his crop with slave laborers. There is rather more information about his son Ephraim, who expanded operations at Mesopotamia. Ephraim Stephenson added 2,000 acres along the Cabarita River to his father's property between 1695 and 1705. When he died in 1726 at the age of sixty, his estate (exclusive of land and buildings) was valued at £5,669. His most valuable asset was a cadre of ninety slaves.[9]

Ephraim Stephenson's probate inventory, taken in August 1727, lists the slaves he held at Mesopotamia by name and value, but unfortunately not by age. Only the two drivers, a distiller, a carpenter, and the ten house servants were identified by occupation. There were forty-one "Negroe" men and boys, forty-eight "Negroe" women and girls, and one nameless "Indian Girle." The list is crude, but circumstantial evidence indicates that these slaves were young, healthy, and durable. Only two men were identified as "old." Sixty-seven of these people reappear on the next surviving Mesopotamia inventory, taken in 1736, and forty-two on the 1744 inventory; and twenty-two of them show up thirty-five years later on the 1762 inventory—the first Mesopotamia listing to assign ages to the slaves.

Several of Ephraim Stephenson's slaves were still living at Mesopotamia well beyond 1762. Parry was listed as a driver of a field gang in 1727, and according to his 1762 age statement he was then about twenty-five years old. Parry was identified in subsequent Mesopotamia inventories as a driver in 1736, 1743, 1744, 1757, and 1762 (when he was said to be sixty years old). He stopped working in 1764 but lived to November 1783, when he died at about eighty-two. Kickery, a field hand about twenty years old with an unidentified baby in 1727, became a domestic and then a midwife; she was said to be fifty-five in 1762 and was still working as a midwife when she died at about seventy-two in May 1779.[10] Love, a girl of about fifteen in 1727, apparently had numerous children, none of whom are identified in the records. She worked in the field gang for some forty years, was a nonworker for another thirty years, and died at about eighty-three in December 1794.[11] Primus, a boy of about ten in 1727, was a mule man in 1736, and from 1744 to 1780 he worked as a distiller; he was said to be forty-five in 1762 and died in February 1795 at about seventy-eight. Phillis, a girl of about five in 1727, was a forty-year-old field hand in 1762 who was disabled by illness in most years from 1764 to 1801; she died at about age eighty. Ralph, the younger brother of Primus and Love, was also a child of about five in 1727; he became successively a field worker, a driver, and a watchman, and died at about eighty-one in March 1802.

The careers of these six Stephenson slaves demonstrate that despite the very high death rate it was possible to survive past age seventy, or even past eighty. We will meet numerous other long-lived Mesopotamians. All of the six except Phillis had active working careers of at least forty years, which was not unusual. All of them except Kickery spent many years in nonworking retirement, another fairly common pattern at Mesopotamia. Primus became blind in old age; Parry was incapacitated for twenty years, Phillis for twenty-two years, and Love for thirty years. Parry, Phillis, and Love were unusually long-term invalids, but it should be emphasized that the Mesopotamia workers who escaped early death often experienced protracted debility in old age.

When the owner of Mesopotamia, Ephraim Stephenson, died in 1726, he had no living children and willed all of his property to his widow, Mary. This lady needed a husband in order to operate Mesopotamia, and by August 1727 she had married a man named Heith. When Heith also died, Mary took a third husband in early 1728, a Jamaica physician named Dr. Henry Barham (1692–1746).[12] Barham was the son of a naturalist also named Henry Barham who was an intellectual of some note. The elder Barham wrote a treatise about Jamaican flora and fauna, corresponded with Sir Hans Sloane, and was elected a fellow of the Royal Society in 1717.[13] The younger Barham was a more pragmatic type who owned Spring Plantation, adjacent to Mesopotamia, and rose to wealth by marrying well. After marrying Mary Stephenson Heith and acquiring control of Mesopotamia, Barham borrowed money from several Kingston merchants between 1728 and 1736 to buy additional land, and he greatly expanded the Mesopotamia labor force.[14] We have no record of when and where Barham acquired his new slaves, nor how much they cost, nor how much money he was making from sugar and rum sales during the 1730s. But the doctor was doing well enough that he could think about retiring from direct management and living in England. He had strong motivation for doing so, because in 1731 his sister Elizabeth died at Mesopotamia, and in May 1735 his wife Mary and her sister Sarah Arcedeckne both died at Mesopotamia.[15] Henry determined to get away from the unhealthy Westmoreland climate before the same thing happened to him. So in April 1736 he departed for England, never to return.

Just before leaving Jamaica, Henry Barham took an inventory of his 248 Mesopotamia slaves—nearly triple the number in 1727.[16] There were 124 males and 124 females listed by name and occupation. Seventy-four of these people were still living at Mesopotamia in 1762, and they form the nucleus of the population we will be examining. Ages are not stated, but this population—like the Mesopotamia population in 1727—must have been young and vigorous. Most of the adults had been purchased between 1727 and 1736, and if they came from Africa they would have been in their teens or twenties on arrival at Mesopotamia. All the slaves had occupations except for twenty-six boys and thirty girls who are described as "not yet fitt to work." If this statement is taken literally, it means that 23 percent of the Mesopotamia slaves in 1736 were under the age of seven or eight—the age at which children were routinely put to work on Jamaican sugar estates. And the statement may be literally true. Twenty-one of these "not yet fitt to work" people were still living at Mesopotamia in 1762, and when their stated ages are projected backward, it turns out that fifteen of them would have been seven or younger in 1736, and eighteen would have been under age ten. So it seems quite likely that there were a great many young nonworking children in 1736—an indicator of natural population increase, and a situation that would soon change.

After 1736 the Mesopotamia slaves very rarely saw their masters. Dr. Henry Barham retired permanently to England, leaving the management of his Jamaican property to his medical colleague Dr. James Paterson. As soon as he reached England, Barham married again, this time to a wealthy widow he had known in Jamaica named Elizabeth Smith Foster Ayscough.[17] This lady had a large family by her first husband, John Foster, who died in 1731 leaving five Jamaica sugar plantations staffed by 768 slaves, valued at £33,958.[18] Her second husband, another Jamaica planter named John Ayscough, died in 1735 or 1736. Barham became stepfather to Elizabeth's children, five Foster boys and two Foster girls, and he kept a town house on Grosvenor Street in London and a country seat at Staines for his new family. He took a special interest in his youngest stepson, Joseph Foster, whose Jamaican inheritance—a newly settled sugar plantation named Island in St. Elizabeth parish—was considerably smaller than the older Foster boys' portions.[19] Barham had no children of his own, so he bequeathed his Mesopotamia estate to Joseph on condition that the boy adopt the surname Barham, with the further stipulation that Joseph's mother have lifetime income from the property.[20] Joseph complied with these requirements. Thus, when Henry Barham died in 1746, his widow, Elizabeth Barham, took over the estate, and Joseph became the putative owner in 1750 when he reached his majority and the full-scale operator six years later at his mother's death in 1756.

Joseph Foster Barham I is a central figure in our story, the first of the four slaveholders who compiled the records upon which this book is based. Unfortunately, most of his personal papers have not survived, so a great many aspects of his life are unknown. Furthermore, the Jamaica correspondence that he did preserve is very one-sided. Unlike his son, Joseph I almost never made copies of the letters he wrote to his attorneys and overseers, though he did keep a good many of their replies. In order to get at his thoughts and motivations, we have to puzzle out his views from the responses that his Jamaica correspondents sent to him. Thus, Joseph I is a more shadowy personage than Joseph II or than either John Tayloe III or William Henry Tayloe.

Joseph I assuredly had a long Jamaica pedigree. His great-grandfather reputedly helped to capture the island from the Spanish in 1655, his grandfather was a pioneer planter in St. Elizabeth parish, and his father, John Foster (1681–1731), operated five Jamaican sugar estates. Joseph was born in Jamaica, the youngest of five sons, was a baby when his father died, and was seven years old when his mother married Henry Barham. His mother and stepfather saw to it that Joseph received an elite education, first at Eton College and then at Trinity College, Oxford. He turned out to be a pious young man, and at age twenty-one he visited Jamaica for a little over a year in 1750–1751, accompanied by his tutor, the Reverend George Downing. Joseph spent considerable time on each of his estates, which were more than fifty miles apart.[21] He doubtless had business discussions with his much older Jamaican brother-in-law Florentius Vassall, who had managed Island estate for him in the 1740s, and Vassall probably urged Joseph to live in Jamaica instead of deputizing attorneys and overseers who would mistreat his slaves and skim off his profits.[22] But Joseph could see that Mesopotamia was producing annual sugar and rum shipments valued at £4,000 Jamaican (£2,850 sterling), and Island was producing another £3,000 Jamaican (£2,150 sterling), which would give him more than enough to live on as an absentee. So he returned to England, attended by a Mesopotamia slave boy named George.[23]

Young Joseph seems to have been quite troubled by the conditions he found at Mesopotamia and Island. It distressed him to see that his slaves received no religious instruction. And he was also concerned about their health and life expectancy. He probably didn't realize when first meeting his Mesopotamia slaves in 1750 that over half of the 248 people who had been inventoried in 1736 were dead or gone. But he certainly noticed the need for replacements, and in 1751 he paid his attorney, Dr. James Paterson, £1,023 for twenty-one new Africans to bolster his workforce. These people can be traced via the Mesopotamia records, and only a third of them proved to be long-term productive workers. Four died immediately during 1751, and five more in the first decade after arrival. Another five, who lived into the 1770s or beyond, held marginal jobs or were invalids from the early 1760s onward. The remaining seven all labored at prime jobs for twenty-five years or more. Four were especially long-lived, dying between 1802 and 1822, and Phylander was perhaps the most valuable person in this group, since he served for many years as a boiler in the sugar works. The mixed record of these twenty-one Africans would be repeated in future years when other groups of replacement slaves were brought to Mesopotamia.

Before leaving Jamaica, Joseph Foster Barham was sufficiently concerned about the high slave turnover at Mesopotamia that he ordered the managerial staff to compile annual slave inventories so that he could keep track of changes in the population. Starting in December 1751, the bookkeepers prepared a lengthy inventory at the close of each year and sent it to him in England. Barham also changed attorneys. He replaced Dr. James Paterson, his stepfather's friend who had managed the estate since 1736, with the Reverend Mr. John Pool, who was the local Anglican vicar.[24] Perhaps he hoped that a clergyman would be a benevolent supervisor for an absentee proprietor. But Barham knew that Parson Pool had no interest whatsoever in converting his slaves to Christianity. So when he returned to England in 1751, he set about looking for missionaries who could minister to his Jamaica slaves.

Back home, Joseph Foster Barham was much affected by the preaching of Moravian brother John Cennick. And he fell in love with Dorothea Vaughan, a devout woman with extensive property in Wales who shared his religious values. Barham and Vaughan joined the Moravian Church, or Unitas Fratrum, a central European radical Protestant sect that maintained a small outpost in England and a very active program of overseas evangelism. In 1753 Barham and his older brother William Foster asked the Moravians to send missionaries to the slaves on their plantations in Jamaica—launching a religious odyssey that will be the subject of Chapter 6. In 1754 he married Dorothy Vaughan, and a few months later he bade a tearful farewell to three Moravian brethren as they set sail from London on the first Christian mission to the Jamaica slaves.[25] One should not exaggerate young Barham's benevolence. Supporting the Moravian mission enabled him to reconcile his religiosity with his status as slaveholder, and he knew that the Moravians would preach a doctrine of heavenly salvation combined with passive obedience to earthly masters. Joseph and Dorothy Barham settled in Bedford, raised a large family, and lived very comfortably on their West Indian sugar sales and Welsh farm rents. During the late 1750s Barham received well over £6,000 Jamaican per annum in sugar shipments from the Mesopotamia estate plus £4,000 Jamaican from the Island estate, for a gross income of more than £7,000 sterling.[26]

Joseph never went back to Jamaica after 1751, but he tried to keep track of his two properties from long distance. He received encouraging reports from the Moravian missionaries, who began to evangelize at Mesopotamia in 1758, but few communications in the 1750s that have survived from his attorneys and overseers. The annual Mesopotamia slave inventories that he had commissioned provided useful information, but not enough to tell what was really going on. From 1751 through 1760 the lists identified each Mesopotamia slave by name and status (man, boy, woman, or girl), but not by age or occupation. There were annual death registers, but no birth registers. These inventories, despite their deficiencies, did plainly show that the Mesopotamia population was not self-sustaining. There were 285 slaves in December 1751, and twelve deaths during that year. The next nine inventories averaged ten deaths per year. Twenty-two new slaves were purchased in 1756 and 1759, but by 1760 the population had dropped to 270.

In 1758 Barham sent a man named Daniel Barnjum, who seems to have been a poor relation, to check up on the situation at Mesopotamia, and when overseer Daniel Macfarlane died in 1760, attorney Pool appointed Barnjum as the new overseer. Joseph Foster Barham complained to Daniel Barnjum that the records he was receiving from him were not full enough. Anxious to keep his job, Barnjum constructed an expanded inventory in December 1761 that listed the age and condition of each slave. When Barnjum found out that this list had been lost at sea, he sent a new one dated July 10, 1762. He told Barham, "I have been as exact as I possibly can in the calculations of their ages, and in regard to their conditions rather on the favourable side."[27] This 1762 inventory—in a format repeated until 1833—identifies 268 men, women, and children by name, age, occupation, and state of health, and adds a birth register to the death register. It is our first fully articulated description of the Mesopotamia slave community.

Daniel Barnjum's inventory of July 10, 1762, now housed in the Bodleian Library, is the key that unlocks the secrets of the Mesopotamia inventories. When I correlated it with the subsequent Mesopotamia inventories for 1763–1833 and with the previous Mesopotamia inventories for 1727–1761, the Barhams' slave population was opened to long-term inspection. But this correlation process required a lot of small mechanical adjustments. Usually the Mesopotamia bookkeepers added a year to the age of every slave at each annual listing, but occasionally they repeated last year's age statements, or decided to add or subtract several years. Thus, Sarah Affir, the heroine of Chapter 2, was listed as sixty-two years old in 1833, when she was actually sixty-six. I have corrected the age statements for Sarah Affir and for all the other slaves year by year when they were obviously erroneous. I have also revised the ages of those young slaves in the 1762 inventory who had recorded birth years in earlier inventories. For example, a girl named Joan, who is listed in the July 1762 inventory as age eight, was a newborn in 1751, so I have corrected her to age eleven in 1762. The adolescent African boys and girls imported to Mesopotamia did not, of course, have known birth dates. But when the bookkeepers added a year or two to their ages, I figured that they had noticed pubertal changes justifying these revisions, and so I accepted their new age statements. I have also attempted to systematize estimated age at death for the many slaves whose birth dates are unknown, by retaining the previous year's age for the people who died between January and June, and adding a year for the people who died between July and December.

Barnjum's inventory of July 10, 1762, enables me, for the first time, to diagram the age structure of the Mesopotamia population. Appendix 2 shows how things looked in 1762. An age diagram of a standard population is pyramidal, with numerous young people at the base gradually, tapering to the few oldest people at the top—males shown to the left and females to the right. But this diagram is not pyramidal. There are more adults in their forties than young children under age ten. Is this because Daniel Barnjum's age estimates are wildly erroneous? A pertinent question, since his age statements for the older people are clearly approximate. Barnjum lumped almost all men and women over the age of twenty-five into five-year categories, labeling them as aged thirty, thirty-five, forty, and so on. In an effort to test Barnjum's figures, I have correlated the 1762 inventory with the fourteen previous Mesopotamia inventories (dated 1727–1759), which often indicate birth years and sometimes birth dates. My conclusion is that his age statements turn out to be generally accurate for the young slaves up to about age twenty, and only a few years off plus or minus for slaves between age twenty and forty. The slaves past age forty are the most difficult to check and also the most problematic, and I believe that Barnjum generally exaggerated their true ages by a few years. So the population pyramid for 1762 is probably more top-heavy than it should be. But there were unquestionably a lot of senior slaves at Mesopotamia. The diagram demonstrates this by identifying the slaves who had lived at Mesopotamia since 1727, since 1736, and since 1744. All in all, Barnjum's estimates were close to correct for the younger people, and as accurate as can be reasonably expected for the entire population.

The Mesopotamia population, which had been youthful and vigorous in 1727 and 1736, had become strikingly elderly by 1762. There were not enough workers in their twenties and thirties for effective sugar production, which requires strenuous manual labor. The gender balance of 1736 had also been lost; by 1762 there were only 116 females as against 152 males, producing a skewed sex ratio of 131:100. Within the diminished female population, 40 percent were beyond the age of forty. The small number of women of childbearing age meant that Mesopotamia was poorly positioned for natural increase. Furthermore, a quarter of the adult slaves were listed as nonworking invalids. In sum, from every point of view the population was in poor shape.

Why had the Mesopotamia population lost the youthful, vigorous character it possessed back in 1727 and 1736? This question cannot be resolved, because the slave records from the 1730s, 1740s, and 1750s are much too incomplete.[28] But the surviving inventories taken during the twenty-six years between 1736 and 1762 show twice as many deaths as births, and a high proportion of the slaves who died during these years were children, teenagers, or young adults. Thirty-five of the fifty-six young children described in 1736 as "not yet fitt to work"—boys and girls who would have been prime workers in their late twenties or early thirties had they survived to 1762—were dead or gone. And many of the adults from the 1730s and 1740s who didn't die had become too old for productive labor.[29] To remedy this situation, the managers of the estate acquired large numbers of new slaves between 1736 and 1762, mostly from the African slave ships, and these new people were generally of prime working age on arrival. The age statements of seventy-one replacement slaves who were members of the Mesopotamia slave gang in 1762 indicate that they were about twenty years old on average when they came to the estate. Though some of these replacement slaves had also died, the majority of the prime-aged workers in 1762 were recent arrivals from Africa.[30]

The extraordinarily large number of deaths among both the estate-born slaves and the newly imported slaves strongly suggests that the managers of Mesopotamia had been overtaxing their workers. Mesopotamia crop accounts, which survive for the years between 1747 and 1761, show that the estate was shipping unusually large quantities of sugar considering the size of the working population. During these fifteen years the slaves produced an average of 201 hogsheads of sugar per annum for export, as against 136 hogsheads per annum during the decade 1762–1771 and 133 hogsheads during the decade 1772–1781.[31] Overwork in the cane pieces had so decimated the field gangs that by 1762 overseer Barnjum said he could muster only fifty effective workers, and had to hire jobbing slaves in order to keep the place going. He complained frequently that there were far too many elderly slaves who could do little or no work on the plantation. If only these people would die, he told Barham, "a Happy Release it will be, for they are so Enfeebled by age, as to be scarce able to help themselves."[32]

An unusual feature of the population change at Mesopotamia between 1736 and 1762 is that the female slaves survived less well than the males. Generally in Jamaica, slave women outlived the men,[33] but at Mesopotamia considerably more females had died. In particular, many fewer girl babies than boy babies survived the first few days after childbirth between 1736 and 1762.[34] In consequence, the gender balance of 1736 had been lost. The population diagram for 1762 shows a severe shortage of young women that not only impeded future childbirth but damaged the social fabric of the slave community. Nearly a third of the Mesopotamia men in their twenties, thirties, and forties could have no wives or sex partners unless they mated with women on neighboring estates. And this sexual imbalance would be difficult to correct, since the Jamaican planters always wanted to have young male workers, and habitually bought twice as many males as females from the African slave ships. In sum, the prospects for healthy family structure and natural population increase were not good.

Whether Joseph Foster Barham recognized in 1762 that the Mesopotamia slaves had been overworked is uncertain. He did not fire his attorney or his overseer. Daniel Barnjum continued to be the overseer at Mesopotamia until 1778, and he kept pestering Barham to buy new slaves. And Barham reluctantly complied. Between 1763 and 1774 he purchased eighty-three new people—all from Africa. They were filtered into Mesopotamia in small groups of ten or a dozen every few years. The great majority of them were males, which did nothing to redress the gender imbalance, but most of these newcomers were strikingly young, aged ten to fifteen, which suggests that Pool and Barnjum were trying to lower the age level of the population. As with the 1762 inventory, one may question the accuracy of the age estimates for these new Africans, but I reckon that even a Mesopotamia bookkeeper could observe the adolescent growth changes of pubescent boys and girls with some accuracy. A very significant age marker for young Africans would be the presence or absence of scars or body marks from tribal pubertal initiation rites that were practiced in West Africa at around age twelve.

Then came the American Revolutionary War, when a series of poor crops led to sugar earnings that barely exceeded expenses. Supplies from North America were cut off, and shipping to and from England was disrupted by American and French privateers. Food shortages were endemic in Jamaica, and the Mesopotamia slaves suffered severely. In 1777, twenty Mesopotamia slaves died during the year.[35] In 1780 a tremendous hurricane struck Westmoreland head on, did huge damage to the slaves' provision grounds, and killed two slaves at Mesopotamia. In 1781–1782 there was serious threat of invasion when the French fleet sailed to the Caribbean after Cornwallis's surrender at Yorktown, but Admiral Rodney came to the rescue by defeating the French at the Battle of the Saintes in April 1782. During the war Barham's oldest son, Joseph, arrived to find out what was going on at his father's two plantations, and he stayed in Jamaica for two years. During his visit young Joseph bought twelve slaves for Mesopotamia from the estate of Daniel Barnjum, who had died in 1778. Otherwise no slave purchases were made during the war, and the population at Mesopotamia shrank from 278 in 1774 to 243 in 1783.

With the close of the war, sugar prices rose significantly and economic prospects suddenly brightened. Joseph Foster Barham, in his fifties now, was changing in a way that reflected the new boom era. His wife Dorothea died in 1781, and without her he became estranged from the Moravian Church. In 1785 he married Lady Mary Hill, the widow of Sir Rowland Hill, in an Anglican ceremony and moved with his new bride to Hardwick Hall, a mansion in Shropshire. Though he left the Moravian community at Bedford, Barham continued to support the Moravian mission at Mesopotamia. And between 1784 and 1786 he bought sixty-five new slaves for Mesopotamia—twenty-four Africans from the slave ships and forty-one seasoned slaves from a Westmoreland estate called Three Mile River. This was the first time that Barham had bought a large group of local slaves, and his attorneys encouraged the purchase because Three Mile River "is quite in your neighborhood," and many slaves on that estate "had formed connexions with some of your Negroes."[36] The slave population climbed to a new high of 303 in 1786, sugar production at Mesopotamia expanded to 300 hogsheads, and gross income from the estate rose to £9,000 sterling per annum. This was the situation when Barham abruptly died on July 21, 1789, of a paralytic stroke at the age of fifty-nine.

In December 1789, five months following Barham's decease, there were 299 slaves on the estate. Appendix 3 shows the Mesopotamia population pyramid in 1789. The changes since 1762 are dramatic. There were many new young adults in their twenties and thirties, and most of the numerous people over age forty in 1762 were dead. The labor force was heavily male-dominated because Barham's attorneys had bought 117 new male slaves and only 43 new female slaves since 1762. The concentration of seventy-one young men in the twenty-five-to-thirty-four age bracket is especially notable. Most of these young men had been imported very recently, between 1784 and 1786. Relatively few had been born on the estate. Most of the Mesopotamia-born slaves were either too young or too old for heavy work, as the diagram demonstrates, and sugar production would have been almost impossible without the imported workers. All in all, the diagram shows how an over-the-hill workforce could be quickly rejuvenated by bringing in a lot of young adult slaves.

From the slaves' perspective, their population was suddenly enlarged in the mid-1780s with sixty-five strangers added to the mix, the new Africans with their very conspicuous tribal body scars speaking a medley of tribal languages, and the new Three Mile River people yanked from their previous Jamaican community—all assembled arbitrarily in order to keep Mesopotamia functioning. But how long would this rejuvenation last? Many of the problems from 1762 persisted. The proportion of youngsters under age twenty was even smaller than in

1762, and the proportion of females was also smaller. The sex ratio had risen from 131:100 to 134:100. The inventory of December 1789 also indicates a dismal health status for nearly all of the older Mesopotamia slaves. Only 14 percent of the 102 men and women past age thirty-five in 1789 were described as "able"; 86 percent were described as sick, weak, or infirm; and 13 percent were so incapacitated that they were listed as invalids unable to work. And there were still two deaths for every birth.

The new master of Mesopotamia in 1789 was Barham's eldest son, Joseph Foster Barham II. Brought up as a Moravian nonconformist, young Joseph was barred from Eton and Oxford, so his parents sent him abroad to be educated at a Moravian school at Barby on the Elbe River in Germany. In 1774, at age fifteen, he was studying Greek, Latin, and German. He declared at Barby that he wished to be received into the Moravian Church, but when he wrote home asking to exchange his coach for a chaise, his mother feared (probably correctly) that he was becoming too worldly and losing his faith. In 1775, at age sixteen, Joseph entered Göttingen University.[37] Four years later, at age twenty, he went out to inspect the family estates in Jamaica—very much as his father had done—but the situation was far more dangerous because he traveled to the West Indies in the midst of the American Revolutionary War.[38] From 1779 to 1781 young Joseph spent most of his time managing Island estate, but he also visited Mesopotamia, and despite the wartime shortages he spent a lot of money renovating the Island and Mesopotamia great houses and installing new furniture. At Island he bought a phaeton, several riding horses, and a £100 harpsichord for himself, and livery for his household slaves.[39] The attorney at Island, John Van Heilen, complained to the senior Barham that his son had done a poor job of managing the estate.[40] But young Joseph, for his part, considered that his time in Jamaica had given him the experience he needed to supervise the family plantations from long distance. He sailed back home in the summer of 1781, but his ship was captured by an American privateer, and he was taken to Boston. His family supposed that he was lost at sea, and when he finally came home via Newfoundland in December 1781 he was greeted as "one raised from the dead."[41]

Joseph II never returned to Jamaica. Instead, he settled immediately into a patrician life in England. In addition to his father's Jamaican estates, he acquired extensive property in Wales from his mother's family in 1803. He conformed to the Church of England, which opened the possibility of a political career. And Joseph II soon proved to be a much more ambitious and worldly man than his father. He took a nine-month continental tour in 1791–1792, then married into the nobility upon returning to England. His wife Lady Caroline Tufton was the daughter of the Earl of Thanet. Lady Caroline was a leading figure in the beau monde, and the Barhams were renowned for the brilliance of their entertainments at their London house in Queen Anne Street.[42]

The younger Barham enjoyed a significant career in the House of Commons. In 1793 he bought one of the two seats in Stockbridge, a notorious rotten borough in Hampshire, and sat for Stockbridge in 1793–1799, 1802–1806, and 1807–1822. He soon had to deal with Wilberforce's effort to ban the Atlantic slave trade. This was an issue that troubled him. While in Jamaica he had witnessed "some disgusting scenes"—perhaps slave auctions or slave floggings—and he came to believe that the continuous influx of new slaves from Africa had a corrupting influence on the sugar planters, who tended to abuse their workers because they could easily buy new ones. In 1792 Barham warned his Jamaica attorneys that the slave trade would soon be ended.[43] And in parliamentary debates in 1795 and 1796 he approved of gradually restricting the British trade. From 1804 onward he fully supported Wilberforce's bill for abolition, which finally passed the House of Commons in February 1807. While this vote put him at odds with his fellow planters, Barham repeatedly voiced his support for the sugar interest in Parliament, opposed the registration of slaves in Jamaica and elsewhere in 1817, and denounced the mischief done (as he saw it) by Methodist missionaries. During his final years in Parliament, Barham warned that the West Indian planters were in desperate financial difficulty, and he tried without success to advance a favorite scheme of his, to bring indentured laborers from India or China to the Caribbean as substitutes for the African slaves.[44] In June 1832, ten years after he left the House of Commons and four months before he died, Barham's rotten borough seat at Stockbridge was eliminated by the Great Reform Act.[45]

When Joseph II assumed the ownership of Mesopotamia in 1789, sugar prices in England were high, sugar production on the estate was double what it had been in 1762, and the slave population was larger than ever before. Barham was in an ebullient mood. He announced to his attorneys that he wished to expand his slave force so that three people could do the work of two in the past, which would "ease the labor of my slaves and make their task light." Recognizing that his father had installed an especially detailed form of bookkeeping, he decided not to change it. He directed that the Moravian missionaries continue to receive salaries and supplementary allowances. And he declared himself an apostle of "amelioration": the policy of improving living and working conditions for the slaves in order to encourage natural increase.[46] So he especially urged the attorneys to treat his slaves humanely. "I carry my ideas a little farther than many on this subject," he added, "and think myself obliged to consult not only their health but their happiness."[47]

Barham's attorneys were eager to implement Barham's expansion plans. In 1791 they bought sixty-one seasoned slaves from Southfield, a farming settlement in Westmoreland, said to be an attractive purchase because there were fifteen children in the group and only two invalids. And in 1792–1793 the attorneys bought 30 young Africans from the slave ships, including 19 adolescent girls, which brought Mesopotamia's population to 383 in December 1792.[48] The new purchases led Barham to believe that he now had enough slaves to lighten the workload on the estate and that he could encourage the young Mesopotamia women to have more babies. And he decided to stop dealing with the African ships. In an undated letter of around 1800 Barham wrote to his attorneys, "I cannot approve the purchase of any negroes from the ships on any account whatsoever." But he asked to be informed "when any eligible purchase should offer of negroes in the vicinity of my estate."[49] Thus, Barham withdrew from the Atlantic slave trade fourteen years before Parliament prohibited it, a span of time (1793–1807) in which his fellow Jamaican planters imported more than 150,000 slaves from Africa.

In 1790 Barham was dismayed to hear that a hurricane had smashed most of the sugar works buildings at Mesopotamia and ruined the slaves' houses. And attorney Wedderburn urged him in 1791 to buy still more slaves. "You say you wish 3 Negroes to do the labour of 2," Wedderburn commented. "From your strength & crops I think your slaves are full hard worked."[50] As it turned out, Mesopotamia sugar production was robust during the next twenty years, and the estate produced a record crop in 1810: 415 hogsheads of sugar valued at £17,092 Jamaican. But with no fresh recruits the slave population steadily declined—

from 380 in 1793 to 332 in 1803 and to 298 in 1813. To bolster the workforce Barham's attorneys bought a large group of 55 slaves from Cairncurran, a coffee plantation in Westmoreland, in 1814, but by 1818 the population had again declined to 309.[51] At this point Barham decided in December 1818, without consulting his attorneys, to buy a small sugar estate named Springfield in Hanover parish, and to transfer the 112 Springfield slaves to Mesopotamia.[52] Though seven of the Springfield people were left in Hanover, this new and largest acquisition brought the Mesopotamia population to a peak total of 421 in December 1820.[53] Barham made no further purchases, and with 125 recorded births as against recorded 230 deaths over the next fourteen years, the slave force shrank to 316 by 1834.

The seasoned slaves who came to Mesopotamia from Three Mile River, Southfield, Cairncurran, and Springfield between 1786 and 1819 had quite different characteristics from the people who were purchased from the slave ships. They arrived at the Barhams' estate already formed into family groups, they ranged in age from infancy to seventy-plus, and the sexes were almost evenly balanced (52 percent male). About half of these people had been born in Africa, half in Jamaica.[54] And since they were forced to leave their families and friends at their former plantations, most of them did not want to come to Mesopotamia. The Southfield and Springfield slaves, in particular, openly resisted, and many of them continually tried to get back to their old neighborhoods, which were a considerable distance away. The Three Mile River and Southfield cohorts included a good many young workers, but few of the Cairncurran and Springfield slaves were in their teens or twenties, and there were lots of people in all four groups who were too old for productive labor.

The slaves who were brought in shackles directly from Africa were of course in an even worse predicament. They also hated coming to Mesopotamia, and many of them ran away. But from the slaveholder's viewpoint, they made better replacement workers. The new Africans were almost entirely in their teens and twenties on arrival, and 68 percent of them were male. Thus (as I shall try to demonstrate in Chapter 4), Barham made a business mistake when he refused to deal with the slave ships. But from a social perspective Barham's policy helped to stabilize the Mesopotamia population by introducing more females and more established families. Though deaths continued to outnumber births, and the population continued to drop, the proportion of young women was increasing, and these young women were producing more children. There was potential for natural growth.

Joseph II was quickly disappointed with his Springfield purchase. By the 1820s he came to believe that the Mesopotamia people must be responsible for their demographic failure. Accepting the arguments of his Jamaican attorneys, Barham judged his laborers to be naturally "dissolute" and incapable of hard work. In 1823 he published a pamphlet entitled *Considerations on the Abolition of Negro Slavery* in which he claimed that the slaves in the British Caribbean were totally unfit for freedom and needed "moral improvement." He proposed a fanciful scheme in which the home government would buy out all of the Caribbean planters and through a massive educational program teach 800,000 Negroes how to become productive workers.[55] Barham reckoned that the government ought to pay the British planters £64 million for their slaves at £80 per slave, plus another £64 million for their Caribbean land and buildings. Had this plan been adopted in 1823, Barham would have personally received about £64,000 from the British government for his Mesopotamia estate. But of course his scheme was not adopted, and he did not live to see the British government's resolution of the British Caribbean slavery problem. Joseph II died on September 28, 1832, at the age of seventy-two. After his death his 329 slaves at Mesopotamia were priced for probate purposes at £20,195 in August 1833—the final full slave inventory for this estate. His total Jamaican estate (exclusive of land and buildings) was valued at £41,093.[56]

Barham's probate inventory taken in August 1833 gives us our last full look at the Mesopotamia slave population. Appendix 4 shows that the age structure was now considerably more pyramidal than it had been in 1762, and no longer distorted by the recent addition of new slaves as in 1789. Despite the fact that 255 slaves had been imported to the estate since 1789, Mesopotamia-born people now dominated the workforce. The proportion of young children on the estate was larger than in 1762 or 1789, although still strikingly small. The proportion of older men and women was also greater. In 1833 there were seventy-nine slaves above the age of fifty, compared with thirty-six in 1789 and thirty-four in 1762. And thirty of the Mesopotamia people—nine men and twenty-one women—were over the age of sixty. This increase reflects the fact that large numbers of senior slaves had been imported, especially in

1819, but it also suggests that health conditions on the estate were improving. From management's perspective, however, the Mesopotamia slave gang by 1833 was encumbered with far too many nonworking people and was a lot less effective than in 1789, when 39 percent of the slaves had been prime-aged workers in their twenties or early thirties—as against 25 percent in 1833. Furthermore, the gender balance had shifted. By 1833 there were only 157 males as against 172 females, for a sex ratio of 91:100, and the women in the prime field gang considerably outnumbered the men. In sum, although Mesopotamia's population structure in 1833 was quite different from the structure in 1762, we once again see a workforce not very well designed for effective sugar production.

In 1832 Joseph II's eldest son, John Foster Barham (1799–1838), inherited Mesopotamia, but he took no interest in his slaves, since they were on the brink of emancipation. He directed his agents to stop taking slave inventories, and he closed the Moravian mission. He was, however, very interested in receiving recompense from the British government for the loss of his property. On August 1, 1834, Barham's agents filed a claim valuing the 316 men, women, and children living at Mesopotamia on Emancipation Day at £13,538—or £43 per slave. Barham was doubtless disappointed when on October 12, 1835, the Office of Commissioners of Compensation awarded him only £5,612—or £18 per slave.[57] It is certainly unfortunate that Barham terminated record keeping at Mesopotamia just before the British government abolished slavery in Jamaica, because there is very little surviving evidence about what happened on this estate during the four-year emancipation period in 1834–1838. But we can be grateful that John Barham's widow brought his family archive with her when she married the Earl of Clarendon, because she thereby preserved for posterity Joseph I's and Joseph II's richly detailed longitudinal record of slave life at Mesopotamia.[58]

Reviewing the population changes at Mesopotamia between 1762 and 1833, we see an extraordinarily abusive slave system at work, in which the high death rate triggered rapid turnover. The Barhams' slave gang was a constantly changing polyglot mix of people brought together arbitrarily to keep the place functioning. Just over half of the 835 new slaves during these seventy-two years were born and raised on the estate, with parents and siblings to relate to. Another third (some African-born and some Creole) had lived previously on other Jamaican estates before they were moved to Mesopotamia in family groups. These people all shared something in common with the Mount Airy slaves who (as we shall see) developed a strongly kinship-based community in Virginia. But at Mesopotamia there were also 137 Africans who came directly from the slave ships between 1763 and 1793—and they had no parallel at Mount Airy. The Barham records provide very little personal information about these Africans; all we know is that they had been stolen from their West African families and tribal societies and brought as captives to Jamaica, and were singles aged ten to twenty-five on arrival at Mesopotamia. Fortunately, the diarist Thomas Thistlewood (who lived close to Mesopotamia and will figure prominently in Chapter 4) has a graphic description of the ten Africans he bought from a Guinea ship in April 1765:

> 1. A Temne man about 19 or 20 yrs. old, 5 ft 5 1/2 ins, several rows of punctures across the belly, & thus on the face.
> 2. A Coromantee man about 20 yrs., 5 ft 6 ins bow-shinned, 3 spots on his right cheek, 4 on his left.
> 3. A boy, about 13 or 14 yrs., 4 ft 10.3 ins., 3 perpendicular scars down each cheek.
> 4. A Susu boy 5 ft 1 in, about 17 yrs, several perpendicular small & one diagonal scar on each cheek.
> 5. A Coromantee or Ashanti boy, about 10 or 11 yrs, 4 ft 5.6 ins.
> 6. A Susu woman, 4 ft 8 ins., about 19 yrs. Two black marks descending obliquely from her eyes, and her belly full of her country marks.
> 7. A Coromantee woman-girl, about 12 yrs old, 4 ft 7 ins. Very black eyebrows.
> 8. A woman, 5 ft 1.9 ins, scar on her left cheek, back & belly full of country marks, holes through her nose. About 16 or 17 yrs old.
> 9. A woman-girl, 4 ft 9.1 ins, her face all over chamfered, about 13 yrs.
> 10. A Chamba woman-girl, on her face 3 long strokes down each cheek, 2 small oblique, etc. Her belly full of country marks, & in arch between her breasts. 4 ft 8.9 ins, about 13 yrs.[59]

Thistlewood identified seven of his ten new African slaves by tribe, and if he is correct, they came from widely separated regions of West Africa: from present-day Sierra Leone, Ghana, Guinea, and Togo.[60] The two youngest children (number 5 and number 7) had not yet received their tribal initiation rites. All of the others had individually distinctive body marks. And as a group they were short in stature. Thistlewood measured their height to the tenth of an inch, and found that the adolescent boys and girls were well under five feet, the mature females were around five feet, and the mature males were five feet six inches. We can be sure that every African at Mesopotamia who was over the age of twelve before being taken into captivity had body scars and country marks equivalent to those described by Thistlewood. In 1762 about half of the Mesopotamia slaves were African-born, so the mélange of tribal identifications must have been very conspicuous. But as time passed the African proportion declined. By 1833 only about 15 percent of the slaves at Mesopotamia were African-born, and most of them were elderly.[61]

Looking ahead, we will observe the slave system in action at Mesopotamia from various angles, and also grapple with the demographic issues exposed in this chapter. Why was the Mesopotamia birthrate, despite continuous changes in the composition of the population, always so low between 1762 and 1833? Why was the death rate always so high? Why did the gender balance shift from male dominance in 1762 and 1789 to a female majority in 1833? And we will also consider how the work experience of the slaves affected their health and life expectancy, and explore the relationship between slave labor and motherhood at Mesopotamia. This is the agenda for Chapter 4.

## *Mount Airy*

Two members of the Tayloe family kept outstanding slave records at Mount Airy: John Tayloe III, the master from 1792 to 1828, and his third son, William Henry Tayloe, who managed this plantation for his father from 1824 to 1828 and then became the owner through the Civil War. The father was a much bigger planter than his son, the reverse of the situation at Mesopotamia, where Joseph Foster Barham II operated on a larger scale than Joseph I. This was because the Barhams practiced primogeniture, passing all property from eldest son to eldest son, whereas John Tayloe III divided his estate among his seven sons. On John's death in 1828 the best Mount Airy farmland and nearly half of the Mount Airy slaves went to William Henry's younger brothers. One need not feel sorry for William Henry. He started out in 1828 with a very sizable inheritance, and by 1863 he held more slaves than Joseph Foster Barham II ever assembled at Mesopotamia, and had converted most of his Mount Airy people into cotton pickers in Alabama.

The Mount Airy inventory series starts in 1808, forty-six years after the first fully detailed Mesopotamia inventory. But since slavery in the Old South continued for thirty-one years after emancipation in Jamaica, both sets of records illuminate the last three generations of slave life in their respective regions. The Mount Airy inventories identify each man, woman, and child by name, age, and occupation—and also by location, which is important because these people were not concentrated in a single village as at Mesopotamia. Some of them lived and worked on the home plantation, whereas others lived and worked on eight farm quarters in three counties strung a distance of thirty miles along both sides of the Rappahannock River. This dispersion makes the Mount Airy inventories more complex than the Mesopotamia inventories, and for other reasons the Tayloe series is more difficult to piece together. There are major gaps, particularly in the 1850s; births and deaths are seldom noticed; young children are often not listed until they are several years old; and—for reasons that we shall soon see—the population is in constant flux. In John III's inventories, people sometimes drop out of the record without explanation, so it is unclear whether they have died or have been moved to an undisclosed new location. In William Henry's inventories, the Mount Airy slaves sent to Alabama were frequently switched from one cotton plantation to another. Thus, the Tayloe slaves are hard to track. But the effort is worthwhile because a population pattern emerges that is fundamentally different from the Barhams' setup in Jamaica.

Appendix 5 tabulates the population changes from 1809 (the first complete inventory) to 1863 (the last complete inventory). It summarizes the births, deaths, and movements of 964 people (517 males and 447 females) who lived at Mount Airy during these years.[62] The table shows that both John III and William Henry enjoyed a healthy surplus of slave births over slave deaths: the opposite of the situation at Mesopotamia. During these fifty-five years there were 502 births at Mount Airy as against 286 deaths. The Mount Airy slaves who were moved to Alabama by William Henry continued the same trend: 134 births versus 57 deaths. On average there were 11.6 recorded slave births and 6.2 slave deaths per annum.[63] As at Mesopotamia, these totals are by no means complete. A large but unknown number of infants and young children were never inventoried, and died without a trace, so it is impossible to establish accurate fertility or mortality rates. It is clear, however, that the population was always growing, which enabled both Tayloes to manipulate their bondsmen to a high degree. It is also notable that despite the constant movement, only fourteen slaves managed to escape from the Tayloes' control, all but one of them during the Civil War, when the Union lines were temptingly close to Mount Airy.

As with Mesopotamia, it is helpful to put the population figures of 1809–1863 in perspective by looking back at the early history of this Virginia estate. John Tayloe III was a fourth-generation slaveholder, the beneficiary of a long process by which his great-grandfather, grandfather, and father built one of the largest slave forces in the Chesapeake, mainly through natural increase.[64] William Tayloe (1645–1710), an immigrant from London, started small. He began to plant tobacco along Cat Point Creek in the bottomlands close to the Rappahannock River around 1680—just when Edmund Stephenson was starting Mesopotamia estate in Jamaica. Tobacco was a much less lucrative crop than sugar, but 1680 was an optimal date for starting a Northern Neck farm, because slave ships were coming into the Rappahannock River for the first time, and planters who could scrape together the money or the credit were switching from white indentured servant labor to black slave labor.[65] Obtaining slaves in Virginia at the close of the seventeenth century was not easy, because only a few hundred Africans arrived per year to service some 10,000 planters (as against 4,500 Africans per year to Jamaica to service a few hundred planters), and most of the slave ships went to the James and York Rivers. But William Tayloe did well in this raw frontier environment. He served as sheriff, colonel of the county militia, and member of the Virginia House of Burgesses, and when he died in 1710 his personal property was valued at £702—the estate of a small rancher in Jamaica, but a rich man's holding in Virginia. His principal asset was a workforce of twenty-one slaves, the nucleus of the Tayloe family's fortune.[66]

In the next generation, John Tayloe I (1687–1747)—a contemporary of Dr. Henry Barham at Mesopotamia—greatly expanded his father's land holdings and slave force. He was in the select group of Virginia planters—led by Robert "King" Carter of Corotoman on the Rappahannock and William Byrd II of Westover on the James—who suddenly emerged as major entrepreneurs in the early eighteenth century. John Tayloe I acquired extensive property through marriage, opened up ironworks on the Rappahannock and on the Potomac, and sent his ship, the

*Tayloe,* with tobacco to England. He served as a factor for a number of slave ships on the Rappahannock, which gave him first choice in buying slaves for himself.[67] Like his father, he became colonel of the Richmond militia and sat in the House of Burgesses, and in 1732 he gained a seat on the Virginia Council. So he became a much more conspicuous figure than Dr. Henry Barham in Jamaica, especially since most of the other Virginia tobacco planters continued to farm with few or no slaves. In Richmond County just two families became major slaveholders: the Carters of Sabine Hall and the Tayloes of Mount Airy. By 1747 John Tayloe I owned over 20,000 acres—developed and undeveloped—in five Virginia counties and three Maryland counties. And he expanded his father's slave force fifteenfold. He left his heirs 328 slaves: 167 at Mount Airy, another 121 on other Virginia work sites, and 40 in Maryland.[68]

John I's 1747 probate inventory gives us an early view of the Mount Airy slave force. It is an imperfect view, since the slaves are identified only by name and value, not by age, occupation, or family. They all have English names, though some of them must have been born in Africa.[69] The sex ratio (145:100) is heavily male, and the population seems to be youthful. While only fourteen slaves are specifically identified as children, nearly half are valued at £5 to £25, and most of these low-priced slaves were probably young boys or girls.[70] Tayloe's inventory was taken sixty-two years before the first fully detailed Mount Airy inventory of 1809, yet several of the oldest men and women in 1809 are possibly traceable back to 1747. A field hand named Tom, who was said to be seventy-one years old in 1809 and lived into the 1820s, could be the Tom who was valued in 1747 at £20, the right price for a nine-year-old boy. Likewise, Old Sucky, who was sixty-seven and an invalid in 1809 (though she lived for another twenty-two years), could be the Sucky who was valued in 1747 at £15, an appropriate price for a five-year-old girl.[71]

In the third generation, John Tayloe II (1721–1779)—a contemporary of Joseph Foster Barham I at Mesopotamia—was a somewhat less hard-driving man than his father, and enjoyed the civic role of wealthy planter. In the 1750s he built the Palladian mansion that the Tayloes still occupy to this day. "Here is an elegant Seat!" exclaimed the diarist Philip Fithian, who visited the house in 1774. Fithian particularly admired his host's twenty-four prints of "English Race-Horses, Drawn masterly, & set in elegant gilt Frames" and the "large, well formed, beautiful Garden, as fine in every Respect as any I have seen in Virginia," embellished with "four large beautiful Marble Statues." Fithian added that he had "some agreeable Conversation" with his host during his visit to Mount Airy, and that "Horses seem to be the Colonels favorite topic."[72] John Tayloe II also built a town house in Williamsburg, which he used frequently because he served twenty years on the Virginia Council, from 1757 to 1776, and then switched over to the revolutionary Council of State. In 1775 Tayloe displayed his patriotic ardor in a letter denouncing Royal Governor Dunmore. "Our Govenor," he told his Maryland son-in-law, "has issued his Proclamation giving freedom to all slaves and servants belonging to Rebells as he calls Us. He is a D——l, but has little in his power, thank God."[73] But John Tayloe II was always more interested in horses than in politics. Newspaper notices in the *Virginia Gazette* reported his triumphs at races in Williamsburg, Yorktown, and Fredericksburg, and Tayloe also advertised annually in the 1770s that his most celebrated steed, Yorick, was available for stud.

Meanwhile, John II by no means neglected his business affairs. Between 1747 and 1770 he acquired an additional 10,000 acres in the Northern Neck as well as new holdings in Maryland, and by 1779 he had doubled his father's acreage.[74] He used the ship *Tayloe* to transport convict servants from the London jails to Virginia, and he employed some of these convicts at his two ironworks and as sailors on his schooner *Occaquan*.[75] And his slave force continued to expand. Though no probate inventory was taken when John Tayloe II died, the 167 slaves at Mount Airy in 1747 seem to have increased to about 250 during his ownership, and by 1779 he held another 250 or so slaves on his outlying properties in Virginia and Maryland. Upon his death he gave £2,000 to each of his eight daughters and bequeathed the rest of his property to his one son, John, who was only eight years old.[76]

The new owner of Mount Airy, John Tayloe III, was a contemporary of Joseph Foster Barham II at Mesopotamia. During his childhood his large estate was managed by uncles, cousins, and brothers-in-law, with brother-in-law Ralph Wormeley acting as his chief guardian. Wormeley had received an elite English education at Eton and Cambridge, and in 1785, two years after the Revolutionary War ended, young John followed suit. He sailed to England at age fourteen and entered Eton College—where Joseph Foster Barham I had been a schoolboy nearly half a century before. In 1789 he matriculated at St. John's College, Cambridge University, where he spent the next two years.[77] Returning home in 1791, John III took possession of his estate when he reached his majority in 1792. He married Anne Ogle from a prominent Maryland family, and they soon began to produce a large number of children.

During John III's minority, county tax lists in 1782–1787 give us another partial view of the Mount Airy slave force. The people are identified by name only, and are grouped in three broad categories: "working" (that is, taxable), "under sixteen," and "old and infirm" (both nontaxable).[78] The lists show that the Mount Airy slave population in 1782 stood at 265, and had zoomed to 316 by 1787—a very notable increase since John II's death in 1779. The population in 1785 was youthful, with about 110 children under the age of sixteen, and only a dozen people too old or infirm to be taxed. The proportion of children at Mount Airy in 1785 was double the proportion of children at Mesopotamia in 1789, reflecting the most fundamental difference between the two communities. The Mount Airy sex ratio was quite even at 108:100, but there were significantly fewer women than men among the working slaves—which makes me suspect that the estate managers had been selling some of the young females, which became John III's policy when he took charge of Mount Airy in 1792.

During young Tayloe's six years in England the Mount Airy slave force continued to grow rapidly, and on his return he found that he had about 350 slaves at Mount Airy, a number that grew to 370 by 1792. This was more manpower than he needed, because he was switching from tobacco to grain production, which was much less labor-intensive.[79] And he could use the profits from slave sales to buy more land or pay off debts. So in September, October, and November 1792 John Tayloe III placed the following concisely worded advertisement in at least five Chesapeake newspapers: "For sale, two hundred Virginia born men, women, and children, all ages and descriptions." Prospective buyers were directed to Tayloe's agents in the town of Fredericksburg for further particulars.[80] The phrase "Virginia born" is telling: none of the slaves for sale were Africans, because the Tayloes had stopped buying from the slave ships at least fifty years before. The big question is, How many slaves did John III sell in 1792? Unfortunately, I have found no sale records, but the county tax lists show that the population at Mount Airy fell from about 370 in 1792 to about 260 in 1794, indicating that John III sold over 100 slaves during these two years. In 1799–1805 he added three new farms to his Mount Airy complex, which almost doubled his Rappahannock acreage and provided work for many more slaves. And he soon had more manpower, because the slave population at Mount Airy rapidly built up again from the low point in 1794. The tax lists show a population of 305 in 1801 and a population of 365 in 1805. The first complete slave inventory in 1809 enumerated 382 slaves at Mount Airy—which was a larger total than before the slave sale in 1792.

John Tayloe began to keep a slave inventory book in 1808, but the opening six pages of the volume have been torn out, eliminating the slave lists from three farm quarters. Therefore, the first complete surviving Mount Airy slave inventory was taken a year later, on January 1, 1809. This inventory lists 382 men, women, and children by name, age, occupation (if any), and value, and identifies the mothers of all children under the age of ten.[81] Unlike the Mesopotamia inventories, there is no effort made in this listing (or in any later Tayloe inventory) to describe the slaves' state of health. Also unlike the Mesopotamia inventories, where the slaves are always grouped together, the Mount Airy inventory of 1809 consists of nine separate lists. There is an inventory for the home plantation, where 106 domestics and craft workers lived with their families, and eight inventories for the widely scattered farm quarters—named Forkland, Old House, Doctor's Hall, Marske, and Menokin (all in Richmond County), Gwinfield (in Essex County), and Hopyard and Oaken Brow (in King George County)—where 276 field hands lived with their families.[82] The inventory for 1809 cannot be tested for accuracy, because we have only the 1808 inventory to check it against. But the age statements on the lists are clearly not wildly exaggerated. There is no indication that the ages of older people are inflated, since only 8.7 percent of the slaves are stated to be forty-five or above—which is less than half the proportion of senior slaves at Mesopotamia at any time between 1762 and 1833. And the ages of the younger slaves are distributed very evenly: for example, there are four men listed as age twenty, five as age twenty-one, four as age twenty-two, five as age twenty-three, four as age twenty-four, and five as age twenty-five. This persuades me that the 1809 Mount Airy inventory is based on previous listings, and that John Tayloe III had begun to register his slaves annually perhaps as early as 1792 in inventory books that have not survived.

It should be noted that, even more than with the Mesopotamia inventories, a lot of small mechanical adjustments are needed when correlating the Mount Airy inventories year by year. The overseers or clerks who compiled the Mount Airy inventories were sloppier than the Mesopotamia bookkeepers. Usually they added a year to the age of every slave at each annual listing, but quite often they repeated last year's age statements, or added several years or subtracted several years. Thus, two shoemakers, Ruffin and Joe, lost a number of years during the annual listings; they were said to be forty and fifty-five years old, respectively, in 1828, when (according to the 1808 inventory) they were forty-four and sixty-one. Similarly, three carpenters, named Charles, Andrew, and Phill, were said to be fifty-four, forty-nine, and thirty-four in 1828, when (according to the 1808 inventory) they were sixty-one, fifty-four, and forty. Coachman Harry, housemaid Franky, and gardener Godfrey had the opposite problem; they were said to be fifty-five, sixty-five, and fifty-one in 1828, when the 1808 inventory shows them to be somewhat younger: aged fifty, sixty-two, and forty-seven. I have corrected the age statements for these people and for all the other Mount Airy slaves year by year when they deviate without explanation from earlier age statements.

Appendix 6 presents Mount Airy's age structure as of 1809. Unlike any of the three Mesopotamia diagrams, we have here a genuine step pyramid. Mount Airy in 1809 had twice as many young children and half as many old people as Mesopotamia at the same date. The Mount Airy pyramid is not perfect, of course. The shortfall of boys aged up to four and of girls aged five to nine may reflect aberrations in recent infant births and early childhood deaths. But most of the other distortions result from John Tayloe III's policy of favoring male slaves over female slaves. The sex ratio in 1809 was 131:100, and hence much more male-dominated than it had been in 1787. The deficit among young women aged fifteen to twenty-four is especially evident, and indicates that Tayloe had recently been selling or transferring some of his teenage girls. Even so, there were more women of prime childbearing age at Mount Airy in 1809 than at Mesopotamia in 1762 or 1789. And the shortage of young women did not seriously impede population growth, since the Mount Airy women bore many more children than their counterparts in Jamaica.

The population diagram for 1809 highlights the cohort of eighty-five traceable slaves who had lived at Mount Airy since 1785. These people were survivors of the big slave sale back in 1792–1794. Almost two-thirds of them were males. And the gender difference is particularly striking if we trace the 110 children who were below age sixteen in 1785. Half of the males in this group were still living at Mount Airy in 1809, but less than one-quarter of the females. Here is another indication that when John Tayloe III staged his sale in 1792 he was much more willing to dispose of young women than young men. The diagram shows numerous males from the 1785 cohort in the forty-to-forty-four age bracket in 1809; these men had been twenty-three to twenty-seven years old in 1792 and at the peak of their strength for slave work, so Tayloe didn't give them up. All in all, despite the slave sale, the Mount Airy gang in 1809 was in a lot better working shape than the Mesopotamia gang ever was between 1762 and 1833.[83]

By January 1809 John Tayloe III was thirty-seven years old and the head of a large family. Like his father, John III was a champion breeder of racehorses, and was recognized as the leading Virginia turfman of his generation. John and his wife, Anne, lived during the summer months at Mount Airy, and during the winter social season in Washington, D.C., at their elegant town house—the Octagon—which they had constructed in 1799–1801 two blocks from the White House. The Tayloes had six sons and four daughters in 1809, with another son and three more daughters to come, which made his family situation very different from the two previous generations. John Tayloe I had had two daughters and one son, and John Tayloe II had had eight daughters and one son—and in both cases the sole son inherited all of the land and almost all of the slaves. Thus, from 1710 onward the Tayloes had been able to keep their estate intact. But since John Tayloe III had seven sons to provide for, all of whom expected to receive a portion of the family land and slaves, he was committed to dividing up his property. So he was not interested in having another large slave sale. To provide each son with an adequate portion, he acquired new farm quarters in Virginia and Maryland and two ironworks in the Blue Ridge. And he moved most of the slaves that he didn't need at Mount Airy to these other Chesapeake work sites.

In 1817 John and Anne Tayloe decided to live year-round at the Octagon in Washington, and in 1818 they placed their eldest son, John IV, in the family mansion at Mount Airy with a retinue of domestic servants and some craft workers. But John III kept control over the Mount Airy farm quarters, and most of the Mount Airy craft workers, until his death. There were 252 recorded slave births and 142 slave deaths at Mount Airy between 1809 and 1828, providing John III with 110 extra slaves. During these years he imported thirty-six people from his other Chesapeake properties and purchased four new slaves, which gave him forty additional people. He utilized this surplus to transfer 109 slaves out of Mount Airy and to sell 44 slaves that he didn't want. During the last twenty years of Tayloe's life 35 percent of the slaves who were living at Mount Airy in 1809 were either sold or transferred off the estate.[84] John III also moved eighty-five slaves from one work quarter to another work quarter within the Mount Airy complex. And every year he moved a dozen house servants from winter duty at the Octagon in Washington to summer duty at Mount Airy—until 1817, when they all stayed year-round with their master and mistress at the Octagon.

It may seem as though Tayloe was shuffling hundreds of people for no purpose, but in fact all of his moves were calculated to maximize the effectiveness of his labor force. The slaves who were shunted around by John Tayloe III had differing group characteristics. The dozen house servants who traveled between Mount Airy and the Octagon twice a year were especially wanted and needed because they were the Tayloes' personal attendants. The thirty-six people imported from outlying properties were mostly young craft workers or domestics who had talents that Tayloe wished to make use of at Mount Airy. By contrast, the eighty-five slaves who were switched from one Mount Airy work quarter to another were fledgling farmhands, boys and especially girls in their teens, who were separated from their parents as soon as they were old enough to work on their own. And the 109 slaves who were transferred to outlying Chesapeake work sites between 1809 and 1828 were mostly males of prime working age. During the 1810s forty-nine of these people were sent to Cloverdale, an iron furnace in the Blue Ridge over 200 miles to the northwest, to work there as founders, woodcutters, and colliers.[85]

All of this movement played havoc with family life. For example, a carpenter named Harry and his wife Agga (a spinner at the textile workshop) had eight children—John, Michael, Kitty, Caroline, Georgina, Ibby, Tom, and George. The parents lived out their lives at the home plantation, Harry dying in 1829 and Agga in 1856. But their children were widely dispersed. John was sent to Doctor's Hall quarter at age ten to become a farmhand, and was sold in 1819 at age twenty-one. Michael joined his father as a carpenter at Mount Airy until he was transferred to Cloverdale in remote western Virginia in 1827 at age twenty-seven. Kitty was sent to the Tayloes' Octagon House in Washington as a housemaid in 1816 at age eleven. Caroline was sold the year before John in 1818 at age eleven. Georgina became a farmhand at Doctor's Hall quarter in 1819 at age ten. Ibby was sent forty miles away to Windsor in King George County in 1822 at age eleven. Only the two youngest boys, Tom and George, who both became carpenters, like their father, remained permanently with their parents on the home plantation.

A farm woman named Winney (1775–1834) at Old House quarter was the victim of even more extreme manipulation. Winney worked as a field hand at Old House until 1817, when she was forty-two, then became a cotton spinner at Mount Airy until she retired in 1832. And while she was a field hand she bore seven children between 1801 and 1813. Her eldest son, Tom, disappeared from the Mount Airy records in 1811 when he was ten years old; he was apparently moved to a distant work site or sold. Patty, the next child, started, like her mother, as a field hand at Old House in 1815, and the next year, when she was thirteen, was sent away or sold, as Tom had been. Daniel, the third-born, started as a miller's boy and continued as a jobber until 1827, when at the age of twenty-one he was sent to the Cloverdale ironworks in western Virginia. Twins Barneby and John came next in 1811; Barneby died at age two, whereas John became a shoemaker at age eight. Eve, the youngest, joined her mother as a spinner until she was sold in 1829 at age sixteen. Five years later, mother Winney died at age fifty-nine, and in this same year her one remaining child, shoemaker John, was sold or moved off the estate at age twenty-three. Winney's family was not broken up; it was completely wiped out.

While he was building labor gangs for his seven sons, John Tayloe III sold forty-four slaves between 1809 and 1828. Thirty-two of them were females, and the older women selected for discard seem to have been in poor health, with a lower monetary value than other women of their age. In 1816 Tayloe dispatched twenty-six slaves for $7,210—or

$288 apiece.[86] This group included five mothers who were sold together with their young children. But two young boys, four-year-old Alfred and six-year-old Bailor, became orphans at Mount Airy when their mothers were taken from them. The most surprising member of this group in 1816 was Peter, a valued house servant who seems to have been sold in punishment for undisclosed misconduct. Tayloe clearly used the threat of sale to keep his slaves in line. When a miller named Reuben defied him by running away in 1817, he was caught the next year and put in the Fredericksburg jail, and sold in 1819.

Appendix 7 presents Mount Airy's population pyramid as of January 1, 1828, the year of John Tayloe III's death. It is instructive to compare this diagram with Appendix 6 for the population pyramid in 1809, because the two are surprisingly similar considering all the movement that had taken place. Less than half of the people who had been inventoried nineteen years before were still living on the estate. Yet the population (378) was virtually the same size as in 1809, the sex ratio was unchanged, and the age structure was fairly similar. To be sure, there were fewer teenagers than in 1809, significantly fewer adults in their thirties and early forties, and more than twice as many senior slaves over the age of forty-five. So the continuous out-migration had depleted the strength of the Mount Airy labor force. But Tayloe was probably not greatly bothered by this, because he had engineered (as in 1809) a Mount Airy cohort of prime-aged slaves with many more male than female workers. And half of the people in 1828 were under the age of twenty, proof that young workers and future workers were in good supply.

When John Tayloe III died in March 1828 he liberated one slave—his faithful body servant Archy (who was fifty-seven years old)—and bequeathed well over 700 other African Americans to members of his family. His eldest son, John IV, had died in 1824, and his second son, Benjamin (known as Ogle), wanted to live in Washington D.C., so third son William Henry inherited the core Mount Airy property in Richmond County with a reduced—but still very sizable—workforce. In addition to the family mansion, William received four farm quarters and 209 slaves.[87] The other four Mount Airy farm quarters, which lay in Essex and King George counties, went to William's younger brothers Edward, Henry, and Charles together with the 163 slaves who lived on these farms. Another nineteen Mount Airy slaves, mainly craft workers, were apportioned to William's brothers and his widowed mother. Thus, nearly half of the people living on the estate in 1828 abruptly disappeared from the annual inventories. The Tayloe brothers took some trouble to keep slave families together, but the division of 1828 permanently broke apart the extended Mount Airy slave community, and the social disruption must have been very great.

The new master of Mount Airy, William Henry Tayloe, had moved into the family mansion in 1824 when his eldest brother, John IV, died. He married his cousin Henrietta Ogle, and they had three children by 1828, with six more to come. After the settlement of 1828 William's Rappahannock domain was a lot smaller than his father's had been, but his workforce had plenty of potential. His 209 slaves expanded to 221 by 1830, and these people were extremely youthful: a full third were under the age of ten, and over half were less than twenty years old. In addition to his Mount Airy holdings, William shared with his brother Ogle the ownership of Deep Hole plantation on the Potomac River close to Washington, with another sixty slaves. But if Tayloe was labor rich, he considered himself to be land poor. The best Mount Airy farms had gone to his brothers, and William had to cope with declining crop yields and depressed grain prices. In the 1810s his father had generally sold about 7,000 bushels of Mount Airy wheat and 2,000 barrels of Mount Airy corn for $15,000 per annum, while William during the 1830s was averaging 1,800 bushels of wheat and 950 barrels of corn for only $5,250 per annum.[88]

To escape from this predicament he engaged in a joint effort with his brothers Henry and Ogle to start cotton production in Alabama. The story of how William moved 218 slaves from Mount Airy a distance of 800 miles south between 1833 and 1862, and what happened to them after they came to his new cotton plantations in Marengo and Perry counties in west-central Alabama, is the subject of Chapter 7. Here we need only briefly describe the step-by-step process that he employed to move these people to his new cotton work sites in Alabama.

The migration to Alabama was made possible by the continuing expansion of the Mount Airy population. In the 1830s William Henry Tayloe needed to maintain full agricultural production at Mount Airy in order to cover the loan he had advanced to his brother Henry to start the venture, so at first he was cautious about sending any of his Mount Airy workers to Alabama. When Henry set forth for Alabama in 1833, William gave him eight Mount Airy slaves. In 1835, after Henry sent encouraging reports, he sent sixteen more. And in 1836, when Henry was falling into debt and calling for quick cash, William sent another fourteen slaves to him with the aim of selling them at high prices in Alabama. This was his riskiest (and most reprehensible) move so far, but he had reduced his Virginia population only slightly; there were still 193 slaves at Mount Airy in 1837. William then stopped sending slaves to Henry, because he saw that his brother was a reckless manager who was moving too fast and was out of control. In 1843 Henry went bankrupt, and the other Tayloe brothers immediately seized his assets in order to protect the family investment. Henry's mismanagement had cost William dearly; thirteen of the Mount Airy people William sent him had been sold, and Henry owed him $21,000. But William now shared ownership of four Tayloe cotton plantations and was receiving about $6,000 in annual cotton sales, which was more than his Virginia grain sales, so he could not pull out.

In 1845 William Henry began to disengage from his partnership with his brothers so that he could operate his own cotton farm. He became the principal owner of Oakland plantation, where fifteen of his Mount Airy slaves were already working, and he sent a large party of forty-five Mount Airy slaves to join them. Nearly half of the slaves in this party were teenagers, of the right age to learn how to pick cotton. William was able to send this many people to Oakland because by 1845 his Virginia population had rebounded from 193 to 212. The migration of 1845 was a pivotal move. It established Oakland as a direct branch of Mount Airy, and the young people he sent down from Virginia soon began to marry and raise families. In 1847 William acquired full ownership of this plantation, and he also became half owner of two other Alabama plantations named Adventure and Walnut Grove. At Mount Airy, the exodus to Oakland dropped the population down from 212 in 1845 to 159 in 1846, but in the next few years it gradually built up again. By 1850, according to the U.S. slave census taken that year, William held 166 slaves at Mount Airy, and in Alabama he had seventy-one slaves at Oakland and sixty-two at Adventure and Walnut Grove.[89]

William Henry Tayloe now felt able to establish a second cotton plantation in Alabama staffed, like Oakland, mainly with his own Virginia slaves. In 1854 he sold his shares in Adventure and Walnut Grove to his brothers and used the money to buy a tract named Woodlawn, a few miles from Oakland. Then he sent another party of thirty-six Mount

Airy people down to Alabama. His Mount Airy population, which had risen from 159 to 182, now dropped to a new low of 151. Despite the loss of manpower, William was increasing his corn and wheat production at Mount Airy by improved husbandry, and Virginia crop prices were rising. In the mid-1850s he shipped 2,800 bushels of Mount Airy wheat and 1,050 barrels of Mount Airy corn per annum, and sold this grain for $9,500, nearly double what he had earned in the 1830s.[90] But his chief focus was on Alabama. In 1858 he moved his Woodlawn slaves to another tract named Larkin, so that both of his plantations—Oakland and Larkin—had ready access to the railroad for shipping. In 1858–1860 William's net receipts from cotton sales were averaging $39,000 per annum—quadruple his agricultural earnings in Virginia.[91] And by 1860 he had more slaves in Alabama than in Virginia. The federal slave census credited him with 169 slaves at Mount Airy and 277 slaves at Oakland and Larkin.[92] In a little over thirty years the Mount Airy population had lost only forty slaves, while William had more than doubled his workforce and had shifted his chief business activities from Virginia to the Deep South.

Then came the Civil War. William Henry Tayloe expected the Confederacy to win this contest, and he looked forward to expanding his slave-based cotton plantations after the war and converting his Virginia farms to livestock ranches. The war forced him to hasten the depopulation of Mount Airy, which was dangerously close to the Union lines. Thirteen of his slaves—ten men living at Mount Airy and three Mount Airy girls who worked for William's daughter in Georgetown, District of Columbia—deserted to the enemy during the war. But William's son Henry avoided greater desertions by moving forty-eight Mount Airy people down to Alabama in December 1861 and another fifty in March 1862. Altogether, William and his son moved a total of 218 slaves from Mount Airy to Alabama between 1833 and 1862. Unfortunately for William, the Union blockade of Confederate ports stopped the export of his cotton crop, and after 1862 he had to convert his cotton fields to food production. The glory days of the 1850s were over, at least temporarily. But William's Alabama slaves were in the geographical center of the Confederacy, as far removed as possible from the Union Army. In 1863 the father in Alabama and the son in Virginia took their last complete slave inventories, and these lists are especially informative, grouping parents with their children and identifying each family by surname.[93]

The combined population of Oakland and Larkin now stood at 391, and only 66 slaves were left at Mount Airy.

Appendix 8 presents the population pyramid for William Henry Tayloe's total slave force in Virginia and Alabama in 1863. This diagram is more complex than the diagrams for 1809 and 1828 since it combines four groups of people: 66 Mount Airy slaves who stayed in Virginia, 187 Mount Airy slaves who had moved to Alabama between 1833 and 1862, 110 children born to Mount Airy slaves in Alabama after 1833, and the 94 non–Mount Airy slaves acquired by Tayloe in Alabama or born on his Alabama plantations. Altogether, 76 percent of Tayloe's Alabama slaves were directly or indirectly from Mount Airy. And the diagram also shows how decisively he had shifted his labor force from Virginia to Alabama. In 1863, 86 percent of William Tayloe's slaves were living in Larkin and Oakland, and Mount Airy was now a nursery for young children and a retirement home for old people. Most basically, the diagram demonstrates that—despite the repeated shocks of sale and transfer and long-distance migration and wartime desertions—the African American population under Tayloe's control in 1863 remained vigorously expansive.

There were surprisingly few old people in the Tayloes' combined Mount Airy–Alabama slave population in 1863: only seven men and thirteen women were over the age of sixty. The total had been higher back in 1828: fifteen men and twelve women were then sixty or older. In fact, despite the huge demographic contrast between Mount Airy and Mesopotamia, William Tayloe in 1863 held considerably fewer old people than could be found at Mesopotamia in 1833, where thirty men and women were over the age of sixty in a smaller population. But if Tayloe's slaves in 1863 were not long-lived, they certainly had plenty of offspring. The sex ratio, at 109:100, was much more balanced than in 1809 or in 1828, and the mothers in 1863 were producing more young children than in 1809 or 1828. A full third of the slaves in 1863 were under the age of ten, and 53 percent were under the age of twenty. Thus, the population retained the strong pyramidal structure and potential for further growth that it had exhibited ever since 1809.

By 1863 the war was beginning to go badly. Grant's capture of Vicksburg in July 1863 and Sherman's capture of Atlanta in September 1864 opened Alabama to invasion, and in January 1865 Tayloe reported to his Virginia family that Union troops were advancing from

Pensacola to Montgomery. His worst fears were soon realized. On April 2, 1865, the Federal army attacked heavily fortified Selma and captured the town in a battle that lasted only two hours. One week later, Lee surrendered to Grant. In June 1865 Tayloe told his daughter-in-law Courtenay Tayloe that his neighborhood in Alabama was now controlled by Union troops.[94] All of a sudden, nearly 400 African Americans in Oakland and Larkin were no longer slaves.

Reviewing the demographic pattern at Mount Airy from 1809 to 1865, we can see that the constant growth of this slave population was a tremendously powerful engine. Between 1809 and 1828 the steady emergence of young Mount Airy laborers enabled John Tayloe III to send numerous workers from the home plantation to the new farms and ironworks that he laid out for his seven sons. Between 1828 and 1860 the continuing population increase at Mount Airy enabled William Henry Tayloe to staff new cotton farms in Alabama while still maintaining his grain farms in Virginia. And both Tayloes made thousands of dollars by selling the Mount Airy slaves that they didn't want. Thus, as at Mesopotamia, we have a slave system with high impact for the bondsmen. The Mesopotamia system led to frequent deaths, the Mount Airy system to frequent family breakups. In both cases there was a lot of turnover in the slave population.

Looking ahead, we will examine in later chapters the impact of the Tayloes' forced movement and/or sale of their slaves on the Mount Airy people. Slave family formation is well documented in the Tayloe records. The annual inventories identify all mothers and most fathers of the Mount Airy children, and they show these families in action over a long time period. Some of the families were mainly composed of field workers, while others were mostly domestics and craft workers. This occupational division was significant, as we shall see in Chapters 5, 7, and 8.

### *Westmoreland Jamaica versus Northern Neck Virginia*

Throughout Jamaica, as at Mesopotamia, the slaveholders were always dealing with population loss. From 1655 to 1807 they bought hundreds of thousands of replacement workers from Africa, and though the price of new Africans kept rising, they managed to enlarge the workforce and expand production until Parliament terminated the African slave trade in 1807. How did the slaveholders then deal with a shrinking labor force? To find out, we will survey the big sugar estates in Westmoreland parish, where Mesopotamia lay, during the years from 1807 to the termination of Jamaican slavery in 1834. Meanwhile, in Virginia most of the slaveholders profited from the expanding size of their African American workforce. They stopped buying from the slave ships before the American Revolution because they were able to supply present and future demands for black labor through natural increase. They soon had enough bondsmen to set up slave-based farm quarters across the interior of their state. And after 1790 they increasingly sold their surplus slaves or moved them beyond Virginia to the frontier West and the cotton South. How did this out-migration affect the Northern Neck of Virginia, where Mount Airy lay? To find out, we will examine black and white population changes in this long-settled Tidewater region between 1790 and 1860.

To start with Westmoreland, there is a great deal of documentation on the structure of the slave population in this Jamaica parish during the closing years of slavery.[95] Annual parish tax returns detail the distribution of slaves among the Westmoreland sugar estates, livestock pens, coffee plantations, provision farms, crossroads villages, and the port town of Savanna la Mar. They show that Westmoreland parish during the final generation of slavery was Mesopotamia estate writ large: less than 2,000 white people exercised mastery over more than 20,000 black people. And over half of the slaves in the parish were attached to some sixty-five sugar estates. Overall, as sugar production expanded, the slave population in the parish rose from 11,155 in 1740 to 16,700 in 1788 and to a peak of about 23,000 in 1807–1808.[96] In 1807, the last year in which new slaves from Africa were shipped to Jamaica, there were thirty-six Westmoreland sugar estates with gangs of 200 or more slaves, and thirteen (including Mesopotamia) with 300 or more slaves.[97] But the termination of the African slave trade immediately reversed this growth, and some of the marginal sugar producers soon shut down. By 1834 there were only twenty-four large estates with 200-plus slaves, though eleven of them (including Mesopotamia) still operated with gangs of more than 300 slaves.[98] And on the eve of emancipation the Westmoreland slave population had dropped by nearly 15 percent, to 20,003.[99]

A printed Westmoreland poll tax list dated March 28, 1814, identifies the 371 tax-paying property holders in the parish, and supplies much useful information about them. Joseph Foster Barham II is credited with 2,448 acres at Mesopotamia, 355 slaves, and 466 head of livestock. His estate was managed by six white men headed by the overseer John Patrickson, who personally owned eight slaves, two horses, and a two-wheeled chaise.[100] When Barham examined this list, he noted that his Scottish friend John Wedderburn held five Westmoreland sugar estates and two livestock pens occupying 16,069 acres and worked by 2,027 slaves.[101] The tax list also showed that nine of the thirty-six biggest sugar estates (those with 200-plus slaves) hired large numbers of extra workers. Fort William estate, which lay several miles east of Mesopotamia, listed 223 slaves, but 98 of them were hired or jobbing slaves. If the tax list is accurate, Mesopotamia and the remaining twenty-six large estates had no hired slaves as of March 1814, probably a positive sign, since this was crop time, and if overseers were shorthanded they would need to recruit extra help for the sugar harvest.

Three years after the Westmoreland poll tax list of 1814 was printed, the Jamaican Assembly was pressured by the British government to conduct a complete census of the slave population on the island. The census takers registered 345,252 slaves in Jamaica as of June 28, 1817. In Westmoreland parish, 22,659 slaves were registered, an increase of 1,642 over the poll tax total for 1817 taken three months previously, which shows that the new registration was much more thorough.[102] The registration also provided a lot more information about the slaves, identifying each individual by name, sex, age, color, and African or Creole birth. Subsequent triennial registrations, recorded in 1820, 1823, 1826, 1829, and 1832, reported on the slave births and deaths on every Westmoreland property within each three-year period. The 1817 census and the five registrations in 1820–1832 make it possible to compare changes in the Mesopotamia population with changes on the other large Westmoreland sugar estates.[103]

The 1817 registration for Mesopotamia, which is copied directly from the most recent inventory on the estate, reports that Joseph Foster Barham II held 321 slaves. Twenty-seven of these people (or 8 percent of the total) are identified as racially mixed or colored: eighteen mulattoes (black and white parents), three quadroons (mulatto and white parents), and six sambos (black and mulatto parents). The registration also identifies sixty-four of the Mesopotamians (or 20 percent) as born in Africa. The Barhams had purchased thirty-one of these Africans from the slave ships and acquired the other thirty-three when they bought groups of seasoned slaves in Jamaica.[104] Compared with the other Westmoreland sugar estates, Mesopotamia had an unusually small proportion of Africans, reflecting the fact that Barham had bought no people from the slave ships since 1793.[105] The other Westmoreland sugar gangs in 1817 tended to have more Africans and also smaller numbers of mulattoes or quadroons. At Cornwall (owned by the Gothic novelist "Monk" Lewis), 38 percent of the slaves were African-born and only 3 percent were colored; at Petersfield, 32 percent were African and 5 percent were colored; at Blackness, 44 percent were African and 4 percent were colored; and at Spring Garden, 41 percent were African and just 2 percent were colored. On the other hand, at Friendship and Greenwich estate (which bordered Mesopotamia), only 11 percent were African, while a very large 17 percent were colored. As we shall see in Chapters 2 and 4, the colored slaves at Mesopotamia were generally fathered by members of the white managerial staff, and it is hard to believe that the whites who operated Spring Garden were less sexually active than the whites at Mesopotamia. The Barhams required payment of £100 from those fathers who wished to free their colored slave children, and it may be that other owners were readier to let colored offspring go.

The 1817 slave registration, followed by triennial birth and death reports taken in 1820–1832 and 1834 (covering the final two years of slavery), shows the steady shrinkage of Joseph Foster Barham's slave force at Mesopotamia. In 1817 Barham had 321 slaves at Mesopotamia, and there were 114 slaves at Springfield (to be purchased by Barham in 1818), for a two-plantation total of 435 slaves. By 1820, after twenty births and thirty-two deaths in three years plus one runaway at Mesopotamia, and six births and seven deaths at Springfield (where the slaves had not yet been moved to Mesopotamia), the total was reduced to 421 slaves. In 1823, after twenty-four further births and thirty-seven deaths, and six Springfield people left behind in Hanover parish, Barham had a Mesopotamia population of 402 slaves. By 1826, with seventeen further births and thirty-three deaths, plus one slave transported and another manumitted, the total was down to 384 slaves. In 1829, with twenty-one further births and a staggering fifty-four deaths, Barham's slave force was reduced to 351. By 1832, with twenty-eight further births as against forty-six deaths and another manumitted slave, Barham now had 332 slaves. Finally, fifteen further births and twenty-eight deaths between 1832 and 1834 produced a final slave tally of 316—a decline of 27 percent in fifteen years.[106] And the slave registration returns show this same pattern on most of the other large Westmoreland sugar estates between 1817 and 1834.

Barry Higman has marshaled much statistical evidence to demonstrate that the slaves on Caribbean sugar estates had shorter lives and fewer children than the slaves on livestock pens or coffee plantations—or the slaves in any other form of labor. The Westmoreland slave registrations of 1817–1832 clearly illustrate Higman's argument. On the livestock pens, roughly equivalent to cattle ranches in the North American West, the slaves tended horses and cattle and other stock, and also planted corn, built fences, and cleared pastures—work that was definitely lighter and more varied than on the sugar estates. In the slave registration returns covering 1817–1832, nine of the pens in the parish with 100 or more slaves reported 509 births as against 484 deaths, and though two of them closed down before their workers were emancipated, four of the remaining seven had larger labor forces in 1834 than in 1817. By contrast, forty-two Westmoreland sugar estates with 100 or more slaves reported 3,314 births and 5,409 deaths between 1817 and 1832.[107] Only two of these estates claimed more births than deaths during this fifteen-year period, and ten estates reported more than twice as many deaths as births. These sugar planters filed compensation claims in 1834 for 1,400 fewer people than they had registered in 1817—a loss of approximately 13 percent in seventeen years.[108]

Appendix 9 compares slave population change at Mesopotamia with change at twenty-six of the other largest Westmoreland sugar estates, via the printed poll tax lists for 1807 and 1814, the slave registration returns for 1817–1832, and the compensation claims filed by the sugar planters in 1834 when slavery was terminated. In 1807 these estates all had gangs of 200 or more slaves, or 316 on average, which places Mesopotamia almost in the center of the group.[109] From 1807 to 1817 these sugar gangs collectively stayed at about the same size, but in order to keep at strength many estates bought or hired extra workers from local slaveholders who were closing down. We know that Mesopotamia acquired fifty-five new slaves in 1814, yet still suffered population loss during this decade, and Black Heath, Blue Castle, Bog, Canaan, Carawina, Fontabelle, Georges Plain, Kings Valley, Mount Eagle, Paul Island, Petersfield, and Shrewsbury must have similarly added many new workers between 1807 and 1817. The evidence of population loss is especially clear in 1817–1832, where twenty-five of the estates reported more deaths than births in the triennial slave registrations. Only Bog and Lenox reported natural increase during these years. Again, we know that Mesopotamia made up for its population loss by purchasing 107 slaves from Springfield estate in Hanover parish in 1818, and *Appendix 9* shows that Bath, Black Heath, Kings Valley, Petersfield, Shrewsbury, and Springfield (a Westmoreland estate, not to be confused with Barham's Hanover estate of the same name) similarly brought in many new slaves between 1817 and 1834. Yet by 1834 the collective slave population on these twenty-seven estates had declined by 18 percent since 1817.

There was also a significant shift in sex ratio on the big Westmoreland sugar estates between 1817 and 1834. Mesopotamia's sex ratio quickly changed from near parity (101:100) in 1807 to a decided female majority (90:100) by 1814. In the slave registration return of 1817, Mesopotamia's ratio remained at 90:100, while the other big sugar estates reported slightly stronger female dominance at 89:100. In 1817 twenty-two of the twenty-seven estates had more female than male slaves. And by 1834 the gender gap had widened further. Mesopotamia's sex ratio shifted to 88:100, while the twenty-seven big estates collectively reported a much greater imbalance of 83:100 in 1834. On the eve of emancipation every one of these big Westmoreland estates except Springfield had a female majority. This was by no means what the sugar planters wanted. For the past 150 years they had been bringing twice as many males as females from Africa in order to sustain male-dominated work gangs. And now they had to rely on women to handle most of the arduous labor of cane cultivation.

We get our closing look at Mesopotamia and the other big Westmoreland estates on Emancipation Day, August 1, 1834, when the sugar planters all filed claims to receive payment from the British government for the slaves they were losing. *Appendix 10* utilizes these compensation claims to compare the slave employment pattern at Mesopotamia with the other twenty-six estates we have seen in *Appendix 9*. When the government announced that it would distribute £20 million to the Caribbean slaveholders, and issued a table of valuations for various categories of workers and nonworkers, the planters filled out forms specifying the number of male and female "head people" (drivers and managers), "tradesmen" (craft workers), field laborers, domestics, children under age six, and elderly invalids on their estates.[110] All of the estates had much the same job distribution. Less than a quarter of the slaves were said to be too young or too old to work. Of the 5,356 males and females categorized as workers, about 5 percent were described as head people, 8 percent were assigned to craft jobs, 6 percent to domestic jobs, and an overwhelming 81 percent to field labor.

The proportion of elderly nonworkers in the 1834 tabulation looks suspiciously small, suggesting that the planters were categorizing some retired people as workers in order to claim more compensation money from the British government. Also, the minimal number of domestic servants indicates that—as at Mesopotamia—the owners of these places were absentees. Only Carawina, Fort William, and Lenox show household staffs that were large enough to suit the style of a resident sugar planter. Mesopotamia had the same job distribution as the other estates, but a higher proportion of old people and more young children than could be found on most of the other estates.

The 1834 tabulation highlights the damaging effects of the shortage of strong young males of the sort regularly supplied by the African slave ships up to 1807. In 1834, as always, most of the head people and all of the craft workers were men, and only a small number of women were employed as domestics, so 89 percent of the female workers labored in the field gangs, and they outnumbered the male field workers by a wide margin. Mesopotamia had fewer female field laborers than most of the other estates, but (as we shall see in Chapter 4) in 1834 more than half of the Mesopotamia first gang workers, who performed the most strenuous physical labor, were women. This was almost certainly the case on the other estates as well. On the valuation forms for compensation claims, field hands were divided into two categories: laborers and inferior laborers, with the "inferior" workers priced lower. And women outnumbered men in the higher-priced prime laborer category on every estate except Lenox and Springfield. The most extreme cases were Fort William, with 101 female and 50 male prime laborers, and Black Heath, with 100 female and 52 male prime laborers. And the women who heavily outnumbered men in most of these sugar slave gangs were producing relatively few young children. Mesopotamia, with twenty boys and twenty-four girls under age six, was well above average. Only seven estates on this list had a higher percentage of young children.

In sum, by 1834 these twenty-seven Westmoreland slave gangs were poorly structured for hard physical labor or for childbearing and population increase.

In the 1820s and early 1830s the big sugar planters in Westmoreland and throughout Jamaica still commanded large gangs of slaves who produced large quantities of sugar and rum. But their long-established system of enforced labor was becoming increasingly dysfunctional. Their slave gangs were more poorly proportioned than in the past, and their workers were more openly discontented and rebellious. The price for Jamaican sugar in the British market had fallen, the planters' profits were shrinking, and some were losing money. Most obviously, they had not solved the problem of a diminishing workforce. While the planters after 1807 made token efforts to encourage women to conceive and nurture their children, and slightly shortened the slaves' working days, they still made their people labor too long and hard and fed them too little, and deaths continued to outnumber births. Joseph Foster Barham II claimed that he cared about his slaves' health and happiness, but his efforts to improve their lives were superficial at best. And this was true for his fellow sugar planters. Their slave system was in serious decline.

Turning to Virginia, the history of the slave population in this state is far less well documented than in Jamaica, probably because the slaveholders didn't care to publicize much statistical information about their human property. The federal manuscript census schedules for Virginia in 1790 and 1800 were destroyed during the War of 1812, and the manuscript census schedules for 1810–1860 lump the slaves into broad age categories rather than listing them (as in the 1817 Jamaica slave registration) by name and age. Most Virginia county tax returns are even less helpful, since the youngest (nontaxable) slaves were often not counted. There is no possibility of comparing Mount Airy with other large Virginia plantations to document population changes and sex ratios over time as in Appendix 9 for the big Westmoreland sugar estates. Nevertheless, the census evidence that we do have shows a very different kind of slave system in action.

In 1810, when John Tayloe III inventoried 347 slaves at Mount Airy, Virginia was 200 years old, and (if the interior region that became West Virginia is included) this state had the largest population in the United States, just ahead of New York. According to the federal census of 1810, there were 983,000 people in the state.[111] About 559,000 Virginians were white and 424,000 (or 43 percent) were black, of whom 31,000 were free and 393,000 were enslaved. This was much unlike Jamaica, where blacks outnumbered whites by more than ten to one. But again unlike Jamaica, the black population in Virginia had doubled since 1776 entirely through natural increase. The Virginia slaveholders had a far larger geographical area to develop than the Jamaicans did, and for a century they had been steadily pushing into the interior in search of fresh farmland for their workers. By 1810 there were many more slaves in the central Piedmont region of the state than in the coastal Tidewater where settlement had first started, and slaveholders had moved across the Blue Ridge Mountains into the Shenandoah valley.[112]

In 1810 there were eighty-three counties in Virginia, excluding the sixteen counties in what is now West Virginia. The federal census taken that year, though very carelessly compiled, permits a county-by-county examination of the slaveholding pattern throughout the state. *Appendix 11* compares Richmond County (the seat of Mount Airy) with five other geographically diverse counties: York, one of the eight earliest-settled Tidewater counties at the mouth of the York River; Amelia, a Piedmont county in the center of the state with one of the highest slave ratios in the Chesapeake; Amherst, in the far western Piedmont, and Fauquier, in the far northern Piedmont, both more recently settled counties with rapidly expanding slave populations; and Frederick, beyond the Blue Ridge in the Shenandoah valley, where slavery was spreading more slowly. Collectively, these six counties had a population that was 54 percent white and 46 percent slave, which was not far from the statewide ratio of 57:43. And when set against each other, the tabulations for these counties show surprisingly small differences between the oldest-settled Tidewater region, the booming Piedmont region, and the frontier Shenandoah valley.

The first thing that a Jamaican would notice about this Virginia tabulation is the scarcity of really big slaveholders. Only thirteen planters in these six counties held more than a hundred slaves, and they possessed a mere 6 percent of the slave force. Nathaniel Burwell in Frederick County with 325 slaves was the leader of this group, and John Tayloe III in Richmond County came next with 225. Tayloe's Mount Airy complex was spread over two adjoining counties, so in 1810 he held an additional 122 slaves on three farm quarters in Essex and King George counties, plus numerous slaves in other parts of Virginia. Some of the other big planters similarly operated farm quarters or ironworks and other industrial projects with enslaved workers in multiple counties. Yet the great majority of the bondsmen in all six counties were owned by smaller planters, especially by the owners of six to twenty-four slaves.

There was a strong contrast between slave-rich Amelia, where 82 percent of the white householders were slave owners, and frontier Frederick, where only 32 percent were slave owners, and other regional differences can be noted. But the commonalities are more striking. Overall, in these six counties 48 percent of the white householders stood outside the slave system. Another 9 percent owned a single slave. A further 17 percent owned two to five slaves, and 22 percent owned six to twenty-four slaves, enough to populate a farm quarter. This group of relatively small slaveholders held 51 percent of the bondsmen, far more than the biggest planters like Burwell and Tayloe. Thus, compared with Jamaica, slaveholding was distributed widely and fairly evenly across the white population. We see here an extensive and deeply entrenched slave system.

Such was the situation in 1810. But new developments were under way with major consequences for the Virginia slave population. As early as the 1790s, when the best land within the state was mostly taken up, large numbers of white Virginians looked for better opportunities elsewhere. They began to migrate to Kentucky and Tennessee, taking some 60,000 Virginia slaves with them between 1790 and 1809.[113] And the exodus accelerated after 1810 when Virginians began to move with their slaves to the cotton states or began to sell their surplus slaves to professional traders who took them to the Deep South. It has been estimated that approximately 440,000 slaves were exported from Virginia between 1810 and 1860—nearly half of the total number of slaves exported from the seaboard slave states during these fifty years. Some 445,000 slaves were imported to the two states of Alabama and Mississippi in 1810–1860, an unknown but huge percentage of them from Virginia.[114] Despite this enormous out-migration, the Virginia slave population continued to grow after 1810, but much more slowly than before, and stood at 490,000 in 1860.

What was the impact of this massive out-migration on the Tayloes' neighborhood, the Northern Neck, the Tidewater region bordered by the broad Potomac and Rappahannock Rivers? There are five relatively small counties in the Northern Neck: Lancaster, Northumberland,

Richmond, Westmoreland, and King George. Mount Airy is situated in Richmond County, but the Tayloes also had slave quarters in King George County to the northwest of Mount Airy, and in Essex County across the Rappahannock River. For many years in this Tidewater region, where the farmland was tobacco-worn and crop prices were low, both white and black people had been moving out, first to the Virginia Piedmont and the Shenandoah valley, then to Kentucky and Tennessee. After 1810 they would increasingly go much further, to Alabama, Mississippi, and Louisiana. Because of this continuous exodus, the Northern Neck from 1790 onward became one of the few regions in Virginia to experience population loss.[115]

Overall, in 1790 the five Northern Neck counties had a white population of 15,989 (43 percent), a slave population of 20,262 (55 percent), and a free black population of 623 (2 percent). By 1860 the totals were 15,318 whites (45 percent), 16,151 slaves (47 percent), and 2,922 free blacks (8 percent). Only Westmoreland County gained in total population during these seventy years, and this was because the number of free blacks in this county grew from 114 in 1790 to 1,191 in 1860—a most unusual development, precipitated by Robert Carter of Nomini Hall, the chief slaveholder in Westmoreland, who decided in 1791 against the protests of his white heirs to free all of his 509 slaves.[116] Thanks to Carter, the free black population in the region expanded vigorously, while the enslaved black population was shrinking by 20 percent.

Richmond County, the seat of Mount Airy, saw striking changes, losing 38 percent of its slave force between 1790 and 1860, while gaining in white population. Alone among the Northern Neck counties, Richmond changed from a black majority to a white majority. But while the Richmond slave force fell from nearly 4,000 in 1790 to under 2,500 in 1860, the county census returns demonstrate that the resident slaves were vigorously reproducing. In 1820, at the close of a decade in which the county lost 514 slaves, the census grouped the Richmond slaves in four age categories: 44 percent were children aged 0–13, 24 percent were young people aged 14–25, 20 percent were older adults aged 26–44, and 12 percent were adults past 45. In other words, two-thirds of the slaves were under the age of twenty-six, signifying a young and expanding population. In 1830, when the county tabulated thirty-four fewer slaves than in 1820, the census grouped the Richmond slaves in six age categories, with much the same result: 37 percent were said to be under the age of ten, and 62 percent were under twenty-four. And in 1840, after another decade's population loss of 267, the census found still greater youth: 65 percent of the Richmond slaves were now categorized as under the age of twenty-four. In short, the Richmond slaves were most certainly not dying off. Rather, they were being taken to work sites outside of the county by their masters or sold to professional slave traders.

We have seen in *Appendix 11* that in 1810 Richmond County's slave distribution was fairly standard, with slightly fewer nonslaveholders, slightly more owners of one to twenty-four slaves, slightly fewer owners of twenty-five to ninety-nine slaves, and more slaves held by the two biggest planters—the Tayloes of Mount Airy and the Carters of Sabine Hall—than the average for the six geographically diverse counties in this appendix. It remains to ask whether the 1810 pattern changed significantly during the next fifty years while at least 2,000 slaves were being removed from Richmond. The manuscript census schedules show remarkably little change. In each census from 1820 to 1860, close to 12 percent of the slaves in Richmond were held by the Tayloes and Carters, another 28 percent by planters who owned twenty-five to ninety-nine slaves, and the remaining 60 percent by small farmers with one to twenty-four slaves.

Digging a little deeper, however, we find considerable fluidity among the substantial householders who owned twenty-five to ninety-nine slaves. In 1810 there were twenty-five people in this category, but by 1820 fifteen of them were gone, replaced by nine people with new surnames, five of whom had also disappeared by 1830, when another seven people with new surnames appeared on the list. And this pattern of changeover among the more substantial slaveholders continued in the census listings for 1840, 1850, and 1860. There were several Richmond families with twenty-five to ninety-nine slaves—the Belfields, Brockenbaughs, Chinns, McCartys, Mitchells, and Settles—who remained in residence from 1810 to 1860. But just as William Henry Tayloe was sending his best young field hands from Mount Airy to Alabama, so many of the more ambitious Richmond farmers were selling their human property for high prices, or picking up stakes and moving with their slaves to better farmland in the West and the South. Meanwhile, the majority of the Richmond slaves belonged to small householders who typically owned five or six slaves apiece. In Richmond County, as everywhere else in Virginia, the slave system penetrated deeply, and the fact that huge numbers of ordinary white Virginians possessed black bondsmen solidified the Old Dominion's decision to go to war in 1861 in order to defend its slave-based way of life.

Thus, we see two radically different slave systems in action. The sugar planters in Jamaica had been making lots of money for many years by treating their workers as disposable cogs in a machine: importing slaves from Africa, working them too hard, feeding them too little, exposing them to debilitating disease, and routinely importing new Africans to replace those who died. It was a shockingly inhumane system, and when public uproar in Britain forced the closure of the transatlantic slave trade in 1807, the planters' game plan had to change if they were going to retain the use of slave labor. But they were unable to change, because the high death rate and low birthrate had undermined their system. And the planters were so dependent on the British army and navy for military protection and on the home market for their sugar sales that when the government offered them £20 million in compensation they were forced to accept the abolition of slavery in 1834.

In Virginia, the continuous growth of the slave population was a compulsive force that enabled the slaveholders to populate the interior of their state with black workers, and to move or sell hundreds of thousands of Virginia blacks to slave-hungry states in the West and the South. The Northern Neck statistics show a shrinking slave population in this long-settled region, but this was no demographic decline as in Jamaica. The slaves in Richmond County were locked into a multistate system that was expanding aggressively. The Virginia slaveholders, dismissing criticism from northern abolitionists, were confident that they had developed a well-functioning labor system. In total contrast to the British Caribbean situation in the 1830s, it would take four years of ferocious warfare, fought largely on Virginia battlefields, before the slaves in the Confederate states could finally be liberated.

# 2

## *Sarah Affir and Her Mesopotamia Family*

When examined year by year, as we saw in Chapter 1, the Mesopotamia and Mount Airy inventories show how each of these two slave populations looked at a particular date, and how they changed in size, in gender balance, and in age structure over time. When examined person by person, they reveal the skeletal biographies of all the 1,103 slaves who lived at Mesopotamia between 1762 and 1833, and all the 973 slaves who lived at Mount Airy between 1808 and 1865. The documentation for most of these 2,076 slaves is admittedly minimal, but the inventories *do* annually track the employment history of every male or female, plus (at Mesopotamia) his or her changing state of health, and (at Mount Airy) his or her removal from one work site to another. The annual entries for the women are especially pertinent, because they disclose childbearing and motherhood, and permit the reconstruction of family histories. And it is sometimes possible to tease out more satisfying information concerning individual men and women. In this chapter and the next, I focus upon two well-documented slave women—Affy (later known as Sarah Affir) at Mesopotamia and Winney (later known as Winney Grimshaw) at Mount Airy—and also track Affy's and Winney's families over a span of four generations.

In Affy's case, two aspects of her life are particularly interesting. First, she illustrates the process of racial mixing at Mesopotamia. Affy was a black woman who had sex with both black men and white men and bore six children. Four of her babies were black, and two, named Robert and Ann, were mulatto. Her mulatto daughter Ann had sex with black men, mulatto men, and white men and (like her mother) bore six children. Two of her babies were black, two were mulatto, and two, named Jane and John, were quadroon. In Jamaican parlance, any black person with some white ancestry was "colored," but everyone was conscious of the gradations between sambo (mulatto-black parentage), mulatto (black-white parentage), and quadroon (mulatto-white parentage). And racial mixing greatly affected one's employment opportunities. Mulattoes Robert and Ann and quadroons Jane and John were given far more attractive job assignments than black Affy because they had white fathers. A second point to notice is that Affy became a member of the Mesopotamia Moravian congregation and was baptized as Sarah Affir. Subsequently, her mulatto son Robert was baptized as Robert McAlpine and her quadroon granddaughter Jane was baptized as Jane Ritchie. All three had close personal contact with the missionaries who maintained a station at Mesopotamia from 1758 to 1836, but (as we shall see) Affy, Robert, and Jane had strikingly different encounters with the Moravian missionaries.

Affy's family tree is presented in *Appendix 12*, and in full form on the website that accompanies this book, *www.twoplantations.com*. Affy appears in sixty-six of the Mesopotamia inventories, Robert in forty inventories, and Jane in nineteen. Here is their story, generation by generation, pieced together from all of the references to them in the Barham plantation papers and the Moravian church records.

## *Sarah Affir*

According to the slave inventory for 1767, three babies were born at Mesopotamia during this year.[1] A female named Affy was born in April. A male child, oddly named Boy—perhaps to indicate the gender of an infant who was not expected to live very long—was also born in April. A second female named Doll was born in October. The bookkeepers failed to notice the arrival date of a third baby girl named Nancy, whose birth month is therefore unknown. The revised total of four births in 1767 is strikingly small, considering that some 80 women and 110 men were living together in the slave village on the estate. One reason for the small number is that the bookkeepers seldom bothered to register babies who died within a few days, and there were many such babies at Mesopotamia. It was extremely common in Jamaica for slave infants to be infected by neonatal tetanus, caused when the umbilical cord was cut by a nonsterile knife or when the baby's umbilical stump wound was not treated hygienically. Neonatal tetanus (termed *lockjaw* in the Mesopotamia inventories) was—and still is—generally fatal. The Westmoreland diarist Thomas Thistlewood reported over and over again that new slave babies on his estate were dying of lockjaw a few days after birth; during the 1770s he recorded seven neonatal lockjaw deaths in a population of about thirty slaves.[2] But the four Mesopotamia newborns who entered the records in 1767 escaped lockjaw. In fact, they all turned out to be quite long-lived. Boy died at forty-eight in 1816, Nancy at fifty-two in 1819, and Doll and Affy were still living on the estate in August 1833, aged sixty-five and sixty-six.

The 1767 inventory reports not only a small birth cohort but a large death cohort. Nine people died during the year, a little below the Mesopotamia average of 10.4 deaths per annum between 1762 and 1833. This is of course the minimal number, excluding unrecorded infants who died within a few days. And it shows the standard pattern: slave deaths outpaced slave births in sixty-three of the recorded seventy-two years at Mesopotamia. Death was ever present in Jamaica. Vincent Brown has written an excellent book on the subject, arguing persuasively that "death structured society and shaped its most consequential struggles."[3] But the death list in 1767 also tells us something about life. Most of the people who died in this year had lived into their forties or fifties, and they had been at Mesopotamia for many years. Hodge was first recorded in 1736 as a sawyer, aged about twenty-four. Oxford was also first recorded in 1736 as a field hand, aged about nineteen. The youngest person, field hand Jack, had been born at Mesopotamia and was said to be in excellent health when he caught pneumonia and died at age twenty-one. On the other hand, Murdock and Richmond were new Africans imported from a slave ship in 1765, and they lasted at Mesopotamia for only two years.

Neither of Affy's parents is identified in the birth registry of 1767. Nor are the parents of two other babies born in 1767: Boy and Nancy. The Mesopotamia bookkeepers only began to record the mothers of newborns in 1774, and they never recorded the fathers. Thus, Affy, Boy, and Nancy are among the forty-four children born in the 1760s and early 1770s whose mothers are not known. By accident, however, we can identify the mother of the fourth baby who entered the Mesopotamia slave record in 1767. Half a century later, in 1817, when all of the Jamaica slaves were registered by the British government, Doll (then fifty years old) was identified as the daughter of a Mesopotamia woman named Lucy (who was then sixty-eight years old). This seemingly irrelevant piece of information will turn out to be helpful, as we shall soon see.

Like all of the other Mesopotamia babies born at this time, Affy was given no surname in the record book. And what about "Affy"? Was it a real name chosen by her mother or a book name imposed by the white record keeper? Some 12 percent of the names given to the 420 slave babies at Mesopotamia between 1762 and 1833 seem unlikely to have been selected by their mothers. Why would a Jamaican slave give her child a British place-name like Aberdeen, Cambridge, Chelsea, Dorset, Dundee, Dublin, Edinburgh, or Windsor? Or a classical name like Brutus, Cato, Dido, Hercules, Juno, Jupiter, Neptune, Phoenix, or Psyche? Or a comic name like Blackemoor, Bunny, Danger, Goodluck, Prince of Wales, or Trouble? These slave names were almost certainly imposed by the overseers or bookkeepers. But most of the newborns received typically English names and nicknames or biblical names, which seem more likely to have been chosen by the mothers—even though the mothers must have known that the cattle on the estate carried these same names. Only about 8 percent of the Mesopotamia babies received African or pseudo-African names such as Affy. In the annual inventories her name was spelled several different ways, somctimes Affie or Affry, but most often Affy. Again, I believe that her mother chose this name. And did Affy actually use it? Probably she did, because when converted to Christianity by the Moravians as an adult she took a baptismal name that was a variant or extension of her book name—and this also happened with other members of her family.

From 1767 to 1773 Affy was annually listed among the other little girls on the estate, with no added comments about her occupation (since she had none) or her physical condition. Then at age seven she was put to work: the common pattern at Mesopotamia. Many of the little boys and practically all of the little girls began employment in the grass gang, which in Affy's time was a squad of about thirty youngsters aged six to sixteen who gleaned the fields and pastures, carried grass to the livestock pens, carried manure into the cane fields, and hoed the young sugarcane after it had been planted. Once these children began as field hands, they continued in the same line of work as adults, performing strenuous gang agricultural labor until their health broke down. For the boys at Mesopotamia, the odds of becoming a field hand were one in two; for the girls, the odds were five in six. But Affy escaped this scenario temporarily, and started to work as a domestic servant instead.

Affy's first employment supplies us with a strong hint as to who her mother was. Doll, also born in 1767, started working as a domestic in 1774, and Doll, as we have learned, had a mother named Lucy. In 1774 Lucy was employed as a seamstress at the Great House. Clearly she was well placed to obtain an indoor job for her little girl. Was there another woman working as a domestic in 1774 of the right age to be Affy's mother? Three of the five other house servants were women in their fifties and hence too old. Washerwoman Silvia (age forty-three) was possible. But head servant Amelia (age thirty-five) was far more likely. She was the housekeeper for overseer Daniel Barnjum, who occupied the Great House in the absence of the owner, Joseph Foster Barham, and was certainly in the best position of any slave woman on the estate to secure a favored house job for her young daughter. In 1774 Amelia was probably two or three years older than thirty-five, for reasons that I will explain below, and was actually about thirty or thirty-one when Affy was born.

Assuming that Amelia was Affy's mother, one might like to conjecture that overseer Barnjum was Affy's father. But this was not so. We *do* know (see Chapter 6) that Barnjum had a slave mistress named Hannah and sired at least one mulatto baby by her, and he may well have had sex with Amelia as well. But Affy was not his child. Mulattoes were always designated as such in the Mesopotamia records, because they were considered to be genetically superior to black children. Affy was never identified as a mulatto, and after her childhood she was never given the privileged job assignments that mulattoes always received. But it remains possible that Affy lived with her mother in the Great House rather than in the slave village as a child, and she undoubtedly came into closer contact—for better or worse—with the white managerial staff than most of the other slave children.

Amelia, whom I take to be Affy's mother, must have had some standout qualities to be chosen as the overseer's housekeeper in a slave system where almost all of the females were put to work in the field gangs, and where colored slaves had first call on the few domestic jobs. Unfortunately, the broken and skimpy series of Mesopotamia inventories taken before 1762 reveals very little about her. It is unclear whether

Amelia was born on the estate or imported as a child, but since hardly any Africans became domestic servants at Mesopotamia she was probably born in Jamaica. Amelia was not listed in the inventories taken in 1736 and 1743, but she did appear in the 1744 inventory, where she was identified simply as a girl, age unstated. Seven years later, on the next inventory, taken in 1751, she was again grouped with the girls, but from 1752 onward she was listed among the women. This suggests that Amelia was about sixteen years old in 1752 and hence born around 1736. In 1756, when all the slaves were priced for probate purposes, she was valued at £65—close to the top figure for a female slave. The annual inventories from 1757 to 1761 supply no new information about her. But in July 1762—when for the first time the Mesopotamia slaves were identified by age, occupation, and state of health—Amelia was described as twenty-two years old (though she was probably closer to twenty-six) and as a seamstress in good health. This was five years before Affy's birth.

Starting to work in 1774, Affy served as a house girl for five years. She probably assisted Amelia, and undoubtedly came into close contact with overseer Daniel Barnjum, who managed Mesopotamia from 1760 until he retired in 1778. During Barnjum's regime the slaves complained frequently to the Moravian missionaries that they were badly mistreated, but he was a much less severe taskmaster than John Graham, who became the overseer in 1778. And Affy quickly experienced the change of management, because she was reassigned to field work in 1779, when she was twelve years old. Amelia continued to be Graham's housekeeper, and stayed in this job until she died in 1784, when Affy was seventeen. But she was unable to protect the girl from sugar labor. Doll had already been sent into the cane fields in 1778, and an older African girl named Monimia was also moved from domestic to field work in 1779.

Between June 1779 and July 1781, when Affy was twelve to fourteen years old, she was undoubtedly presented to her future owner, Joseph Foster Barham II—along with all of the other Mesopotamia slaves—on several occasions. The younger Joseph Barham came out to Jamaica at age twenty to inspect the two sugar estates he would inherit from his father, and he stayed for two years. Though he spent most of his time at his Island plantation, he also visited Mesopotamia, and in Jamaica when an absentee master came to inspect his estate the whole slave force would be ceremonially assembled to greet him.[4] Unlike her children and grandchildren, Affy was in the small minority of slaves living at Mesopotamia in 1762–1833 who actually got to see a real live Barham.

The 1779 inventory lists Affy as a field worker without specifying what sort of agricultural labor she performed. There were 125 field workers at Mesopotamia in 1779, constituting half of the slave population, and they were organized into three large labor gangs, supervised by black drivers who kept the people in each squad working in regimented lockstep. About fifty-five men and women labored in the Great Gang or first gang: these were the prime field workers, who performed the most strenuous tasks. The second gang was of somewhat smaller size and was composed of teenagers not yet strong enough for first gang labor and adults in their thirties and forties who were beyond their working prime. The youngest hands were organized into the grass gang.[5] At age twelve Affy was probably assigned to the grass gang. If so, she was one of the older workers in this squad, and was bossed by a forty-eight-year-old woman named Peg, who was the children's driver.[6]

After a year or so of field labor Affy contracted a bad case of the yaws, a bacterial infection of the skin, bones, and joints that produces disgusting ulcerating skin lesions and often causes disabling bone damage and permanent disfigurement. At least 15 percent of the Mesopotamia slaves were seriously afflicted by the yaws during Affy's lifetime, and seventeen people died of it, according to the estate death registers. Youngsters were particularly susceptible; in over three-quarters of the recorded cases at Mesopotamia, boys and girls were first recorded as having the ulcers and scabs characteristic of the disease when they were between the ages of two and seventeen.[7] Affy suffered from this highly contagious disease for at least four years, and may have spent some time in the plantation yaws house, which was situated on an isolated hill so that patients in the worst shape could be quarantined. By January 1, 1785, when she was seventeen years old, Affy was finally on the mend, and the bookkeeper noted "Yaws Recovering" in the health status column after her name.[8] From age eighteen to thirty she was considered to be in good health, and was always categorized as "able" or "healthy" in the annual inventories.

During the early 1780s Affy was moved from the grass gang to the second field gang. Here she was drilled by Cuffee Tippo, an African-born man in his fifties who probably came from the Gold Coast. About half of her fellow workers on the second gang were female, and about half of them were teenagers. Affy was now performing more strenuous tasks than in the grass gang, cleaning the pastures and weeding canes, and during the sugar harvest she hauled the discarded cane tops to the cattle for fodder and the pressed canes to the trash house to be dried and used as fuel. Then in the late 1780s, when Affy was about twenty, she was transferred from the second to the first field gang. The policy at Mesopotamia from 1801 to 1833, when every field hand was annually identified by gang, was to place almost all of the "able" hands in their twenties and thirties on the first gang, and to relegate almost all of the "weak" or "sick" hands in this age cohort to the second gang. If the same procedure was followed in Affy's case, she worked on the first gang for about a decade, from 1787 through 1797.

Two drivers managed the Great Gang during Affy's time: a pair of men named Francisco and Hector. Francisco, the head driver, took charge of this gang in 1779 at the young age of twenty-seven and remained in command for twenty years. Shortly before Affy joined the first gang, Francisco complained to the Moravian missionaries on the estate that the Mesopotamia overseer, John Graham, "wanted more work of him & the Negroes than he was able to make them do, & more indeed than was right, & more than they were able to do. He said that they worked harder than any other Negroes, on any Estates round about."[9] And Francisco added that Graham expected him to "swear & teare at the Negroes, particularly when he came to look at them."[10] But however much Francisco disliked his job, he continued as the head driver until he died in 1798. His fellow driver Hector had a remarkable forty-one-year tenure, serving as driver from 1775, when he was thirty, to 1816, when he died at seventy-one. And Hector—like Francisco—must have been a formidable field boss. In 1797, when Affy was probably still working in the first gang, Hector was charged with murdering a fellow slave. He scuffled so strenuously with a field cook named Cudjoe that Cudjoe burst a blood vessel in his brain and died.[11]

The women and men in the first gang were required to perform the most taxing and hazardous labor on the estate. Working six days a week from sunrise to sunset, they did everything in unison and by hand, with no reliance on draft animals or on labor-saving tools. The hardest work of all came during the fall months, when they planted eighty to one hundred acres of cane by excavating four-foot-square holes with hoes, and placing cuttings from old cane stalks in these holes. This was known as holing. The field hands worked in pairs, each couple digging up to one hundred holes per day, and the two drivers had the job of making sure that every pair kept pace with the work of the others. Once the cane was planted, they replanted the cuttings that failed to sprout, and weeded and manured the rows of young shoots. Fortunately for the first gang workers, less than half of the Mesopotamia cane pieces were dug up and replanted in an average year. In the remaining cane pieces the field hands cut the stalks just above the root so that they could resprout and produce a second or third or fourth crop. This was called ratooning; ratooned cane required much less labor and matured more quickly than new cane, but produced much less sugar.

In early 1778, about ten years before Affy joined the first gang, overseer Barnjum sent to England a list of "canes to cut" for the upcoming Mesopotamia crop. He reported that six newly planted cane pieces totaling eighty acres were expected to produce eighty-five hogsheads of sugar, or 56 percent of the crop. Five cane pieces planted with first ratoon cane totaling fifty-eight acres were expected to produce thirty-one hogsheads of sugar, and six cane pieces planted with second to fourth ratoon cane totaling seventy-five acres were expected to produce thirty-five hogsheads of sugar.[12] Newly planted cane took up to fifteen months to ripen, while ratooned cane matured in about a year. Crop time stretched from January into May or June, and Barnjum had to figure out a planting schedule that would enable the field workers to harvest seventeen cane pieces one at a time in 1778. If too many sugar fields ripened simultaneously, the cutters would be overwhelmed. It was always essential to cultivate the cane pieces so that they would be ready for harvest in a sequential pattern over a four- to six-month period.

During crop time the first gang hands cut the ripe cane—now towering three feet above their heads—with sharp curved knives called bills. And having trimmed the cut cane, they hauled it in bundles to the mill, where they fed the canes through the mill rollers by hand in order to extract the cane juice from the stalks. This was dangerous work, where accidents and serious injuries were common. Then they stoked the furnaces that boiled the cane juice into crystallized sugar. During crop time, the first gang hands labored up to ninety hours per week, including three nights of work per week, because the Mesopotamia mill and boiling house furnace were running twenty-four hours a day.

A good many of the Mesopotamia slaves decided that this arduous agricultural regimen was intolerable. The records identify 140 men and women who fled from the estate at least once between 1762 and 1831; the great majority of them were field workers, and seven managed to escape permanently. A boy named Sam, who had worked in the overseer's kitchen when Affy was a house girl, began to run away as soon as he was put into field work. After seven years of capturing and recapturing him, Barham's agents decided to get rid of Sam and sold him to the colony government for £40 to be transported off the island as an incorrigible. But this was a drastic step: only three other slaves were sold off the estate during Affy's lifetime. And Affy herself was never identified as a runaway—perhaps because she had a family of young children at Mesopotamia to care for and to love.

Affy was the mother of six children, and she probably conceived and bore the last five of them while she was a member of the Great Gang. Her first child, Princess, was born in February 1785 when Affy was only seventeen. Then Hagar followed in October 1788, Davy in November 1790, Robert in April 1793, Ann in March 1796, and Rodney in February 1798. In her pregnancies for Hagar and Davy, Affy was cutting sugarcane during the first trimester, and in her pregnancies for Robert and Ann she was holing cane pieces during the first trimester. When the overseer discovered that a cane worker was pregnant, he would move her temporarily to the second gang, so that Affy escaped a good deal of the most strenuous labor as her six babies came to term. But she must have been an exceptionally strong young woman. Her babies were spaced at intervals of forty-four, twenty-five, twenty-nine, thirty-five, and twenty-three months, which suggests that she very likely lost another child between Princess and Hagar. Four of her children were sick with the yaws when young, but all six of them lived to adulthood, which was an unusual achievement at Mesopotamia.

Two of Affy's babies were mulattoes, and the other four were black, reflecting a common pattern of biracial sex on Jamaican plantations. Twenty-six slave women at Mesopotamia bore forty-five children fathered by white men between 1762 and 1833—constituting 12 percent of the total births. The Mesopotamia plantation birth registers never disclose the paternity of these biracial babies. In Affy's case, the white father of her son Robert can be identified, and the white father of her daughter Ann can be guessed at. But unfortunately the father or fathers of Affy's four black babies cannot be traced. So while we know that Affy was pursued sexually by her white masters, we do not know whether she formed a steady marriage alliance with a slave husband.

Affy's four black children all followed their mother into field labor at a tender age. Princess was ill with the yaws from age three to five, but nevertheless joined the grass gang when she was eight years old, the second gang at sixteen, and the first gang at nineteen. Hagar was sickly from birth and afflicted with the yaws at age eight, but she was put into the grass gang at ten, spent a year minding the fowl house (a job reserved for feeble workers), and at eighteen joined the second gang, where she spent the next two decades of her life. Davy, despite being afflicted with the yaws at age five, was in the grass gang at eight, moved on to the second gang at sixteen, and entered the first gang when he was twenty-one. And Rodney (who escaped the yaws as a young child) was in the grass gang at age six, served for two years as a houseboy and eleven years as a cattle boy, and eventually at twenty-three joined the first field gang, where he stayed for the next decade or more.

In sharp contrast, Affy's two mulatto children were automatically exempted from their mother's arduous toil as a field hand. The Jamaican planters assumed that slaves of white/black parentage were genetically superior to slaves of black/black parentage, and treated them accordingly. None of the fifty-two mulattoes and quadroons who lived at Mesopotamia between 1762 and 1833 ever had to hoe a cane piece or chop down a stand of mature cane. Both Robert and Ann were put to work a little later than their siblings, Robert at age eleven and Ann at ten. More important, Robert entered employment as a houseboy in the Great House and went on to become a skilled artisan in adulthood, while his sister Ann entered domestic service in 1806 and spent the rest of her recorded career working either in the Great House or in the overseer's house at Mesopotamia.

As her six children were entering the workforce, Affy herself was wearing down in strength and health. In 1798, at age thirty-one, she was described as "weak" for the first time, and from this date until 1823 Affy was almost always listed as being in a "weak" or "weakly" state of health. In 1798 or 1799 she was probably shifted back to the second gang, and by 1801, when all field hands were identified by gang for the first time, Affy and her teenage daughter Princess were both working on the second gang. Princess would soon to be moved to the first gang, while Affy was no longer strong enough for first gang labor. In 1803 Affy was removed from field work, and at age thirty-six became a washerwoman at the Great House, one of three women employed in cleaning the household linen and clothing of the eight or nine white people who lived on the estate. Interestingly, another of these washerwomen was a slave we have already met named Doll, whose experiences at Mesopotamia paralleled Affy's to an almost uncanny degree. Both women were born in 1767, started work together as domestics at age seven, spent two decades in field labor, and were now consigned to the same laundry job.[13] Like a great many other women at Mesopotamia, Affy and Doll went through a cyclical labor pattern in which they started doing relatively easy work as children, worked desperately hard as young adults, and sank gradually back into the equivalent of child labor status as their health broke down.

Affy worked as a washerwoman for the next six years, from 1803 through 1808. There were two buildings at Mesopotamia where laundry was done: a combined cook room and washhouse measuring twenty by forty feet servicing the Great House, and another cook room and washhouse measuring sixteen by forty feet that serviced the overseer's house and the bookkeepers' barracks.[14] The cooks and washerwomen worked together because they could share a stove or stoves when heating their kettles and tubs. Thus, Affy's workplace must have been oppressively hot.

It was during her washerwoman years that Affy sought entry into the Moravian Church. Since 1758 Moravian missionaries had been operating a station at Mesopotamia and teaching the slaves the United Brethren's mystical religion of the heart—to love and adore Our Saviour (as they addressed Jesus Christ) by identifying emotionally and physically with his blood, his wounds, and his sufferings on the cross.[15] From 1801 to 1808, Brother Joseph and Sister Rachel Jackson managed the slave congregation at Mesopotamia. About forty men and women were baptized members, and a good many other slaves attended chapel services, though the Jacksons complained that attendance was very spotty during crop time. To judge by the letters Brother Jackson wrote to his patron, Joseph Foster Barham II, in England, he did his best to help the slaves without openly challenging Barham's attorney and overseer.[16] On Sundays he preached a doctrine of future rewards, urging the slaves to accept their present subjugation and to look forward to compensatory heavenly bliss in the next world. The Jacksons reported that they were baptizing a few slaves every year, and fortunately we can identify the men and women they brought into the church. During their tenure at Mesopotamia they kept a fascinating conference book—so called because the missionaries recorded in it their spiritual conferences with Jesus Christ.[17] Every month or so the Jacksons met with the Saviour by praying and singing hymns to him, after which they proposed various slave men and women as candidates for baptism or communion. During 1808 the Jacksons proposed four men and six women for baptism, and Affy was in this select company.

The Jacksons' ten baptismal candidates ranged in age from thirty-seven to fifty-six, and were mostly—like Affy—in poor health and marginally employed. Five had been born in Africa. And only one of them was actually baptized in 1808. Six waited for several years to be baptized, and three never became members of the Moravian Church. This was because the church admission process at Mesopotamia was determined by lot, which the Moravians interpreted as the voice of the Saviour. Their standard practice was to place three papers in a tube, one with a favorable message, another with a negative message, and a third that was blank. After mixing these papers, they drew out one of them, which was the Saviour's choice. When the Jacksons proposed Affy's candidacy on June 5, 1808, they drew out a blank or negative message; the Saviour said no to Affy.[18] Two months later the Jacksons proposed her a second time, probably because she seemed very eager to join the congregation. But once again the answer was no. On October 23, 1808, they tried a third time and drew out a positive paper. Affy was now a candidate for baptism; the next step was to ask the Saviour for permission to baptize her. But the Jacksons never got this far, because Brother Jackson fell sick and died in December 1808. His death stopped everything at the Mesopotamia chapel: church services, meetings with the baptismal candidates, and spiritual conferences with the Saviour.

Shortly after Brother Jackson's death, Affy was given a new job. In 1809 overseer Patrickson assigned her to operate a Mesopotamia-style day care center with two other women, looking after twenty little boys and girls aged two to five while their mothers were at work. Affy had no youngsters of her own to mother in 1809; her six daughters and sons (ranging in age from twenty-four to eleven) were all working, and as yet she had no grandchildren. The children under her charge were "regularly brought up to the house two or three times a week to be inspected and each have check [cloth] to make frocks for them," according to the estate agent.[19] So part of Affy's job was to sew up the clothing that these little children wore. During 1810 Affy was sent back to the wash-house, where she continued to launder her white masters' linen for the next fourteen years, until 1824. Soon after she resumed her laundry job, calamity struck. In June 1811 Affy's oldest child, Princess, died very suddenly of the dry bellyache or dry gripes, an abdominal disease caused by lead poisoning which produces excruciating cramps in the stomach and bowels. Lead pipes and coils were used in the distillation of Jamaican rum, and Princess was probably poisoned by drinking lead-contaminated rum. She had been in apparent good health and was working in the first field gang when she died.[20] Princess was only twenty-six years old and had no children.

In this time of loss Affy was able to turn once again to the Moravian Church. A new pair of Moravian missionaries had finally arrived in April 1811: Brother John Samuel Gründer and Sister Sarah Gründer. After spending considerable time getting acquainted with the communicants, the baptized, and the candidates for baptism on the estate, the Gründers began holding spiritual conferences with the Saviour, as the Jacksons had done, and on May 19, 1812, they asked if Affy could receive Holy Baptism. The answer was no. They asked again on July 14, 1812, on February 24, 1813, on April 27, 1813, and on March 22, 1814, with the same result. The Gründers reported to the Saviour that Affy was one of the baptismal candidates who "attend regularly to their speaking" (that is, who came regularly to the missionaries' weekly instructional classes), which is doubtless why they kept on proposing her. And finally, when they asked for the sixth time on May 17, 1814, the answer was yes.[21] So Affy joined the Moravian Church at age forty-seven after six years of trying. She chose Sarah Affir as her baptismal name, or perhaps the Gründers chose this name for her. Sarah, being the wife of Abraham and the mother of Isaac, was one of the more popular baptismal names at Mesopotamia; at least four other women were baptized as Sarah when they entered the church.

During the Gründers' seven-year term of service at Mesopotamia, they proposed twenty-eight candidates for baptism to the Saviour and fourteen candidates for communion. They never proposed Sarah Affir as a candidate for communion, probably because the Saviour had voiced such repeated doubt about her baptism. Perhaps she was no longer a faithful attendant, but this seems most unlikely, because several other members of her family followed her into the Moravian Church. Between 1818 and 1833 her sons Davy and Robert, her granddaughter Jane, and her grandson John all became baptized Christians.[22] Surely the Christian promise of redemption was comforting to this victimized woman. But the Moravian mission was not in a healthy state. The Mesopotamia congregation was shrinking in size, and the missionaries complained that those slaves who attended their services showed little religious ardor. Sister Gründer died in 1815, Brother Gründer in 1818, and their successor, Brother Ward, in 1819. For the next eleven years there was no resident Moravian missionary at Mesopotamia.

In June 1817 Sarah Affir was noticed for probably the only time by a record keeper from outside Mesopotamia. The British government directed the Jamaican authorities to compile a complete census of the slaves on the island, so Barham's attorneys provided the name, sex, age, color, and origin of each slave living at Mesopotamia at this date. Whereas most slaveholders could only guess the ages of their people, Barham's attorneys had solid documentation, which they copied from the most recent inventory of January 1, 1817. But even the Mesopotamia age statements were not completely accurate, since the bookkeepers had neglected to add a year to everyone's age in 1800 and again in 1805. In the Jamaica slave registration return for 1817, Sarah (identified as Affy) was stated to be a Negro woman born in Jamaica, aged forty-seven, whereas she had actually just turned fifty. She was one of 321 slaves registered in Mesopotamia, one of 22,659 slaves registered in Westmoreland parish, and one of 345,252 slaves registered in Jamaica.[23]

By the mid-1820s Sarah Affir had spent almost fifty years in drudge labor at Mesopotamia, mostly in the fields and in the washhouse. In 1824, at age fifty-seven, she became an invalid, incapable of any further work. She had contracted the ugly disease scrofula, or tuberculosis of the lymphatic glands, and her neck was grossly swollen. In the Mesopotamia inventories, Sarah was said to be afflicted with "the king's evil"—the popular name for scrofula because it was believed that the king of England could cure this disease by touching the affected person's swollen glands. But the king of England was very unlikely to visit Mesopotamia.

Sarah's black children were also wearing down. Princess, her eldest, had already died in 1811. Sarah's second child, Hagar, was the next to go. While she had never been strong enough to work in the first gang, Hagar spent twenty-four years as a member of the second gang. From 1817 onward she was listed as "weak" rather than "able" in the annual inventories, and starting in 1824 she began to run away from the estate at frequent intervals. Probably Hagar didn't go very far, just into the wilderness slopes of the Hanover Mountains, where she could hide out for a few days or weeks and then return to receive a flogging and resume her labors. In January 1830 she was discovered dead in a Negro hut. Her white supervisors suspected suicide, which would have looked very bad to absentee owner Barham, and also to the abolitionist critics of Jamaican slavery had they found out about it. So a jury considered the case and concluded soothingly that Hagar's death was "occasioned by the visitation of God."[24] She was forty-one years old, and like her sister Princess was childless.

Sarah's two black sons failed more slowly. Davy worked in the first gang for sixteen years, starting in 1811, but at the close of 1825, when he was thirty-five, he was described for the first time as "weak," and after another year was removed from field work. In 1827 he was switched to the less physically taxing job of mason, and was still working as a mason in 1832. Although he was described as "Weakly & Diseased," in the final inventory at age forty-two he was valued at the rather high price of £120—only £20 below the top valuation for a Mesopotamia slave.[25] And Davy had become a member of the Moravian Church. His Christian name, listed for the first time in 1833, was David Reid, which suggests that his black father adopted the surname Reid. The only other Mesopotamia slave to use the name Reid in the inventories was Claret, son of Quasheba and eight years younger than Davy, whose baptized name was Thomas Reid Foster. So Davy and Claret may well have had the same father, but who this man was I cannot tell.

Finally, Davy's younger brother Rodney, although still a member of the first gang in 1832, was characterized as "Weakly & Hipshot"—meaning that his hip joint was dislocated. Nevertheless, in 1833 Rodney was valued like Davy at £120 at age thirty-five. Rodney did not have a baptismal name, and, most regrettably, the records do not reveal whether he or Davy had a wife and children.

And Sarah Affir herself? Although she became a permanent invalid in 1824, she was still alive nine years later in August 1833, listed as sixty-two years old but actually sixty-six. Her state of health was "weakly," and she had a valuation of £0. In the eyes of her white owners she was totally worthless. But she was in good company. Forty-six other elderly or feeble slaves in the 1833 inventory were valued at £0—which was 14 percent of the Mesopotamia slave population in 1833.[26] And although the appraisers considered Sarah to be a burden on the estate, they did at least validate her spiritual worth by recognizing her Christian name. Back in 1819 some of the baptized slaves had apparently told the overseer that they wished to be known by their Christian names rather than their book names. And Barham's attorneys objected. To add these new names to the inventories "will give more writing in making out the lists—an unnecessary measure," they told Joseph Foster Barham II. And Barham agreed with them that adding the new names was "uncalled for altogether."[27] But eventually, as slaves throughout Jamaica began pressing for Christian baptism, the attorneys relented. On January 1, 1832, and on August 23, 1833, they listed the slaves by their "original" or "former" names and also by their "Christian" names in parallel columns. On both lists close to 20 percent of the Mesopotamia slaves bore Christian names. In 1832 and again in 1833 Sarah was listed both as "Affie" and as "Sarah Affir."

Here her story abruptly breaks off. We know almost nothing about the inner life of this woman. Nor do we know what happened to her—presuming that she was still alive in 1838 at age seventy-one—when she became free at last. But through the year-by-year skeletal outline of Sarah's career at Mesopotamia we can observe slavery in action on this sugar estate and sense its harsh configurations. Nothing came easy for Affy, including admission into the Moravian Church. Yet she hung on to life. She was a survivor.

## *Robert McAlpine*

To turn from mother to son, the Mesopotamia birth register for 1793 contains the following entry: "April 11. Affie Deliver'd of a Mulatto Boy named Robert."[28] There were ten births recorded during this year, compared with sixteen deaths, and three of the babies were mulatto. The number of mulattoes was strikingly high considering that 116 black men between the ages of eighteen and forty-five were living on the estate at the time of Robert's conception, and only 7 white men. Though the birth register does not identify Robert's father, the boy evidently knew who he was. When he joined the Moravian Church as an adult, he chose "Robert McAlpine" as his baptismal name. And there was a white man by the name of McAlpin living at Mesopotamia in the summer of 1792. One of many Scots who came to Jamaica in the late eighteenth century, Andrew McAlpin worked on the estate as a bookkeeper from 1790 to 1796 at the modest starting salary of £30 per annum.[29] Interestingly, another Scot named Robert McAlpin (who was probably Andrew's brother) visited Mesopotamia for four months in 1793, arriving on April 30 and leaving on August 30.[30] But Robert McAlpin cannot have been the father, for he arrived a month after the baby's birth. Perhaps it was Andrew rather than Affy who chose the name Robert for this child. He was probably the clerk who entered the baby's birth into the Mesopotamia records. Nor is there any surviving evidence that Andrew McAlpin tried to manumit his slave child. Several of the other white fathers at Mesopotamia purchased the freedom of their sons and daughters, but McAlpin probably could not afford to do this; the standard fee to the estate was £100 per child.[31] So Robert lived into the 1830s as a slave.

Affy had just turned twenty-six when she bore Robert, and she was working in the first field gang. Her fourth child was a sickly little boy with a very bad case of the yaws, which persisted from age six through age nine. He was quarantined from the other slaves during 1801 and 1802, and lived with six fellow patients (five of them young children) in the plantation yaws house on an isolated hill.[32] Here he was nursed by an African-born woman named Lydia, who died so suddenly in November 1802 (when Robert was probably still under her care) that a jury was convened to consider the cause and declared that her death was "a visitation by God." By the time he turned eleven in 1804, Robert was restored to health and given his first job, which was to wait on William Rogers, who was Joseph Foster Barham II's attorney or estate agent and lived in the Mesopotamia Great House. Here, if not earlier, Robert encountered the facts of life about white power. Nine white people were living on the estate in 1804: the attorney, the overseer, the distiller, the pen keeper, the carpenter, two bookkeepers, and a Moravian missionary couple. Thirty-three slave servants attended them, and this number would have been even larger had the absentee owner been resident. Seven of the slave servants were men, twelve were women, ten were boys, and four were girls. Seven of the servants were mulattoes, mainly children like Robert. At the Great House he joined a staff of seven domestics headed by sixty-one-year-old Hannah, who had been overseer Barnjum's mistress when Affy was a child. And Robert must have noticed that he was starting out with a far more attractive job than his mother, since in 1804 Affy was also working at the Great House in the behind-the-scenes position of washerwoman.

In 1805 Robert had a new job description: "waiting upon the Doctor." He now attended the white doctor who physicked the slaves at Mesopotamia for an annual fee of 6s. 3d. per head. The name of this physician is unrecorded, but one hopes that he was more successful than his predecessor, John Horsley, who amputated the leg of a man named Abner, operated on the thorax of a man named Greenwich, and tapped the abdomen of a woman named Love—all three of whom died immediately after his procedures.[33] The doctor who replaced Horsley was living at Mesopotamia in 1805, which explains why he needed a waiting boy. Robert served him for a year, and then in 1806 attended the overseer, a man named James Brodie Rose. At the overseer's house he joined four other slave domestics: three mulatto children managed by a black housekeeper named Pysche.

In March 1806, just before his thirteenth birthday, Robert had a violent fight with a slightly older black boy named Tamerlane, a fourteen-year-old who was starting his apprenticeship training to become a cooper at Mesopotamia. Nothing is known about the circumstances of their quarrel, but Tamerlane (who had started working in the grass gang at age seven) was very likely resentful of Robert's favored treatment. Both boys knew that the fifteen coopers and carpenters held the choicest slave jobs at Mesopotamia, because their skilled craft work was far less strenuous than field labor and permitted some measure of individuality. Tamerlane got his chance to become a craft apprentice through family connections (his older brother Richard was a carpenter), and he very likely worried that mulatto Robert was going to steal his job away. Robert for his part probably saw Tamerlane as an inferior rival. Having served the white managerial elite at Mesopotamia for several years, he may have felt that he really belonged with them rather than with his black mother and black siblings. When he was eight years old, two mulatto children at Mesopotamia had been manumitted by their father, David Forrester, a former overseer, and when he was eleven, two more mulatto children had been manumitted by Patrick Knight, the white carpenter. Knowing of these manumissions, did Robert feel a special rage against his own white father for deserting him and thereby condemning him to a life of perpetual bondage? Whatever the motivation, when Robert and Tamerlane had their fight, Robert kicked Tamerlane in the intestines so hard that he killed him.[34]

The jury verdict on Tamerlane's death was that Robert had committed manslaughter, not murder. He had acted unlawfully, but without premeditated malice: it was not a capital crime. But Robert was undoubtedly flogged for destroying valuable property. Slave floggings in Jamaica were frequent and severe. The Westmoreland diarist Thomas Thistlewood, who lived a few miles from Mesopotamia, recorded that he flogged most adolescent and adult male slaves in his charge at least once a year, and many of the females as well. Thistlewood routinely ordered fifty lashes for a female slave who stole food, and one hundred lashes for a male slave who committed a similar minor misdemeanor.[35] On this scale Robert could have been given more than a hundred lashes for killing Tamerlane.

But the new overseer at Mesopotamia, John Patrickson, was evidently interested in rehabilitating the boy, because he assigned him to attend the Moravian missionaries Joseph and Rachel Jackson at the plantation chapel, and there Robert stayed for the next two years. The Jacksons received a salary of £50 per annum and had the use of five slaves: three women, a girl, and a boy. In 1807 the women were all former field workers in broken health, and the girl, a twelve-year-old named Wonder, would soon be shifted into field labor. Robert, now fourteen, was the only mulatto in this group. Two of the chapel women were baptized, and Robert's mother, as we have seen, was proposed as a candidate for baptism in 1808. But the Jacksons almost never brought teenagers into the church, and they did not propose Robert. Brother Jackson may indeed have taught Robert to read; his successor, Brother Gründer, reported in 1818 that he was keeping school at Mesopotamia for some of the mulatto children.[36] But if Joseph Jackson took an interest in young Robert, his support was brief, because he died in December 1808.

At Jackson's death the chapel staff was broken up, and Robert became a houseboy again in 1809. By 1810, at age seventeen, he was deemed ready for adult employment, and started his apprenticeship as a cooper—the craft he would practice for the remainder of his recorded career at Mesopotamia. When Robert started his apprenticeship there were ten coopers at Mesopotamia: eight men and two boy apprentices. It was an experienced crew. Head cooper Redriff had been making casks for twenty-seven years, mulatto Robin for seventeen, Chelsea for ten, and Plato for nine. The coopers shared a workshop with the carpenters, of the same dimensions (forty by sixteen feet) as the washhouse and cook room where Robert's mother worked.[37] Out of crop the coopers spent some of their time constructing and repairing the fences on the estate, but they never did field labor. And their craft required real skill. Using oak staves imported from North America, Robert and his colleagues constructed huge hogsheads and smaller tierces for sugar containers, and puncheons for storing and shipping the plantation rum. They made hogsheads out of red oak staves and puncheons out of white oak staves.[38] They steamed and bent the staves to curve properly, cut and beveled them to fit snugly, fastened the staves into grooved cask headings, and bound the assembled casks with iron hoops. In 1810, Robert's first year, Mesopotamia produced a record crop, and he helped to construct 361 hogsheads, 45 tierces, and 164 puncheons to hold the sugar and rum shipped to England that year.[39] The average Mesopotamia hogshead weighed a little over a ton when filled with sugar, the average tierce had two-thirds the capacity of a hogshead, and the average puncheon held a little over one hundred gallons when filled with rum. These bulging casks had to be built strongly so that they would not split open when carters and dock workers rolled, tumbled, hoisted, and dropped them fully packed. And they had to be sealed tightly so that their contents would not leak out or spoil.

We catch a quick glimpse of Robert in action in January 1815, a few months after his mother's baptism into the Moravian Church. Sister Gründer, the Moravian missionary who had befriended his mother, had become gravely ill, and Brother Gründer begged overseer Patrickson to send a messenger to the closest Moravian brethren (who lived fifty miles across the mountains at the Bogue in St. Elizabeth parish) to tell them of her plight. Patrickson ordered Robert, now twenty-one years old, to go. He set off on this errand of mercy riding a mule and carrying a letter from Brother Gründer asking that Brother Lang or Brother Becker come quickly. Robert was gone three days, but he accomplished his task, for he returned with Brother Becker a few hours before Sister Gründer died. Brother Becker conducted the funeral, "to which a few white Gentlemen came and many Mulattoes & Negroes"—probably Robert and Sarah Affir among them.[40]

Robert worked mostly as a barrel maker in the 1810s and 1820s. He was switched temporarily to the carpenters' crew in 1812–1813, but from 1814 onward he was back with the coopers. If he worked alone, he must have constructed nearly a thousand casks in these two decades. Head cooper Redriff died in 1816, and the other senior coopers also soon died or retired, so that Robert's crew was suddenly much younger and less experienced. In 1817 the eight coopers at Mesopotamia (half of them mulattoes) were on average twenty-five years old, with only five years' experience at cooperage. No one immediately replaced Redriff as supervisor. Robert was evidently considered unsuitable for the post, and when the overseer finally appointed a new head cooper in 1822 he chose a twenty-five-year-old mulatto named William Prince. Robert henceforth worked under the direction of a boss who was four years his junior and had started his craft training three years later.

In 1817 Robert was reported in the Jamaica slave registration return, like his mother. Identified by first name only, he was stated to be a mulatto man born in Jamaica, aged twenty-one, whereas he had actually just turned twenty-four.[41] Unhappily, neither the Jamaican slave registration returns nor the Mesopotamia records tell us anything about Robert's personal life when he was in his twenties. We do not know whether he was married or had children.

In 1826 Robert turned thirty-three. The spring and summer of this year were a time of exceptionally wet, humid weather and debilitating sickness at Mesopotamia. In May and June more than one hundred slaves were hospitalized with influenza, pleurisy, and bowel cramps—a third of the population—and the cane fields were nearly deserted, so that crop time dragged on much longer than usual. Whether Robert took sick is unknown, but nineteen slaves were buried in 1826, including the head driver and the head boiler. A watchman named Mingo, who had been running away for years, was tried at quarter sessions and ordered transported off the island as an incorrigible deserter. And one of the carpenters fell fifty feet from the coconut tree he was climbing and broke his neck.[42] In July of this dismal year, Robert had another violent fight, this time with head distiller Peter, a thirty-seven-year-old black man who held one of the most responsible jobs on the estate. The two men quarreled over a debt of two bits that Peter owed Robert and had failed to pay. A bit was a Spanish coin worth approximately fourpence. Two bits was not a vast sum of money, even for a Mesopotamia slave. For example, the diarist Thomas Thistlewood generally paid one or two bits to a slave woman after fornicating with her—one bit to a woman from his own estate and two bits to a neighbor's slave. A pregnant woman at Mesopotamia was paid a bonus of £1—thirty times the value of two bits—if she delivered a live healthy baby. Robert and Peter probably both earned several pounds a year in tips or for extra work. So it seems likely that Robert's rage was triggered by something other than Peter's two-bit debt. Perhaps both this quarrel and his clash twenty years earlier with Tamerlane were fueled by mulatto-versus-black hostility.

Robert attacked Peter with the cooper's handsaw he used to cut and trim barrel staves, struck him three times on the arm before he could be restrained, and appeared ready to kill the man. Barham's attorney reported that Peter was laid up in the crowded estate hospital, and his injuries were so severe that he had to quit his distillery job and shift into the second field gang. Robert was incarcerated after this vicious assault and brought to trial before the Westmoreland magistrates, who sentenced him to a flogging of thirty-nine lashes and a month of hard labor in the parish workhouse.[43] He was no doubt flogged publicly, in the presence of the entire Mesopotamia slave community. And at the workhouse he was probably ordered to walk on a tread wheel, designed to force the prisoner to climb as the wheel turned. But this would have been small consolation to Peter, who never recovered from his injuries, and died in 1827. For the second time Robert had caused the death of a fellow slave.

One wonders what Peter's brother, a black carpenter named Dundee, said to Robert after he was released from the workhouse, or how he was greeted by his eight fellow coopers, seven of whom were black in 1826. Morale was deteriorating on the Mesopotamia coopers' crew. Cooper Toswell, who had run away for nine months in 1824, disappeared permanently in 1827. In this same year cooper James also absconded for a while, and the bookkeeper noted that James was "ill disposed." Robert was never reported to be ill disposed by his white masters, and there is no indication that he ever ran away, but his workload was becoming increasingly burdensome in the late 1820s. No new young apprentices had been recruited into the coopers' crew for many years, and by 1831

Robert and his colleagues were forty-four years old on average, whereas they had been only twenty-five years old back in 1817. All of them were listed in "able" health as late as 1826, but four were sick or weak by 1829. Robert himself, having enjoyed continuously good health throughout his working career, began in 1827—at age thirty-four—to develop the skin ulcers and sores that had plagued him as a little boy. Once again he was coming down with the yaws.

In November 1827 a forty-five-year-old black field hand named Flora gave birth to a girl named Sarah McAlpin, who must have been Robert's child, since Robert was the only adult on the estate with this surname. Flora, who was nearly eleven years older than Robert, had been an attendant in the Moravian chapel as a girl and a church member since 1814. She was a tough woman who worked in the first field gang for over twenty years and produced two other daughters without listed surnames: Dido, who died at age two in 1812, and Ancilla, who was eleven years old in 1833. Sadly, when Flora gave birth to Robert's child she died of dropsy immediately after delivery, so Robert had a motherless newborn to deal with. He must have enlisted his mother to care for little Sarah, especially since he named the baby after her.[44] But Sarah McAlpin lived for only six months before she expired of a bowel complaint in May 1828.

Sometime between 1818 and 1832 Robert was baptized into the Moravian Church, perhaps encouraged to do so by Flora or by his mother. There is no surviving record of his baptism, but he was listed among the "Christian" slaves in the final inventories of 1832–1833. He was identified by the "Original Name" of "M° Robert" and the "Christian Name" of "Robert McAlpine." Thus Robert announced his claim to the bloodline of the white father who had deserted him long ago. But doubtless he felt envious of Henry Patrickson, a mulatto carpenter twelve years his junior whose father had been an overseer at Mesopotamia when he was born. Around 1830 Patrickson's father left his son a bequest of £100, and Henry used this money to purchase his freedom from Joseph Foster Barham II in 1831, while Robert McAlpine remained a slave.[45]

At last report, on August 23, 1833, Robert was said to be thirty-seven years old (actually he was forty) and was in his twenty-fourth year of employment as a cooper. He was valued at £100, though he suffered from ulcers and sores. And he had fathered a second child. In April 1832 a thirty-seven-year-old black field hand named Abigail gave birth to a girl named Elvelina McAlpine. Abigail, who was not a church member, was two years younger than Robert and much less robust than Flora had been. She was incapacitated with the yaws in her teens and again in her thirties, and worked mainly in the second field gang when she was able to work at all—though, oddly, in 1833 she was reckoned to be worth £110, which was higher than Robert's valuation. Elvelina McAlpine was her only child. The baby was one year old in August 1833, valued at £10.

And here the story of Robert McAlpine abruptly breaks off. The narrative as we have it shows that this mulatto man had a far easier pathway than his black mother. He spent his boyhood in personal attendance on his white masters, and in adulthood was given one of the most tolerable employments on the estate. Yet Robert reacted to his condition with acute frustration and blind rage. And small wonder. For even the mulattoes at Mesopotamia were trapped in a dehumanizing life of exhausting labor, debilitating disease, and demeaning social relationships.

### *Jane Ritchie*

To move to the next generation one must start with Affy's mulatto daughter Ann, who was born in March 1796 when Affy (at age twenty-eight) was probably still working in the first field gang. This child was very likely fathered by Andrew McAlpin, like her older brother Robert, because McAlpin was still working at Mesopotamia as a bookkeeper in 1796. But Ann's paternity cannot be proved because she never adopted a surname that surfaced in the estate records. Unlike her brother, Ann was a healthy child, and in 1806, at age ten, she began working as a domestic in the Great House, where attorney John Blyth was in residence. Mulatto Ann was employed at the Great House for five years, and then at age fifteen she was moved to the overseer's house, where she continued (except for one year back at the Great House) to work as a domestic servant for the next twenty-three years—for the rest of her recorded Mesopotamia career.

Mulatto Ann had her first baby when she had just turned eighteen. In May 1814 she bore a girl named Jane, who was categorized as a quadroon—meaning that one parent was mulatto and the other white. When Jane was later baptized into the Moravian Church, she adopted the surname Ritchie, which indicates that her white father was Dr. Ritchie, who was the medical practitioner at Mesopotamia at the time she was conceived. Dr. Ritchie was a partner in the firm of Goodwin, Ritchie & Distin, and he physicked the Mesopotamia slaves for about five years. In 1816–1817 he earned £101 11s. 3d. for attending 325 slaves, plus £12 10s. for attending three white bookkeepers (two of whom died under his care). In this year Ritchie was also paid £10 13s. 4d. for amputating fowl keeper Cretia's leg—a successful procedure, since one-legged Cretia was still tending chickens in 1833 at age sixty. And he further received £10 for treating a young woman named Luna who had bad sores—a less successful procedure, since Luna never returned to work and died four years later.[46] All in all, Dr. Ritchie earned a good income from Mesopotamia, and unlike Andrew McAlpin he probably had the means to manumit his enslaved daughter, but he chose not to do so.

In October 1816, at age twenty, Ann had her second child, another quadroon, named John. The estate accounts show that Ann was paid £1 for her delivery of this baby. Once again, John's white father can be identified. When John was later baptized, he adopted the surname Bell, which tells us that his father was James Bell, who in 1816–1817 was paid £85 to supervise the slave distillers and also do some bookkeeping.[47] I do not know how long James Bell worked at Mesopotamia, but it is evident that like Dr. Ritchie and Andrew McAlpin he abandoned his child to lifelong bondage.

A variety of men had sex with Ann. She was impregnated by Dr. Ritchie and James Bell when she was young and at her most attractive. After the birth of quadroon John she had no further children for nearly five years, then produced two babies sired by a mulatto man or men, and two babies sired by a black man or men. This was the sequence: in May 1821 she bore a mulatto girl named Mary; in March 1824 she bore an unnamed mulatto boy who died when nine days old; in March 1825 she bore a black girl named Selena; and in August 1828, when she was thirty-two, she bore a black boy named Peter Foster, naming him (as a number of other Mesopotamia mothers did) in honor of her master, Joseph Foster Barham, whom she had never seen.

Turning to Ann's first daughter and Sarah Affir's first known granddaughter, Jane (born in May 1814) was identified in the Jamaica slave registration of June 1817 as an island-born quadroon who was two years old, though she had actually just turned three. She appears to have enjoyed good health as a young child, and started work at age nine. By the 1820s there was a surplus of young colored girls on the estate and not enough suitable domestic jobs for all of them. In August 1823 a vacancy developed at the overseer's house when a mulatto girl employed as a house girl caught a fever and died, so at age nine Jane took her place and started working as a domestic under the tutelage of her mother. Ten people staffed the overseer's house: seven adults and three children. Jane's mother, mulatto Ann (age twenty-seven), was probably the head housekeeper, though black Kate (age thirty-seven) had worked at the overseer's house longer than Ann and had been the mistress of a previous overseer. There was a black cook and two black washerwomen, but the other six domestics were all colored.[48] Jane worked at the overseer's house for four years, then moved to the Great House in 1827, and back to the overseer's house with her mother in 1829. From age eleven to fourteen she suffered from the yaws, but she was never as ill as her uncle Robert had been in childhood, and she spent no time in the plantation yaws house.

When Jane entered domestic service she got to know a mulatto lad named Edward Knight, seven years her senior, who sometimes served at the overseer's house as a waiting boy and, in 1823 when he was sixteen, was employed at the Great House. Edward's father was Patrick Knight, a white man who had supervised the slave carpenters at Mesopotamia for more than thirty years until his death in 1817, and had fathered children by at least three Mesopotamia slave women.[49] Edward's black mother was a slave named Batty (1773–1830) who had been Knight's mistress. Between 1797, when she was twenty-three, and 1811, when she was thirty-seven, Batty had four mulatto children, named Annie, Peggy, Edward, and Mary, all conceived with Knight. From 1801 through 1815 Knight hired Batty as his personal servant. He manumitted Batty's first two mulatto children, Annie and Peggy, in 1804, but chose not to manumit mulatto Edward (born in 1807) or mulatto Mary (born in 1811). And in 1816 he stopped hiring Batty, who was now weak and diseased. Knight himself died in early 1817, leaving an estate comprising twenty-five slaves whom he probably hired out as a jobbing gang to work at sugar estates that needed extra hands. Knight's slaves were appraised at £3,075, and he also left £400 worth of furnishings, carpentry tools, a horse, and a pony.[50] His mulatto daughter Mary, identified in the estate records as Mary Knight, died as a Mesopotamia domestic at age twelve in 1823. And in 1824 Barham's attorney told him that a mulatto boy named Peter Edward was anxious for his freedom, and that his friends were willing to pay £140 to manumit him. In 1825 the executor of Knight's estate, William Cooke, paid this sum, and Peter Knight (as he was now called) became a free man at the age of eighteen.[51]

Jane was only eleven years old when Edward (alias Peter Edward or Peter Knight) was manumitted, but she certainly knew that he had achieved the freedom she craved for herself. Edward continued to live near the estate, and he was a member of the Moravian congregation, as demonstrated by his baptismal name, Peter Edward. Very likely he encouraged young Jane to join the church. There is no surviving record of her baptism, but when it happened she took the name Jane Ritchie. In 1830 Jane turned sixteen, and she and Edward wanted to get married with an official Moravian church wedding, so they asked a visiting Moravian missionary named Brother Jacob Zorn to marry them. The Mesopotamia attorney William Ridgard obtained Joseph Foster Barham II's approval, which was needed since one of his slaves was marrying a free colored man. Then Brother Zorn published the banns in July and married these two young "brown people," as he described them in September. The chapel was filled to capacity as the couple walked in procession with their friends and relatives, the bride in a new white muslin dress and the groom in a new coat and pantaloons. Brother Zorn tells us that Edward and Jane went "to considerable expense in dress, wedding cake, etc." and that after the ceremony they invited all of their colored friends to dinner.[52] One wonders whether Jane's scrofulous black grandmother, Affy, and Edward's black mother, Batty (who died of dropsy that September), were among the guests invited to this feast.

One month after Jane's wedding, in October 1830, Brother Peter and Sister Sarah Ricksecker came to live at Mesopotamia—the first resident missionaries since 1819. The Rickseckers needed chapel attendants, and William Ridgard supplied them with three slaves: blacks Kitty and Prudence (Christian name Mary Barham), who were sickly and had previously been field hands, and newlywed Jane Ritchie. Some of the Moravian missionary couples were easy to get along with, like Brother and Sister Jackson, who had befriended Jane's grandmother Affy and her uncle Robert back around 1808. But the Rickseckers were fussy and thin-skinned. They complained that the members of the Moravian congregation at Mesopotamia lacked religious ardor, they complained that they had no stable in which to keep a horse, and they complained about the slave servants assigned to them by Ridgard. They wanted to have a houseboy, and they objected to the fact that Mary Barham and Kitty were subject to epileptic seizures.[53] But their prime complaint was with Jane Ritchie, who was insufficiently deferential to them. Jane clearly did not like her new job. Since she had more white than black blood and had just married a manumitted mulatto, she probably disdained her black fellow chapel attendants, and she certainly wanted freedom for herself. And if she had to serve the whites, she would much rather wait on attorney Ridgard in the Great House or join her mother, Ann, in attending overseer William Turner than slave for the Ricksecker s. But as we have seen, there were already too many colored females working in the Great House and the overseer's house, and Ridgard probably figured that a girl who had just celebrated a Moravian marriage was a very suitable servant to the missionaries.

Very quickly friction developed within the mission household. In April 1831 Brother Ricksecker, during a conference with other Moravian missionaries, voiced unspecified "grievances" that he and his wife had "concerning the house servants supplied to them by Mr. Barham's attornies."[54] By September 1832 it became evident that their prime grievance was with Jane Ritchie. "Our house servant," Brother Ricksecker wrote, "a brown girl belonging to the Estate, who had distressed us already so much, became so very impudent that we had to send her back to the Overseer & rather chose to be without her, till the Attorney came."[55] So Jane Ritchie had dared to cross the line of acceptable slave behavior. She had become openly insolent to her white bosses. This young, attractive girl had decided that she could no longer tolerate having to fawn on and toady to her master and mistress.

Had Jane acted aggressively toward the Mesopotamia overseer, she would have been in big trouble. But she was clever enough to sense that the attorney and the overseer were hostile to the Moravian missionaries. For the past two years Jane had been telling tales about the Ricksecker s to the other "brown people" on the estate, and now she presented accusations to Ridgard and overseer William Johns (newly appointed in 1832) that these white men could make use of. When Johns learned from Jane that the Ricksecker s had excluded an unusually large number of slaves from church meetings for unchristian behavior, he went to these people and asked "what the parson had said & done to them."

Then, according to Peter Ricksecker, "we had a visit by the Attorney & the Overseer who brought with them the above mentioned girl." The Rickseckers gave an account of Jane's bad conduct, but Ridgard and Johns countered by charging the missionaries with holding meetings at night when this was strictly against the law. And Johns confronted the Rickseckers with complaints from the church members they had excluded. "Tho we explained every particular mentioned," wrote Brother Ricksecker ruefully, "still they seemed to be prejudiced against us. We have learned since, that the above mentioned girl had falsely represented to the brown people at the great houses whatever we said unto the negroes even from the first year of our being here, who relates their stories to the Overseer."[56]

In 1832 there were eleven "brown people" employed as domestics at Mesopotamia, including mulatto Ann and quadroons Jane and John. Allying herself with the other colored domestics, Jane had effectively undermined the missionaries' reputation. Once it became known throughout the slave community that the white managerial staff strongly disapproved of the Rickseckers, attendance shrank at chapel services. "In this way," wrote Peter Ricksecker, "we lost the necessary confidence & respect" of the Mesopotamia congregation.[57] By the end of the year hardly anyone—black or mulatto—was coming to church. The Rickseckers were discredited, and they departed from Mesopotamia in February 1833.

When the last slave inventory was taken at Mesopotamia on August 23, 1833, four of Sarah Affir's six children and six identifiable grandchildren were living on the estate. Her mulatto daughter Ann, now thirty-seven, was valued at £80 and probably was still working in the overseer's house, though the 1833 inventory does not list occupations. Ann's two youngest children, black Selena and black Peter, were eight and five years old, respectively, and not yet working. They were each priced at £30. Ann's older children, twelve-year-old mulatto Mary and sixteen-year-old quadroon John, were both domestics and were valued at £50 and £80, respectively. And nineteen-year-old quadroon Jane—no longer a member of the chapel staff—was considered to be worth £90.

It is interesting to note that while the final Mesopotamia inventories of January 1, 1832, and August 23, 1833, listed over sixty of the slaves by their baptismal or Christian names in addition to their book names, there were significant differences between the two listings. Affy was "Sarah Affir" on both inventories, and "M° Robert" was also "Robert McAlpine" both times, but Affy's oldest son had no baptismal name in 1832, whereas in 1833 he became known both as "Davy" and as "David Reid." Davy had evidently been baptized after January 1, 1832. And the same was true of Affy's quadroon grandson John. He had no Christian name in 1832 but in the final inventory was identified both as "John" and as "John Bell." The reverse happened with John's quadroon sister. In 1832 she was both "Quad Jane" and "Jane Ritchie," but by 1833 Jane no longer had a Christian name. Together with four other "brown people" who had taken her side in the quarrel with the Rickseckers, she had renounced her baptism or had been excluded by the missionaries from the Moravian Church.

Jane had other things than religion on her mind in the mid-1830s. One year after the final Mesopotamia inventory, slavery was terminated at Mesopotamia and everywhere else in the British Caribbean. On August 1, 1834, Jane Ritchie and all the other Mesopotamia slaves over the age of six were upgraded to a new temporary status of apprenticed laborer, while children under six were fully freed. Jane, being one of the fourteen domestic apprentices at Mesopotamia, could expect to be manumitted after four years of continued service, on August 1, 1838, while the 258 agricultural apprentices at Mesopotamia were told that they had to work for six years until they were manumitted on August 1, 1840.[58]

In September 1834, shortly after Emancipation Day, Jane Ritchie (now age twenty) gave birth to a daughter named Frances Johns, who was listed as one month old in an amended compensation claim for the 316 Mesopotamia slaves filed on October 14, 1834.[59] The baby's surname is a surprise. Frances's father was not Peter Knight, whom Jane Ritchie had married with much ceremony in 1830, but William Johns, the overseer of Mesopotamia. Thus, Jane Ritchie, like her mother and her grandmother, was impregnated by a white man, and she became one of the only two slave women to produce an octoroon baby at Mesopotamia, a child who was only one-eighth of African descent. Another new biracial baby, born in July 1834, was William B. Ridgard, who received the surname of the Mesopotamia attorney. William's mother was Fanny Fisher, the head domestic at the Great House where Ridgard lived, so interracial coupling was clearly a continuing factor at Mesopotamia on the eve of emancipation. Jane Ritchie's white sex partner,

William Johns, served as overseer at Mesopotamia from 1832 to 1835, and he later became the attorney as well as the overseer at Mesopotamia from 1839 until at least 1853.[60] There is no way of knowing whether Jane went back to Peter Knight or whether she became William Johns's mistress—and perhaps his wife. But she bought her freedom in 1836 by paying the estate the modest sum of £18 for the valuation of her apprenticeship, and if William Johns helped to arrange this low-priced deal, it was the only time at Mesopotamia in which the birth of a mixed-race child led to freedom for the child's mother. Jane was also the only member of her family to gain manumission before the official termination of apprenticeship in Jamaica on August 1, 1838. Thanks to the emancipation process, in 1834 her baby Frances was already free at birth.[61]

Viewing the generational changes within this particular Mesopotamia family, the contrast between the sufferings of Affy and the rages of Robert and the achievements of Jane is strong and clear. Robert was able to harm only his fellow slaves, while Jane succeeded in inflicting lasting damage on a pair of white missionaries. To be sure, the Ricksecker s were the most vulnerable white people on the estate, but Jane had the talent to figure out how to challenge them successfully. Shrewdly parlaying her colored network to full advantage, she not only bested the Moravian missionaries but married a freed mulatto, shared a bed (at least temporarily) with the white overseer, and secured her emancipation ahead of time. Yet Jane was an extreme rarity at Mesopotamia. In fact, she was the sole person I have found among the 1,103 slaves living on this estate between 1762 and 1833 who outmaneuvered her white masters in a major way.

# 3

## *Winney Grimshaw and Her Mount Airy Family*

When I started to explore slave life at Mount Airy, I expected that the richly detailed documentation compiled by the Tayloes for bookkeeping purposes could be readily converted into a comprehensive series of slave biographies. Yet reconstructing the lives of the Mount Airy people turned out to be more challenging than working with the Mesopotamia biographical data. While hardly any of the Mesopotamia people permanently left the estate until they died, the Mount Airy slaves were frequently shifted from one work site to another, and many were moved or sold from the Virginia plantation and suddenly disappeared from the inventories—often without explanation. Since the Tayloes kept no birth or death registers, demographic information is generally defective. And there is no equivalent at Mount Airy to the observations by outsiders like the Moravian missionaries about slave conditions at Mesopotamia. However, in important respects the Mount Airy data is more satisfying than that at Mesopotamia. Some of the inventories during John Tayloe III's ownership, and most of the inventories compiled by William Henry Tayloe, identify the fathers as well as the mothers of most slave children, making family reconstitution much more comprehensive than at Mesopotamia. And the Tayloes sometimes have interesting things to say about individual slaves in their family correspondence.

This chapter focuses on Winney Grimshaw and her family, who are among the best documented of the Mount Airy people. Appendix 13 presents her family tree, and a fuller version is displayed on the website that supplements this book, *www.twoplantations.com*. By great good fortune three letters to or from Winney's father and sisters survive among the 28,000 documents in the Tayloe Papers at the Virginia Historical Society, and these letters—reproduced in this chapter—provide special insight into the actions and opinions of the Grimshaw family. The plantation records also provide interesting information about two dozen members of Winney's family, including her maternal grandparents, Harry and Winney Jackson; her parents, Bill and Esther Grimshaw; her sisters Elizabeth (known as Lizza), Anna, Juliet, and Charlotte Grimshaw; her brothers James and Henry Grimshaw; her husband, Jacob Carrington; and her children John Carrington and John, Julia, Lizza, Willie Anne, William Henry, and Thornton Grimshaw.

Winney Grimshaw makes a good pairing with Sarah Affir of Mesopotamia. The records from the two plantations enable us to track members of Affy's family and members of Winney's family for four generations. Affy, being a field worker, and Winney, being a craft and domestic worker, illustrate both sides of a major occupational division common to Mesopotamia and Mount Airy. Both of these women bore mulatto children, opening a window into the quite different ways in which racial mixing was handled in Jamaica and Virginia. Another big difference was that Winney Grimshaw lived in close proximity to her master, William Henry Tayloe, who knew her well, whereas Affy and her family had no personal contact with the absentee Barhams. To illustrate this point, Jane Ritchie could never have dared to challenge Brother and Sister Ricksecker had Joseph Foster Barham II (patron of the Moravians) been resident at Mesopotamia in 1832. A big part of Winney's story, as we shall see, is that John Tayloe III and William Henry Tayloe had major personal impact on her and other members of the Grimshaw family.

## *Winney's Grandparents*

To supply context for Winney Grimshaw's story, we need to go back two generations to her grandparents, Harry and Winney Jackson. They first appear on the earliest surviving Mount Airy slave inventory, taken on January 1, 1808.[1] At this date their owner, John Tayloe III, was thirty-six years old and at the height of his powers—one of the largest slaveholders in Virginia. He was an imperious and aggressive planter who in 1808 held 375 African Americans at Mount Airy. His slave force was divided into two parts: 105 people—domestics, craft workers, and their young children—lived at the home plantation, and about 270 people—agricultural workers and their young children—lived on the eight farm quarters. The craft and farm workers all labored for long hours six days a week, with their only free time on Sundays and a few holidays. Many of the domestics attending the Tayloe family must have worked on Sundays and holidays as well.[2]

Winney Grimshaw's grandparents, Harry and Winney Jackson, were both domestic workers. Indeed, they were among the Tayloes' favorite personal retainers—and thus belonged to a category of enslaved servitors with no equivalent at Mesopotamia. In 1808 Harry Jackson was thirty years old and served as one of the two coachmen who drove the Tayloes' carriages. His wife Winney (for whom Winney Grimshaw would be named) was twenty-eight years old, the mother of three children, and one of the two chambermaids who waited on Ann Ogle Tayloe. The Jacksons' master and mistress lived at Mount Airy from mid-April to mid-October, and during the other half of the year they moved with their children to the Octagon, their elegant new town house in Washington, built in 1799–1801 at the intersection of 18th Street and New York Avenue, two blocks from the White House. To handle this situation, John Tayloe III employed a staff of about forty domestic slaves, divided into three groups. In 1808 twenty of these people worked year-round at the Mount Airy plantation house. Another ten or a dozen domestics (the number is uncertain because they were not listed in the Mount Airy inventories) were employed at the Octagon. And eight servants, including Harry and Winney, traveled back and forth a distance of about a hundred miles with their master and mistress, spending the summer months at the country mansion and the winter months at the city mansion.

The Jacksons worked in two of the most impressive houses in the Chesapeake region. The Mount Airy plantation house has already been described in some detail.[3] The Octagon, in a very different way, is an equally distinctive building. Designed by William Thornton, the original architect of the U.S. Capitol, this three-story house cost over $28,000 to construct—which was a lot of money in 1800.[4] Despite its name, the Octagon is a six-sided structure, and it has an eccentric footprint, combining a circle with two rectangles and a triangle. Mount Airy and the Octagon, though very different in architectural design, both celebrate the authority and the hospitality of a large-scale slave-holding plantation master. Mount Airy, an imposing Palladian villa built of brown sandstone with twin dependencies connected by curving passageways, spreads grandly across an open park, whereas the wedge-shaped red-brick Octagon fits into a tight city lot.[5] In both houses, a spacious central entrance hall (circular at the Octagon) leads to a drawing room and dining room for the entertainment of guests. But the Octagon has one feature not found at Mount Airy. Behind the elegant oval staircase that flows up from the ground floor to the third floor is a hidden flight of narrow back stairs designed for the use of the Tayloes' slaves, so that they could get from the basement kitchen to the public rooms on the first floor and up to the bedrooms on the second and third floors via a series of discreet service doors. In this way the numerous black servants at the Octagon were minimally visible to the white occupants and white visitors in the house. The Tayloes employed an elaborate bell system to summon their slaves as needed. Winney and Harry Jackson must have climbed up and down those narrow back stairs thousands of times.

In 1808 the other six domestics who traveled with the Tayloes back and forth from Mount Airy to the Octagon were John III's manservant Archy; Betty, who joined Winney in waiting on Ann Tayloe; Gowen, who shared the coachman's duties with Harry; Peter and John, who were house servants; and a young cook named Billy. Harry and the other male attendants were decked out in livery: blue coats, red vests, and white stockings.[6] At both Mount Airy and the Octagon the domestics lived in slave quarters to the rear of the house that have long since been torn down. The 1808 inventory gives valuations for all of the Mount Airy slaves, stated in British rather than American currency.[7] Pricing ranged from £15 to £120 depending on age, occupation, and gender—with males generally valued at £10 more than females of the same age and occupation. Harry was said to be worth £100 (equivalent to $333) and Winney £90 (or $300). The coach that Harry drove was reckoned to be twice as valuable as he was; it was appraised at £200.[8] The Tayloes' phaeton was valued at £150, and their seven carriage horses were priced collectively at £525.

In 1808 Winney and Harry Jackson had three young children: Betsy, who was ten in 1808 and valued at £60; Esther, who was nine and valued at £60; and Henry, who was seven and valued at £40. The Jackson children lived at Mount Airy year-round, and so were separated from their parents for the winter months. They were probably supervised by

Winney's elder sister Phillis, a textile spinner who lived at Mount Airy. Young Betsy entered training as a spinner in 1810, then was sent to Old House farm quarter at age thirteen to become a field hand. Her sister Esther (the future mother of Winney Grimshaw) began to work as a spinner at age thirteen in 1812 and continued in this line of employment at Mount Airy for the next thirty-four years. Their brother Henry started to work as a stable boy in 1812.

During these years the senior Jacksons appeared only intermittently in the Mount Airy slave inventories, which were taken on December 31 or January 1, since they spent their winters at the Octagon. They were listed on January 1, 1808, and again on January 1, 1809, because in those two years the Tayloes spent Christmas at Mount Airy. But from 1810 to 1814, when the Tayloes celebrated the holiday season in Washington, Harry and Winney disappeared from the records. Young Henry Jackson was also absent in 1813–1814; he had evidently joined his father as a stable boy at the Octagon. The three Jacksons resurfaced at Mount Airy on January 1, 1815, because President James Madison was occupying the Octagon. When the British invaded Washington in 1814 and burned the White House, John Tayloe III lent his town house to President Madison, and the Tayloe family spent the winter of 1815 at Mount Airy.[9] But in 1816 the three Jacksons spent the winter in Washington. Then in 1817 John and Ann Tayloe decided to live year-round at the Octagon; so they took with them a number of Mount Airy servants, including coachman Harry Jackson (now age thirty-nine), housemaid Winney Jackson (age thirty-seven), and their stable boy, son Henry (age sixteen). In 1824 Harry Jackson returned to Mount Airy, but his wife and son now lived permanently at the Octagon, and pass out of our sight since they never reappeared in the Mount Airy records.

### *Winney's Parents*

Our attention now turns to Esther Jackson, Winney Grimshaw's mother. This young woman was eighteen years old in 1817 and had been working as a spinner for five years. From 1817 onward she saw very little of her immediate family, since her parents and brother lived in Washington and her sister Betsy lived at Old House farm quarter several miles away. Esther probably shared a cabin with her aunt Phillis, who was also a spinner in John Tayloe's textile production unit at Mount Airy.

Before the American Revolution the Tayloes had imported all their cloth from England. But in the 1770s, when commerce with the mother country was cut off, John Tayloe II set up a spinning and weaving shop at Mount Airy for the local manufacture of woolen, cotton, and linen cloth, and John III expanded this cloth production. In 1808 he had two ginners, ten spinners, and two weavers. By 1812, when Esther started to work, he was employing thirteen spinners, and in 1828 there were twenty spinners, partnered with one ginner and three weavers.[10] In the generation before the Civil War machine-powered textile mills spread rapidly in the northern states, but slave women continued to spin yarn and weave cloth at Mount Airy until 1865.

Using cotton and flax grown at the farm quarters and wool sheared from the plantation sheep, Esther Jackson and her fellow Mount Airy textile workers produced a good deal of the fabric needed for slave clothing on the estate. Two males ginned the cotton to separate lint from seed. The spinners—all women or girls—then combed or carded the cotton, flax, and wool, and spun these fibers into skeins of yarn. Two weavers, Israel and his daughter Eliza, wove this yarn on hand looms into lengths of cloth. Most of the spinners produced one and a half pounds of cotton yarn or five and a half pounds of woolen yarn per week. By this measure Esther Jackson was spinning sixty-three pounds of cotton yarn and twenty-two pounds of woolen yarn each year.

In 1820, when Esther was twenty-one years old, her situation changed for reasons characteristic of slave life at Mount Airy. Her master, John Tayloe III, having decided to live year-round at the Octagon, needed to delegate the management of the plantation house to one of his sons, so he settled his eldest son, John Tayloe IV (who was twenty-seven years old), at Mount Airy with a retinue of forty household and craft slaves. He gave the young man six of the seventeen spinners on the estate, including Esther and her aunt Phillis. But he was willing to spare only one of his fourteen carpenters. So he moved an eighteen-year-old named Bill Grimshaw from his Neabsco ironworks in Prince William County, some sixty miles away, to serve as John IV's second carpenter at Mount Airy. From 1820 to 1824 Esther Jackson and Bill Grimshaw both worked for John Tayloe IV. And at some point in the early 1820s Esther married Bill. Their first child, a girl named Elizabeth or Lizza, was born in 1824.

We know very little about Bill's background. Grimshaw was an uncommon surname in early nineteenth-century Virginia. An Englishman named Samuel Grimshaw who came to Henrico County in 1795 and operated a tavern there in the 1820s may possibly have owned members of Bill's family. Bill's mother was named Letty, and she appears on slave inventories taken in 1825 and 1828 at the Neabsco ironworks. In 1825 Letty was the only domestic servant among the sixty-seven slaves living at Neabsco, and was listed as forty-six years old. Three years later she was still a house servant but must have looked a lot older, because she was said to be fifty-five. Bill's father was apparently named James, and in 1825 there was a ship carpenter called Jim at Neabsco who was listed as fifty-one years old and therefore of the right age and occupation. But this Jim was identified in two inventories of that year as Jim German or Garman, not Grimshaw. And he was gone from Neabsco or dead by 1828.[11]

In 1824, the year in which Esther and Bill's first child, Lizza, was born, their master, John Tayloe IV, died. There was debate within the Tayloe family as to who should replace him as manager of Mount Airy, and since the second son, Benjamin Ogle Tayloe, didn't want to live there, the third son, William Henry Tayloe (age twenty-five), took charge. So Esther became one of his six spinners, and Bill became one of his two carpenters. In 1825 William Henry Tayloe moved Bill and Esther to Old House farm quarter, along the bottomlands adjoining Cat Point Creek, several miles from the family mansion, and there Bill and Esther stayed until 1829. Esther's older sister Betsy lived at Old House, and Tayloe may have sent Esther to Old House to take care of her, because Betsy had been sickly for several years; in 1824 the plantation slave doctor had visited her three times. But if Esther was the nurse, she couldn't save Betsy, who died sometime in 1825. She bore her only child (who lived less than a year) when she was just thirteen years old, and she died at the early age of twenty-seven.

Probably in the early months of 1826 at Old House quarter Esther gave birth to her second child, Winney Grimshaw, the chief figure in our story. There was no birth register for the slaves at Mount Airy, so we cannot tell exactly when Winney was born. She was reported to be one year old when she appeared for the first time in the Old House inventory of January 1, 1827, listed as Esther's daughter Winney.[12] Unlike the situation at Mesopotamia, the children listed in the Mount Airy inventories were never given book names imposed by the white masters; their names were all chosen by the slave parents. And while in the

1820s almost all of the Mount Airy people were identified in the inventories by a single name or nickname, they were keenly aware of their lineages, and gradually William Henry Tayloe took note of family connections and of surnames. In the 1830s he started to identify slave fathers as well as mothers; in the 1840s he started to supply family names; and by the 1860s he was according almost every black person he owned the same two-name dignity as white people. From 1827 onward Tayloe identified Winney as Esther's child. In 1835 he noted in the record book that she was one of "Car. Bill and his wife Esthers children." In 1845 he referred to Winney's father as "Billy Grimshaw." And in 1862—when Winney was thirty-six years old—he finally wrote her name down for the first time as "Winney Grimshaw."[13]

In 1824, two years before Winney was born, her grandfather Harry Jackson returned to Mount Airy. William Henry Tayloe needed a coachman and ostler, so he asked his father to send Harry Jackson back to Mount Airy. Harry was now forty-six years old. And when John Tayloe III died in 1828, coachman Harry became the property of William Henry Tayloe and lived at Mount Airy until he died in 1838 at age sixty, so young Winney must have known her maternal grandfather very well. But her grandmother Winney and probably her uncle Henry remained in Washington after 1828, the property of the widowed Ann Ogle Tayloe, who was given life use of the Octagon with a coach and carriage and twelve house servants.[14] Winney Jackson lived and worked at the Octagon with her mistress until her death in 1847. What happened to Henry Jackson (who would have been twenty-seven in 1828) is unclear; most likely he succeeded to his father's post as coachman and ostler at the Octagon.

Meanwhile, Esther and Bill Grimshaw were raising a sizable family of seven children. The sequence: Lizza born in 1824, Winney in 1826, Anna in 1827, Juliet in 1829, James in 1831, Charlotte in 1834, and Henry in 1837. No birth dates are recorded for these children, but it is evident that Esther's first five babies were born a little less than two years apart and her last two babies about three years apart. Charlotte died in 1840,[15] but none of the other six children died young. However, none of them stayed very long at Mount Airy, for reasons that we shall see.

The Grimshaws, not being domestic servants, had many fewer personal dealings than the Jacksons with their owners, but William Henry Tayloe certainly knew this family well. Bill Grimshaw was one of the most skilled and versatile workers on the estate, and he served as Tayloe's head carpenter from 1832 to 1845. An inventory taken in 1839 demonstrates his skills. The four carpenters he worked with were each supplied with eight or ten basic items: axes, hammers, saws, chisels, planes, and not much else. Bill, on the other hand, was equipped with twenty-five tools of his trade. He had at his disposal two axes, two hammers, two planes, two augers, one adze, one bevel, two gimlets, two handsaws, one cross saw, one compass saw, one drawing knife, a pair of compasses, and seven chisels. He was the only craftsman on the estate entrusted with finished carpentry.[16]

Several Mount Airy work logs from the period 1805–1814 detail the daily tasks assigned to the carpenters throughout the year, though Bill Grimshaw's routine in the 1830s and 1840s was probably rather different, because his master, William Henry Tayloe, employed only five carpenters, whereas John Tayloe III had had a crew of sixteen back in 1805–1814.[17] According to the work logs, in January and February the Mount Airy carpenters collected timber for barrel staves and fencing, and cut rails and posts at the plantation sawmill. During March and April they constructed new fencing in the several farm quarters. In May and June they joined with almost all of the other workers on the estate to harvest the wheat and oats crops. William Tayloe noted in June 1825 that Bill was one of the eight cradlers who performed the most skilled harvest labor by cutting the ripe wheat so that other workers could rake and bind it into sheaves.[18] Throughout the summer the carpenters made corn barrels—parallel work to the sugar barrels and rum puncheons constructed by Robert McAlpine at Mesopotamia. In 1834, for example, they must have assembled hundreds of casks, because Tayloe shipped 568 barrels of corn from the estate that year.[19] During the fall months they spent much of their time constructing, repairing, or reshingling buildings on the estate, or they were sent to do carpentry at other Tayloe properties. In September 1828 Bill Grimshaw worked at the Octagon, and he was given permission to call at Neabsco ironworks for a day or two on the way home to see his mother and his old friends.[20]

Winney's mother Esther continued to work as a spinner. Her fellow cloth makers seem to have preferred this line of work to agricultural field labor, because at least four of the spinners in 1826 had their daughters working along with them, and two other spinners were the daughters of former spinners. But William Henry Tayloe evidently regarded textile production as marginal. By the 1830s half of the spinners on the estate were elderly semi-invalids, and few young girls were being recruited into the craft. Yet in 1834 they manufactured 310 yards of shirting cloth, enough to supply the field hands on two farm quarters with five or six yards apiece.[21] Thus, Winney's parents were both skilled artisans, contributing significantly to the self-sufficient economy of the Tayloes' Rappahannock plantation.

## *Young Winney*

Bill and Esther Grimshaw's second child, Winney, had, as far as we can tell, a relatively uneventful early life. Yet she must have experienced considerable family tension throughout her girlhood. When she was three years old, in 1829, she stayed with her mother at Old House (renamed Landsdown), where William Henry Tayloe had set up a spinning house. At the same time, Winney's father moved to the home quarter at Mount Airy to be with the other carpenters. A spinner could set up her wheel almost anywhere, and Tayloe allowed several other textile workers who were married to field hands to spin or weave at their husbands' farm quarters. But for some reason he separated the Grimshaws by several miles. From this time onward Esther and Bill probably lived apart much of the time, though they had three more children.

Young Winney lived at Landsdown quarter with her mother and siblings for more than a decade, from age three until age fourteen. Perhaps she envied her older sister Lizza, who went to Washington as a small child in 1830 to live with their grandmother Winney Jackson and stayed on to become a domestic at the Octagon. But Winney had plenty of company at Landsdown, a large farming unit inhabited by about sixty slaves, nearly half of them children. She did not yet have an assigned job in 1839 when she was thirteen years old, by which age over 90 percent of the girls and boys on the estate were already working. She may have been sickly, because there was a lot of illness in her family during the 1830s. Medical bills from Dr. William G. Smith, who treated the Mount Airy slaves, show numerous visits to Esther and her children. Winney's mother became seriously ill in 1834 when she was eight and again in 1839 when she was thirteen, and Esther became a semi-invalid thereafter.[22] Dr. Smith also prescribed medicine for Winney's grandfather Harry Jackson during his final illness in 1838. On hearing of old Harry's death, a friend wrote to William Henry Tayloe: "I am sure you will miss him greatly as an ostler. I understand Ralph has ruined one of your carriage horses allready."[23] Ralph Ward was the new slave ostler referred to in this letter, and despite his inauspicious start he handled Tayloe's horses into the 1860s.

In 1840, at age fourteen, Winney had her first job. She was sent with two of her sisters—twelve-year-old Anna and ten-year-old Juliet—to work for the wife of a local clergyman, Mrs. William N. Ward. William Henry Tayloe was not hiring these girls out; he was lending them to Mrs. Ward, who was expected to supply their food, clothing, and housing for a year. Most years he lent out several slave children to his neighbors; it was a convenient way to train them into domestic service. Probably he sent out the Grimshaw sisters together so that they could keep one another company. Their three younger siblings—James, who was eight, Charlotte, who was five, and Henry, who was two—stayed with their mother, and it was during this year that young Charlotte died of unexplained causes.

In 1841–1842 Winney returned to the plantation to work in the overseer's house at Doctor's Hall quarter. For the first time in her life she was completely on her own, separated from family and close relatives, and living on a farming quarter inhabited by forty slaves who were almost all strangers to her. But this was a common experience at Mount Airy, where girls and boys were typically moved away from their parents in the early to mid-teens when they were old enough to start regular jobs. And at Doctor's Hall Winney did not have to labor in the fields with the other girls her age. Her assignment was to nurse Mrs. Monday, the ailing wife of the white overseer. Winney Grimshaw was becoming schooled in the slippery art of waiting on the white folks.

In 1843, at age seventeen, Winney embarked on an adult career pattern that was strikingly reminiscent of her mother's. She started to spin yarn and weave cloth with six older women in the Mount Airy spinning house. In 1843 and 1845 Winney was listed as a spinner; in 1844 she held the more demanding job of weaver. And around 1844 she married a twenty-year-old youth named Jacob Carrington, who had been working at the plantation gristmill since he was a boy of nine. Her husband had an unusual parentage. Jacob's mother was a slave field hand named Criss, but his father, David Carrington, was a free mulatto, one of the very few emancipated African Americans who lived in the vicinity of Mount Airy.[24] This, however, did Jacob no good at all. Since Criss was a Tayloe slave, all of her ten children by David Carrington were Tayloe slaves also. Perhaps the stigma of enslavement struck Jacob with special force. By 1844 when he had married Winney, his mother, Criss, was dead, and his father, "Freeman David," had moved away. He had six living brothers and sisters, and four of them were working as field hands at Fork quarter. Jacob himself was described by William Henry Tayloe in 1841 as "a smart lad," experienced at grinding and at dressing the millstones.[25] In 1844 he was put in charge of the mill although he was only twenty years old, with two boy assistants to help him, thirteen-year-old Alfred Lewis and twelve-year-old Cornelius Ward. Jacob and his young wife lived at the home quarter, and in 1844 Winney gave birth to her first child, a son named John.

## *A Family Crisis*

In 1845 an event occurred that changed the whole direction of Winney's life. Sometime during that summer her father, carpenter Bill Grimshaw, had an altercation with one of his white supervisors and was whipped. Floggings were rare at Mount Airy, and Bill was so incensed by his treatment that he ran away. William Tayloe advertised for him in the *Alexandria Gazette* in November 1845 and offered $100 in reward money for his capture.[26] But he never caught Bill Grimshaw, who was well used to traveling around the Virginia countryside on his own, doing carpentry jobs at the Octagon in Washington and at other Tayloe properties. This time, he may have headed for a nearby Virginia port and somehow secured passage (or was hired as a sailor) on a ship heading north. Or he may have traveled north on foot. All we know is that he eventually reached the town of Saint John in New Brunswick, Canada, on the Bay of Fundy. Bill was the only Mount Airy slave between 1808 and 1860 who managed to abscond permanently.

William Henry Tayloe did not tolerate runaway slaves, and his policy—like his father's—was to take revenge by breaking up the offender's family. First he crossed Bill's name off the 1845 slave list, and then he took action to discourage any others from following the carpenter's example. "Sent this family away for misconduct of the parents" was Tayloe's cryptic comment on the Grimshaws in his inventory book.[27] Evidently he laid blame on Esther Grimshaw as well as on Bill, but what her "misconduct" consisted of he did not say. Probably she had managed to conceal her husband's getaway until Bill had too much of a head start to be captured. In 1845 Esther was living at the Landsdown farm quarter where Tayloe had placed her in 1829. He wanted to sell her, but had trouble finding a buyer, since Esther was middle-aged and in poor health. So he kept her temporarily at Landsdown while he proceeded to separate her from all but one of her children.

Bill and Esther's seventeen-year-old daughter Anna had been employed as a domestic for several years by the Reverend Mr. Ward, and in the fall of 1845 Tayloe sold Anna to Ward in exchange for a girl of the same age. Bill and Esther's sixteen-year-old daughter Juliet had been similarly employed for the past two years by another white neighbor, Dr. Tyler, and Tayloe made the same arrangement with Tyler, selling Juliet to him in exchange for another teenage girl. In October 1845 he took a more drastic step, sending Bill and Esther's nineteen-year-old daughter Winney, her infant son John, and her fourteen-year-old brother James to faraway Oakland plantation, his new cotton estate in Alabama—a journey that we will follow shortly. Tayloe didn't need to do anything about Bill and Esther's oldest child, twenty-one-year-old Lizza, who belonged to his widowed mother, Ann Ogle Tayloe, and had been working at the Octagon for many years. By the close of 1845 only Bill's wife and their youngest child, Henry (who was eight years old), were still living on Tayloe's Mount Airy property.

In the spring of 1846 young Juliet Grimshaw responded to this crisis by sending a letter to her sister Lizza in Washington. The formal styling of this letter, the accomplished penmanship, and the almost flawless spelling indicate that Juliet dictated her message to an amanuensis, perhaps to her new mistress, Mrs. Tyler. If Mrs. Tyler composed the letter for her, it would explain why Juliet claims to be "satisfied" with her sale to a new owner. But if the wording is not exactly Juliet's, and if her statements sound guarded, the letter nevertheless tells us what was going on among the Grimshaws and conveys Juliet's deep distress at her family's fractured situation. Here is her message to Lizza:

March 27 [18]46

Dear Sister

I would not have postponed writing so long but circumstances which I could not control have prevented me: this is the first letter you have received from me I have been truly sorry it should be so. You I know are too well convinced of my affection for you to let that circumstance make you doubt its sincerity. Our dear mother looks better than I ever saw her Master has moved her on the farm. She is much opposed to Ann's marr[y]ing—and I also am much averse to it, I think she had better wait until she is older.[28] Ann is very well, she had lately written to Sister Winny, who is in Alabama, but has not received an answer yet.[29] Brother Henry is well. You ought to write to Sister Winny and try to get an answer. We have not heard from father since he left. You do not know how anxious we are to see you and your little boy, do try to visit us soon. I feel very anxious to see my little nephew Armsted says you and he are engaged[30] is it so? I think if it is so you ought to let us know You ought to give your mother and sisters your confidence. Armsted says you are to marry him in June is that so? Are you coming down to Master Charles' to live or not?[31] Aunt Nancy is well and has moved to the farm also.[32] Mother and Aunt N are much better satisfied now. I suppose you know both Ann and myself are sold Ann to Mr Ward and I to Dr Tyler we are much better satisfied than when at home. Aunt Betsy is very well but she has lately lost one of her sons, she seems to bear the loss very well.[33] Dear Sister you can not imagine how lonely I feel sometimes when I think how our family is scattered, my father I know not where nor how he is whether dead or alive, one sister in Alabama mother at home alone with out any of her children with her, and we never hear from you. This is enough to make me low spirited is it not? But we have one consolation we will hope one time to meet in heaven never to part again Aunt Betsy sends her love to you and says you must give her love to her Brother. Mother and Ann join me in love to you. Either visit us soon or write. Believe me your affectionate sister

Juliet Grimshaw[34]

What comes through loud and clear in this letter is young Juliet's devotion to her parents and siblings, her anxiety about what has happened to her runaway father, and her grief at the breakup of her family. She was upset by Lizza's detachment from the family crisis. Also evident is her sustaining religious faith, as well as her belief in the serious business of marriage: sister Ann at seventeen was too young to marry. Unfortunately for Julia, the Grimshaw family situation deteriorated still further in the months after she wrote to Lizza. In July 1846 William Henry Tayloe finally got rid of Esther and her young son Henry for the low price of $300. "Sold," he noted after her name on the Landsdown list, "or rather given away." There were now no Grimshaws left at Mount Airy. Only Winney, her baby John, and her brother James still belonged to Tayloe, and they were 800 miles away in Alabama.

Probably a year later, in a partially dated letter that seems to have been written in March 1847, Juliet sent another message to Lizza. But was this message sent by Juliet? The scrawled initials at the close of this second letter may be "J G" but could also be "S P" or "J P" or several other combinations. This document is puzzling in many ways. It is written in a different hand from Juliet's letter of March 1846, is much less polished, and reads as though the author was Lizza's close acquaintance rather than her sister—referring to "your" rather than "our" mother and grandmother, and signing off as one of Lizza's "friends." If Juliet *was* the author, she had been resold by Dr. Tyler to Robert Wormley Carter II and was now living at Sabine Hall, the Carters' mansion two miles distant from Mount Airy in Richmond County.[35] More likely the letter was written by an African American friend of Lizza Grimshaw who lived at Sabine Hall. The slaves at Mount Airy and Sabine Hall were closely connected, and some of the Tayloe people had Carter husbands or wives. The only thing certain is that the writer was replying to Lizza, who wanted to get in touch with her mother, Esther, to tell her that grandmother Winney Jackson had died in Washington and to ask how to dispose of her belongings. The answer to Lizza's plea is as follows:

> Sabine Hall Mar 22nd
>
> It is now a month, my dear Lizzy since I received yours of the twenty-second of february I would have Answered yours immediately as you requested if I Could have seen your Mother I have been waiting day after day and week after week in hopes of Seeing her I had com to the conclusion to write to day Wheather I saw her or not She Came over yesterday to see me, and to hear about you[36] I read your letter over to her She sends you too dollars as a present for your Boy She is very Sorry that it is not more She Wishes you to sell all belonging to your Grand Mother[37] but her Clothes and keep the money for your own use I hope this letter will not put you back in any of your plains I am sorry that I have so little in my power to assist you but what little I have is at your service I will be Responsible for what you owe to A [??][38] and feel pleasure that I can bee of that much service to you your letter gave me both pleasure and pain Excuse this short and imperfect letter I hope you will not be disapointed my dear Lizzey in your expectation if all thing[s] end well w[h]ich I hope will be the case you will not fail to lett your friends know
>
> Yours truly
> J G [??][39]

Juliet Grimshaw, or whoever responded to Lizza Grimshaw, was evidently able to read and write. Thus, in contrast to Juliet's dictated letter of March 1846, we have a document apparently written by a slave. The letter shows the difficulties that slaves had in communicating by long distance. Lizza, who was evidently very anxious to reach her mother, did not write to her directly, perhaps because Esther was illiterate. But Lizza may also not have known the name and address of her mother's new owner. And Esther for her part took a month to visit Sabine Hall and receive Lizza's message, probably because she had to get permission from her new master. The letter also raises a series of teasing questions. Why was Lizza going to be disappointed with her mother's direction to sell grandmother Winney's belongings and keep the money? Evidently Lizza was breaking up with her boyfriend, Armistead Carter, but what was her new "expectation"? What were her "plains"? Regrettably, we have no answers for these questions, because Lizza—along with Esther, Juliet, Ann, and Henry Grimshaw—all disappeared from the Mount Airy records after Bill Grimshaw ran away.

Bill himself, however, did resurface by long distance in William Henry Tayloe's file of correspondence. When he fled in 1845 he kept heading north until he reached Canada. He settled in Saint John, New Brunswick, where he met an abolitionist named William Francis. Bill asked Francis to write to William Tayloe on his behalf, to say that he wanted to buy his freedom from Tayloe. Apparently Francis sent several letters in the late 1840s but got no reply. In August 1851 he tried again, with the following message:

> Saint John NB Aug 14th 1851
>
> Sir
>
> Not knowing whether you have received the letters I have written to you, I again comply with the wishes of William Grimshaw, by once more writing to you concerning him. According to his request I wrote to you in regard to purchasing his body, but have not as yet received any answer. He seems to be quite anxious to know for what amount you will give him a bill of sale of his body, that he may be at liberty to go or come where his business may call him. He says however he knows he is safe and free from any danger of being again taken back to the South, but he is no less willing to make some recompence to you for the labor of which he is well aware you were deprived at the time he ran away from you. He says he never should have ran away from you if he had not been whip'd, for in no other way had he the slightest reason to

complain. If you wish to sell his person and receive the cash for it you can by writing to me and stating for what amount you will give a satisfactory bill of sale.

The money is ready at any moment, that the bill of sale may be produced in Boston or New York as there are persons there who will transact the business for him. He wishes to be remembered to your family with his best respects, and to all with whom he was acquainted who may enquire about him.

If you conclude to write to me concerning him you would very much oblige me by doing so as soon as you could make it convenient as I expect to leave Saint-John on a visit to the United States for some length of time; but if you write within a couple or three weeks I will receive it, before I leave and will be able to know how to proceed.

P.S. Please direct your letter to William Grimshaw to the care of Mr Wm Francis, King Street, Saint John, N.B. And believe him your most obedt

William Grimshaw[40]

William Francis's letter was posted from Canada to Virginia at a cost of ten cents. Bill's amanuensis clearly did the runaway carpenter no favor by phrasing his request so disparagingly: asking Tayloe for a "bill of sale of his body." Bill Grimshaw had repeatedly asked Francis to write on his behalf, so he really did want to purchase his freedom from Tayloe. With a written declaration of settlement from his former master he could move about more freely in the northern United States, especially after passage of the Fugitive Slave Act of 1850, which required that all runaway slaves when captured be returned to their masters. Also, Bill may have wished to get away from New Brunswick, where the Saint John city charter specifically excluded black people from practicing a trade or selling goods.[41] Whether Bill had earned enough money to pay Tayloe's asking price, or whether he was dependent upon his abolitionist friends, I do not know. It seems very unlikely that William Henry Tayloe replied to this letter. But he did keep it, writing "Grimshaw" on the cover. It would be fascinating to find out whether Bill Grimshaw stayed in Canada or moved back to the States, but I have not been able to track this talented and venturesome man past August 1851.

## *Winney in Alabama*

By 1846 most members of Winney's family had disappeared from the Mount Airy records. Winney, her baby boy, John, and her brother James were the only Grimshaws still held in bondage by William Henry Tayloe, and they had been exiled to the Deep South. In October 1845 Winney, John, and James were consigned to a large party of forty-five Mount Airy slaves who were being sent to Tayloe's new cotton plantation named Oakland, situated in what is now Hale County in the Canebrake region of west-central Alabama, north of Demopolis and west of Selma. Winney's husband, Jacob Carrington, was not a member of this group. William Tayloe was unwilling to send his best miller to a plantation in Alabama where he had no mill. Two of Jacob's brothers did go to Oakland in this party, and also the mill boy Alfred Lewis, but Jacob himself stayed on at Mount Airy grinding grain until the Civil War. Winney probably never saw her parents again, and she didn't see her husband, sisters, or youngest brother for at least twenty years.

Winney's party trekked the 800 miles from Mount Airy to Oakland accompanied by five horses and mules and a heavy wagon, and camped each night in tents. There were five little children in the party (including Winney's baby), and the children may have ridden in the wagon, but all of the other slaves walked the entire way. They left Mount Airy on October 14, 1845, went from Richmond through western Virginia across the Appalachian Mountains to Knoxville, Tennessee, then cut down to Summerville, Georgia, and passed through a series of small Alabama towns until they reached Marion in the Canebrake region, and traveled from there a few miles further to Oakland plantation. This route required close to forty days, so they probably reached Oakland before the end of November.[42]

Winney, trained as a spinner and weaver, and James, trained, like his father, as a carpenter, were not obvious candidates for workers on a cotton farm; they had been sent to Oakland because William Henry Tayloe wanted to show the other Mount Airy slaves that bad things happened to families who defied him. The Grimshaws' fellow travelers to Alabama in 1845 were mostly young field hands of the right age to walk the long distance to Alabama and to learn how to cultivate and pick cotton. Their median age was sixteen. Only three members of the group were over the age of thirty. There was one complete family of eight: Alfred and Sinah Lewis and their six children (the oldest of whom was Alfred Jr., the mill boy). There were also several mothers with babies, like Winney. But most were young teenagers migrating without their parents, sometimes in the company of a brother or sister, and sometimes all alone. Winney's Carrington brothers-in-law were in this category. Austin Carrington was a twenty-year-old scullion from the Mount Airy kitchen, and his brother David was a fourteen-year-old field hand from Fork quarter. Their mother was dead and their free black father was gone, but the Carrington boys left three brothers, a sister, a brother-in-law, a sister-in-law, and four little nephews back home at Mount Airy.[43]

Once these transplanted Virginians finally arrived at Oakland plantation, they found themselves on a raw new cotton farm of 600 acres. The newcomers must have seen immediately that they were in for hard times. There were not enough cabins to accommodate the migrants, and new cabins were not built for them until early 1847. Almost all of the Virginians who were old enough to work were put into the cotton fields, but Tayloe treated Winney and James Grimshaw differently. He directed that James, who had started training as a carpenter at Mount Airy just before his father ran off, be continued in this craft at Oakland. Winney posed more of a problem, since there was little demand in Alabama for textile workers. Tayloe purchased all of the cloth needed at Oakland for slave clothing from the New England textile mills, and kept only one spinner on the plantation to make a little yarn or cloth. So Winney became the housekeeper for Tayloe's new overseer, a man named Richard H. Donnahan. In 1846 and 1847 she also had a sickly son to nurse. When Tayloe visited Oakland in May 1847 he listed young John, then nearly three years old, among the slaves. But John died in 1848.

Winney Grimshaw Carrington was a young and evidently attractive woman in the bloom of life. Her supervisor, Richard Donnahan, had a wife and children, but this did not stop him from pursuing Winney. By 1848, if not earlier, she became his mistress. In 1849 she bore him a mulatto son and named the infant John in memory of the black boy she had just lost. Sex with slaves was a taboo topic in the antebellum South, and Winney is the only Tayloe slave woman that I know about who is identified as a black mistress in the family correspondence. Racial mixing certainly occurred. At least ten of William Henry Tayloe's slaves were listed as mulattoes in the 1860 slave census.[44] But Tayloe himself was a pious, abstemious, and sober-minded family man who was very much opposed to interracial sex. And he was upset at seeing Winney's child when he next came to Oakland.

William Henry Tayloe became more exercised when Winney bore a second mulatto baby, named Julia (probably after her sister Juliet), in 1850. Mrs. Donnahan, in Tayloe's view, was part of the problem: she "was a delicate little woman very helpless—she could not get on without Winney and countenanced the two families under her eye."[45] So Tayloe intervened to break up this Oakland ménage à trois. First, in 1852, he tried to get Donnahan to move away from Winney to Walnut Grove, a plantation that Tayloe had an interest in, and manage Oakland from several miles away. But Donnahan said that this was impractical and refused.[46] Since Richard Donnahan was the best overseer he had in Alabama, Tayloe capitulated temporarily. Then a third mulatto baby, named Lizza (after Winney's elder sister), appeared in 1853. Deciding to put pressure on Winney instead, Tayloe took her mulatto son John away from her when he was four years old and presented the child to his widowed niece Lucy Tayloe as a New Year's gift on January 1, 1854.[47] Next, in 1854, he dealt with Donnahan more firmly: "I moved him to Woodlawn to separate him and Winney, but in my absence he made other arrangements."[48] Woodlawn was another cotton plantation that Tayloe owned some miles from Oakland, and as soon as Tayloe went back to Virginia Donnahan brought Winney to live with him at Woodlawn. Here in 1855 she bore a fourth mulatto child named Willie Anne. And when Tayloe visited Woodlawn in April 1857, he discovered that yet another birth had taken place. "Winny has an infant," he reported to his son in Virginia, "which I am told is 'much like the others.' "[49]

What Tayloe didn't mention was that this fifth mulatto child was named William Henry. Why, one may ask, did Winney choose to name this latest infant for her intrusive master? Possibly she did it to mock or shame him. Possibly Winney had genuine affection for William Henry Tayloe despite what he had done to her, and named the baby in his honor. More likely she chose the name in the hope that Tayloe would not take a boy named after him away from her. If so, her strategy worked, for little William Henry Grimshaw stayed with his mother.

While Winney was serving Donnahan, her brother Jim Grimshaw worked as a carpenter first at Oakland and then at Woodlawn. From 1845 to 1857 he probably shared a cabin with Winney and her children. Jim clearly had some sexual adventures or misadventures as a young man, because during 1857 at the age of twenty-five he received eighteen treatments for syphilis over an eight-month period from March to November.[50] By 1858 he was back at Oakland, apparently in an improved state of health, and there he found and married eighteen-year-old Arabella Ward, who had been sent to Alabama from Mount Airy in 1854. James and Arabella had undoubtedly known each other in Virginia as young children, and James had worked in Alabama with two of Arabella's brothers. The Wards were a family very similar in slave status to the Grimshaws. Arabella's father, Ralph Ward, was the groom at Mount Airy, the man accused in 1838 of ruining one of Tayloe's carriage horses when he replaced James's grandfather Harry Jackson as coachman and ostler. Arabella's mother, Eliza Ward, was a spinner who had worked with James's and Winney's mother, Esther Grimshaw, in the Mount Airy and Landsdown spinning houses for twenty years. Arabella had also been a spinner at Mount Airy, and she continued to ply this craft at Oakland, the only spinner listed on the estate inventories from 1855 through 1863. When William Tayloe visited Oakland in March 1858 he remarked to his son, "You know Jim Grimshaw married Arabella and has got her to breeding."[51] But if Arabella was pregnant when Tayloe wrote this letter, she either miscarried or her child died in infancy. Four years later Jim and Arabella successfully started their family. Their first recorded child was a daughter, Sylvia, in 1862, followed by a second daughter, Eliza (named for her maternal grandmother), in 1864.

Meanwhile, back at Woodlawn, William Tayloe was glad to hear that overseer Donnahan was talking of moving to Texas, for "I think it is time for him to separate from us."[52] In early 1858 he finally broke up Donnahan's liaison with Winney by dismissing him. It was at this time that Winney (now thirty-two years old) fell seriously ill: William Tayloe paid Dr. G. W. Browder $80 to visit her daily for nearly a month, from February 10 through March 7, 1858.[53] In this same year Tayloe sold Woodlawn plantation and moved his slaves to another cotton farm called Larkin in Perry County. On the eve of secession he owned and operated two Alabama cotton plantations—Larkin and Oakland—with a combined population of 277 slaves. In 1860 Jim Grimshaw and his wife lived at Oakland, while Winney lived at Larkin with three young children, occupying one of the thirty-four cabins in the slave quarters. Her first son John had died, her second son John had been taken away, and her daughter Julia had also died, but she still had Lizza (age seven), Willie Anne (age five), and William Henry (age three). In 1861 Winney began keeping house for the new Larkin overseer, a man named John W. Ramey, who had a wife, a daughter, and two teenage sons.

Soon after the Civil War broke out, Tayloe moved most of the able-bodied workers still living at Mount Airy to Alabama, in order to keep them from being captured or deserting to the Union Army in Virginia. By 1863, 250 people were crowded into the slave quarters at Larkin, with another 141 at Oakland. Many families that had been torn apart during the 1830s, 1840s, and 1850s were suddenly reunited. Austin and David Carrington, for example, who had been separated from their siblings in 1845, were once again living in the same neighborhood with their brothers Godfrey and Israel and their sister Becky. They also had twenty-two nephews and nieces among the slaves on Tayloe's two Alabama plantations. But one member of the Carrington clan was conspicuously missing. The man Winney had married in 1843, the Mount Airy miller Jacob Carrington, had deserted to the Yankees in December 1861 when he saw that his master was planning to move him to Alabama. Freedom was more important for Jacob than reunion with his lost wife.

As the war got under way, William Henry Tayloe took up residence at Oakland plantation in order to keep an eye on his Alabama operations, leaving Mount Airy in the hands of his son Henry Augustine Tayloe. He also frequently visited Larkin plantation, but felt uncomfortable with overseer Ramey's family because he took an intense dislike to young Charles Ramey, who was eighteen years old in 1861. This youth had managed to get quickly out of Confederate Army service, and according to Tayloe he spent his time at Larkin shooting squirrels instead. He "took Winney's son [William] Henry to tote them for him, and he visited the girls." Soon Ramey's black housekeeper was pregnant yet again. When Winney's baby was born in 1862, the infant—named Thornton by his mother—was unmistakably another mulatto child. Mrs. Ramey was less compliant about this sort of thing than Mrs. Donnahan had been, and when she saw Winney's new baby "a bomb exploded" in the Ramey household. Tayloe reported that "Winney wanted clothes for the Baby. I had scolded about the D's—so she kept out of my way. I never saw Baby nor asked any questions."[54] The situation was indeed an awkward one. Tayloe wanted to get rid of the Rameys, but he knew that it would be extremely difficult to find a replacement overseer during the war. And Ramey was unwilling to quit the Larkin job because he could keep his sons out of the Confederate Army by claiming that they were doing essential plantation work. So Tayloe and Ramey kept an uneasy truce over the next three years.

When the Yankees captured Vicksburg in July 1863, Tayloe sensed that the Confederacy was likely to lose the war. Doubtless aware that his slave ownership was now in peril, William Henry Tayloe made a thorough appraisal of the people that he had gathered together at Oakland and Larkin plantations. He listed all of the slaves by families, indicating who lived in each cabin, and he calculated a monetary value for each person. For example, he reckoned that Jacob Carrington had a market price of $1,500—a value wiped out by his getaway. James Grimshaw was priced even higher, at $1,800, as a skilled carpenter, and Jim Grimshaw's wife Arabella was also valued at $1,800, the highest price for a female on Tayloe's list. Housekeeper Winney was reckoned to be worth $1,000, and her children were valued at $800, $700, $600, and $200—the $200 child being young Thornton. Tayloe's slave valuations in 1863 were about four times as high as the values assigned to his father's Mount Airy slaves back in 1808.[55] Tayloe also recorded the ages of his slaves in 1863, but his age statements are not very accurate. He stated that Winney was thirty-two years old in 1863, whereas she was actually thirty-seven.[56]

On April 2, 1865, the Union forces captured Selma, and a week later Lee surrendered to Grant at Appomattox. Soon the Federal army was flooding into west-central Alabama. Tayloe's slaves, suddenly liberated, stopped working at Oakland and Larkin and collected in the local towns—Selma, Uniontown, Demopolis—to talk with the Yankee soldiers or to consult among themselves.[57] There was great confusion because the freedmen and freedwomen were hoping for land of their own, which they didn't get, and instead were told that their only employment choice was to keep working for their former masters. At least they now had the option of moving about in search of better work sites.

At Oakland plantation most of the former slaves decided in 1865 to stay on for a while, and contracted to work for Tayloe's overseer as hired cotton hands. Jim and Arabella Grimshaw were still at Oakland in early 1866: both are entered on an Oakland employment and housing list for this year. But they evidently dropped out after about three months, since Jim was paid only $33 and Arabella was paid $22, whereas the men who worked for a full year in 1866 received about

$115, and the women about $75.[58] Once they quit Oakland, Jim and Arabella disappeared from sight; I have not been able to find them in the U.S. Census of 1870. But quite a few of their Mount Airy companions did stay on at Oakland, as we shall see in Chapter 9. Thirty-eight sharecroppers born at Mount Airy—twenty-two men and sixteen women—continued to work at Oakland through 1868, when the surviving employment records for this plantation terminate. And in 1870 twenty-six of the twenty-seven black workers' cabins at Oakland were occupied by families of freed people who had formerly been Tayloe slaves.[59]

In contrast to Oakland, there was violent upheaval at Larkin plantation in 1865. William Tayloe went to Washington after the war was over to acquire a presidential pardon, leaving overseer Ramey in charge. After Tayloe returned in October he found the plantation in disorder and accused Ramey of stealing from him and trying to entice his best workers to leave. So Tayloe fired Ramey, who left with about a third of the Larkin freedmen—including many of Tayloe's best hands. By the end of the year another third or more of the freed people also departed to work on other plantations. In January 1866 only a small remnant of Tayloe's workforce was left at Larkin. But Winney was still there with her four mulatto children, now aged twelve, ten, eight, and three.[60] Indeed, Winney may have given birth to one more baby, because on a Larkin list dated November 1865 she is credited with five children.[61]

By 1870 many of the Tayloe freed people who left Larkin in 1865 had come back. The census shows that twenty-five black families formerly enslaved by the Tayloes were living at Larkin. But Winney Grimshaw was not among them. She evidently left Larkin in 1866 or 1867, perhaps remarried, and (like her brother Jim) disappeared from sight. More than half of her companions at Larkin and Oakland can be located via the U.S. Census of 1870 either in Alabama, Virginia, or the District of Columbia, but Winney and her children cannot be traced.[62] She was forty years old in 1866. With no education and four (or five) small dependent children, this freedwoman had meager chances for a better sort of life.

Winney Grimshaw, unlike Sarah Affir at Mesopotamia, was born into a favored slave family. But favorable status could be instantly dissolved. When Winney's father dared to defy William Henry Tayloe, her family was deliberately torn apart. Her mother, youngest brother, and two sisters were sold to three new owners, and Winney and her other brother were packed off to Alabama. The Tayloe records reveal disappointingly little about this woman's personal life. We need a letter from her comparable to her sister Juliet's letter of March 1846. But the narrative of her slave experience reveals quite a bit about power relations between whites and blacks. Subversive interracial sex cuts like a knife through Winney's story. She was manipulated to an extraordinary degree. She had to cope with continual exploitation by her white masters—and cope she did. The last reference to her that I can find comes from her former owner, William Henry Tayloe. Reminiscing about his experiences as a slaveholder in 1868, Tayloe remarked that "Winney Grimshaw could fill a volume with interesting events, if [she] could write."[63]

If she could write . . .

# 4

# "Dreadful Idlers" in the Mesopotamia Cane Fields

THIS BOOK OFFERS many opportunities to observe the fatuous opinions of the white planters about Afro-Caribbean (or African American) slave character. Joseph Foster Barham II provides a classic example in his 1823 pamphlet entitled *Considerations on the Abolition of Negro Slavery*, where he warned that the Caribbean slaves could not be made to perform satisfactorily as free wage laborers should the British government decide to emancipate them. Black people, in his view, were morally deficient and lacked any sort of work ethic. "The Negro race," Barham declared, "is so averse to labour that without force we have hardly anywhere been able to obtain it, even from those who have been trained to work." And he characterized the people who toiled for him at Mesopotamia as "dreadful idlers" by nature.[1]

Barham's own records supply much evidence to support alternative explanations for the labor problems at Mesopotamia. His slave inventories, designed to keep a running check on the status of every man, woman, and child, provide a precisely detailed account of the Caribbean slave labor system in action from 1762 to 1833. The inventories during these years not only report the annual occupation of each worker, but also categorize the healthy workers as "able," the less healthy workers as "weak" or "sickly" (often listing their ailments), and the nonworkers as "invalids." When the occupational and health reports are correlated, it becomes clear that most of the jobs assigned at Mesopotamia had a measurably negative impact on the slaves' life expectancy, and that some job assignments were particularly damaging to the workers' health.[2] The

Mesopotamia records also highlight the differences between the jobs assigned to males and females, as well as job differences between Creoles and Africans. And for the female workers they show the impact of field labor on childbearing.

Unfortunately, the Mesopotamia records provide very little information about how the white attorneys, overseers, and other staff members managed the slave laborers day by day. The diaries kept by the Moravian missionaries at Mesopotamia are of some help here, for the missionaries recorded many accusations from the slaves about bad treatment. But the Moravians always focused on the slaves' spiritual state, and avoided commentary on slave life at Mesopotamia as much as possible except to lament continually that their converts were hindered from attending divine services during crop time. To get a more direct feel for how the slaves were actually treated, I draw upon the enormously detailed diary of Thomas Thistlewood, an overseer and small planter who lived within ten miles of Mesopotamia from 1750 to 1786. Throughout his long Jamaican career Thistlewood usually wrote in his diary every day, and his thirty-seven unpublished manuscript volumes add up to 9,300 pages.[3] To be sure, Thistlewood is a controversial figure. Some historians see him as completely unrepresentative, and certainly he had repellent character traits. But Thistlewood was a practical man, and I believe that he used the standard playbook for slave management in Jamaica. There is certainly some hazard in using him as a proxy, since he spent no time at Mesopotamia as far as I have discovered (though I have only read several years of his huge manuscript), and the slave gangs he managed were much smaller than those at Mesopotamia. Thistlewood rarely had more than one hundred slave workers under his control, so he probably kept a closer eye on them than Barham's overseers could do. Yet his account of his day-to-day activities matches well with surviving information about the behavior of the overseers and bookkeepers and other white staff members at Mesopotamia.

In the following pages we will observe the Mesopotamia labor system from a variety of angles. First we track the productivity of the laborers on this estate: the annual sugar and rum output at Mesopotamia from 1751 to 1832, and the shifting relationship between crop size and slave population. Next we examine the pattern of male and female employment, and the impact of work assignments on the slaves' health and longevity. Then we observe Thomas Thistlewood's brutal style of slave management, which I believe was also practiced at Mesopotamia. Next we look for the causes of the especially high male mortality rate at Mesopotamia and the reasons for the unusual pattern of childbearing on this estate. And finally we compare the occupational histories of four distinctly different groups of Mesopotamians: the mulattoes and quadroons fathered by the white managerial staff, the black Creoles born on the estate, the blacks (both Creole and African) who were brought to Mesopotamia from elsewhere in Jamaica, and the Africans who came directly to Mesopotamia on the slave ships.

## *Sugar Production at Mesopotamia*

Year after year the Mesopotamia laborers produced very sizable quantities of sugar and rum for export to London. The estate records, supplemented by crop reports for Mesopotamia in the Jamaica Archives, provide a comprehensive account of annual sugar production from 1751, the year in which Joseph Foster Barham I inherited the estate, through 1832, when Joseph Foster Barham II died. There are shipping totals for all eighty-two years in the Barham Papers, and more detailed Mesopotamia crop reports for sixty-six of these years in the Accounts Produce series at the Jamaica Archives. These records show that during Joseph I's ownership (1751–1789) the enslaved workers produced 6,821 hogsheads of sugar and 4,547 puncheons of rum valued at approximately £251,099 Jamaica currency—generating a median annual gross value of £6,438. During Joseph II's ownership (1790–1832) they produced 10,135 hogsheads of sugar and 4,553 puncheons of rum valued at £396,516—generating a median annual gross of £9,221. In eighty-two years the Mesopotamia people produced 16,956 hogsheads of sugar and 9,100 puncheons of rum valued at approximately £647,615 Jamaica currency.[4] Enough hogsheads to supply 445,000 modern Americans with sugar for a year; enough puncheons to fill four million bottles of rum; enough Jamaican pounds to buy 12,000 slaves at 1750s prices or 6,000 slaves at 1820s prices.

All of the Mesopotamia sugar was shipped to London except for a hogshead or two reserved each year for estate use. Most of the rum was also shipped to London, but in some years a large number of puncheons were sold locally. Mesopotamia also had a big herd of 300 to 500 livestock used as draft animals and to manure the fields. These animals, managed by the slave stock keepers, were culled almost every year, sometimes with significant sales, as in 1801, when eight horses, thirty-four steers, sixteen cows, three heifers, and eight calves fetched £2,356. From 1785 onward there was a further source of Mesopotamia income: slave laborers cut down logwood trees on the estate and chipped off the bark, so that the heartwood (a source of brilliant red dye) could be sold. In 1787, for example, one hundred tons of logwood fetched £404; forty-four tons were shipped to London, and the remaining fifty-six tons were sold locally.[5] All in all, Mesopotamia earned about £54,000 in livestock sales and £8,000 in logwood sales between 1751 and 1832.

The Barham Papers at the Bodleian Library provide a very useful though incomplete account of expenditures on slave labor at Mesopotamia. Expenses in Jamaica, such as the purchase of new slaves or the hiring of jobbing gangs as extra slave laborers, were recorded in Jamaican currency, while expenses in England, such as the cost of the slave clothing, slave food, and slave tools and utensils that were shipped annually to Mesopotamia, were recorded in pounds sterling, with £1.4 Jamaican equivalent to approximately £1 sterling. The Bodleian records demonstrate that Mesopotamia slave labor costs rose dramatically between 1751 and 1820. A balance sheet for Joseph I's first twenty-seven years of ownership, 1751 to 1777, reports that he spent on average about £270 Jamaican / £193 sterling annually for new slaves and £140/£100 for hired jobbing gangs, plus £644/£460 annually in slave supplies shipped from London. An expense account for 1777–1788 records that the estate was now paying £220/£157 per annum for jobbing gangs; a second expense account for 1798–1808 itemizes payments for jobbing gangs at £410/£293 per annum; and a third expense account for 1814–1819 raises the payments for hired labor to £920/£657 per annum. The cost of slave supplies also climbed steeply. A balance sheet kept by the Barhams' sugar brokers, Plummer & Co., shows that in 1777–1797 Plummer shipped supplies priced at £1,134/£810 per annum to Mesopotamia during this twenty-one year period—almost double the figure for 1751–1777.[6] I have found no figures on the pricing of slave supplies after 1797, but by 1820 these shipments probably cost well over £1,400/£1,000 per annum.

Even so, the proportion of the Mesopotamia budget allocated to slave expenses was always comparatively modest. A tabulation of local running expenses in Jamaica for 1814–1820 shows that the estate expended

£14,305/£10,218 in seven years on the maintenance of more than 300 Mesopotamia slaves. This included £6,083/£4,345 for the purchase of fifty-five new slaves from Cairncurran, £5,860/£4,186 for the hire of extra jobbing gang slave laborers, and around £2,000/£1,429 for local food purchases and medical attention. During these seven years the estate expended £11,848/£8,463 on wages and butcher's meat for the six whites who supervised the 300 Mesopotamia slaves.[7]

The Barham Papers also supply a good deal of information about Joseph I's and Joseph II's net profits from Mesopotamia sugar and rum year by year. The balance sheet for Joseph I's first twenty-seven years of ownership shows that income always exceeded expenses between 1751 and 1777, with a median net profit of £3,760/£2,686. The balance sheet kept by Plummer & Co. for 1777–1797 indicates that the estate continued to make money during these twenty-one years, though with great fluctuations. In 1778 the net profit was a mere £57/£41, whereas in 1796 the estate netted £16,127/£11,519. Plummer generally sold Mesopotamia sugar and rum in London for about two-thirds of the Jamaican evaluation, but London sugar prices suddenly surged in 1790–1792 when the chief French sugar island of St. Domingue was convulsed by revolution, and three Mesopotamia crops valued collectively at £27,905 in Jamaica sold for £31,315 sterling in London.[8] This was an era of major combat between Britain and France, and the Barhams lost a total of ninety-three hogsheads of sugar and fifty-six puncheons of rum at sea to privateers, though some of this loss was covered by insurance. Between 1777 and 1797 father and son bought 172 new slaves for Mesopotamia; the annual bills from Jamaica averaged £2,450/£1,750, and the annual supplies sent from England cost an additional £1,134/£810, yet the bottom line was that in these twenty-one years Plummer & Co. paid the Barhams net earnings of more than £119,000/£85,000, or £5,677/£4,055 per annum. And the good times continued. The tabulation of Jamaican running expenses for 1814–1820 does not include slave supplies shipped to Mesopotamia from England, but if these are estimated at the generous figure of £2,000/£1,429 per annum, then total estate running expenses came to about £7,200/£5,143 annually during these years as against £12,000/£8,571 in yearly sugar shipments and cattle sales. Joseph II's bank books with Messrs. Thomas Coutts show that Plummer & Co. deposited £47,583 sterling into his account between 1805 and 1825.[9]

Appendix 14 charts Mesopotamia's productivity and the Barhams' sugar income against the changing size of the slave population during the years 1751 to 1832. The chart confirms that Joseph II operated on a decidedly larger scale than Joseph I. Over a span of thirty-nine years Joseph I's slave population averaged 267, whereas in forty-three years Joseph II's population averaged 349 and produced 50 percent more sugar with 60 percent more gross profit. Both father and son experienced sharp swings in sugar and rum production from year to year and fluctuations in the market value of the crop. As sugar prices rose over time the swings in market value became more pronounced. But the most intriguing feature of the chart is the relatively close correlation between the size of the slave gang and the size of the crop until 1774, and the much smaller degree of correlation between population and production from the American Revolution to 1832.

The chart begins in 1751, when young Joseph Foster Barham I came to Jamaica to take possession of his inheritance. Finding that his work gang was depleted, he bought twenty-one new slaves, which brought the population up to 285. When Joseph I went back to England, his attorney, the Reverend John Pool, did what managers of absentee properties often did when armed with new laborers: he drove the Mesopotamia slaves to the maximum in order to obtain large crops and generous returns for his employer and extra profits for himself. Between 1754 and 1761 the slaves produced 209 hogsheads of sugar per annum, a total not seen again until 1784. But by 1762 (when Joseph I obtained his first fully detailed slave inventory) the slave population was in such poor shape that production declined to an average of 156 hogsheads during the next decade. Eighty-three new slaves were added between 1762 and 1774, and sugar production began to rise during the early 1770s. But during the American Revolutionary War only ninety-nine Mesopotamia hogsheads of sugar were shipped annually in the years 1776–1781. In October 1780 a great hurricane lacerated the Mesopotamia cane, leading to a record low sugar shipment of fifty-nine hogsheads in 1781. The war years also brought food shortages and extra suffering to the Mesopotamia people, and by 1783 the population had shrunk to 243. But immediately after the war Joseph I bought 67 new slaves, which increased the population to 303 in 1786. With a greatly expanded number of young laborers (as illustrated by Appendix 3 in Chapter 1), Mesopotamia in the last two years of Joseph I's ownership shipped, respectively,

262 and 313 hogsheads of sugar, and the £11,823/£8,445 market value of the crop in 1789 was almost double the average value during Joseph I's tenure.

Buoyed by his father's success, Joseph Foster Barham II purchased 91 more slaves for the estate in 1791–1793 and expanded the population to 383. But although there was ample acreage along the Cabarita River for new cane fields, sugar production actually declined a little. Having shipped 231 hogsheads per annum in 1784–1789, Mesopotamia shipped 227 hogsheads per annum in the 1790s, and 216 hogsheads per annum in the 1800s. Joseph II told his attorneys that he wanted his slaves' workload lightened, and perhaps the attorneys followed his instructions, but if so they did not prevent a population drop from 376 in 1794 to 307 in 1810. The death toll was particularly high among the 102 people brought to Mesopotamia from Three Mile River estate in 1786 and Southfield estate in 1791. In 1810–1813, when the population was back to the size it had been in 1786, the Mesopotamia people harvested four unusually large crops averaging 314 hogsheads at a market value of £13,279/£9,485. In 1810, the peak year, they shipped 415 hogsheads of sugar and 164 puncheons of rum priced at £17,092/£12,209—more than triple the average valuation in 1751–1779. And they achieved this result with a shrinking workforce: there were only 146 prime workers in 1810 compared with 188 in 1793. Joseph II's attorneys unctuously assured him that the bumper crop of 1810 was produced without any extra exertion by the Mesopotamia slaves, who "work well, and go cheerfully, and contentedly through their work."[10]

The highly profitable crops in 1810–1813 encouraged Joseph II to purchase fifty-five slaves from Cairncurran (a coffee plantation) in 1814. Meanwhile, his attorneys were supplementing the estate labor force with much heavier use of jobbing gangs than ever in the past. After four more years of good returns in 1814–1817, Joseph II bought another 112 people from Springfield sugar estate in 1818, and after 105 of them were moved to Mesopotamia, he had 421 slaves in 1820. But this large expansion did not translate into still greater sugar output. On the contrary, Mesopotamia sugar production slid from an average of 280 hogsheads per year in the 1810s to 210 hogsheads during the 1820s, and with a decline in sugar prices the annual crop valuation dropped much further—from £11,729/£8,378 in the 1810s to £6,317/£4,512 in the 1820s, which was close to the valuation back in the 1760s. Thus, the productivity chart concludes with striking variance between the size of the slave gang and sugar production.

Joseph II, as a new owner in 1789, had announced to his attorneys that he wanted above all to promote his slaves' health and happiness. But the steep decline in sugar production and sugar profits during the 1820s changed his mood. By 1825 Barham was calling on his attorneys to deal more energetically with the runaways and other refractory slaves on his estates and to punish the women who aborted their babies. In 1829, when he heard that several of the Mesopotamia women had recent stillbirths and that others were faking pregnancy during crop time in order to get out of work, he announced that this "malpractice" and "vicious conduct" must be checked, either by shortening the women's supplies, by making them work harder, or by hiring them out as cane-hole-digging job laborers.[11] By the end of his life Barham had completely abandoned the position he started out with in 1789.

## *The Labor Pattern at Mesopotamia*

The men and women who produced 17,000 hogsheads of sugar and 9,000 puncheons of rum for the two Joseph Foster Barhams between 1751 and 1832 lived in the Westmoreland plain of western Jamaica, a low-lying region that was very hot, very rainy, very humid, and infested by mosquitoes. The Mesopotamia slaves worked an estate of 2,448 acres close to the center of the plain, laid out along both banks of the Cabarita River, five and a half miles inland from the port town of Savanna la Mar. Mesopotamia was bordered on three sides by other large sugar estates: Black Heath to the west, Friendship and Greenwich to the east, and Blue Castle to the south. Many of the Mesopotamia men and women must have found mates among slaves from these neighboring estates, but we have absolutely no record of such unions—or of the enslaved black fathers from Black Heath, Friendship, or Blue Castle whose children were born to Mesopotamia women. We do know that the Mesopotamia people had to trek into the mountain land on the northern perimeter of their estate in order to reach the provision grounds where they grew most of their food. The closest of these provision tracts were laid out about three miles from the slave village, while the more remote tracts were five or six miles away.

The prime cane land where the Mesopotamia field hands worked six days a week stretched along the Cabarita, with 400 acres divided into some forty cane pieces, about half of which were harvested each year. The sugar works also stood at the river's edge, centered on the water mill (powered by a wheel eighteen feet in diameter and ten feet wide) into which the field hands fed ripe cane stalks at crop time. Close by was the boiling house where the boilers ladled the cane juice into a series of red-hot kettles of decreasing size (from 450 to 80 gallons in capacity), heating, clarifying, and evaporating the juice into crystallized sugar. In the curing house other slaves packed the sugar, still full of molasses, into hogsheads with pierced bottoms to drain and dry out. Meanwhile, in the still house the distillers fermented the molasses extracted from the sugar into spirits, then aged the rum in huge cisterns. The Barham Papers contain detailed descriptions of the sugar works, and also of the white people's accommodations at Mesopotamia: the Great House, which was often occupied by the Barhams' attorneys, the overseer's house, the bookkeepers' barracks, and the chapel and dwelling house for the Moravian missionaries.[12] But the Barhams' record keepers never bothered to describe the number and size of the slave huts in the Negro village, which was located close to the chapel. The standard practice in Jamaica was that the slaves constructed their huts with wattle and daub walls and thatched roofs, and each hut had an adjoining small garden.[13]

Far worse than the absence of information about the slave huts is the absence of information about who lived with whom inside these huts. For the Mount Airy people, there is plentiful evidence (discussed in Chapters 5, 7, 8, and 9) that most of these Virginia slaves lived in tight nuclear family units, and that more-extended kinship ties were also extremely important. For the Mesopotamia people, family relationships are much harder to determine. The Barhams' attorneys claimed that the Mesopotamia slaves were promiscuous and that there were few slave marriages or long-term stable relationships between men and women. The only family ties singled out in the Mesopotamia records—and in almost all other Jamaican plantation records—were between slave mothers and their children, which has prompted commentators to assume that the Caribbean slave family was matrifocal and severely stunted. Barry Higman challenges this view. Employing a list of "families and their dependents, if any" living in 253 households on three properties owned by Lord Seaford in 1825, he shows that male/female couples lived in about 100 of these households, and that half of these couples (or 20 percent of all householders) lived with children in nuclear family units. In another seventy households mothers lived with their children, without an adult male, in matrifocal settings. The remaining households were mainly occupied by single men or women, plus a few single-sex groups of men and women.[14] This mixed picture of conjugal, matrifocal, and solitary living arrangements fits well with what little is known about family life at Mesopotamia. The slaves purchased from Three Mile River, Southfield, and Cairncurran were listed on arrival at Mesopotamia in a similar mix of nuclear families, female-headed families, couples without children, and solitaries. And the Moravian missionaries, who came to Mesopotamia with European notions about family structure, identified quite a few married couples among their baptismal candidates, but also found much evidence of promiscuity and polygamy to complain about. In sum, conjugal family structure appears to have been always substantial at Mesopotamia, and it probably developed more strongly over time. But family relationships remained fragmented in many ways. The nuclear family at Mesopotamia never became the vital unifying force that we find within the Mount Airy community.

If family life is poorly recorded at Mesopotamia, the labor pattern is exceptionally well documented. As on all Jamaican sugar estates, full-scale employment began at a very early age. Every child above the age of six or seven had a job unless incapacitated by disease. We have seen in Chapter 2 how Sarah Affir, her children, and her grandchildren were routed through a three-stage occupational system: the training jobs they held as children, the primary jobs they held as adults, and the secondary occupations they were switched to when they were no longer able to perform their primary jobs. The critical point in this cycle came at about age sixteen, when the bookkeepers generally promoted "boys" to "men" and "girls" to "women" in the inventories, and when the overseers assigned these new "men" and "women" to the jobs they would hold during their prime working years. Boys and girls whose first childish jobs were in field work, cutting grass or carrying dung, almost always continued as field workers after age sixteen. Boys who began as cattle boys or hog herders also generally continued after age sixteen as stock keepers or carters. But boys who started work as domestics, waiting on the white managerial staff, were usually converted to apprentice craft workers by age sixteen, while most girl domestics became field hands. Whatever their line of adult employment, the sixteen-year-olds at Mesopotamia were not yet adult in a physical sense. Teenage girls began their adolescent growth spurts at about thirteen and boys at about fifteen, considerably later than in most current populations. The Barham Papers provide no information on the growth pattern and stature of the Mesopotamia people, but slave registration returns from Trinidad, Saint Lucia, and Berbice in Guyana show that Afro-Caribbean boys and girls were approximately the same height at age sixteen, averaging just over five feet. Women would grow another inch or so to achieve mature stature, and men would reach five feet four inches by age twenty-two. Caribbean slaves were thus notably shorter than slaves in the antebellum South, where men reached a median height of five feet seven inches and women five feet three inches.[15]

A total of 1,103 slaves—602 males and 501 females—were recorded as living at Mesopotamia between July 1762 and August 1833. About 20 percent of these people—117 boys and 109 girls—either died in childhood or were under the age of sixteen when the last Mesopotamia inventory was taken. This leaves 877 people—485 males and 392 females—whose primary occupations can be studied. Appendix 15 tabulates their job assignments. The nineteen drivers had the responsibility of keeping the field gangs working in unison and on schedule. The twenty-six boilers and distillers (three of whom also worked as drivers) possessed essential technical skills, since it was by no means easy to make acceptable sugar and rum. Among the craft workers (designated as "tradesmen" throughout the British West Indies), the thirty-one coopers had to construct leak-proof hogsheads and puncheons, while the twenty-seven carpenters, eight masons, and five blacksmiths kept the plantation buildings and machinery in working order. The twenty-nine stock keepers and twenty-one carters, all of whom worked with the animals on the estate, enjoyed more autonomy than most of the other laborers. Seventy-six percent of the stock keepers were Mesopotamia-born, which suggests that young Creole insiders sought out and secured these jobs. The twenty-two domestics had occupations that were physically less taxing, but they had to deal directly with the often abusive white managerial staff on a day-to-day basis.

The 561 field workers, by far the largest contingent, were initially divided into three gangs, and by around 1800 into four gangs. During the years 1802–1832, the members of the several gangs were all annually identified. In 1804, for example, the fourth (or hogmeat) gang had twenty-three members, the third gang had thirty-nine members, the second gang had twenty-three members, and the first (or Great) gang had sixty-two members. The youngest child workers were in the hogmeat gang, and their chief job was carrying grass to the cattle. The older children in the third gang also performed comparatively light tasks such as hoeing the cane shoots. The members of the second gang were given harder work; they weeded the cane pieces, cleaned the pastures, and assisted the first gang at crop time. The prime workers in the first gang performed the hardest physical labor, digging the deep, square cane holes in which to plant new shoots, and cutting the ripe cane stalks at crop time.

Almost all of the more responsible and skilled jobs, as well as occupations that offered some degree of independence, were held by the men. Women could be drivers, but only of the third or fourth gangs, and they were excluded from craft work, transport, and all stock work except for fowl keeping. So they had little chance of escaping field labor: 85 percent of the adult Mesopotamia female workers were field hands. To be sure, the women in the West African societies from where these slaves came did most of the agricultural work, but certainly not the gang labor system practiced at Mesopotamia. Significantly, 131 men and women (15 percent of the adult slaves) in the years 1762–1833 were marginal workers or nonworkers. But this does not mean that these people had never had primary occupations. Half of them were aged forty and above in 1762 and had already labored for many years at Mesopotamia, and most of the others were elderly people acquired from Three Mile River, Southfield, Cairncurran, or Springfield. The great majority of the eighty-one marginal workers and fifty nonworkers in this tabulation had probably worked in field gangs when they were younger and stronger.

*Appendix 16* demonstrates that the 877 people over age sixteen who lived and worked at Mesopotamia during this seventy-two year period had short careers and short life spans. Overall they put in 17.5 years of adult plantation labor; they were in good health just 45 percent of the time, weakly 44 percent of the time, and unable to work 11 percent of the time; and the 617 people who died had lived on average for only 46.6 years. Among the women, it is difficult to assess the impact of occupation on health and longevity, since practically all of them—94 percent of the prime workers—were field hands. The 17 female domestics worked longer at their prime and secondary jobs than the 305 female field hands and enjoyed better health, but their number is so small that the comparison may be irrelevant. Among the men, however, job comparisons are more meaningful. When the 256 male field hands are set against the 71 male craft workers and 29 male stock keepers, the field hands clearly fared less well. They toiled at their prime jobs for less than twelve years, whereas the craft and livestock workers labored for more than seventeen years. The field hands' total working time at Mesopotamia was more than five years shorter per man than that of the stock keepers and two years shorter than that of the craft workers. The field hands were in "able" health only 51 percent of the time as against 68 percent for the craft workers, and they died three years earlier than the stock keepers.

This tabulation also shows significant differences between male and female longevity. While the adult males at Mesopotamia had slightly longer working careers than the women and were in somewhat better health, they died on average 3.5 years sooner. In consequence, although the Barhams imported sixty-nine more male than female slaves in 1762–1833, the population shifted from a male majority of thirty-six in 1762 to a female majority of fifteen in 1833. The gender differential is most noticeable when we compare the adult male and female field workers. The 305 females continued in their prime gang labor jobs for 2.6 more years than the 256 males and lived 2.9 years longer. Another difference is that forty-one of the men (16 percent of the total) lasted for five years or less in the field gangs, while only twenty-seven of the women (9 percent) were removed from gang labor within five years.

But one may ask: Was there a gendered division of field labor, with the men working in the first field gang, while the women were placed in the less strenuous second field gang? Fortunately, the Mesopotamia inventories supply most of the answer to this question. From January 1, 1802, to January 1, 1832, the inventories annually identify every member of each field gang.[16] Examination of these 1802–1832 inventories shows a clear pattern that can be applied to the years 1762–1801 when all cane laborers were lumped into a single "field" category. Both male and female field hands from age sixteen to about twenty almost always worked in the second gang. Males and females in their twenties, thirties, and forties, when listed in "able" health, were almost always assigned to the first or Great gang, while the "weak" or "sickly" hands almost always stayed in the second gang or moved back into it. Marginal old people of both sexes, together with the young teenagers, were placed in the third gang, while the youngest field workers aged six to ten were placed in the fourth or hogmeat gang.

Having found the key to Mesopotamia's field employment system, I tracked the proportion of women among the total workers, the adult field workers, and the members of the first field gang at ten-year intervals from 1762 to 1832. Appendix 17 shows the results. For three decades (1762–1792) the female proportion of the population remained steadily at about 44 percent, the female proportion of the workforce at about 40 percent, and the female proportion of the field gangs at close to 50 percent. Then things changed. Between 1792 and 1802 the overall gender balance shifted to near-equality, and there were suddenly many more women than men in the field gangs, with a strong female majority in the first field gang. As late as 1808 there were slightly more males in the total population, but in 1809 the females outnumbered the males for the first time since the 1730s. Thereafter the population stayed at about 52 percent female until 1832. From 1802 to 1832 women continued to outnumber men in the Mesopotamia field gangs by a sizable margin, but they had a smaller proportion of workers in the first gang in 1822 and 1832 than in 1802 and 1812.

My estimates for the number of men and women in the first field gang in 1762, 1772, 1782, and 1792 are solidly based and cannot be very far wrong. I have no doubt that during these years there were always a good many women in the first gang, and that they were probably always in the minority. Then came a dramatic shift as the first gang declined from eighty-five members in 1792 (52 percent male) to only fifty-one by 1802 (41 percent male). This puzzled me, so I searched through the inventories to find out what had happened to the forty-four males who by my calculus had worked in the first gang in 1792. Only seven of these men were still there in 1802, while twelve had died, three had become drivers or craft workers, twenty had been shifted to secondary jobs, one had been transported, and one had run away. Even with eight new male members in their early twenties, in January 1802 the first gang could only muster twenty-one men, joined by thirty women. By June 1802 the first gang had been enlarged to twenty-six men and thirty-three women and was now 56 percent female.[17] For the next thirty years there were always more women than men in the first gang. And since they worked in pairs, each couple compelled by the drivers to keep pace with all the other couples, there must have been a number of female pairs, engaged in the same kind of strenuous labor as the men.

As we have seen in Chapter 1, Mesopotamia's trend toward a heavier reliance on female field workers was the standard pattern in western Jamaica. The close of the African slave trade in 1807 shut off the planters' external supply of new young male laborers, and in the years between 1817 and 1834 the twenty-seven large sugar estates in Westmoreland parish charted in Appendix 9 went from a male/female sex ratio of 89:100 to 83:100—a more decisive change than at Mesopotamia. By 1834, when the plantation owners filed their compensation claims, these same twenty-seven sugar estates claimed to have 2,824 "prime" field laborers, of whom 1,686 were women, while only 1,138 were men. Thus, 60 percent of the prime field hands on these estates were female, solid evidence that in the closing years of Jamaican slavery women performed most of the hardest physical work.

What were the health consequences of Mesopotamia's production of unusually large sugar crops from the 1790s to the 1810s and of the increased use of women in the first gang to produce these crops? Martin Forster and Simon D. Smith, a pair of cliometricians who have employed survival analysis techniques to conduct an investigation of the mortality pattern at Mesopotamia, demonstrate that the damage was very considerable. Intrigued by the detailed information in the Barham slave inventories at the Bodleian Library, Forster and Smith have linked the 1762–1832 inventories (very much as I had done) in order to quantify the relationship among labor, health, and mortality for these people.[18] Comparing all of the slaves first observed during Joseph I's tenure with all of the slaves first observed during Joseph II's tenure, they found that the mortality rate for the males at Mesopotamia did not change appreciably between 1762–1789 and 1789– 1832. But the female slaves who were first observed during Joseph II's tenure faced a considerably greater hazard of death than the females observed during 1762–1789. Forster and Smith agree with me that the labor regime at Mesopotamia exerted a large and quantifiable effect on slave survival, and they argue that the increase in female mortality was most likely caused by the women's greatly expanded enrollment in cane field labor from the 1790s onward. Increased exposure to field labor, they claim, reduced a woman's survival time by about 30 percent.

The Mesopotamia inventories show that while the sex ratio on this estate switched fairly rapidly from male to female between 1792 and 1812, it did not continue to change, and instead held fairly steady at about 52 percent female until 1832. And in the field gangs there was a similar pattern in more exaggerated form: a dramatic expansion of women workers in 1792–1812, followed by a shift back toward closer gender equality in 1822–1832. Thus, when the Mesopotamia women were assigned a much larger share of first gang labor in the years between 1792 and 1812, their health and survival rate deteriorated. As Forster and Smith point out, while Joseph Foster Barham II set out to ameliorate conditions at Mesopotamia for the slave women, he actually made things considerably worse for them.[19]

### *Jamaican Slave Management as Illustrated by Thomas Thistlewood*

Orlando Patterson observes that "there is no known slaveholding society where the whip was not considered an indispensable instrument."[20] Mesopotamia account books list the purchase of "a Gag for a Negro" and "a patent iron mouthpiece," which sounds like another gag, and by 1760 the estate was equipped with a set of stocks used to immobilize and humiliate slaves who had committed minor offenses.[21] But I have found no mention of whips in the Mesopotamia records. Indeed, I realized early in my research that I was going to have great difficulty in discovering much about day-to-day working conditions, let alone methods of punishment, on this estate. The Barhams' attorneys and overseers sent hundreds of letters to their absentee employers, but their comments about their black laborers were invariably brief, bland, and uninformative. And they never wrote in concrete detail about how they treated the slaves. So when I learned from Kenneth Ingram's bibliographical survey of manuscript sources of Jamaican history[22] that an enormously detailed diary by an overseer and small planter in Westmoreland parish was available for inspection, I went to the Lincolnshire County Archives in 1977 and read Thomas Thistlewood's diary volumes for 1748–1750, 1751, and 1765, as well as his nephew John Thistlewood's diary of 1763–1765.[23] Little or nothing had been published about Thistlewood at that point, so I came to him with no preconceptions—and was fascinated by what I found.[24] I have not attempted to survey all thirty-seven volumes, but since 1977 I have read further in the diary on microfilm and have benefited from Douglas Hall's and Trevor Burnard's contrasting biographies of Thistlewood.[25] And I am firmly convinced that this diarist helps us to understand what was going on at Mesopotamia.

Thomas Thistlewood's unvarnished reporting of his daily routine, unencumbered by explanation or interpretation of what he did or saw, is a far cry from the unctuous missives that the Mesopotamia attorneys sent to the Barhams. Especially in the opening volumes of the diary the reader is struck by the matter-of-fact way in which Thistlewood narrates the lurid events that he witnesses or hears about, and his frank description of his daily actions—such as the large number of lashes he administers in slave punishments, or his frequent sexual intercourse with numerous slave women, or his careful inspections of his penis when coping with venereal disease. At first I saw him as an unusually violent and sex-driven man. Then I found that during his last two years in England (1748–1750) Thistlewood recorded that he fornicated about twenty times with a mix of prostitutes, single women, and married women—which is not an excessive number for a bachelor in his twenties.[26] So the interesting point is that as soon as he got to Jamaica his sexual encounters accelerated from 10 to about 200 a year. And as for violence, a careful reading of his diary shows that the many other white men who populate his Jamaica narrative generally behaved even more extravagantly than he did.

When Thistlewood was employed as an overseer, he recorded numerous occasions in which two or three white males came to his plantation, got roaring drunk, and commandeered their favorite slave women for sex. In August 1761, for example, "Mr Cope [the owner of the plantation], Mr John Dorward, Mr Alexander McDonald dined here, also supp'd and got very drunk, disturb'd me sadly in the Night, but I would not let them into my house. They had Eve and Margaritta, amongst them." Thistlewood also reported that the white men employed as his assistants sometimes punished the slaves for no apparent reason. In January 1761 he observed "John Groves like a Madman amongst the

Negroes, flogging Dago, Primus, etc. without much occasion." When he reprimanded Groves for this behavior, the next morning "John Groves went away, because he might not flogg the Negroes as he pleased: very stubborn & resolute."[27] Thus, in the Jamaica context, Thomas Thistlewood comes through as more moderate than many other members of the white master class.

Immediately after he arrived at Savanna la Mar on May 4, 1750—the same year Joseph Foster Barham I came to Jamaica to inspect his property at Mesopotamia and Island estates—Thistlewood was confronted, as Barham must have been, by the vicious character of Jamaican slavery. Staying at first at William Dorrill's plantation, he noted in his diary: "Many runaway Negroes took up, Mr. Dorrill had his own whipt severely, and rubb'd with pepper, salt, and lime juice." A few days later he reported that "a runaway Negro being dead and buried while we were away, Mr. Dorrill caus'd him to be took up, his head stuck upon a pole, and his body burnt."[28] Armed with these evidences of master/slave relations on the island, Thistlewood contracted on June 29 with Florentius Vassall to manage Vineyard pen for a year at a salary of £50. Vassall was Joseph Foster Barham's brother-in-law, and on July 1 Thistlewood rode to Vassall's Friendship estate, which adjoined Mesopotamia, to receive his instructions and initial payment.

Thistlewood's first Jamaica job was a testing assignment that required him as a single white man to manage forty-two blacks armed with knives and guns. Vineyard pen was situated on an isolated tract in St. Elizabeth parish about five miles from Barham's Island estate. During his year at Vineyard, Thistlewood sometimes saw no other white person for two or three weeks. Thirteen of the Vineyard slaves were new Africans, likely to be particularly difficult to handle. And all of the Vineyard slaves were short of food and very hungry because their provision grounds were parched by drought.[29] Thistlewood arrived at Vineyard on July 2, and Vassall followed on July 11 to show Thistlewood how to maintain white supremacy in these challenging circumstances. Vassall ordered the lead Vineyard slave, driver Mulatto Dick, bound to an orange tree and given nearly 300 lashes "for his many Crimes and negligencies," which sent poor Dick to his hut for the next nine days.[30]

During Thistlewood's year at Vineyard, eight of the slaves ran away in order to get out of work, some repeatedly, and some for several weeks at a time. They were always caught or came back voluntarily, but while on the run they were a big problem, because they broke into the slave huts or lurked around the provision grounds and stole food. And the slaves who didn't run away were killing sheep and chickens in search of something to eat. Thistlewood himself was in no danger of starvation, because Vassall supplied him with four barrels of beef, which the slaves never could get at. To feed the hungriest slaves, he distributed plantains, corn, flour, herring, and spoiled salt pork. And he very quickly decided that the Vineyard Negroes were all "a nest of Thieves and Villains." On July 21, 1750, he administered his first whipping, giving Mimber fifty lashes for stealing corn. And on August 4 he had sex for the first time with Marina, who became his Vineyard favorite.[31]

During the next eleven months Thistlewood applied the lash freely. According to his diary, he whipped eighteen of the twenty-four male slaves thirty-one times, and seven of the eighteen females eleven times—50 lashes per whipping at a minimum, and sometimes 100 or 150 lashes. His highest total was 250 lashes administered to a Negro stranger who was caught inside the pen. Among his own workers, Thistlewood particularly went after the newly arrived African males, and spared only a few of the older men. By March 1751 he felt able to leave the pen unsupervised for five days while he went on a hunting expedition with two senior slaves, driver Dick and Guy. During the year he had sex regularly with Marina, and on forty-four recorded occasions fornicated with nine of the other seventeen Vineyard females. Shortly before the end of his contractual term, Thistlewood paid driver Dick to build a new house for Marina, and attended her house-warming party, at which everyone became drunk and merry. When he left in early July 1751, he distributed presents to all of the Vineyard slaves. He probably still thought of the Vineyard people as thieves and villains, but knew that he had learned how to manage them successfully.[32]

Moving ahead to 1765 we find Thistlewood as the overseer at Egypt estate, a small sugar plantation six miles southwest of Mesopotamia, with about ninety workers, including eleven of his own slaves.[33] Egypt was a marginal operation compared to Mesopotamia and produced a much smaller sugar crop.[34] By 1765 Thistlewood had been the overseer at Egypt for fourteen years and had settled into a somewhat less violent system than at the Vineyard. He continued to be sexually very active. In 1765 Thistlewood recorded having intercourse with his slave mistress Phibbah one hundred times, but also fornicated with twenty-three other slave women (including three from neighboring plantations whose names he didn't know) an additional fifty-five times. In this same year he punished about a third of the men and women under his command. Seventeen males and eleven females were whipped a total of forty-nine times—slightly more whippings than at Vineyard, but administered to a much larger slave gang. During the year Thistlewood had the head Egypt driver flogged twice, first for being saucy to his white supervisors, and again for his "carelessness and stupidity." Thistlewood also made use of the bilboes, forcing five men and two women to lie down in the dirt with their legs hoisted up and their ankles shackled to an iron bar. Several of them were kept immobilized in the bilboes for two or three weeks at a time.

In 1765 at Egypt, as in 1750–1751 at Vineyard, the slaves were rebellious for very good reasons: they stole food because they were hungry, and they ran away because they hated the work they were forced to perform. Being underfed all year round, the Egypt slaves took sugar and molasses from the boiling house, stole corn or grain used to feed the cattle, purloined fruit and vegetables from their fellow slaves' provision grounds, and cut and ate sugarcane growing in the fields. And during 1765 ten of them ran away, mostly for only a few days to avoid working. Two inveterate runaways, Derby and Adam, were more determined to escape, and though Thistlewood clamped iron collars on them they continued to get away whenever they could. The chief rebel was Plato, who disappeared for over three months, and when he reappeared Thistlewood taught him a lesson by bringing him to trial at Savanna la Mar, where he was sentenced to one hundred lashes and had his right ear cut off. After that, Plato made no more trouble for the rest of the year.

Plato was by no means the only Egypt slave who received particularly nasty treatment. Nine years earlier, Thistlewood had experimented with an unusually disgusting and degrading method of punishment. In January 1756, after catching Derby eating sugarcane, "Had Derby well whipped, and made Egypt shit in his face." In May Derby was again caught eating cane: "Had him well flogged and pickled, then made Hector shit in his mouth." In July Thistlewood went one step further. Port Royal had run away, and after whipping and pickling him, Thistlewood "made Hector shit in his mouth, immediately put in a gag whilst his mouth was full & made him wear it 4 or 5 hours." Also in July, a female slave, Phillis, was given the defecation treatment twice (though without a gag). In August Thistlewood reversed his January punishment, and made Derby evacuate into Egypt's mouth. In October 1756, after a total of seven slaves (including Hector) had been given "Derby's dose," Thistlewood abandoned this sadistic mode of punishment, perhaps because it had proved to be totally ineffective.[35]

Another way of viewing Thomas Thistlewood is to see how he treated his son John, a mulatto borne in 1760 by Phibbah, Thistlewood's long-term slave mistress and effective common-law wife. John was the first and only child for whom Thistlewood claimed paternity out of nearly 4,000 recorded acts of sexual intercourse. The father clearly valued his biracial son, for he worked hard to obtain John's manumission in 1763, while mother Phibbah remained enslaved. And he trained him up in the hope that he would amount to something. Thistlewood imported children's books for John to read and sent him to boarding school in Savanna la Mar. In 1775, when John was fifteen, Thistlewood apprenticed him to a carpenter so that he could earn his living as a craftsman. But John wanted to be a soldier, not a carpenter, and during the next three years he kept running away from his master, which instigated several floggings from his father. On Christmas Day 1778 Thistlewood gave Phibbah seven bottles of rum and John a pair of new shoes. Four days later, on discovering that John had not gone back to work at the carpentry shop, he sent out three slaves to find him, and when they brought him home he put the boy in the bilboes overnight. The next morning he "flogged John pretty well, pulled off his shoes, and put them up," then sent him back to his master shoeless, like a slave, "with his hands tied behind him." Would Thistlewood have treated a white son in this manner? Poor John had a brief moment of glory in 1779 when the militia was mobilized in expectation of a French invasion; he joined the Free Negro Company in his red uniform and camped for a month near Kingston. Then in 1780, back to his dreary apprenticeship, John suddenly fell ill and died, just twenty years old.[36] Had mulatto John kept a diary, one wonders what this hapless youth would have said about his father.

In January 1759 Thistlewood noted that when he called at Salt River plantation he met "mr. McFarthing Barham's Overseer," but he didn't know him well enough to get his name straight. In a later volume of the diary that I haven't read Thistlewood may well have encountered Daniel Macfarlane's successor, Daniel Barnjum, who was the overseer at Mesopotamia from 1760 to 1778. I see Barnjum as an operative very much in the mode of Thistlewood. He is the best documented of the Mesopotamia overseers because he corresponded frequently with the elder Barham, who had hired him in 1758 to go to Mesopotamia for five years and report back on the state of affairs on this estate.[37] Like Thistlewood, Barnjum came to Jamaica as a penniless young man with no prospects in England. When Macfarlane died in 1760, he was suddenly thrust into the role of overseer, and—like Thistlewood—he held this job for a long time because he quickly learned how to manage slaves. Both men used their earnings to assemble their own small slave holdings. Barnjum held twenty slaves when he died in 1778, and Thistlewood had twenty-eight slaves when he left Egypt in 1767 and became an independent small planter.[38] Barnjum also had an active sex life. In 1759 a young distiller, Augustus, complained to the Moravians that his wife (whom I cannot identify) was being used as a whore at the Great House where Barnjum was living.[39] And Barnjum, like Thistlewood, kept a long-term slave mistress—Hannah, the daughter of head driver Matt—by whom he had several mulatto children. In 1761 Hannah bore him a mulatto son who quickly died, and Hannah was probably the mother of two mulatto boys—Tom, born in 1762, and Jack, born in 1764—whom Barnjum manumitted in 1774. In 1778 Hannah bore another mulatto son named Matt, who was never manumitted, perhaps because Barnjum died immediately after his birth.

Certainly Barnjum faced Thistlewood's chief managerial problems: trying to control slaves who were often extremely hungry and hated their forced labor and kept running away. As we shall see in Chapter 8, there was a great deal of unrest at Mesopotamia in the 1760s. Many slaves ran away, several were jailed, and two were transported. What we don't know is how often Barnjum used the whip or the bilboes to keep order. My guess is that he didn't resort to "Derby's dose." But since he managed three times as many people as Thistlewood, I figure that Barnjum administered a hundred or more whippings per year at Mesopotamia.

### *Male Mortality at Mesopotamia*

Women outlive men in most times and places, of course, but the male mortality problem at Mesopotamia was extreme, and it reflected the general trend in Jamaica. Indeed, the history of the Jamaican slave trade was shaped in large part by the relatively brief lives of the male slaves. The planters wanted as many male laborers as possible, and since their enslaved men died faster than the women, they needed male-heavy shipments of new workers from Africa. The editors of the Trans-Atlantic Slave Trade Database calculate that 61.8 percent of the Africans disembarked in Jamaica were male, so that some 500,000 of the approximately 810,000 Africans acquired by the Jamaican planters between 1655 and 1807 were men and boys.[40] Yet when the slave population peaked at about 355,000 in 1808 at the close of the slave trade, the males held only a slight majority (50.9 percent) with an island sex ratio of 104:100. The close of the African trade produced two quick changes. The Jamaican slave population, which had been growing thanks to the trade, experienced a 15 percent decline to 310,000 by 1833, and during these twenty-five years the island sex ratio changed to 95:100, or 48.7 percent male.[41] And as we have seen in Chapter 1, the gender balance on twenty-seven large sugar estates in Westmoreland parish saw a more dramatic shift, from a sex ratio of 89:100 in 1817 to a quite lopsided ratio of 83:100 (only 45.3 percent male) by 1834.[42]

Scholars who have tried to explain the high mortality rate for Caribbean slaves—both male and female—generally emphasize three factors: the inadequate slave diet, the lethal disease environment, and the brutal and exploitive labor regimen.[43] All three of these factors unquestionably contributed to the high death rate at Mesopotamia, but I don't believe that they fully explain why the male slaves died more quickly than the females. The labor regimen statistics in *Appendix 16* show that the 485 Mesopotamia men worked a little longer than the 392 women and were in a somewhat better state of health during their working careers. Furthermore, 161 of them (33 percent) had less strenuous jobs than the vast majority of the women. Thus, male slaves would seem to be positioned to live longer, yet they died three and a half years sooner. And it was after they stopped working that the Mesopotamia men most quickly succumbed. Once the women became too weak or sickly to work, they continued to survive as invalids for 3.3 years on average—or 2 years longer than the men. What is the explanation for this two-year retirement differential?

Turning to the slave diet, the managers of Mesopotamia—like all Jamaican estate operators—economized as much as possible on food for their workers. The slaves were expected to grow their own plantains, yams, cassava, fruit, and other vegetables, supplemented by minimal weekly rations of salt herring, with extra beans or flour doled out sparingly in emergencies. A watercolor by William Berryman of a Jamaican slave provision ground around 1810, presents an enticing scene of tropical fruitfulness. And the Mesopotamia managers always claimed that the slaves were amply provided for. In 1802, for example, they described a new plantain walk that supplied 1,500 to 2,000 extra plantains per week to the slaves, with 660 of these plantains distributed weekly to twenty-two invalids and orphaned children who were unable to grow their own food. Eleven nursing mothers received a quart of oatmeal and a pint of sugar weekly, and eleven especially favored slaves—the drivers, the chief craftsmen, and the female doctor who dispensed herbal medicine—received weekly rum rations.[44] Every time a new group of slaves was brought to Mesopotamia, a jobbing gang would be hired to do the hard work of preparing new provision grounds for them by chopping down trees and planting food crops. In 1814 hired laborers spent 1,512 days (at a cost of £223) to clear and plant the provision grounds for the fifty-five new Cairncurran slaves.[45] Left unsaid in the 1802 and 1814 reports (though observed by the Moravian missionaries) was the fact that the slave provision grounds were all laid out in mountain land on the edge of the estate, a laborious three- to five-mile walk from the slave village. Indeed, it appears that the Barhams' managers deliberately made the slaves waste time getting to and from distant provision grounds to discourage them from producing extra food that they could sell in the Negro market at Savanna la Mar.

The Mesopotamia people worked from dawn to dusk six days a week, except for some free Saturdays out of crop time, so during most of the year they could cultivate their food crops only on Sundays. This method of feeding the population was very different from the Mount Airy situation, where the slaves produced food—corn, wheat, and pork—for the market, and ate a portion of this food. And the Mesopotamia method was much less satisfactory in many respects. First and foremost, the slaves needed more than one day a week to grow enough food to live on, especially since they had to spend several hours just getting to and from their allotments. During crop time (January to May) the slaves had especially little time to plant adequate crops for summer harvest, with the result that when summer arrived they had far too little to eat. Even in good times, as the Moravian missionaries frequently observed, the Mesopotamia people ran short of provisions every summer and suffered acutely from malnutrition.[46] The crops they grew were high in carbohydrates and low in protein. The plantain, in the banana family, was their staple food, and the tall, treelike plants that grew this fruit were easily toppled in a high wind. When the Mesopotamia plantain walk was wiped out in the hurricane of October 1780, and again in a great storm in October 1812, it took nearly a year for a new set of plantain plants to bear fruit. In these emergencies the Barham management never came up with adequate rations to compensate, and the Moravians reported—as at Thistlewood's Vineyard and Egypt—that the Mesopotamia slaves resorted to stealing food from the estate storehouses, or from the missionaries, or from one another.[47]

Mesopotamia's annual death registers shed some light on another mortality factor—the disease environment. Cause of death is reported for 414 of the 432 males who died between 1762 and 1832, and for 307 of the 319 females. These Mesopotamia death statistics match up well with Michael Craton's findings at Worthy Park estate in Jamaica, where cause of death statements are recorded for 401 slaves between 1792 and 1838.[48] *Appendix 18* presents the death reports from the two estates, employing the medical terminology of the day, ranked according to the frequency of stated causes at Mesopotamia. The white doctors who reported these fatalities undoubtedly misdiagnosed many of the terminal illnesses, confusing symptoms with causes. And they also contributed to the death totals. Addicted to humeral medicine, and ignorant of the germ theory of disease, they habitually bled, blistered, and purged their slave patients, killing more people than they cured.[49]

This disease tabulation shows that many of the white doctors' diagnoses are almost useless. A total of 105 Mesopotamia men and women, almost all over the age of fifty, are said to have died of "old age." Another eighty-five mostly younger people succumbed to "debility" or "distemper" or "disease" or "decline." And twenty-one people, again mostly elderly, expired because they were "invalids." Thus, we have no medical explanation for more than a quarter of the reported Mesopotamia deaths or a third of the Worthy Park deaths. Another thirty-four Mesopotamia fatalities and twenty-six at Worthy Park are attributed to "fever," with no distinction drawn between mosquito-borne yellow fever and malaria, or high temperatures symptomatic of quite different infections. Viral and bacterial diseases such as smallpox, whooping cough, and measles, however, were more easily identified. A major epidemic of smallpox hit Mesopotamia in 1768, when 30 slaves caught the disease and 10 died, while another 115 people were inoculated and escaped infection. Only seven Mesopotamians are said to have died of hookworm or other parasitic diseases, probably an understatement. But the absence or near-absence of major modern killers like heart disease and cancer is almost certainly accurate.

The four principal killers at Mesopotamia were stated to be tuberculosis, dropsy or edema, dysentery, and ulcerous skin and bone diseases.[50] Both TB and edema can signify poor nutrition, and dysentery is closely associated with unsanitary living conditions. Many of the other ascribed causes of death could also indicate malnutrition, vitamin deficiencies, and/or bad hygiene. The seventeen reported fatal occurrences of dirt eating or geophagy at Mesopotamia are particularly interesting, because dirt eating was widespread in Jamaica. The planters and doctors were mystified and alarmed by the fact that many slaves had a craving for clay (which they dug up, baked, and ate); they tried hard to stop this seemingly suicidal practice. Modern commentators now contend that the slaves' appetite for clay was actually triggered by deficiencies in calcium and iron in their diet, so the clay they ate was a remedy for malnutrition, not a disease.[51]

There were interesting differences between the ascribed causes of male and female deaths at Mesopotamia. Dropsy, yaws, and other ulcerous skin and bone lesions, pleurisy, apoplexy, lung disease, and epilepsy were all much more common afflictions among the males. Thus, the men seem to have been more susceptible than the women to the most hideously destructive tropical afflictions. Those with advanced cases of yaws and kindred ulcerous diseases suffered excruciatingly. Frederick, an African field hand who ran away frequently, was described as "covered with ulcers" and "rotten" when he died at age forty-eight. Barnet, a mulatto carpenter from Cairncurran, was "deranged with dreadful ulcers" when he died at fifty. London, an African watchman, was "rotten" with ulcers when he died at forty-four. Three other male ulcer victims, Chelsea, Cato, and Cumberland, were also described as "rotten" at death. A pen keeper named Tom had a different though equally painful problem: he was diagnosed with opisthotonus, a state of extreme hypertension in which his body was arched rigidly and his head was bent backward, when he died at thirty-one.

Males also suffered three times as many fatal accidents as females. The most horrific accident occurred in January 1809 when a driver named Bristol slipped and fell into a copper cauldron of bubbling red-hot cane juice in the boiling house. Another man, Duke, was fatally scalded in the boiling house in 1802, and Page died of burns in the distillery in 1792. Two men (Fuller and Frank) were killed when they fell while climbing coconut trees, and mulatto John had a fatal fall while thatching his roof.[52] Three men (Peter, Jupiter, and Tom) drowned while swimming in the Cabarita River, and two men (Hazard and Tony) died in the hurricane of 1780.

Jamaican planters generally agreed that newly arrived Africans from the slave ships had great difficulty in adjusting to Caribbean slave life, and they noted that a significant number failed to survive their initial "seasoning" period of two or three years. Some observers claimed that a quarter to a third of the new Africans died within three years.[53] At Mesopotamia, it was a lot less than that. But the new African males *did* have much more trouble with seasoning than the new African females. Of the forty African women and girls who were brought from the slave ships between 1762 and 1793, only two died within three years. But fifteen of the ninety-seven African men and boys who arrived during this same period died within three years, fourteen within two years, and nine within one year. Three of the African males who failed to survive the seasoning period reportedly died of yaws, two of apoplexy, two of pleurisy, and two of breast complaints, and another two were described as dirt eaters. Some of these people were in dire mental and psychological condition. Cesar, who arrived at age twelve in 1772, died in a state of "lethargy" in 1773, and Othello, who arrived at age twenty in 1765, died "sullen" in 1766. Seemingly they stopped wanting to live.

The disproportionate number of male seasoning deaths strengthens my opinion that the high male mortality rate at Mesopotamia cannot be fully explained by the brutal working conditions, the semistarvation diet, the toxic disease environment, or the collective impact of these three lethal factors. I see additional and more hidden factors at work—factors more psychological than physical that may collectively be labeled as "emasculation." The dictionary synonyms for "masculine" include "bold," "virile," "potent," "aggressive," "powerful," "macho," and "two-fisted"—all male characteristics that are incompatible with enslavement, especially with Jamaican enslavement. Most males through time have been taught to suppose that they belong to the dominant sex, and I believe that the totally humiliating and hopelessly degrading nature of Jamaican bondage was deeply distressing to the male slaves at Mesopotamia—especially to the African newcomers. Every aspect of life at Mesopotamia challenged their manhood. They were categorized as subhuman because of the dark color of their skin. They were branded (on the shoulder) like cattle to identify them as property should they try to escape. They were ordered to obey their white masters submissively, and were flogged for every real or imagined fault—the males much more frequently and severely than the females. Even the head people, the drivers and leading craft workers, could not escape whippings that demeaned them. And if they resisted or rebelled or ran away, they were subjected to torture and imprisonment. All of these insults to their manliness can be seen as forms of psychological castration.[54]

Close to 40 percent of the 601 Mesopotamia males during our period were born in Africa, as against about 30 percent of the 502 females,[55] and the African men must have been profoundly shocked when they were forced into exhaustive and endless robotic manual labor. They also had much difficulty in sustaining their West African social and sexual practices. They could drum and dance and sing every Saturday night, to the great annoyance of the Moravian missionaries, but finding and keeping the women they wanted was much more of a challenge. The Moravian missionaries frequently reported that some of the Mesopotamia men were polygamists and that others pursued serial sexual relationships. This meant that many of the other sexually active men had to visit neighboring estates to find mates, and some of them must have gone without, since there were fewer women than men in the Mesopotamia (and Westmoreland) slave population through the 1790s. Those men who did form families had to stand by helplessly while the whites freely fornicated with their wives and daughters, and they were totally unable to protect their women from corporal punishment. Those men who had no mates and lived singly had no one to comfort them when they were depressed, and no one to care for them when they were sick or old, while many of the women were supported in old age by their children, which surely helps to explain why the elderly female invalids lived two years longer than the elderly male invalids.

But if the women at Mesopotamia survived somewhat better than the men, they too were cruelly victimized by the Jamaican slave system, as we shall now see.

## *Motherhood at Mesopotamia*

The most striking feature of female slave life at Mesopotamia was the pattern of childbearing. Of the 501 females who lived on the estate between 1762 and 1833, 299 women were of childbearing age (between sixteen and forty), and they produced 420 recorded children. The exact number of Mesopotamia mothers is unknown, because before 1774 the bookkeepers almost never listed the mothers of newborns, and they missed a few thereafter. Thus, 44 of the 420 children had unrecorded mothers; 38 of these "motherless" babies were born in 1762–1773, and six in 1774–1780. From 1774 onward the picture becomes much clearer. The annual inventories identify 129 women who gave birth to 376 children between 1774 and 1833, averaging just under 3 children per mother. If the same ratio is applied to the forty-four "motherless" children, there were fifteen additional mothers. However, eight of the twenty-eight females of prime childbearing age with no listed children in 1762–1773 *did* have recorded births after 1774, which makes them likely candidates for motherhood before 1774. Therefore I estimate that there were about ten additional mothers, rather than fifteen, who bore most of the "motherless" babies. If so, we have a grand total of 139 Mesopotamia women (46 percent of the females of childbearing age) with recorded live births in 1762–1833. Even if there were twenty additional unidentified mothers in 1762–1780, more than half of the Mesopotamia women were childless.[56]

The contrast with motherhood at Mount Airy is striking. On the Tayloes' Virginia plantation there were 449 females in the total population between 1809 and 1865, but only 191 of them were between the ages of sixteen and forty. The childbearing cohort at Mount Airy was notably smaller than at Mesopotamia because the Tayloes sold twenty-eight girls before they reached the age of sixteen, and assigned another forty-nine young Mount Airy girls to other family owners, most of them in 1828–1829 after the death of John Tayloe III. The 191 women of childbearing age at Mount Airy bore considerably more recorded children than the 299 Mesopotamia women. While 44 of them had no live births in 1809–1865, the other 147 (77 percent of the total) had 636 recorded children, 502 born at Mount Airy and 134 in Alabama. So while there were 108 fewer potential mothers at Mount Airy than at Mesopotamia, they gave birth to 216 more children over a shorter (fifteen-year) time span. The contrast is stark and epitomizes the difference between the two populations.

Both the Mesopotamia and Mount Airy birth statistics clearly understate the total number of mothers in both populations, because many of the older "childless" women in 1762 and 1809 had babies before the records started, while many of the younger "childless" women in 1833 and 1865 had babies after the records stopped. I estimate that about 90 percent of the 191 Mount Airy women and up to 60 percent of the 299 Mesopotamia women became mothers. Both sets of records also understate the number of live births. At Mesopotamia the bookkeepers supplied an exact birth date for each recorded child, but many of the infants who died within a few days were not recorded. For example, Joseph Foster Barham II's attorneys told him that three newborns died of lockjaw (tetanus) before they were nine days old in 1808, and "these births were not entered into the Plantation Book."[57] At Mount Airy the Tayloes did not keep a birth register, so very few of the 636 listed children have exact birth dates, and the newborns who died between inventories were never noted. The Tayloes occasionally neglected to list small children until they were several years old, though the mother of each listed child was always identified, and sometimes the father. Thus, the number of babies in both places who died before they were recorded can only be guessed at.[58]

To demonstrate the contrast with Mount Airy we need to focus on the 129 Mesopotamia mothers who bore 376 children in 1774–1833. *Appendix 19* compares them with the 147 Mount Airy mothers who bore 636 children in 1809–1865. The two time spans are equivalent: sixty years versus fifty-seven years. Far more of the Mesopotamia women were childless in these years, and the Mesopotamia mothers bore 2.9 children on average, while the Mount Airy mothers averaged 4.3 children. Forty percent of the Mesopotamia mothers had only a single child, compared with just 16 percent at Mount Airy. Twenty-three percent of the Mesopotamia mothers had four to seven children, compared with 37 percent at Mount Airy. And 5 percent of the Mesopotamia mothers had eight or more children, compared with 14 percent at Mount Airy. A few of the Mesopotamia women did, however, have surprisingly large families. Clarinda had eight children between 1799 and 1813; Judy had nine children between 1791 and 1815; Sally had ten children between 1796 and 1820; Matura had eleven children between 1797 and 1820; Cooba had twelve children between 1805 and 1829; and Minny had fourteen children between 1784 and 1814. These six women, plus the thirty who bore four to seven children each, account for 59 percent of the 376 births. In short, at Mesopotamia relatively few women had most of the babies.

There is a close correlation between the low Caribbean slave birthrate and the high Caribbean death rate. The three factors that we have already examined—the disease environment, the inadequate slave diet, and the brutal labor regimen—clearly also contributed to the low birthrate. On the issue of diet, the biological demographer Rose Frisch argues that the female adolescent growth spurt, the onset of menarche, and the maintenance of regular menstrual function are all associated with body weight and fat storage. Poorly nourished women experience delayed menarche and irregular menstruation, bear infants with low birth weight, produce less breast milk to feed their babies, and reach menopause earlier than well-nourished women.[59] Frisch's critics contend that her nutritional arguments are exaggerated and that fertility studies of modern populations demonstrate that malnutrition has only slight impact on childbearing. They point to the large families produced by malnourished, hard-working women in poverty-stricken regions of the world today.[60] However, Frisch's critics admit that frequent starving times (as at Mesopotamia) would cause some women to stop ovulating and others to have very irregular menstrual cycles. Sexually transmitted infectious diseases such as syphilis and gonorrhea would increase the chances of miscarriage or stillbirth. And malnutrition would lead to low birth weight and greater risk of death in infancy. One further point: the estate records show that the Mesopotamia women faced great obstetrical hazards while giving birth, since eight of them died in childbed.

The estate managers kept assuring Barham that they took special care of pregnant women. The estate had a maternity room where a white midwife delivered the babies at a payment of £2 per child. The new mothers were given £1 apiece as a reward, together with a suit of baby clothes.[61] In 1811 attorney John Blyth told Barham that he moved pregnant women in the first gang to the second gang and assigned them light work, recommended by the medical gentlemen as positively conducive to their health. He sent them home six weeks before delivery and allowed them to tend their newborns for a month after birth. When the mothers returned to work, he claimed, the estate midwife cared for their infants. Blyth added that women with five living children were now exempted from labor.[62] This was a new policy at Mesopotamia, first instituted in 1805 for housekeeper Minny. By 1817 seven members of the first gang and one member of the second gang, most of them still young and in "able" health, were allowed to retire from field work in order to attend their young families.

During the four years 1823 to 1826, the Mesopotamia attorney William Ridgard became concerned about pregnancies that resulted in miscarriages and stillbirths or that turned out to be nonpregnancies. In these four years twenty women were described as pregnant as of December 31, the date of the annual inventory, but only eleven of these pregnant women bore live babies the following year. Four of them had stillbirths, according to the inventories, and the other five had no birth record. And during these four years the inventories recorded a total of nine stillbirths as against twenty-nine live births. This is an unusually high number of stillbirths, but the bookkeepers may have labeled miscarriages or abortions as stillbirths. Ridgard complained to Barham that the Mesopotamia women were practicing "artful" tricks in order to avoid the heaviest field labor. At the beginning of the 1825 sugar harvest four women—Clara, Dorinda, Phoenix, and Sophia—had claimed to be pregnant, and when the slave midwife Kickery examined them and confirmed their condition, they were excused from work during the whole of crop time. But, says Ridgard, this turned out to be a sham, and the four women went back to work as soon as the harvest was over.[63] Fustina, Kitty, and Dido also claimed to be pregnant in December 1824 but may not have known for sure, since Fustina had her baby in September 1825, and Kitty in August 1825. Dido produced stillborn twins in August 1825 and was sent to the workhouse by Ridgard in punishment.[64] Not surprisingly, Dido was subsequently said to be "ill-disposed and refuses to work," and her disposition did not improve when she was put into solitary confinement for miscarrying in 1827. In

October 1830 Dido did have a live birth, but the child died after three months. At this point Dido ran away, doubtless fearing a further stint at the workhouse.

These stillbirths and false pregnancies raise an important issue. Some historians contend that Caribbean slave women resisted motherhood, not wishing to bring any more slaves into the world, and that they deliberately aborted their fetuses. Or they practiced infanticide by "overlaying" or dropping their newborn babies.[65] I have no doubt that some of the Mesopotamia women faked pregnancy in order to get out of work and that others handled their newborn infants very carelessly. And it is certainly possible that some of them aborted their fetuses or deliberately killed their babies. But I am inclined to suspect this line of argument, since the only evidence comes from overseers and attorneys like Ridgard who were looking for ways to blame the slaves for the low birthrate. As we have seen, nearly half the Mesopotamia women were mothers with one or more children, and a few of them had very large families. What needs explaining above all is the huge number of childless women. Is it credible that they were all practicing abortion? I prefer to look for alternative options. One possibility is that many of these childless women had no sexual partners, or infrequent sexual partners, or very irregular opportunities for sexual intercourse. Another is that they and their sexual partners were biologically impaired by infectious disease and malnutrition. And there is a further possibility that can be tested via the Mesopotamia evidence: that they and their sexual partners were physically drained and damaged by continuous sixty-hour weeks of heavy sugar labor.

In previous essays I have argued that the sugar labor performed by the Mesopotamia women in their prime childbearing years was the main cause of their low birthrate.[66] Now, after examining motherhood at Mesopotamia more closely, I still see sugar labor as a key factor, but place more emphasis than before on diet, disease, and the social environment. Women at Mesopotamia always did much of the grueling field work, and from 1802 onward the majority of the workers in the first gang were women—many of them in their twenties, when they were most likely to bear children. And it is surely pertinent that Minny, the mother with the largest number of children, was a house servant who performed much less taxing work. However, five of the six most prolific mothers, with eight to twelve children—Clarinda, Judy, Sally, Matura, and Cooba—were field workers who started their large families while they were members of the first gang. Clarinda, who was African-born, gave birth to six of her eight children while she worked in the Mesopotamia first field gang for sixteen years. Judy, born in Mesopotamia, gave birth to five of her nine children while working in the first gang for fourteen years. Sally, also from Mesopotamia, had five of her ten children while working in the first gang for twelve years. Matura, another African, had five of her eleven children while working in the first gang for eleven years. Cooba, another Mesopotamian, produced the first five of her twelve children while working in the first gang for eleven years.

Obviously there were some strong women at Mesopotamia who bore large families while performing the hardest physical labor. These women, however, constituted a small minority. *Appendix 20* demonstrates this by looking at the 119 females who were sixteen or older on January 1, 1802—the first year in which membership in field gangs was identified—and tabulating all of their 224 recorded births. Because of Minny, the four domestics who worked in the overseer's house and the bookkeeper's house were the leaders in fertility. But the thirty women who worked in the first field gang in 1802 produced almost half of the 224 children; the first gang had many more mothers than the second gang, and the first gang mothers had larger families.[67] In 1802 six members of the first gang gave birth. Eve at age forty had a stillborn child, but Marina, Matura, and Judy produced live infants during the 1802 sugar harvest season, while Polly had her child in August, and Sally in November.[68] Twenty-seven of the first gang women in 1802 were listed in "able" health, six of them toiled in the first gang for twenty-five years or more, and few of them became invalids later in their careers. Clearly the estate managers had selected very tough women for the first gang, women who could best endure the rigors of the sugar labor routine.

The remaining eighty-five women in 1802—the second gang members, the marginal workers, and the invalids—had fewer children and were generally in sad shape. Fourteen of the second gang women were listed as "weak" or "sickly." They had shorter working careers than the first gang women, and they died younger. And the women with marginal jobs and the nonworking invalids in 1802 were mostly elderly and in poor health. One of the childless women with a marginal job was Bella, a forty-two-year-old grass cutter who had lost her hand in a sugar mill accident in 1793 but continued to cut grass and cook for the field workers until 1817, when she was retired at age fifty-eight. Another was Rose, a fifty-year-old nurse whose arm had been amputated after another mill accident in 1765 and who served as a water carrier for the next thirty-two years, balancing the container on her head.[69] The great majority of the marginal workers and invalids had previously worked in the field gangs, more often in the second than the first gang, and the small number of recorded children they bore is just one of many signs that the sugar labor routine had beaten them down.

The Mesopotamia and Mount Airy inventories make it possible to construct family trees for three or four generations on both plantations. Once more, the differences are striking. When we examine family life at Mount Airy in Chapter 7, we will find mothers with huge extended families. Sally Thurston (1780–1843) had thirteen children, forty-two traceable grandchildren, and twenty-four known great-grandchildren by 1863. And Franky Yeatman (1766–1852) had six children, thirty-five recorded grandchildren, fifty-nine great-grandchildren, and eleven great-great-grandchildren by the end of the Civil War. At Mesopotamia there were no families of anything like this size, and many family trees can be devised for only two generations. This is partly because familial relationships were more poorly recorded than at Mount Airy, with fathers of children rarely identified, blocking out half of a family pedigree. But also far too many Mesopotamia offspring died young.

For example, Minny (1770–1826), the housekeeper who bore fourteen children, had a very large family that was ripped apart by death. Her stunted family tree is displayed on the website *www.twoplantations.com* that accompanies this book. Minny had more children than any of the Mount Airy mothers. She bore her nameless first child (who quickly died) when she was fifteen, and her last (called Joseph Foster Barham in honor of her absentee owner) when she was forty-five. But four of her babies died in infancy, and five more died between the ages of eleven and thirty-seven. One daughter, mulatto Susannah, was manumitted at age fourteen, so that only four of Minny's children—Daniel, Kitty, Fanny, and Wellington—were still living at Mesopotamia in 1833, aged thirty-two to twenty-one. Since none of Minny's five sons who reached adulthood were identified as fathers, only Kitty and Fanny had recorded children, and Kitty's babies all died, leaving William Allen, age three, as Minny's sole identifiable grandchild in 1833.

In contrast with Minny's broken family, a couple named Betty and Qua had an unusually large and well-documented Mesopotamia slave family traceable for four generations. Their family tree is also displayed at *www.twoplantations.com*. Betty and Qua's children were not born at Mesopotamia, but at Three Mile River estate. In 1786, when Betty was forty-six and her husband Qua was forty-five, they were moved from Three Mile River to Mesopotamia, together with their seven children, who ranged in age from twenty-two to seven.[70] Qua came as a field driver, and the two older sons, Charles and Chelsea, were craftsmen, so they were among the "head people," in Jamaican terminology, who held responsible and skilled jobs. Qua died shortly after arrival, but Betty spent twenty-five years at Mesopotamia as a children's nurse and lived to age eighty-three. Her second son, Chelsea, became a long-term field driver in Mesopotamia and succeeded Qua as the family patriarch. Starting with Chelsea and his wife Charity in 1802, ten members of this family joined the Moravian congregation in Mesopotamia and were noticed in the missionaries' records—as will be discussed in Chapter 6. All seven of Betty's children lived into their forties, and three were listed in the 1833 inventory, aged sixty-two to fifty-five. Betty had fifteen recorded grandchildren and twelve recorded great-grandchildren, and though there were many sad deaths along the way, twenty-one members of this extended family were living at last report in 1833.

There was a good deal of racial mixing in both Minny's and Betty's families. Six of Minny's children were mulattoes (white father), three were sambos (mulatto father), and five were black. The four mulatto boys—Alexander, Thomas, Andrew, and Daniel—all started out as houseboys and became carpenters in adulthood, while mulatto daughters Susannah and Fanny became domestics. But Minny's three sambo children—Kitty, Wellington, and Joseph Foster Barham—having a smaller admixture of white blood, were consigned, like her black children, to field labor. Obviously a variety of men had sex with Minny, but only two of them can be identified. In 1800 the white carpenter on the estate, Patrick Knight, manumitted Minny's mulatto daughter Susannah, and in 1802 overseer David Forrester tried to manumit her mulatto son Alexander. But either Forrester didn't offer enough money or

Barham was offended by his promiscuous behavior, because Alexander remained a slave until he died at age twenty-five in 1813.

In Betty's family, none of her children were mulatto, but there was considerable racial mixing in the next two generations. Two of Nancy's four children, the first of Bessy's seven children, and all three of Sophia's children were mulattoes, fathered by unknown white men.[71] Sophia was very possibly the mistress of a staff member living at Mesopotamia between 1798 and 1804 who could not or would not afford the money required to manumit any of his three enslaved children. In the next generation, Tabia (the daughter of Chelsea and Charity) bore a mulatto son named James, and mulatto Nancy (Sophia's younger daughter) had two quadroon children, Eliza and William, whose father was evidently William Parrott, a bookkeeper and supervisor of the distillery at Mesopotamia, since these children were listed as Eliza and William Parrott in the final 1833 inventory.[72] Thus, Minny's and Betty's families highlight the intrusive role of the white managerial staff in helping to increase the estate's population by fathering slave babies.

Between 1762 and 1833, twenty-six Mesopotamia slave women gave birth to thirty-seven mulattoes and six quadroons, accounting for 10 percent of the 420 recorded slave births. Observers of Caribbean slavery have long emphasized the avidity with which the white masters (who generally constituted about 5 percent of the adult males at Mesopotamia) pursued the slave women. Thomas Thistlewood records having about one hundred sexual encounters a year with slave women. He fornicated with these women at all times of the day or night, sometimes in his house, sometimes in outbuildings, sometimes in the open fields—and this style was likely followed by the Mesopotamia overseers and bookkeepers as well. His biographer Trevor Burnard, who has examined the entire 9,300-page diary, states that Thistlewood engaged in 3,852 acts of sexual intercourse with 138 slave women during his thirty-six-year Jamaican career.[73]

The whites who fathered the forty-three mulatto and quadroon children at Mesopotamia were clearly attracted to the younger women on the estate. Nine of the girls they impregnated were in their teens. Minny had her first two mulatto children when she was sixteen and eighteen. A mulatto house servant named Nelly was only fifteen when she gave birth to her only child, a quadroon named John, in 1798. A young field hand named Margretta had just turned sixteen when she bore a mulatto boy named William in 1797. Most of the mothers of miscegenated children had black children as well, which undoubtedly affected family harmony. Of course the unanswerable question is, How consensual were these black-white liaisons? Did the Mesopotamia women *want* to have sex with their white masters? Did they *want* to bear mulatto or quadroon babies?

If we can believe Thomas Thistlewood's nephew John, who joined his uncle at Egypt plantation in 1764 and kept his own diary, a Mandingo slave named Lettice from a neighboring plantation came to him on February 3, 1765, "to perswaid me if possible to ly with her for she wanted to have a Child for her master and she was very certain that she never should have one by him" because he was impotent.[74] John agreed to meet Lettice the following night in the boiling house, but failed to report on what happened next. We learn from uncle Thomas's diary that John was in hot pursuit of Little Mimber, the wife of Johnnie, the black driver at Egypt, and that on the night he was supposed to lie with Lettice he locked the driver in the boiling house and slept with Little Mimber instead. During the next two months John Thistlewood and Little Mimber were repeatedly discovered having sex together. When driver Johnnie complained bitterly to Thomas Thistlewood, Little Mimber responded by tearing her husband's coats into pieces. Thomas Thistlewood flogged Little Mimber three times and put her in the bilboes. He also reprimanded his nephew at least five times. On March 30, 1765, when Thomas was away in Savanna la Mar, John Thistlewood paddled onto the river alone in a canoe to go shooting. The canoe was soon seen drifting upside down, and Thomas got two slaves to dive into the river to search for John. The next day they found his floating body—probably an accident, perhaps a suicide, possibly even a murder.[75] Thomas Thistlewood was devastated, especially when a few nights later he heard "two guns fired with a loud huzza after each, on the river against our Negro houses for joy that my kinsman is dead I imagine."[76]

It is reasonable to suppose that much of the white-black male sexual competition on Egypt estate was also simmering at Mesopotamia, and that some of the Mesopotamia women, like Lettice when she approached John Thistlewood, wished to have mulatto babies in order to strengthen their bonds with their white masters. They all knew that any mulatto children they bore would escape from field labor and be assigned to the most favored jobs, the females to housework and the males to craft work. They could even hope that these children might be manumitted by their white fathers. But in fact only nine of the forty-three mulattoes and quadroons born at Mesopotamia were freed between 1762 and 1833. And manumission was never an option for the mothers, all of whom remained slaves.

One such mother was Batty (1773–1830), whom we met in Chapter 2 because her son Edward Knight married Jane Ritchie. Batty, a Mesopotamia field hand, bore four mulatto children and was the long-term mistress of carpenter Patrick Knight, who lived on the estate and fathered at least two children born to other Mesopotamia slave women. In 1793 she had her first child, a black girl named Martha (father unknown), who died in three weeks. Batty's second child, born in 1797, was a mulatto girl named Annie, fathered by Knight, as was another mulatto daughter named Peggy in 1800. In 1801 Batty was switched from field labor to work in the bookkeeper's house where Knight lived, and in 1802 Knight hired Batty as his personal attendant, and also asked Joseph Foster Barham II for permission to purchase the manumission of Annie and Peggy. In 1804 the two girls were manumitted, while their mother remained a Mesopotamia slave. Batty continued as Knight's servant/mistress from 1802 to 1815, and during these years she had two more mulatto babies, Edward in 1807 and Mary in 1811, both fathered by Knight, although he did not free them. By 1814 Knight may have stopped having sex with Batty, because her last child was a black daughter named Hannah. In 1816 he sent Batty back to the Mesopotamia work pool, and at age forty-two she became a washerwoman. In 1823 Batty's mulatto daughter Mary died at age twelve, and two years later (after Patrick Knight had died) the executor of Knight's estate paid Barham £140 for mulatto Edward's manumission. Poor Batty now had only one of her six children still living with her at Mesopotamia, her black field hand daughter Hannah. She was afflicted with leprosy and dropsy, but continued to work as an attendant at the Mesopotamia yaws house until she died at age fifty-seven in 1830.

Predatory white males seem to have avoided having much sex with the African-born females at Mesopotamia. Perhaps they were put off by the African women's filed teeth and the body scars or country marks that covered their bellies, though nearly half of the forty-one females imported to Mesopotamia directly from the African slave ships between 1762 and 1793 were listed as eleven or twelve years old on arrival and were probably too young to have experienced puberty initiation rites in West Africa.[77] There was only one mulatto child with an African mother at Mesopotamia. A woman named Coumba, who arrived on a slave ship in February 1773 at about age twenty, was quickly impregnated and bore a mulatto boy named Captain in March 1774; he was afflicted with the yaws from infancy, served intermittently as a houseboy, and died at age fifteen.

More significantly, few of the African women bore *any* children. The forty-one females who came to Mesopotamia via the slave ships in 1763–1793 had only forty-nine recorded live births. Just seventeen of them became mothers, and thirteen had small families. It has been argued that the African women in the Caribbean produced fewer children than the Creole women because they nursed their young for upward of three years, resulting in wider spacing between conceptions,[78] but the Mesopotamia evidence does not support this contention. Nine of the seventeen African mothers bore only a single child, so spacing was not an issue, and the others spaced their babies just as the Creole women did, becoming pregnant optimally about every two years. In fact, the African women shared the same fertility pattern as the other females on the estate: a very few women produced most of the babies. Four women with five to eleven children accounted for 61 percent of the recorded African slave births.

Matura had the largest family of any African woman from 1762 to 1833. She came to Mesopotamia in a slave ship in 1792 at age thirteen, and produced eight boys and three girls in twenty-four years. Her babies were spaced very regularly at two-year intervals, arriving in September 1798, unknown month 1800, May 1802, May 1804, June 1806, August 1808, January 1810, December 1811, December 1813, February 1816, and November 1821. By 1808 she had six youngsters, all of whom were living, and she was released from the first gang in order to raise her family. But she experienced a distressing time between 1813 and 1817, when four of her little boys died, and in the 1820s one daughter died, another was childless, and the third, Dido, was punished repeatedly (as we have seen) for failing to produce babies that lived. By 1833 Matura had only three recorded grandchildren, two of whom had died. Her one identifiable surviving granddaughter, ten-year-old Georgiana, was working in the grass gang. Matura herself was a healthy fifty-four-year-old, valued at £0 because she was not doing plantation work. Her four surviving sons and two daughters were all field workers, but Dido, Anthony, and Jarvis had become chronic runaways.

Matura's Dido was one of thirty-six Mesopotamia women and girls listed in the annual inventories between 1762 and 1833 as habitual runaways. Thirty-five of them were field workers. Only two women got away permanently, but many disappeared for lengthy intervals year after year. The sixteen female runaways born in Mesopotamia started young: seven began to flee in their teens, and five in their twenties. But the most determined absconders were females who had been brought unwillingly to Mesopotamia from other Jamaican estates. Nine women purchased from Springfield estate ran away repeatedly in the 1820s in order to get back to their relatives and friends in their old Hanover neighborhood, and six of these women were in their forties and fifties. The most aggressive female absconder was Harriet, purchased from Southfield pen in 1791, who was seriously afflicted with the coco bays (a disease akin to leprosy) and clearly hated Mesopotamia. She first ran away in 1808 when she was thirty-nine years old, fled again in 1815 and 1816, and in 1817 disappeared into the wilderness backcountry, where she was captured by the Maroons. Harriet was sent to the parish workhouse in 1816 and 1817, but this did not deter her. She fled again in 1818, and though overseer Peter Hogg caught her one more time, she escaped permanently that same year at the age of forty-nine.[79]

Harriet was childless, and it seems significant that thirty-two of the other thirty-five female runaways were also childless. I believe that many of the Mesopotamia mothers would also have run away except that they had loved ones to tie them to their wretched jobs. The three runaway mothers all had young children whom they left behind. Lizzie was the mother of six children; in 1825 she dropped her infant daughter, who died, and in 1830 she ran away, leaving children aged one and four in the care of a thirteen-year-old daughter, but returned and bore her sixth child in 1832.[80] A Southfield woman named Mary also ran away temporarily at age twenty-nine in 1814, gave birth to a girl named Amba in December 1818, and fled again in 1819 until the Maroons caught her and brought her back. Amba died at the age of one in July 1820, as did Mary two months later. The third runaway mother, Matura's daughter Dido, had twice been punished for bearing stillborn twins and miscarrying, and when her three-month-old daughter died in January 1831 she fled to avoid further punishment. Dido left behind her eight-year-old daughter Georgiana, but Matura could presumably care for this girl, and Dido was back at Mesopotamia before the end of the year. Mother love was clearly a powerful deterrent against running away.

## *Four Labor Patterns: Colored, Mesopotamian, Jamaican, and African*

One of the most distinctive features of the Mesopotamia workforce, when compared with Mount Airy, is its extreme diversity. The Mount Airy workforce was composed of relatively homogeneous families that had been living together for generations—families continually torn apart through forced separation and sale, but renewed through new marriages and new births. The Mesopotamia workforce was far more heterogeneous, made up of four distinct groups flung together arbitrarily: (1) the mulattoes and quadroons, known because of their white fathers as colored or brown people; (2) the Creole blacks who were born and raised at Mesopotamia; (3) the blacks (some of them Jamaica-born Creole and others African) who were brought to the estate from local Jamaican plantations; and (4) the African blacks who were purchased for the estate directly from the slave ships. We have no way of knowing how well or poorly these four disparate sets of people got along together, though there seems to have been considerable tension between the colored and black people, as also between the Creole and African people. The slave inventories do enable us to compare the productivity of these four groups, and in particular to see whether the Mesopotamia-born laborers had longer or shorter working careers than the slaves imported from Africa and from local Jamaican estates.

We can also test the Barhams' change in slave-buying policy in the 1790s. Joseph Foster Barham I acquired most of his 160 new slaves directly from the African ships, making only one major purchase of "seasoned" Jamaican slaves when he bought forty-one people from Three Mile River estate in 1786. Joseph Foster Barham II reversed this policy. He bought thirty new Africans in 1792–1793, but then stopped, fourteen years before the frenetic close of the slave trade in 1807. In a directive to his attorneys he declared, "I cannot approve the purchase of any negroes from the ships on any account whatsoever."[81] Instead he acquired the replacement workers he needed from local Jamaican estates:

61 from Southfield in 1791, 55 from Cairncurran in 1814, and 105 from Springfield in 1819. The Mesopotamia slave records enable us to ask, Was this a wise business decision?

Appendix 21 tabulates the statistical differences between the four groups of Mesopotamia workers in order to compare their productivity. Unlike the appendixes discussed earlier in this chapter, it excludes all of the 199 adult slaves listed in the July 1762 inventory, and considers only the 678 people who entered into adult (age sixteen) employment at Mesopotamia in the seventy years between 1763 and 1832. There are two reasons for eliminating the adults working in 1762. First, it is impossible to tell for sure how many of the older slaves in 1762, who can be traced back to the 1727, 1736, and 1743–1744 inventories, were Africans and how many were born at Mesopotamia or elsewhere in Jamaica. Second, the population in 1762 was in particularly bad physical condition, with a quarter of the adults identified as long-term nonworking invalids, and we get a much clearer picture of worker performance if we look instead at the men and women who entered the adult workforce from 1763 onward. Appendix 21 probably overstates the productivity of the 135 Africans in comparison with the 251 Mesopotamia-born blacks and the 265 "seasoned" Jamaican blacks, since 90 percent of the Africans were dead by 1833, whereas only 52 percent of the Mesopotamians and 58 percent of the Jamaicans had died, and many of the younger people in these two groups had long working lives ahead of them. However, this tabulation *does* establish the years of labor that the Barhams' attorneys and overseers were able to extract, on average, from the individuals within each of the four groups of slaves during a seventy-year period.

The twenty-seven mulatto and quadroon men and women who entered the adult workforce between 1762 and 1833 had the shortest working careers among the four groups, averaging only 11.1 years of prime labor and 12.1 years of total labor.[82] The 265 adult blacks imported from other Jamaican estates were more than thirty years old on average when they joined the Mesopotamia workforce, so they too had short careers at Mesopotamia, averaging just 9.1 years of prime labor and 14.1 years of total labor. Both the 251 Mesopotamia-born blacks and the 135 Africans from the slave ships averaged many more years of productive work. The Mesopotamians averaged 15.8 years of prime labor and 21.2 years of total labor, whereas the Africans averaged 14.9

years of prime labor and 22.3 years of total labor. The men and women imported directly from the slave ships were in poorer health than the Mesopotamia-born people, but worked a little longer and overall were just about as productive. On average, the Africans worked nearly six more years at their prime jobs than the slaves acquired locally from Jamaica, and it thus appears that from a business viewpoint Joseph Foster Barham II should have patronized the slave trade until it closed in 1807.[83]

It may seem surprising that the mulattoes and quadroons, who all escaped field labor and were assigned to less strenuous jobs—the females working as domestics and the males as carpenters, coopers, masons, and smiths—had the shortest careers at Mesopotamia. But according to Barry Higman, colored slaves in Jamaica were generally reputed to be feeble and sickly.[84] At Mesopotamia there were only three mulattoes in 1762, but during the next seventy-one years forty-three slaves with white fathers were born on the estate, and another seven were imported from Three Mile River, Southfield, Cairncurran, and Springfield. Of these fifty-three colored people, twenty-seven were dead by 1833 and nine had been manumitted, leaving twelve colored adults and five colored children on the eve of emancipation. Affy's son Robert McAlpine, whose career was charted in Chapter 2o, was one of the more robust Mesopotamia mulattoes, in his twenty-fourth year of employment as a cooper at age forty in 1833. Robert's lethal fights with a black boy named Tamerlane in 1806 and with black distiller Peter in 1826 suggest the animosity between the black and brown people at Mesopotamia, fueled by the common knowledge that colored boys and girls were always assigned to favored jobs.

Despite their small number, the colored slaves held a quarter of the craft jobs and half of the domestic jobs at Mesopotamia. Indeed, by the close of our period there were barely enough domestic positions to accommodate the colored female population. As of January 1832, the last inventory to list occupations, there were eleven adult domestic servants, and seven of them—six females and one male—were colored. In addition, four colored girls and two colored boys were in training as domestics, and room would soon have to be found for their employment as adults. Among the craftsmen, there was much more room for enslaved blacks because of the high mulatto manumission and death rate; in 1832 only five of the twenty-one carpenters, coopers, masons, and blacksmiths were colored. Even so, Barham's attorneys thought there were too many colored people among the tradesmen, so when several young mulatto men tried to buy their freedom during the last decade of Joseph Foster Barham II's ownership, the attorneys supported them. But by the 1820s Barham demanded a steep payment of £140 or £160 for manumission. Colored carpenter Henry Patrickson was a victim of this policy. In 1823, when Henry was nineteen, his white father, John Patrickson, formerly of Mesopotamia and now the overseer at neighboring Blue Castle estate, offered to buy Henry's manumission for £140, but the offer was refused or delayed, and John Patrickson died in 1826 without many assets.[85] In 1831 Henry Patrickson (now twenty-seven) came up with £100 to purchase his liberty, but this was not considered sufficient, and as of December 1832 his freedom payment was being held in escrow.[86]

The second group under consideration, the 251 black adult workers born at Mesopotamia, tended to die at an early age, particularly the women. But this was largely because they were exposed to the brutal labor regime earlier, all entering into prime Mesopotamia labor at age sixteen—two and a half years younger than the 135 slaves imported from Africa, and fourteen years younger than the 265 slaves acquired from other Jamaican estates. Despite their short life spans, the Mesopotamia-born blacks were the most productive laborers on the estate. Although they constituted only 37 percent of the new adults employed in these years, they performed 45 percent of the prime labor on the estate, and 43 percent of the total labor. Half of the new field drivers were Mesopotamia-born blacks, as were half of the recorded sugar mill hands, a third of the craft workers, 56 percent of the stock workers, 39 percent of the field workers, and a third of the domestics.

From the Barhams' point of view, there were never enough Mesopotamia-born workers, and they bemoaned the low birthrate. But the 124 Mesopotamia-born black women had statistically higher rates of reproduction than the other slave women. Over half of them bore children during these years—a higher percentage than the African women and a much higher percentage than the women from other Jamaican estates. And the Mesopotamia-born black women had larger families than the other 179 adult women, producing over 60 percent of the children born on the estate between 1762 and 1833.[87] By contrast, less than 30 percent of the black women imported from other Jamaican estates had babies during these years, since many of them were well beyond childbearing age when they were brought to Mesopotamia.[88] In fact, the forty-one African women had a higher percentage of mothers, and more children per mother, than the 129 Jamaican women. Thus, Joseph Foster Barham II probably exacerbated the fertility "problem" on the estate with his decision in the early 1790s to turn from the Africa slave trade to Jamaica for his replacement slaves.

It appears that some of the parents of the Mesopotamia-born slaves were able to help their children avoid the least desirable occupations. For example, when Camilla was elevated from field hand to driver in 1792, she had two young sons; Jamante was switched from cattle boy to houseboy in 1793, and Harry was switched from hog herder to houseboy in 1795, and both sons then became carpenters. Sabina was a lowly field hand, but her unknown husband may have had influence, because her sons Dundee and Peter both became skilled workers; Dundee was a carpenter, and Peter became the head distiller. Likewise, Coco, another field hand with an unknown mate, produced two skilled sons: Chelsea was a cooper, and his brother Jeffrey was the head boiler. And apprentice cooper Tamerlane, the boy mulatto Robert killed in a fight in 1806, would likely have followed his older brother Richard and become a carpenter. It is also probably no accident that the majority of the jobs that involved working with draft animals—cattle boy, pen keeper, stock keeper, plowman, and carter—went to the Mesopotamia-born slaves. These were less debilitating occupations. The boys and men who worked with animals at Mesopotamia were in effect pursuing the same line of work as on the numerous livestock pens in Westmoreland; they enjoyed a degree of independence and escaped the dreary routine of gang sugar labor.

Turning to the 265 adult blacks imported from other Jamaican plantations, only 36 of the 136 males and 24 of the 129 females in this group entered the adult workforce at Mesopotamia as sixteen-year-olds. The great majority (77 percent) of these people had been prime laborers on their previous estates, and sixty-four of them (24 percent) were over the age of forty on arrival at Mesopotamia. Those who were already prime laborers came to the Barham estate with much the same range of work assignments as the Mesopotamia-born blacks. The chief occupational differences were that fewer Jamaicans were trained as stock keepers and carters, and none were boilers or distillers. Since many of the Jamaicans were past prime working age, 21 percent of the women and 20 percent of the men came as marginal workers or as invalids and never held prime jobs at Mesopotamia. Also, some of the Jamaicans had short stints in the occupations they came with. For example, three masons—Joe, Exeter, and William—all arrived from Cairncurran in 1814 and were immediately switched to harsher assignments. Joe (age forty) was moved to the first field gang, where he died of incurable ulcers in 1817. William (age forty-five) was also moved to the first gang, where he worked until 1821, when he became a watchman; he died in 1829. Exeter (age fifty-five) was head mason for one year, then worked as a stable hand, but by 1824 he was almost blind and a permanent invalid, and though he was still alive at seventy-four in 1833 he was valued at £0.

Most of the slaves imported from Three Mile River, Southfield, Cairncurran, and Springfield seem to have come very unwillingly to Mesopotamia. They were forced to abandon the living quarters, gardens, and provision grounds that they had established in their previous communities, and they hated to be separated from family members and friends who were left behind, not to mention the graves of deceased kinfolk. In addition, while Three Mile River and Springfield were sugar estates, Cairncurran was a coffee plantation and Southfield was a mixed farming settlement, and at both of these places the labor requirements were less onerous than at Mesopotamia. Yet the Barhams' managers extracted valuable work from many of the people uprooted from all four places. The Three Mile River and Southfield people, arriving many years earlier, worked longer than the Cairncurran and Springfield people; ten men and ten women from these two estates spent twenty years or more as prime hands at Mesopotamia. Chelsea, who came from Three Mile River as a fifteen-year-old houseboy, had one of the most remarkably long and varied job histories of any slave at Mesopotamia. He worked eleven years as a domestic, four years as a cooper, twenty-three years as the first or second field driver, and ten years as head watchman—and after forty-eight years of service Chelsea was still employed as a watchman in 1833, aged sixty-two and valued at £40.

The most discontented of the slaves transferred from Three Mile River, Southfield, Cairncurran, and Springfield ran away, and many of them tried to get back to their old living quarters. There were numerous Southfield fugitives in the 1790s, and a larger number from Springfield in the 1820s. The Springfield slaves were especially distressed by their forced removal from their former plantation in Hanover parish, and they turned out to be the least productive workers of any group at Mesopotamia. Fifteen of the fifty-three adult Springfield men and thirteen of the forty-nine adult Springfield women never had a primary job at Mesopotamia. By 1833 only sixteen Springfield men were assigned to primary jobs, eight held marginal jobs, one had been transported off the island, and twenty-eight were dead. Only nine Springfield women held primary jobs by 1833, thirteen held marginal jobs, five were nonworking invalids, one had run away, and twenty-one were dead. And just seven of the Springfield women gave birth (to fourteen children) between 1819 and 1833. But this is not the whole story. Eight Springfield men and seven Springfield women worked at prime jobs for all fourteen years of their recorded service at Mesopotamia. And some of the Springfield people stoutly defended Joseph Foster Barham's property. During the slave rebellion of 1831–1832 head mason Richard and blacksmith Edward, who both came from Springfield in 1819, were key figures in protecting Mesopotamia from rebel attack, as we shall see in Chapter 8.

The 135 people imported from the African slave ships between 1763 and 1793, the last of our four groups to be considered, were overwhelmingly consigned to the least desirable and most taxing jobs at Mesopotamia. All of the African women and 81 percent of the men became field workers. At least five of the African men did become boilers or distillers at Mesopotamia—skilled jobs of great importance that were frequently held by Africans on this estate.[89] But the new Africans were largely blocked from other skilled jobs or stock work, and none became domestics. Since the Barhams' agents selected these people for their apparent youth and vigor, they were almost all forced into arduous labor on arrival at Mesopotamia. Just 2 of the 135 African adults never held prime jobs. Yet life was brutal and short for these people. The 121 new Africans who died by 1833 lived on average only 40.5 years.

We can follow the fortunes of a group of twenty imported Africans who were purchased for Mesopotamia in March 1792. Joseph II's attorney reported that he traveled to Montego Bay to buy from a Guinea slave ship: "I went there and picked for you 11 very fine young boys and 9 girls which I hope will turn out well."[90] Two of the boys were estimated to be eleven years old, and nine were thirteen, while seven of the girls were listed as eleven, and the two older girls were thirteen and fifteen. If these ages are correct, the nine eleven-year-olds had probably not yet undergone their pubertal initiation rites in Africa, and so would not have scarification or country marks, whereas the eleven who were thirteen or older would have conspicuous facial and/or bodily scars. Six of the boys were named for British or European towns or places, and most of the girls were given equally inappropriate feminine-sounding names. Bell served as a domestic for a year, but all the other girls and boys were immediately put into the grass gang. Fancy died in 1793 and Granby in 1794, while the others were moving into the second field gang. By 1801, when they were in their early twenties, Bacchus, Major, Paris, Port Royal, Ralph, and Smart among the males and Belinda, Clarinda, and Matura among the females were already working in the first gang. Devonshire was a pen keeper, and the other three young men and five young women remained in the second field gang. As the years passed, nine of the eleven males and six of the nine females spent time in the first gang at Mesopotamia. These Guinea people grew up in close company with their shipmates, and doubtless as they matured they formed sexual and familial relationships—though we do not know who lived with (or mated with) whom. Over time the females clearly survived better than the males. By 1815 (when those still living were in their mid-thirties) Bell, Fidelia, and Susannah had died, while five more males—Bacchus, Cubbena, Glasgow, Major, and Smart—were dead, and Ralph had absconded. By 1833 only six of these twenty Africans were left. All in their early to mid-fifties, they had survived at Mesopotamia for forty-two years: Belinda, Clarinda, Juliet, and Matura out of the nine females, and Devonshire and Port Royal out of the eleven males.

Just three of the nine Guinea females had recorded children. Clarissa had a single boy named Cupid who burned to death accidentally at age three, while Clarinda had eight children, and Matura had eleven—the largest family produced by an African woman from 1762 to 1833. Matura's exceptionally large family has already been discussed. But it is worth adding that in 1802 she named her second daughter Fancy in memory of her shipmate who had died nine years before, suggesting a powerful bond between these African girls. Perhaps Matura had a less burdensome life than any of her African companions. She stopped working in the cane fields in 1808, and by 1833 she had been tending her younger children for twenty-five years, and was surrounded by four surviving sons, two daughters, and at least one granddaughter. But if

Matura had children to support her, most of her nineteen African shipmates did not. It is painful to contemplate the traumatic experiences that these Guinea people lived through. They had been snatched by slave raiders from their families in West Africa when most of them were preadolescent boys and girls, trekked in coffles to the seacoast, sequestered in shore-side barracoons, and loaded in chains onto slave ships. They had been forced to endure the brutality and fright of a long voyage across the Atlantic, were pawed over by slave buyers on arrival in Jamaica, were branded on arrival at Mesopotamia, and found themselves—all still in childhood—living among 300 strangers in the Mesopotamia slave gang and working twelve hours a day in the cane fields. Doubtless they found lovers and dear friends in Jamaica, but few of them had families, and four of the males—Bacchus, Devonshire, Major, and Ralph—turned into deeply alienated runaways.

Though the four groups of workers had differing backgrounds and labor histories, all of them were victims of a vicious labor regimen. Far from being "dreadful idlers," they all toiled at Mesopotamia to the limits of their endurance. But they were trapped in an environment in which death was more common than birth. Death was omnipresent from conception to old age. Scores of slave babies died within a few hours or days after birth and never entered the Mesopotamia record book. Of the 420 boys and girls whose births were recorded between 1762 and 1833, 37 died in infancy, another 21 at age one, another 23 between ages two and five, another 17 between six and ten, and another 7 between eleven and fifteen—for a total of 105 deaths in childhood, a quarter of the recorded births. Among the 678 slaves who reached the age of sixteen after 1762, 421 (or 62 percent) were dead by 1833, at an average age of forty-two. Death also stalked the white managers of the slaves. Many of the Mesopotamia bookkeepers, pen keepers, and other white staff members died within months of arrival on the estate. Few whites survived at Mesopotamia for more than a few years. Fifteen of the Moravian missionaries were buried in the Mesopotamia cemetery. Vincent Brown, in his powerful account of the culture of death in Jamaican slave society, sees the Grim Reaper as a gardener as well as a harvester. "Death," he says, "tended and nurtured the activities of the living, cultivating their understanding of the world and their struggle to shape it."[91] Death indeed was the common currency for all the peoples of Mesopotamia, black, brown, and white.

# 5

## "Doing Their Duty" at Mount Airy

When John Tayloe III completed his education at Eton and Cambridge and returned to Mount Airy in 1791 at the age of twenty, he was one of the ten largest slaveholders in Virginia.[1] No census of his slaves survives for this date, but county tax records indicate that in 1791 he held about 350 slaves at Mount Airy and many additional slaves at ironworks and farm quarters elsewhere in the Chesapeake.[2] Had Tayloe allowed his slave population to grow unchecked at its natural rate, he would have owned close to 1,000 enslaved people at Mount Airy by the time he died in 1828[3]—the opposite of the situation faced by the Barhams in Mesopotamia, where the slave population would have shrunk from 268 in 1762 to less than 100 by 1833 if allowed to decline at its natural rate. But Tayloe pursued a different strategy. Figuring that he held more slaves at Mount Airy than he could profitably employ, he advertised in the newspapers of Richmond, Norfolk, Fredericksburg, Alexandria, and Annapolis in 1792 that he was putting 200 men, women, and children up for sale—more than half the total population at Mount Airy. Although Tayloe's sales records are lost, his tax returns indicate that he did sell about 110 of his Mount Airy slaves between 1792 and 1794. He used the proceeds to buy several thousand acres of additional farmland, and his slave force was rapidly restored through natural increase.[4] By 1809 there were more slaves at Mount Airy than in 1791, and Tayloe now had enough Rappahannock acreage so that he could utilize all of his black workers. While the Barhams were purchasing hundreds of replacement slaves to keep Mesopotamia in operation, John Tayloe III's naturally expanding slave force was the engine for his business success.[5]

In the year of his big slave sale John III married Anne Ogle, the daughter of a leading Maryland planter-politician, who bore him fifteen children, thirteen of whom lived to maturity. Needing to provide for seven sons and six daughters, Tayloe devoted his career to maintaining and expanding his large estate. He became the most aggressive entrepreneur among the four slave owners in our story, a considerably more venturesome American gentleman than his British contemporary, Joseph Foster Barham II. Tayloe opened the Mansion Hotel on the site of the present Willard Hotel in Washington, D.C., built other property in the new capital city, and operated three ironworks, a fishery, numerous farm quarters, and several taverns and stores in rural Virginia and Maryland. In 1799–1801 he employed William Thornton, the first architect of the United States Capitol, to build an elegant town house, the Octagon, on 18th Street and New York Avenue, close to the White House, so that he and his wife could spend the winter months mixing with the leaders of the national government. John and Anne spent the summer months at his Mount Airy mansion, the center of his agricultural domain. Starting with five Rappahannock farm quarters in 1791, Tayloe added three more, and by 1805 he was operating eight farms on 8,000 acres in three adjacent counties along both sides of the Rappahannock River. In addition, he ran an extensive congeries of craft shops on his 1,700-acre home plantation. The Mount Airy home plantation was and is situated in Richmond County. Five of the farm quarters—Old House, Doctor's Hall, Forkland, Marske,[6] and Menokin—were also in Richmond County. Gwinfield was in Essex County across the Rappahannock, and Hopyard and Oaken Brow were in King George County thirty miles upriver from the home plantation. Tayloe's "Mount Airy Department" (as he called it) was valued in 1808 at £98,054 or $326,520. On the last page of his 1808 inventory, John Tayloe III scribbled a note. "I wish the above Valuation could be realized," he remarked. "Its certainly much too great."[7]

Tayloe lived among his enslaved workers, unlike the Barhams, but he was always imperiously remote in his dealings with them—as also with the numerous white overseers who supervised their labors. He seldom identified any of his "people" (as he termed the slaves) in his correspondence, but he did have a consistent message for them that he frequently repeated. In 1794, when two slaves ran away, he placed a newspaper advertisement in which he urged the runaways to "return to a sense of their duty."[8] In 1801 he told his manager at Neabsco ironworks to press the overseers on the farm quarters: "You must not spare these Fellows, but push 'em on & make 'em do their duty."[9] In 1809 William Holburne, the manager at Mount Airy, reminded the Hopyard overseer of Tayloe's dictum: "His People . . . are to be well taken care of in sickness and in health, particularly the former, but to be made [to] do their duty."[10] In 1815 Tayloe concluded his instructions to his wheat harvesters with the same litany: "In a word, each overseer and all hands are expected to do their Duty."[11] In 1824 he told his son William Henry that "your Garden and House enclosure ought to be kept in perfect order by your lazy gardners, were they made to do their duty."[12] And in 1827 he wrote to his manager Benjamin Boughton, "Make *all hands do their duty—Black and White.*"[13] In Tayloe's view, his enslaved workers were not being forced to work for him. They had a moral *obligation* to work for him.

This chapter examines slave life at Mount Airy during John Tayloe III's tenure, with prime focus on the employment pattern. In some ways the Mount Airy system resembled the Mesopotamia system. In both places the slaves were divided into three occupational groups: domestics, craft workers, and field hands. In both places over 60 percent of the prime workers were field hands. In both places the women had fewer employment options than the men and were largely relegated to the most burdensome jobs. In other respects there was a strong contrast with the Barhams' sugar labor regimen. The Mount Airy field hands grew corn and wheat and raised cattle and pigs for the market—a system of mixed agriculture that was practiced on white family farms throughout the United States. Cultivating grain and raising livestock required plenty of long, hard work, but the Mount Airy slaves escaped the grueling and debilitating gang labor at Mesopotamia. They also had a much ampler and more nutritious diet, and experienced no starving times. What these Mount Airy people *did* experience was continuous family breakup, as Tayloe sold his surplus slaves to new owners or moved them to new jobs in distant places.

## *The Labor Pattern at Mount Airy*

John Tayloe's documentation for slave life and labor at Mount Airy differs in important respects from the Mesopotamia documentation in

Chapter 4. The annual Mount Airy slave inventories track each man, woman, and child living on the estate by name, age, occupation, and location, and thus reveal the great mobility of the population. But the inventories provide no data on the slaves' physical condition, nor statistical evidence showing the impact of particular jobs on slave health and longevity. Information about annual crop production and livestock sales at Mount Airy is much less complete than for sugar and rum production at Mesopotamia. And I have found no diarist equivalent to Thomas Thistlewood to shed light on the day-by-day management of the Mount Airy slaves. To be sure, three diarists living in the Northern Neck—Landon Carter, who wrote from 1756 to 1778, Philip Fithian, who wrote in 1773–1774, and Robert Carter of Nomini Hall, who wrote from 1773 to 1792—are famous commentators on Virginia slavery in the previous generation.[14] But all three are very poor proxies for John Tayloe III. Philip Fithian was an outsider from New Jersey who detested slavery. Landon Carter was a notoriously intrusive, irascible, and inept slave master, with none of Tayloe's remote, imperial demeanor. And Robert Carter was inspired by his Great Awakening conversion experience to manumit all 509 of his slaves—the last thing in the world that Tayloe would consider doing. Besides, Tidewater Virginia in the early nineteenth century had changed a lot from the revolutionary era. Tobacco was no longer the cash crop, the slaveholders were now living on a more dynamic national stage, and increasingly they were moving their black workers to the interior of Virginia or to the new western and southern states, or they were selling the slaves they did not need to professional traders.

The lack of an outsider observer equivalent to Thistlewood is regrettable, because the Mount Airy documentation of slave life and labor from 1791 to 1828 is entirely from slave master John Tayloe III's perspective. However, one very important set of Tayloe slave documents with no counterpart at Mesopotamia does open up a view of the Mount Airy slave force in full action. An assortment of minute books and shop books, dating from 1805 to 1819, record the labor assignments of the Mount Airy craft workers and field hands. These work logs were compiled by two of Tayloe's estate managers, William Holburne and Benjamin Boughton, and when pieced together they reveal a great deal about the character of slave labor at Mount Airy. The key volume is a 190-page minute book, running from January 1 to December 7, 1805, which describes the daily assignments of Tayloe's craft workers and the weekly assignments of his field hands during this forty-eight week span.[15] Five subsequent minute books similarly chart the daily craft and weekly farm labor routines for lengthy periods in 1806,[16] 1807,[17] 1811–1812,[18] and 1814,[19] and for six mostly complete years in 1813–1818.[20] Unfortunately, the minute books for 1806 and 1813–1818 are in such a poor state of preservation that they are mostly unreadable. The minute books for 1807, 1811–1812, and 1814 are, however, in much better shape, and while considerably less detailed and informative, they confirm the picture established in the 1805 work log. Tayloe also kept shop books tabulating the productivity of his craft workers, several of which have survived: a blacksmith book for 1793–1794, spinning books for 1805–1806, 1806–1807, and 1816–1819, a mill book for 1810–1813, and a shoemaking book for 1816–1817.[21] Collectively, these work logs and shop books capture the variety and the seasonal rhythm of labor at Mount Airy.

While John Tayloe's Mount Airy plantation records are richly detailed for the years 1805–1828, nearly all of the documentation for the first fourteen years of his management has been lost. This is exceedingly unfortunate. The sale that Tayloe advertised in September–November 1792 of "two hundred Virginia born men, women, children, all ages and descriptions" is the most significant event by far in his slaveholding career, and the details of this very large transaction have not been preserved. The tax records for Richmond, Essex, and King George counties indicate that the Mount Airy slave population dropped in 1793 from about 370 to about 290, and dropped again in 1794 to about 260, for a total loss of at least 110 slaves. The sale was organized by Tayloe's estate manager William Holburne, who supervised the Mount Airy farms and craft shops, and was the white man best acquainted with the individual abilities of all of the slave workers. There is no way of identifying the 110 or so people who were taken away or of knowing how many slave families were ripped apart. We cannot tell whether the Mount Airy slaves were dispersed individually or in large groups, to big planters, or to small planters, or to slave traders. We *do* know that Tayloe sold more females than males. This is demonstrated by the Mount Airy population pyramid for 1809 diagrammed in Appendix 6, which shows a distorted sex ratio and a severe shortage of women aged fifteen to forty-nine—of females who would have been thirty or younger at the time of the sale seventeen years previously. If Tayloe and Holburne followed the pattern of later sales, they mainly sold people from the farm quarters, along with some craft workers, but very few domestics. They probably avoided selling complete families, and focused instead on individual members of large families, or mothers with their youngest children.

The massive breakup of families and friendships in 1792–1794 must have been etched indelibly into the consciousness of every Mount Airy slave who remained on the estate after 1794. The sale proved that no individual or family on this estate had true security, and what had happened once could certainly happen again.

The county tax records show that the Mount Airy population began to increase annually immediately after the sale ended in 1794. There were about 290 slaves at Mount Airy by 1797, 320 by 1803, and 350 by 1806.[22] The first surviving complete slave inventory shows a population of 382 in 1809. Meanwhile, soon after his big slave sale, Tayloe bought three farms, which almost doubled his Rappahannock agricultural acreage. He expanded Hopyard farm quarter in King George County in 1795 and added Menokin plantation and Doctor's Hall farm quarter in Richmond County in 1799 and 1801, and Oaken Brow farm quarter in King George County in 1805.[23] Now he was in a position to use some of his surplus Mount Airy people to increase Rappahannock farm production.

On this new land Tayloe grew wheat and corn in place of tobacco, the export staple that had mainly occupied the labor of his grandfather's and father's slaves since the early eighteenth century. The European market for Chesapeake tobacco was collapsing in the 1780s and 1790s, owing to the disruptions of the American Revolution, postwar British commercial policy, the French Revolution, and the European wars of the 1790s. By 1802 American tobacco exports to Britain were only 25 percent of what they had been in 1773.[24] When Tayloe returned from England, tobacco was still being grown on four of his five Rappahannock farms: between fifty and sixty hogsheads were shipped from Hopyard, Gwinfield, Old House, and Forkland to England annually from 1789 to 1792. Production continued at Hopyard from 1793 to 1796. Then it stopped.[25] Tayloe continued to ship tobacco from some of his other Chesapeake farm quarters as late as 1801,[26] but the Mount Airy slaves no longer grew any tobacco from 1797 onward.

During the years 1805–1828 John Tayloe III always had a Mount Airy population of around 350 slaves at his disposal, and a workforce of

225 to 250 farmhands, craftsmen, and domestics. The cessation of tobacco production meant that he needed fewer agricultural laborers per hundred acres, since wheat and corn require less intensive and arduous manual labor than tobacco. Tayloe might have liked to convert to cotton, another labor-intensive crop that was just beginning to be profitable in Carolina and Georgia, but Virginia was too far north for commercial cotton farming (although cotton was grown at Mount Airy for homespun cloth making, as we shall see). The only practical choice was to expand grain and livestock production. So in the early nineteenth century the Mount Airy field hands grew very large crops of wheat and corn, as well as hay, oats, rye, peas, and potatoes, and tended herds of cattle, horses, hogs, and sheep. According to the work logs, they spent about two months per annum on wheat (Tayloe's principal cash crop), about five months on corn, and the remaining five months cultivating the minor crops, as well as grubbing, manuring, and fencing the fields. Meanwhile, the craft workers were given frequently changing assignments depending on the season of the year.

The first complete Mount Airy inventory, taken on January 1, 1809, shows that 106 slaves lived on the home plantation and 276 slaves lived on the eight Mount Airy farm quarters, for a total population of 382.[27] As *Appendix 22* demonstrates, the people on the home plantation had quite different occupations from the people on the farm quarters. And at Mount Airy (unlike Mesopotamia) the two groups of laborers were physically separated, so much so that the domestics and craft workers on the home plantation saw very little of the agricultural workers on the three most distant farm quarters: Gwinfield, Oaken Brow, and Hopyard. About two-thirds of the Mount Airy slaves had jobs: 153 were farm laborers, 59 were craft and specialty workers, and 36 were domestics. Eleven old people had no stated employment—only 3 percent of the total population, while at Mesopotamia the elderly and sickly nonworkers constituted 8 percent of the population in 1809.[28] And 123 young children at Mount Airy (32 percent of the total) were not yet working, whereas at Mesopotamia only 43 children (13 percent) were unemployed in 1809, including 6 of working age who were hospitalized with the yaws. There were twice as many young children at Mount Airy as at Mesopotamia, and they entered the listed workforce three or four years later: the boys when nine to twelve and the girls when ten to thirteen.[29] On both plantations the boys and girls who were field or craft or domestic workers in their teens almost invariably continued in the same line of employment when they became adults. The occupational division was perhaps stronger at Mount Airy, because the children of the domestics and craft workers frequently entered into their parents' occupations, while the children living on the farm quarters had no alternative to field work.

In the occupational tabulation for 1809, the workers listed as "Men" and "Women" were eighteen or older, while the "Boys" and "Girls" were seventeen or younger. Mount Airy had a youthful population in 1809, as *Appendix 6* has demonstrated. Altogether, forty-eight of the eighty slave women living on the estate were mothers with young children: a total of 104 boys and girls under the age of ten (as against 46 at Mesopotamia). At the home plantation there was an apprenticeship system, in which boys and girls were placed in every skill position except blacksmith and sailor. And overall, the male workers decidedly outnumbered the female workers: 72 percent of the artisans and domestics and 58 percent of the field hands were men and boys.

The nine Mount Airy inventories dated 1808–1816 attach a value to every man, woman, and child. Overall, the average price of a male worker above the age of sixteen in 1808 was £83 ($276), while the average price of a female worker above the age of sixteen was £65 ($216).[30] Farmhands were priced almost exactly the same as artisans and house servants, except that several of John and Anne Tayloe's personal attendants were given the highest valuations. Three of the four male slaves priced at £120 in the 1808 inventory were house servants Archy, John, and Peter; the fourth was head carpenter Charles. On the female side, house servant Ibby was the only woman to be priced at £100. In general, a Mount Airy slave's worth was determined by age, gender, and condition, not by sentiment or occupation or skill. Over half of the adult male slaves were priced at £90 or £100, and 87 percent of the adult women were rated at £60 or £70. Among the children, infants of both sexes were valued at £15 or £20, a ten-year-old at £50, and a teenager at £60. A further index of the Tayloe scale of values is that the twenty-one carriage, riding, and race horses stabled at Mount Airy were collectively priced at £2,510 or $8,358—a higher figure than the £2,375 valuation for the forty-three African Americans who lived and worked at Oaken Brow quarter.

John and Anne Tayloe maintained a very large number of personal attendants at Mount Airy, and John's craft system was designed to promote the economic self-sufficiency of the plantation. The artisans lived in the slave quarters adjacent to the mansion house, as did the domestics, and worked in craft shops on the estate. Using cotton grown and ginned on the estate and wool sheared from Mount Airy sheep, the textile workers—fourteen spinners, two weavers, and a ginner—made coarse cloth for slave apparel and fine cloth for household linen. The four shoemakers tanned and dressed leather from Mount Airy cattle to make coarse shoes for the slaves and custom shoes for white clients (who paid Tayloe for this footwear), as well as harnesses for the horses, mules, and oxen. The five blacksmiths and four joiners built and repaired wagons, plows, and hoes, and the smiths also shod horses. The eleven carpenters and three masons moved about the estate erecting and repairing buildings, while the miller ground Mount Airy grain for food consumption on the estate.

From 1809 until John Tayloe's death in 1828, the number and proportion of slaves enrolled in domestic service, craft work, and farm work at Mount Airy was remarkably stable. In 1827, as in 1809, 27 percent of the slaves lived on the home plantation and 73 percent in the farm quarters. But during the years 1829 to 1865, John's son William Henry Tayloe cut down the proportion of domestic servants and expanded the proportion of field hands. *Appendix 23* takes a broad look at the Mount Airy labor pattern by tabulating the primary occupations of all the 542 adult slaves (eighteen and older) who lived at Mount Airy between 1809 and 1865. It then compares this overall Mount Airy labor pattern with the overall Mesopotamia pattern, previously presented in *Appendix 15*. Clearly there were strong occupational parallels. The men in both places held all of the most highly skilled jobs, and women were particularly exposed to heavy manual labor. Sixty-eight percent of the women workers at Mount Airy and 85 percent at Mesopotamia toiled in the fields. But the differences were also significant. In 1809, 17.7 percent of the Mount Airy adults were domestics, a figure that decreased over time to 11.6 percent, but at Mesopotamia only 2.5 percent of the adults were domestics. The proportion of craft workers (which did not change greatly from 1809 to 1865) at Mount Airy was double the proportion at Mesopotamia. On the other hand, the proportion of field hands (which rose over time) was always lower than at Mesopotamia. Since the field hands on the Mount Airy farm quarters were also stock keepers and carters, 61.8 percent of the Tayloe slaves were agricultural workers, as against 69.7 percent of the Barham slaves. And only 4.2 percent of the Mount Airy adults were marginal workers or nonworkers, as against 14.9 percent at Mesopotamia.

In sum, forced labor at Mount Airy was considerably less onerous and debilitating than at Mesopotamia by every index of measurement. But John Tayloe's work logs demonstrate that the Mount Airy slaves also toiled long and hard throughout the year.

## *Agricultural Labor, 1805–1828*

It was impossible for the Mount Airy field hands to produce grain crops that could match the sugar and rum profits generated at Mesopotamia, but they made plenty of money for John Tayloe. From 1805 onward, when the grain fields were in full operation on all eight farm quarters, the field hands annually grew some 7,000–12,000 bushels of wheat and 4,000–6,000 barrels of corn, which produced a gross profit of about $15,000 for their master.[31] Tayloe paid between 7 and 8 percent of the wheat and corn harvested to his seven overseers, used about 15 percent of the wheat to seed the next year's crop, and retained 2,000 barrels of corn to feed his slaves and stock. But this still left him a lucrative surplus. In an average year, he shipped about 7,000 bushels of Mount Airy wheat and 2,000 barrels of Mount Airy corn to Baltimore or Alexandria, and received $8,000 to $9,000 for his wheat and another $6,500 to $7,500 for his corn. In addition, he shipped and sold some of his neighbors' grain. In June 1809 Tayloe announced that he expected to market 10,000 bushels of wheat at $1.25 per bushel.[32] Livestock production added further to his profits. In 1809, according to the annual inventory, the field hands on the eight quarters tended twenty-eight horses, sixty mules, eighty-seven oxen, 354 milk and beef cattle, 413 sheep, and 353 pigs, valued collectively at £4,570 or $15,218. The horses, mules, and oxen were work animals, the cows produced milk for the estate, the heifers were sold, the sheep produced over a thousand pounds of wool per annum,[33] and a third of the pigs were butchered every year for consumption on the estate.[34]

Susan Dunn argues that the Virginia planters in the early nineteenth century were so hobbled by their dysfunctional slave labor system that they became ruinously indolent and incompetent farmers.[35] John Tayloe III does not fit this description. He was an innovative agriculturalist who invested in the latest plowing and harvesting techniques, and his slave laborers produced impressively high crop yields. He operated a pair of schooners, one based on the Rappahannock and the other on the Potomac, and used both ships to transport his Mount Airy corn to Alexandria and his wheat to Baltimore. The plantation records show that Tayloe's estate manager and his seven overseers effectively synchronized the individual labors of some 200 farm and craft workers to keep an elaborate enterprise in continuous motion.[36] The work logs detail how the overseers on each farm quarter ordered labor assignments each day, supervised the field work, and made weekly reports to the estate manager.[37] The work log for 1805, supplemented by the later Tayloe work logs, also documents the cyclical character of the agricultural year as the calendar slowly revolved through the seasons, each with its appropriate and interlocking set of tasks.[38] The Mount Airy artisans and field hands generally labored at separate jobs, except for two weeks in late June and early July, when everyone joined in the grain harvest, but the work assignments of the artisans and farm laborers were always closely coordinated.

During the coldest six weeks of the year, from mid-December to the end of January, when the previous season's crops had all been harvested and the winter wheat was in the ground, the field workers and jobbers shucked and beat corn, cut and hauled timber to the sawmill for fence rails and posts, and cut ice from Rappahannock Creek for storage in the Mount Airy icehouse. This was a big project. On January 3–5, 1805, they hauled 229 cartloads and 108 wagonloads of ice—"all good Ice"—and filled the icehouse to capacity.[39] Here they must have used most if not all of the seventeen wagons and carts inventoried on the five Richmond County farm quarters, and perhaps the nine tumbrels as well. The carpenters, meanwhile, operated the sawmill, while the smiths and joiners repaired and sharpened some of the 207 plows in use at Mount Airy for spring planting. In February the field hands worked with the carpenters and joiners in putting up fencing, with the jobbers in grubbing and manuring the fields for plowing, and with the sailors in loading the previous year's surplus corn for shipment and sale. John Tayloe changed his pair of cargo ships every few years. In 1805 the *Federalist* (signifying the owner's politics) carried corn to Alexandria in March, and the *Henrietta* (named after his daughter) carried wheat to Baltimore in September and November. In 1811 the *Saragossa* transported Mount Airy corn in April, while the *Virago* took on loads of Mount Airy wheat in July, August, and September. These vessels were all manned by four slave sailors.

Meanwhile, in late February the spring plowing began, and after the soil was broken the field gangs planted oats in March, corn in April, and cotton and peas in May, while the smiths and joiners sharpened the 186 hoes used to till and weed the new plants. In mid-April, just before the Tayloe family arrived for the summer from the Octagon in Washington, seventy laborers from the five closest Rappahannock farms came to Mount Airy to dress up the mansion lawn.

In late May, while the field workers began to weed the corn, the smiths, joiners, and carpenters were making and mending rakes and cradle scythes for the coming harvest. In mid-June the wagoner went to Kinsale, a nearby town, to fetch five barrels of whiskey, and Tayloe also laid out extra rations of bacon, pork, and herring for his harvesters.[40] The wheat crop was harvested in a two-week period starting in late June.[41] Parallel gangs were deployed on each of the quarters to cut, rake, bind, and stack the grain. In 1815 John Tayloe III wrote the following memorandum for his harvesters: "Give a dram at 6 OClock AM, a Drink of Grog at 12 noon & at 4 PM to the rakers. The cradlers have their ½ pint given to them to use as they please. All hands 'tis expected to dine together in the Field at One OClock—The Overseer to have his Dinner also sent him There, as it is expected he is never to be absent except in case of sickness—for fear of accident. Their throats should be well gargled & they should throw a little water on their foreheads before Drinking—in a word, each overseer and all hands are expected to do their Duty."[42]

Tayloe's key harvest operative was the cradler, who cut the ripe wheat with a cradle scythe, a formidable tool whose long blade is fitted to a large framework of curved rods which gather the grain cut by the scythe so that the cradler at the end of his cutting stroke can drop the wheat into a row or swath. Cutting with a cradle was very hard work; Tayloe's cradlers routinely cut swaths seven feet wide.[43] Twenty-three cradle scythes were stored at the home plantation for use on the nearby wheat fields at Old House, Doctor's Hall, Forkland, and Menokin, while another twenty-two cradle scythes were stored at Gwinfield, at Hopyard, and at Oaken Brow for harvest use in those distant quarters.[44] Interestingly, Tayloe chose his artisans rather than the regular field hands to do most of the cradling. Every summer twenty to thirty experienced blacksmiths, jobbers, carpenters, masons, joiners, shoemakers, sailors, and house servants were assigned to cut the grain, along with a few strong men from the farm quarters. The second big harvest job was to rake the cut wheat and bind it into sheaves, and here again Tayloe employed twenty to thirty young male artisans and female spinners, weavers, and housemaids as rakers and binders.[45] Tayloe also hired slaves from neighboring planters as supplementary cutters and rakers,[46] and recruited a few free blacks to join them at $3 each.

Tayloe's system is illustrated by comparing the lists of cradlers and rakers for Doctor's Hall and Forkland quarters in June 1816, as shown in *Appendix 24*.[47] At Doctor's Hall, a 695-acre farm that produced about 1,100 bushels of wheat, the cradlers were all carpenters who were used to working together, and four of the rakers and binders were female textile workers who were also close colleagues. None of the seven rakers were field hands. At Forkland, a 320-acre farm that produced only 600 bushels of wheat, the cradlers were more miscellaneous, three of them being young and inexperienced, and Fork Tom was a field hand at Forkland quarter. Four of the Forkland rakers and binders were again female textile workers, while Marske George was a field hand. Four years earlier, on June 30, 1812, it was noted with some surprise that Fork Tom, who was then a twenty-nine-year-old field hand, cut wheat occasionally during the Forkland harvest, and cut "very well"—so well that he was used as a cradler in succeeding harvests.[48] But Tom was an exception. Most of the field hands played a secondary role in the harvest; their job was to stack the wheat sheaves after the grain was cut, raked, and bound.

On the evening of June 18, 1801, John Tayloe III sat down to write a letter, protesting to his correspondent that he was "just from my harvest field and fatigued to death."[49] Tayloe was not fatigued from handling a cradle scythe or a rake; he was worn out from riding through his fields to supervise the pace of harvest work. No doubt the men and women who did the real work were also pretty tired. In 1805 a thirty-three-year-old blacksmith named Lewis and a twenty-year-old joiner named James each cut the ripe wheat by hand with cradle scythes for eight straight working days, from June 22 through July 1, first harvesting the fields at Forkland, then moving on to Old House, and finishing up at Menokin.[50] There are eight surviving harvest lists spanning the years 1811–1827; Lewis was a cradler in 1811 and 1812, and James was a cradler in 1813, 1814, 1815, 1816, 1824, and 1827. Lewis continued to work as a smith at Mount Airy until his death in 1832 at age sixty, and James worked as a joiner even longer, until the year he died in 1852 at age sixty-seven.

In July the seasonal pressure continued, as most of the hands joined for another week to harvest the oats crop. During the next few weeks the field gangs worked mainly in the cornfields, hilling and hoeing the plants. The Mount Airy farm laborers spent a lot of their time hoeing, and they were well supplied with the tools for this work. Two hundred hoes were inventoried on the eight farm quarters in 1809: 102 weeding hoes, 84 hilling hoes, and 14 grubbing hoes.[51] In August the field hands turned from hoeing the corn to cutting the hay and threshing the wheat. The jobbers helped with the hay, the carpenters made grain barrels, and the smiths and joiners worked as before on plows and wagons. In September, and again in November, the wagoner and sailors helped the field hands to load the *Henrietta* with wheat for the market in Baltimore. Now it was time to pick apples, to gather fodder to feed the livestock during the winter, and to start the long process of seeding the winter wheat for next year's crop. As the fall racing season approached, the smiths set to work on shoeing Tayloe's racehorses. Every year John Tayloe held races on his Mount Airy course, and many of the slaves from the home plantation and the nearby farm quarters must have been in attendance to watch Top Gallant and other Tayloe thoroughbreds compete against rival steeds.

Shortly after the races, the Tayloes departed for Washington, and the craft workers could now make necessary repairs on the mansion house—as in 1805, when the carpenters, joiners, jobbers, and masons worked for a month reshingling the mansion roof under the supervision of a hired white builder.[52] In October the spinners joined the field hands at picking cotton. In late November and from mid-November to mid-December, after the wheat fields were seeded, the field hands harvested the corn and hauled the stalks to the mill. Shortly before Christmas, during the slackest work period of the year, the slaves performed public work for Richmond County. In 1812, for example, the field hands from Old House and Doctor's Hall hauled bricks and stone to the Richmond County Court House, after which the Mount Airy masons, carpenters, and jobbers used these materials to repair the building. In other years the Mount Airy slaves spent a week or two in December repairing the county roads. After all of this yearlong labor they were allowed a five-day Christmas holiday, which normally ran from December 25 to December 29.

While John Tayloe maintained a dignified distance, the overseers on his farm quarters had direct management of the field hands. They assigned the daily tasks, supervised and disciplined the slaves at work, and reported every week to Tayloe's general manager William Holburne or his successor Benjamin Boughton. Tayloe remarked, "I have but few of that class [i.e., overseer] are worthy of [trust]," and Holburne kept a close eye on the quantity and quality of the crops they supervised. In November 1794, for example, Holburne wrote to Ben Baber, overseer at Hopyard, "I wish your wheat got out as soon as possible, of which send me a sample by the first opportunity."[53] Tayloe's condescending treatment of his overseers is illustrated by a letter he wrote in 1813 to Ephraim Beazley, who had managed his biggest farm, Gwinfield quarter, for the past eleven years. "You know I wish not to part with you," Tayloe wrote. "I therefore expect you will stay with me another year, tho so many wants your place, but I shall expect more attention to my farm than you have lately bestow'd, for to be candid with you, you have been more negligent and careless than you out [*sic*] to be and the place by no means improves or looks as neat as is wished by yours, etc. John Tayloe."[54]

Ephraim Beazley and Tayloe's other overseers, for their part, were motivated to push the slaves hard, because they received a 7 to 8 percent share of the wheat and corn crops, a significant addition to their salaries. There was frequent overseer turnover, particularly on the small farm quarters with meager income. A Mount Airy account book identifies twenty-seven men who served as overseers on seven of the farms between 1790 and 1815, but the total was probably larger, because there are many gaps in this listing. Six of the twenty-seven overseers were promoted from a small farm to a larger one.[55] Hopyard had seven or more overseers between 1790 and 1815, Old House had eight or more, and Forkland had ten or more. The biggest farms, Oaken Brow and Gwinfield (where Ephraim Beazley was employed in 1813), had the least turnover.[56]

There is very little mention in John Tayloe's correspondence of slave punishments. In 1794 a "wench" named Sukey belonging to a neighboring planter, Dr. Brockenbaugh, was accused of "vilanous" conduct at Mount Airy, and Holburne wanted her brought to Mount Airy to receive her punishment, since "such wayward condu[ct in] slaves, is not to be born." Interestingly, Sukey's villainous, wayward conduct is not specified, and the number of lashes ordered for her in this letter has been obliterated![57] But another example suggests that whippings at Mount Airy were less frequent and less severe than at Mesopotamia. In March 1809 a twenty-two-year-old Hopyard field hand named Nancy, the mother of three young children, came to Mount Airy and complained to Holburne that the Hopyard overseer, George Gresham, forced her to work when she was sick and generally mistreated the slaves on this quarter. Holburne told Gresham that he did not believe Nancy's story, but reminded him that "when correction is necessary to give it in moderation and with no other weapon than a Hiccory Switch." Holburne sent Nancy back to Hopyard, instructing Gresham to "set this wench to her work immediately and forbare any punishment." But he soon learned that Gresham whipped Nancy "contrary to my directions," and he told Gresham, "Mr. Tayloe is informed of this and will judge of it accordingly."[58] In 1810 Gresham was replaced by a new overseer at Hopyard.

In 1801 the overseer at Forkland quarter—a man named John Cannady—suddenly disappeared in the middle of the wheat harvest, leaving thirty slave workers unsupervised. John Tayloe was outraged by Cannady's negligence, and reacted decisively. In a letter bristling with haughty disdain, he told the man that he was fired and that his pay was stopped; thus, Cannady would not receive his share of the Forkland wheat crop, which probably amounted to about forty bushels, worth $44. "Your conduct," Tayloe wrote, "has been so extraordinary in absenting yourself, particularly in the midst of harvest, that I have no further use for you as an overseer. You will therefore prepare yourself to leave the plantation tomorrow. My Waggon will be ready to convey your furniture etc. any distance within 10 miles."[59] Tayloe's wagoner was a twenty-seven-year-old slave named Barnaby, who continued to labor at Mount Airy until he died in 1825 at age fifty-one. Barnaby carted all manner of heavy goods during the course of the year—ice, coal, and wood in the winter, fence rails and manure in the spring, harvest whiskey in the summer, stone and bricks in the fall. Hauling away the overseer's furniture would be a comparatively light task for him, and perhaps an amusing one.

There were not many slack times for the Mount Airy slaves. Tayloe's managerial staff kept the workers continually busy by shifting their job assignments, and the work logs list a multitude of tasks performed by each individual agricultural and craft worker during the course of the year. In 1805 the smiths, joiners, masons, and jobbers were each given thirty to forty different assignments, and the field workers had an even greater variety of jobs. At Old House and at Doctor's Hall, the field hands labored at more than fifty separate tasks, and seldom worked for more than two or three days on any given assignment. From the slaves' perspective, the labor routine at Mount Airy was surely taxing. They were required to work six days a week throughout the year. In 1805 the jobbers worked a total of 300 days, the smiths, masons, and joiners a total of 303 days, and the field hands a total of 306 days. Apart from Sundays, the field hands had only seven days of released time: Easter Monday, one free Saturday in July, and the five-day Christmas break in December.

The field hands at Mount Airy put about 60 percent of their labor into corn and wheat production. For example, in 1805 the field hands at Old House quarter worked for 128 days on the corn crop, fifty-four days on the wheat crop, twenty-seven days cutting rails and building fences, twenty-three days on the oats crop, sixteen days on the cotton crop, sixteen days on fodder, sixteen days grubbing, eight days on hay, seven days on peas, and eleven days on other miscellaneous tasks.[60] The corn crop was much more labor-intensive than the wheat crop, and required attention every month of the year except September and October.[61] At Old House in 1805 corn tasks included plowing (ten days in February), manuring (six days in February), planting (twenty-seven days from March to May), replanting (four days in May), weeding and thinning (sixteen days in May and June), hilling (nine days in June), hoeing (four days in July), harvesting (twenty-one days from August to December), shelling (six days in December and January), measuring (five days in December), carting and hauling the cornstalks to the mill (thirteen days in December and January), beating the dried corn (five days in March), and delivering about 450 barrels of the 1804 crop to the *Federalist* for shipment and sale (two days in March).

Labor on the wheat crop was much more concentrated. Again, to take the example of Old House in 1805, since the wheat was planted in the fall and sprouted in the spring, the crop needed no attention until mid-June, when the field hands spent two days "cockling" the wheat—weeding the cockle plants that had invaded the wheat field. The wheat harvest was spread over ten days at Old House, from June 22 to July 1. The slaves then spent fifteen days intermittently from August into November treading and cleaning the harvested grain. They delivered the wheat crop for sale in two batches, loading about 600 bushels onto the *Henrietta* in mid-September and 400 bushels for the next voyage of the *Henrietta* in late November. Meanwhile, they were starting work on the 1806 crop. The Old House hands spent twenty-five days planting—a task that stretched with interruptions from early September to mid-November, when they finished seeding 258 bushels of wheat.

The other crops—oats, hay, cotton, and peas—required far less attention. In 1805 the Old House hands spent twenty-three days cultivating and harvesting the oats crop: six days of plowing and manuring in February, seven days of planting in March and April, five days of harvesting in July, and five days of treading the oats from August to November. Sixteen days were spent on the cotton crop: four days of plowing and planting in May, three days of weeding in June, two days of hoeing in July, and seven days of picking the cotton in October and November. The pea crop took only two days for planting in May and five days to pick the peas in August and September. And work in the hay field was about the same: three days to plow the fallow ground in November and five days to harvest the crop in August. A bigger annual task was repairing or rebuilding the fences that were designed to keep the cattle, sheep, and pigs out of the grain fields. Tayloe's Rappahannock farm gangs spent about a month every year on fencing. This was cold work, done mostly in January and February. And it was heavy work, for the slaves had to cut fence rails, haul logs to the sawmill, dismantle the old fencing, and set up new posts and rails.

By the standard of the day, Tayloe's field workers seem to have been exceptionally productive. Lorena Walsh has measured the per capita productivity of Chesapeake farm laborers in the early nineteenth century, and by her reckoning the Mount Airy slaves were at the top of the scale.[62] According to Walsh, the planters in the lower Rappahannock region (the neighborhood of Mount Airy) averaged ten barrels of corn per laborer in the 1730s, fifteen barrels per laborer in the 1760s, and thirty-three barrels per laborer by the 1810s. John Tayloe's Mount Airy slaves were doing better than this. In 1809, for example, the total corn crop was 5,303 barrels, and when 422 barrels are deducted as the overseers' share, the net crop amounted to 4,881 barrels. In this year Tayloe employed 143 field hands: 57 men and 40 women who were (according to Walsh's methodology) full shares, and 24 boys and 22 girls under nineteen years of age who were half shares, for a total of 120 shares. Tayloe's farm workers thus produced nearly forty-one barrels of corn per laborer in 1809.[63] And in wheat production the Mount Airy slaves performed even more strongly. In the lower Rappahannock region, Walsh reports that planters obtained about twenty bushels of wheat per laborer in the 1760s and improved to nearly fifty bushels per laborer in the 1810s. Tayloe's field hands far surpassed this level. In 1811 the net wheat crop at Mount Airy amounted to 8,010 bushels. The eight farm gangs were now smaller than in 1809 because a number of slaves had been moved to new work sites, and they totaled 102 ½ shares. By this measure Tayloe's farm workers produced seventy-eight bushels per laborer in 1811.

Lorena Walsh and her colleague Lois Carr also contend that as the Chesapeake farmers improved their wheat and corn outputs and modernized their production techniques during the period 1760–1820, female laborers were forced into increasingly burdensome work routines. The new skill jobs like plowing, harrowing, and carting were taken up by the slave men, while the drudge jobs like hoeing, weeding, and grubbing were left to the slave women. The plantation records at Monticello and Mount Vernon show that the majority of Jefferson's and Washington's field hands were women, and that the Mount Vernon field hands were divided by sex into two work groups, with the men having the skilled assignments.[64] Unlike Jefferson and Washington, John Tayloe III deliberately constructed a male majority on his farm quarters by importing men and selling women. But he very likely divided his field hands into work groups by sex as Washington did. A typical work log entry for March 23, 1805, reads: "Old House. Ploughs Fallowing all the week. Tumbrels Hauling out manure, one cart hauling post & railing for Carpenters, the other hands delivering corn to Fed[eralist], leaveling ditches, Grubing etc." Though no slave inventory survives for 1805, we can estimate from the 1808 inventory that about eight men and sixteen women and children were at work on that quarter in March 1805. Assuming that the fallow plowing was done by men, and that the tumbrels and cart were also operated by men, then the women and children must have been doing most if not all of the ditching and grubbing. But the gender division was not complete. At harvest time, as we have seen, men did all of the mowing, and women did most of the raking and binding, but Tayloe also employed several men as rakers and binders. And women at Mount Airy had somewhat more variegated employment opportunities than at Mount Vernon or Monticello, because Tayloe employed a considerable number of female cloth workers and domestics—as we shall now see.

## *Craft Labor, 1805–1828*

The differences between craft and agricultural labor at Mount Airy are significant. The craft workers were skilled (or semiskilled) artisans, whereas the field workers mainly performed drudge manual labor tasks such as grubbing and weeding and hauling manure. On the other hand, Tayloe's farmhands produced for a competitive market and used the latest agricultural equipment (such as newly designed plows) to achieve high crop yields, whereas Tayloe's textile workers employed preindustrial handiwork techniques to make cloth, a product that was readily obtained from machine-powered factories in England or New England. From the slaves' perspective, craft jobs seem to have been more appealing, or less unattractive, than farm jobs. Over two-thirds of the 110 craft and cloth workers listed in *Appendix 23* were males, who generally started their apprenticeships at about age ten and often learned their skills from their fathers and older brothers. They escaped much of the harassing supervision by white overseers that the field hands were always subjected to, though it is clear from the work logs that Tayloe's manager or his clerk was checking up six days a week on what each craft worker was doing, Monday through Saturday. Still, the craft workers could take some personal pride in their artisanal expertise.

The carpenters and the sailors were able to get away from the slave regimen at Mount Airy on a regular basis. The carpenters constructed and repaired buildings at many of Tayloe's farms and ironworks in Virginia and Maryland, and also did carpentry at the Octagon in Washington. In 1801 John Tayloe told the manager of his Neabsco ironworks in Prince William County to push the carpenters to finish their job at Neabsco: "I want them [at Mount Airy] as quickly as possible." In 1809 William Holburne sent two carpenters to Oaken Brow to build a fisherman's hut, expecting the other carpenters to pick them up shortly "on their way to the city."[65] The carpenters often traveled in a group, apparently without white supervision. The sailors were even more mobile and independent, as they plied Tayloe's schooners laden with plantation cargo down the Rappahannock and up the Potomac to Alexandria, or sailed into Chespeake Bay to Baltimore. In an average year, Tayloe's pair of cargo ships made four or five trips from Mount Airy to Alexandria or Baltimore. In 1808 there were four slave sailors based at Mount Airy: Rolly (age thirty-four), Sandy (thirty-two), Bill (twenty-four), and Sam (eighteen). The first three of these men were valued at £100, and Sam at £90. Since the sailors were constantly on the move, they appeared only intermittently in the slave inventories. But Rolly can be tracked over a long span of time; he was still listed as a sailor in 1836, when he was sixty-two years old.

John Tayloe III's spinners and weavers and shoemakers were craft workers in a seventeenth-century mold instead of a nineteenth-century mold. Tayloe resisted buying machine-made cloth imported from the English cotton and woolen factories, and instead used fifteen or twenty of his slaves to produce cloth in the old-fashioned way, manufacturing by hand. In 1808 he had two ginners, ten spinners, and two weavers. By 1818 he was employing sixteen spinners, and in 1828 there were twenty spinners, partnered with one ginner and three weavers. Tayloe's two male ginners in 1808 were sixty-two-year-old Cato, who died in 1814, and twenty-four-year-old Lame Sam, who continued in this job until 1833.[66] Using Eli Whitney's cotton gin or a variant machine, Cato and Lame Sam removed the seeds from the raw cotton fiber so that the spinners could twist the fiber into yarn at their spinning wheels.[67] The ten spinners in 1808 were all women or girls, and mostly quite young, seven of them in their teens, twenties, or thirties. Five of these women were still spinning yarn at Mount Airy into the 1830s. By 1828 the three hand-loom weavers were also female, leaving ginner Lame Sam as the only male in a twenty-four-person production unit.

The Mount Airy work logs seldom report on activities in the spinning and weaving shop, but fortunately a spinning book has survived that records weekly production during a two-year period, January 1806 to December 1807.[68] This book shows that during the course of a year the spinners spent forty-two weeks spinning cotton, four weeks spinning wool, three weeks harvesting grain, and two weeks picking cotton. The spinning proficiency of each worker is recorded week by week. Most of the women produced one and a half pounds of cotton yarn or five and a half pounds of woolen yarn per week. The fastest workers occasionally spun up to a maximum of two pounds of cotton yarn in a week, but as a group they seem to have aimed for a lower work pace of one and a half pounds per week. There were four women in 1806–1807 who consistently fell below the one and a half pound level: Sylva, who was dead or gone by 1808; Beck, age forty-eight, who was switched to a dairy job by 1808; Nanny, age forty-nine, the wife of ginner Cato, who was relegated to the laundry by 1808; and Judy, who was thirty-seven in 1806 and listed as sickly two years later. When Judy's productivity slumped to half a pound of cotton yarn per week in 1806–1807 she was characterized as "good for nothing," but she continued to spin at Mount Airy for another twenty-eight years, until 1834, when she was sixty-five years old.

The Mount Airy weavers in 1808 were Israel, age forty-three, and his daughter Jane, age nineteen. By good fortune we can learn quite a bit about their family connections, which are discussed in Chapter 7, and charted in full detail in the Yeatman family tree at *www.twoplantations.com*. Israel (who worked as a weaver until his death in 1821) was married to Franky, a house servant at Mount Airy, whose sisters Agga and Jenny were both spinners. Franky and Israel's daughter Jane was born around 1788, and she had five siblings: older brother Dick (born around 1786), who became a gardener at Mount Airy; Nancy (born around 1791), who became a spinner in 1819; Tom (born around 1800), who became a blacksmith; Fanny (born around 1805), who worked as an apprentice weaver in 1817–1819 before she became a field hand; and Eliza (born around 1807), who also started as a weaver in 1819 and switched to spinner in 1820. Thus, Jane (who worked as a weaver until 1828) came from a domestic-cum-craft family, and six of her close relatives were fellow textile workers. This pattern continued into the next generation. Jane's mate cannot be identified, but we know that she had four children: Kitty (born around 1807), who started work as a weaver in 1818; Will (born in 1810), who started work as a ginner in 1824; Mary (born in 1813), who became a spinner in 1825; and John (born in 1817), who was not yet working in 1828. But during the 1820s Jane and her children were dispersed. Kitty at age fourteen was given to Tayloe's second son, Benjamin Ogle Tayloe, in 1822 and sent to Windsor plantation, and when the Mount Airy slave force was divided up at John Tayloe's death in 1828, Jane and her two younger children were given to Henry Augustine Tayloe and sent to Gwinfield. Jane returned in 1839 without her children and worked as a spinner from 1843 to 1855, then lived in retirement with her brother Dick until she died around 1859, age seventy or seventy-one.

The weavers produced about eleven and a half pounds of cotton cloth and fifteen to twenty pounds of woolen cloth per week, but this was by no means enough fabric to clothe the slaves.[69] In June 1806, 807 ½ yards of cloth were distributed to the eight farm quarters, including 215 ½ yards of "Virginia cloth," 112 ½ yards of oznabrig linen woven from flax, which was not grown at Mount Airy, and 479 ½ yards of "brown rolls." Tayloe paid $102.75 for the brown rolls and $297.31 for the oznabrig, so only the Virginia cloth could have been made on the estate.[70] Similarly, in November 1807 the slaves received for their winter clothing 140 yards of blue plains cloth valued at £28, 719 yards of cotton cloth valued at £107 17s., and 1,211 yards of oznabrig valued at £80 14s. 8d.[71] The cotton and blue plains may have been made by the Mount Airy weavers, but not the oznabrig. Another bit of evidence comes in February 1809, when Holburne sent to the overseer of Hopyard quarter thirty-seven and a half pounds of sole leather to mend the people's shoes, and twenty yards of cotton and wool Virginia cloth for the children most in need of clothing. If this cloth was made on the estate, it is worth noting that Holburne apologized to the Hopyard overseer for the small quantity, saying that he had not another yard of cloth on hand.[72]

Tayloe's other preindustrial handicraft workers were his four shoemakers, three of whom in 1808 had the same name: Joe (age forty-one), Big Joe (age twenty-two), and Little Joe (age ten).[73] Their labor routine is particularly well documented. The shoemakers spent 95 percent of their time making shoes—though this included such ancillary tasks as dressing hides, cutting tan bark, tanning leather, cleaning the tan vats, and constructing lasts. The 1805 work log shows that in this year they logged 110 days making coarse shoes for the field workers, 48 days making custom shoes for the Tayloes' white neighbors, 36 days making shoes for the slave artisans and domestics, and 17 days making shoes for the Tayloe children. On June 19, 1805, "Joe made a pr. of shoes for Master Edward."[74] Joe was a thirty-eight-year-old African American who would continue to make shoes at Mount Airy until he died in 1829, age sixty-two. Master Edward was John Tayloe III's fourth son, Edward Thornton Tayloe—future student at Harvard College—who was two years old at the time.

The shoemakers' workshop was equipped with (among other items) two hammers, four rasps, two awls, four knives, two pairs of pinchers, and one hundred lasts.[75] The three Joes and their colleagues appear to have been quite accomplished craftsmen. During a four-month period in 1812 they turned out 231 pairs of shoes. At this pace, each of the four cobblers produced about 170 pairs of shoes per annum, or more than one shoe per working day.[76] Furthermore, their work must have been of pretty high quality. A shoemakers' shop book for 1816–1817 indicates that most of the leading planters in Richmond County ordered shoes from them. Three members of the Carter family from nearby Sabine Hall ordered twenty-two pairs of Mount Airy shoes. All in all, Tayloe's slave shoemakers made 142 pairs of shoes for white customers during this two-year period.[77] But their chief job was to make shoes for the Mount Airy slaves, as documented in November 1807, when 179 pairs of shoes were distributed to the people at Gwinfield, Old House, Doctor's Hall, Forkland, Hopyard, Oaken Brow, Menokin, and Marske. These shoes were listed at six shillings a pair, for a total value of £89 6s.[78]

The other Mount Airy craftsmen all worked in direct support of the field hands. This was particularly true of the jobbers, who were not really craftsmen but semi-agricultural laborers. Their chief work assignments were cutting ditches and constructing or repairing fences. In 1805 they spent 154 out of 300 working days ditching at Old House, Doctor's Hall, Forkland, and Marske, as well as at the Richmond County Court House, and they spent 34 days cutting rails, digging post holes, and putting up fences. They also spent fifty-seven days in 1805 on the wheat, corn, oats, and hay crops. In May they sheared the sheep, and in December and January they cut ice.

There were seven jobbers in 1808: six men—Tom (aged forty-two), Ben (thirty-eight), Urias (thirty-eight), Joe (twenty-four), Prince (twenty-three), and Jacob (twenty-two)—plus Marcus, an eleven-year-old boy. Tom died in 1809, Jacob was sent to Cloverdale ironworks in western Virginia in 1810, and Ben died in 1816, but the other four were still working as jobbers at Mount Airy in 1827, the year before John Tayloe III's death. By 1827 they had been joined for some years by three replacement workers—Thomas in 1816, Moses in 1817, and Daniel in 1821—so the size of the jobbing gang at Mount Airy was unchanged, a good example of Tayloe's ability to keep each of his specialty work groups at the size he wanted. In 1827 Urias was much the oldest of these workmen at fifty-seven, and the others ranged in age from Joe (forty-three), Prince (forty-two), Marcus (thirty), Moses (thirty), and Thomas (twenty-six), to Daniel (twenty-one), who had begun to work as a jobber when he was a boy of fifteen. In 1828, upon John Tayloe's death, the seven jobbers were portioned out among Tayloe's sons; all of them were sent to outlying family work sites, and none remained at Mount Airy.

During 1808–1828 there were two groups of woodworking artisans at Mount Airy, the carpenters and the joiners, with similar skills but distinctly different work assignments. Altogether, twenty-six men and boys worked either as carpenters or as joiners in these twenty years, with some of the carpenters frequently identified as coopers and four of the joiners switching to carpenter. In 1808 John Tayloe III employed twelve carpenters and four joiners, in 1818 fourteen carpenters and five joiners, and in 1828 fifteen carpenters and two joiners. Harry was the senior carpenter in 1808 at age forty-two. He was married to spinner Agga, and he had four daughters and four sons, three of whom became carpenters at a very young age: Michael in 1813 at thirteen, Tom in 1825 at eleven, and George in 1829 also at eleven. Harry broke his leg in 1805, and he mashed his arm in 1807, but he kept working as a carpenter at Mount Airy until he died in 1829 at age sixty-three. Many of the other carpenters and joiners (like the jobbers) put in long years of service. Nine of these craftsmen worked as carpenters or joiners at Mount Airy for the entire twenty-year span 1808–1828. A carpenter named Tom (not Harry's son) was the clear champion; he put in at least fifty-six years, since he was a fourteen-year-old carpenter's boy in 1808 and was still carpentering at Mount Airy in 1864, the final year of his enslavement, at age seventy.

The joiners spent nearly half of their time—126 days in 1805—constructing or repairing the forty farm wagons, carts, and tumbrels on the estate, or making wheels for these vehicles. In 1805 they also put in sixty-six days making and repairing plows and plow stocks, and twenty-one days making or repairing doors and gates. On four days they made coffins. The carpenters were far more perambulatory than the joiners. Every few weeks they would move to a new location, sometimes to make repairs at the Tayloes' town house in Washington, other times to do carpentry in the Tayloes' outlying plantations and ironworks in Virginia and Maryland. In 1805 Harry was the only carpenter in residence at Mount Airy for the full year. In that year the other carpenters were absent in January and most of February; they surfaced in the work log from February 26 to April 10 when they cut posts and railings for fencing at Old House quarter; they then moved to Oaken Brow on April 11, to Hopyard on May 2, to Neabsco in Prince William County on May 4, and in late May or June to the Octagon in Washington. Meanwhile, Harry remained at the carpenters' shop making and repairing cradle scythes for the wheat harvest. His fellow carpenters returned to Mount Airy from Washington on June 24 to cradle the wheat harvest, and spent July in the farm quarters adjacent to Mount Airy harvesting oats and repairing the gardeners' house; then they went back to Oaken Brow in August and from there to Nanjemoy in Charles County, Maryland. At Mount Airy Harry coopered barrels to ship the plantation corn, but he broke his leg on August 19 and was laid up for three weeks, returning to work on September 11.[79] On October 23, carpenters John and Abram and joiners James and Joe arrived from Nanjemoy, and the four of them joined Harry in dressing shingles and then reshingling the mansion house roof, a job that lasted from October 26 to December 4.

No early nineteenth-century American farmer could function without the aid of a blacksmith, and Tayloe needed multiple smiths for his large operation. He employed seven blacksmiths in 1808, six in 1818, and seven in 1828. These men ranged widely in age and experience. In 1808 James was fifty-eight years old, Lewis thirty-six, Jerry seventeen, and Jarrett only thirteen; and by 1828 Lewis was fifty-six, while newly trained Will and Enoch were eighteen and nineteen. There was considerable turnover during this twenty-year span, with three of the men moving to new work sites and two dying, but Lewis, Jerry, and Jarrett worked continuously at the forge from 1808 to 1828. The smiths performed a wide variety of tasks. An account book entry in March 1807 reads: "to smith's work for a pair of hand cuffs for a Mr. Woodford . . .

£0 4s 6d."[80] And a work log entry in 1812 summarizes activity at the smithy during the month of January as follows: "Smiths shop. Making & repairing plows for the plantations, making axes for Ditto & the Carpenters, shoeing the mules to haul wood to Mt. Airy & the different plantations in Ice, repairing their burned tools, making a new Bellows, cutting coal wood etc."[81]

The smiths' principal task was to service the plow teams on the farm quarters. During 1805 they spent ninety-six days, or the equivalent of four months, making and repairing plows. Plowing was a key element in Tayloe's system of cultivation; during plowing season his farmhands on each quarter ran up to ten plows simultaneously, so Mount Airy had to be well stocked with these implements. In 1809 over 200 plows in seven different designs were listed in the eight farm quarter inventories: 78 half share plows, 69 bar share plows, 25 skimmer plows, 12 trowel plows, 8 carry plows, 8 "small" plows, and 7 new ground plows. And there were eighty-seven oxen and twenty-eight horses on the farm quarters to draw these plows. Here again the smiths provided vital assistance; in 1805 they spent fifty-nine days making horseshoes and shoeing the horses and mules on the estate.[82] They also spent thirteen days making nails for the horseshoes (and probably for the carpenters as well), twenty-seven days repairing wagons and wagon wheels, and forty-seven days making or repairing other farm equipment such as chains, hoes, harrows, garden tools, and gridirons.

There were three mills at Mount Airy, gristmills at Hopyard and Menokin operated by the millers, and a sawmill operated by the carpenters. The millers are rather hard to trace because they were sometimes omitted in the inventories, but there were always three or four of them at work. Altogether, eight men and boys worked at the Mount Airy gristmills between 1808 and 1828, often for only a few years. The millers in 1808 were Ralph (age forty-six), Phil (age forty), and Wayman (age eighteen). Phil and Wayman were soon moved to other jobs, so that only Ralph worked continuously as a miller throughout this twenty-year period, though Phil returned to the mill in 1816 and Wayman in 1835. Ralph had a daughter named Eve who became one of the most fecund women at Mount Airy, with eleven recorded children, none of whom became millers. Wayman had a son James who worked briefly at the mill with his father in 1835–1836. The millers spent most of their time grinding grain for Tayloe's neighbors. The work log for

1805 tracks production at the Menokin mill; here the millers put in 250 days of custom work, as against 4 days spent grinding Mount Airy wheat, 24 days grinding Mount Airy corn, and 4 days installing new gudgeons at the mill. The mill was inactive during twenty days that year, partly because the water in the mill race was too low.[83]

John Tayloe III generally employed two masons at Mount Airy.[84] In 1805 the masons spent about half their time raising stone for their building projects and cutting wood to fire their brick kiln and lime kiln. In this year they also built three chimneys, which took fifty-two days, and put in nineteen days plastering and whitewashing. Two men, Billy and Adam, served as masons from 1808 to 1828, and both of them stayed on at Mount Airy when a great many of the other craft workers were taken away in 1828 by William Henry Tayloe's brothers. Billy was thirty-two in 1808. He had been hired out in 1806–1807, was hired out again in 1812, and in 1815 was sent to work in Washington, perhaps to repair masonry damaged by the British attack on the capital. But otherwise he labored every year as a mason at Mount Airy until his death in 1832 at age fifty-six. Adam had a similar career. He was seventeen in 1808, briefly became the Tayloes' jockey in 1809–1810, returned to masonry in 1811, was hired out for a year in 1812, and joined Billy in Washington in 1815; otherwise he worked at Mount Airy as a mason from 1811 to 1836 and after that as a jobber until 1842, when he became too feeble to work any longer. Adam lived in retirement for another fourteen years, dying in 1856 at age sixty-three. A third slave named Edward joined Billy and Adam in 1826 at age eleven as an apprentice mason. After working as a jobber in 1827–1828, Edward served William Henry Tayloe for thirty-three years as a combined mason/jobber until he seized the chance during the Civil War to escape from slavery—one of thirteen Mount Airy slaves who fled to the Yankees. In March 1862, at age forty-seven, Edward made his way across the Northern Neck to get to the Union lines. But his escape ended in tragedy. Having reached the Potomac, he drowned while attempting to cross the river.

### *Domestic Labor, 1805–1828*

John and Anne Tayloe's domestic servants at Mount Airy are poorly documented. Many of the household slaves show up very infrequently in the January inventories, because they spent their winters with the Tayloe family at the Octagon in Washington. Only the gardeners are recorded in the work logs. But this is a very interesting group, the Mount Airy slaves in closest personal contact with their master and mistress.

Anne Ogle Tayloe (1772–1855), the mistress of Mount Airy, supervised the domestic staff. The daughter of Benjamin Ogle, a large planter and the governor of Maryland in 1798–1801, she was as accustomed as her husband to elite social life. Brought up in Annapolis, the capital of Maryland, she enjoyed spending the winter social season at the Octagon in Washington, and she wanted many servants to take care of household chores, to attend her numerous children, to prepare meals, and to wait on her visitors. By 1808 Anne was the mother of nine children aged fifteen to two, with a tenth baby born during 1808, so she probably required extra servants to care for her younger children. The Tayloes in 1808 employed twenty-eight slaves—seventeen men and boys and eleven women and girls—as domestics at the plantation mansion house.[85] There were eleven house servants, three cooks, two dairymaids, a laundry maid, two coachmen, three grooms, four gardeners, and two women who tended the henhouse. They ranged in age from sixty-one (house servant Marcus) to thirteen (gardener John and housemaid Lucy). In addition, about a dozen domestics whom I cannot identify were based at the Octagon in Washington. The inventories tell us that a core group of the most-needed Mount Airy servants accompanied the Tayloe family back and forth between Mount Airy in the summer months and the Octagon in Washington during the winter months, and it appears that a smaller number of the most-needed Octagon servants also spent the summer months with the Tayloes at Mount Airy. So Anne and John Tayloe were always surrounded by a large number of black servitors.

In Chapter 3 we met eight of John and Anne Tayloe's personal attendants who accompanied the family back and forth between the Octagon in the winter and Mount Airy in the summer. The slave inventories of January 1, 1808, and January 1, 1809, identify them as (1) Archy (age thirty-seven in 1808), who was John Tayloe's manservant; (2) Betty (age fifty-eight), who was Anne Tayloe's senior chambermaid; (3) Harry (age thirty), who was the Tayloes' senior coachman and the grandfather of Winney Grimshaw; (4) Winney (age twenty-eight), who was Anne Tayloe's junior chambermaid and the grandmother of Winney Grimshaw; (5) Gowen (nineteen), who was the Tayloes' junior coachman; (6) Peter (thirty-one) and (7) John (twenty-three), who were house servants; and (8) Billy (eighteen), who was a cook. All eight of them then disappeared from the records for five years because they were with the Tayloes in Washington for the winter season. All except Betty resurfaced at Mount Airy in the slave inventory of January 1, 1815, because President James Madison was living at the Octagon after the British invaded Washington in 1814 and burned the White House, so the Tayloe family was spending the winter at Mount Airy. At this point four additional young servants who ordinarily lived and worked at the Octagon entered the 1815 Mount Airy inventory: fourteen-year-old houseboy Little Archy who was probably Archy's son; fourteen-year-old stable boy Henry, who was coachman Harry's son; fifteen-year-old housemaid Nancy; and sixteen-year-old housemaid Maria. Probably senior chambermaid Betty stayed behind at the Octagon to help Dolley Madison's black servants adjust to their temporary quarters, while Little Archy and the other three Octagon teenagers came to Mount Airy because there was no room or need for them in Washington.

After 1815, John and Anne Tayloe always spent the winter in Washington, and in 1817 they decided to live year-round at the Octagon. That fall they traveled from Mount Airy to the Octagon with a retinue of sixteen slaves enumerated by Tayloe as "Archy, Harry, John, Lewis, Betty, Henry, Winny, Marian, Sirah, Kitty, Godfrey, George, Travis, Hosena, John and Winny."[86] I parse this group of people as follows: (1) Archy, Harry, John, Betty, and Winny were five of the eight domestics who had been accompanying the Tayloes year-round at Mount Airy and the Octagon since at least 1808; Betty returned to Mount Airy, but the other four became permanent residents at the Octagon in 1817. (2) Henry, Kitty, Godfrey, the second John, and the second Winny were Mount Airy domestics who were sent to the Octagon in 1817; Godfrey and John, both gardeners, quickly returned to Mount Airy, and Kitty returned a few years later, but Henry (who was Winney Grimshaw's uncle) and the second Winny moved permanently to Washington. (3) Lewis, Marian, Sirah, George, Travis, and Hosena were all Octagon servants who had spent the summer at Mount Airy in 1817, but were not domiciled at the Virginia plantation.

The departure of John and Anne Tayloe greatly reduced the size of the domestic staff at Mount Airy. Twenty-eight slaves had worked at the plantation mansion in 1808, and twenty-two in 1815. But in 1818, when young John Tayloe IV and his wife Maria were put in charge of the house, there were only twelve domestics at Mount Airy: six house servants, a groom, a cook, a dairymaid, and three gardeners. By 1822, when John IV and Maria had several young children, the number of domestics expanded to eight males and seven females: five house servants, a chambermaid, a nurse, two cooks, two dairymaids, a laundry maid, and three gardeners. And there may have been more. John IV's probate inventory after his early death in 1824 shows that he owned eight slaves not listed in the Mount Airy inventories, and these people may have supplemented the domestic staff at Mount Airy.[87] By 1825, when William Henry Tayloe and his wife Henrietta had taken over the management of Mount Airy, the number of domestics was down to thirteen: six house servants, a cook, two dairymaids, three gardeners, and an ostler. William, who was passionately interested in horses, kept his main stable at Old House quarter in 1825, staffed by two ostlers. By 1828 William and Henrietta had trimmed their Mount Airy staff a little further; they now had six male and six female domestics: four house servants, a cook, a dairymaid, a laundry maid, three gardeners, a coachman, and an ostler.[88]

There was a lot of turnover on the domestic staff during these years, with about fifty slaves working at the big house or in its gardens between 1808 and 1828. Generally there were three or four child workers who stayed for a year or two and then were moved to other jobs or to the Octagon. Some of the adult domestics were commandeered by Benjamin Ogle Tayloe, John III's second son, when he set up his own household staff in 1820. Only three slave domestics—gardener Godfrey, gardener Dick, and housemaid Franky—served continuously at Mount Airy throughout these twenty years.

A Mount Airy groom and ostler named Sam was fifty-one years old in 1808; he had probably worked with the Tayloes' horses for more than thirty years, and would continue as a groom until 1816, when he either died at age fifty-nine or was moved to the Octagon. In 1808 Sam was in charge of John III's valuable string of three racehorses. Top Gallant was appraised at £600, while Pavilian was valued at £300, and

P.M. General at £200.[89] Sam himself was valued at only £70. But as a horseman he was given the same sort of independence that the carpenters and sailors enjoyed. In February 1809, manager William Holburne dispatched Sam on a journey of some 130 miles to Frederick County in the Shenandoah valley for the upcoming racing season, with a note explaining that Sam was riding Top Gallant, one of John Tayloe's very best horses. Sam had been ill, the manager said, and his health was still precarious—but he was apparently expected to spend many months racing Top Gallant in Frederick County. Sam is bringing "a full suit of blue and red dress clothes," Holburne reported, "and a linen sheet for summer wear."[90] Five years later, in October 1814, Sam was sent on another journey of sixty miles, to Neabsco in Prince William County. On this occasion he was given thirty-seven and a half cents for his expenses.[91]

The Mount Airy gardeners are much better documented than Sam and his fellow stable hands. The 1805 work log records their activities week by week throughout the year. The gardeners were able to work in the winter months because Tayloe had an elaborately heated greenhouse, valued at the very high price of £4,000.[92] In December, January, and February they mostly worked inside the greenhouse and hothouse, where they watered plants, set young trees in the nursery, prepared hotbeds, and made beanpoles and pea sticks. Outdoors they shoveled snow and helped to cut ice. In March and April they planted onions, carrots, beets, potatoes, and melons in the greenhouse, grafted apple trees, repaired the cucumber frame, and tended the asparagus and artichokes. They also dug and weeded in the kitchen garden, cleaned the gravel walks, transplanted trees and plants from the greenhouse to outdoor sites, rolled the bowling green, and dressed the flower beds. In May and June they planted more vegetables outdoors, weeded the kitchen garden, repeatedly mowed the lawn and rolled the bowling green, and helped with the wheat harvest. In July and August they repeated most of these chores, and also constructed a lime kiln, harvested vegetables, trimmed hedges, and planted cedar trees. In September they began to move plants back into the greenhouse and laid down a cistern in the garden. In October they picked apples, helped pack up the Tayloes' belongings for their return to the Octagon, prepared asparagus beds, and cleaned the hothouse flue. In October and November 1805 they dug post holes and helped put up scaffolding for the reshingling of the mansion roof. They also planted plum, apricot, and fig trees at the bottom of the garden and shrubs along the serpentine walk, tended the raspberry bushes and the strawberry bed, covered the cabbages, and prepared the kitchen garden for the winter. And at all times they carted manure in wheelbarrows and made compost.

Between 1808 and 1817 Tayloe generally employed four gardeners, two men and two boys, then from 1818 to 1828 he reduced the number to two men and one boy. In 1808 the two men were twenty-seven-year-old Godfrey and twenty-two-year-old Dick, and both of them were still working at this job in 1828. Dick was the son of weaver Israel and the brother of weaver Jane, and he tended the mansion house grounds for more than fifty years, until 1861, when he retired; he died a freeman in 1866 at age seventy-nine. The junior gardeners in 1808 were Horace, age seventeen, and John, age thirteen. Horace soon went to the Octagon, and John became a house servant; they were replaced by a succession of boys including Dick's sons George and Richard, most of whom put in short terms of service. But Richard, starting to garden in 1829 at age eleven, continued to work alongside his father for the next thirty-two years until he was moved to Alabama during the Civil War in 1862, while old Dick was left behind at Mount Airy.

The pairing of gardener Dick and his gardener son Richard, and weaver Israel and his weaver daughter Jane, show how some craft and domestic jobs were passed directly from parent to child. More broadly, the children of craft and domestic workers were usually channeled into craft and domestic employment. Here are three examples. Spinner Grace, who operated a spinning wheel at Mount Airy from 1808 to 1828, had two daughters who died in infancy and three sons with long careers at Mount Airy. Emmanuel worked as a houseboy and carpenter for twenty-five years until he died in 1852, Edward worked for thirty-six years as a mason until he deserted in 1862, and Armistead worked as a house servant for thirty-four years until he was emancipated at age forty-four in 1865. Another spinner, Elsy, had one son who died young and five other children who became craft workers or domestics: Jenny and Betsy were spinners like their mother, Thomas and Peter started as gardeners and became jobbers, and Marcus was a house servant. Our third example, dairymaid Nancy, was an exceptionally prolific mother, giving birth to eight boys and five girls.[93] Six of Nancy's children died in early childhood, two from scarlet fever in 1832. Another two of her children were sold or sent away before they reached working age. But the remaining five all had domestic or craft jobs at Mount Airy: Richard became a carpenter, Billy a cook, Becky a spinner, Martha a housemaid, and Jane a laundry maid. It is a striking fact that none of the thirteen children of Grace, Elsy, and Nancy who joined the Mount Airy workforce became field hands.

There was a parallel pattern in the eight farm quarters, where almost all of the children—if they lived to working age—*did* become field hands. At Doctor's Hall quarter, for example, there were twelve children not yet working in 1808. Three of them died in early childhood, while eight who reached working age—four boys, named John, Billy, Ralph, and James, and four girls, named Rose, Sucky, Winney, and Anny—all started as field hands at ages ranging from nine to twelve, and continued to be farm workers as long as they stayed at Mount Airy. The one exception in this group of twelve was Travers, who was moved to the home plantation in 1814 at age seven and became an apprentice joiner, the craft he pursued until he left Mount Airy in 1828. The probable reason? Travers was treated differently because his mother, Mary, was not a field hand but a domestic; she was the housemaid for the Doctor's Hall overseer.

### *The Transfer and Sale of Mount Airy Slaves, 1808–1828*

As we saw in Chapter 1, John Tayloe III manipulated his slaves to a high degree. Between 1808 and 1828, he transferred 109 Mount Airy people to distant Chesapeake work sites, and he sold forty-four Mount Airy slaves. These tactics especially affected the farm workers, because most of the slaves who were moved or sold were field hands. The most striking example of Tayloe's method—which in many ways anticipated his son's decision to send more than 200 Mount Airy slaves to Alabama between 1833 and 1862—was his large-scale transfer of Mount Airy people to Cloverdale, an iron furnace in Botetourt County in the Blue Ridge Mountains, which lay 200 miles to the northwest of Mount Airy. In 1810 Tayloe bought Cloverdale, which had been in operation since the 1780s, and between 1810 and 1827 he sent forty-nine Mount Airy slaves to work there—thirty-nine field hands and ten craft workers.

Tayloe started the exodus to Cloverdale in a big way by dispatching thirty-six people in 1810—twenty-four males and twelve females.[94]

*Appendix 25* identifies these slaves. There were thirty-one field hands from Doctor's Hall, Old House, Forkland, Hopyard, Oaken Brow, and Gwinfield farm quarters, a smith and a jobber from the home plantation, and three nonworking children. The group included Doctor's Hall Cambridge and Gwinfield Phil, who had been sent by manager William Holburne to Oaken Brow in 1809 to become seine fishermen. When Holburne was told by the Oaken Brow overseer that Cambridge and Phil needed shirts and trousers, he called them "slovenly scoundrels" for wearing out their clothes too fast.[95] Another Cloverdale migrant was Hopyard Joe, who had been whipped by Holburne for misconduct in 1809.[96] Most of the migrants were young. Seventeen were in their teens, and six of the girls were only thirteen or fourteen. These young girls were sent to Cloverdale in the expectation that they would soon mate with some of the boys. Only one conjugal family was sent to the iron furnace in 1810: Charles from Oaken Brow with his wife Milly and their three young children—Joe, Milly, and Fanny. The most unfortunate person was Nancy from Doctor's Hall. She was listed in the January 1810 inventory as living with three young children, but William and Fanny died just before Nancy was sent to Cloverdale,[97] and her ten-year-old daughter Sucky was put into field labor at Oaken Brow and sold five years later. So Nancy was suddenly childless. Bridget was in a similar plight, because her two-year-old son Daniel had died at Doctor's Hall in early 1810. All of the other migrants in this group were listed as singles, though some of the older men, like Tom, George, Prince, and Ralph, were almost certainly, like Nancy, separated from Mount Airy mates and children who were not identified as their kin in the Tayloe records.[98]

None of the thirty-six people who left for Cloverdale in 1810 ever came back to Mount Airy, and as far as the records show they never saw their parents or other family members again. Meanwhile, at their new work site, William Gordon, the Cloverdale manager, reported that he was having trouble keeping the furnace in full operation. Gordon needed more wood choppers, he told John Tayloe in August 1817, and the wood cutters and haulers that he had were ranging further in search of trees to cut down. Also, he had hired a mason to rebuild the furnace stack, which "will require a great many wagon loads of stone to rebuild it, [and] in raising the stone, will take the wood choppers from cutting wood."[99] In other words, the Cloverdale slaves were engaged in very heavy stone and timber work. They were almost certainly pressed harder than they had been at Mount Airy, because work at early nineteenth-century furnaces and forges was notoriously demanding and dangerous, with round-the-clock labor when the furnace was in blast.

A slave inventory taken at Cloverdale in December 1817 tells us that five of the males and six of the females sent from Mount Airy in 1810 were no longer living at the furnace, but the other twenty-five were heavily involved in the iron-making business. In fact, as shown in *Appendix 25*, they held most of the key jobs at the furnace.[100] The migrants from Mount Airy were valued in 1817 at about double their appraisal prices in 1810, and almost all of their ages were stated incorrectly, though seldom greatly distorted. Three of the Mount Airy migrants were woodsmen who felled, chopped up, and hauled the timber needed for the furnace; four were colliers who turned the wood into charcoal; two were wagoners who carted charcoal to the blast furnace; one was a wheelwright who serviced the wagons; two were fillers at the blast furnace; one stoked limestone into the furnace mix; one was the keeper who monitored the blast's progress; one was a gutter man who channeled the liquid iron; one was a bank hand who gathered and dumped slag; and two were smiths who worked the pig iron into bar iron.[101] Four of the women worked in the fields, as they had at Mount Airy, growing food crops for the Cloverdale iron workers.

While most of the young girls sent to Cloverdale in 1810 were no longer there seven years later, family life was definitely taking root at the furnace. In 1817, 35 of the 124 slaves at Cloverdale were nonworking young children. The only identifiable couple sent from Mount Airy in 1810, Charles and Milly, still had their three children with them. Charles held the responsible position of keeper, Milly was a spinner, son Joe was a wheelwright, daughter Milly was a field hand, and nine-year-old Fanny would soon start to work. Four of the other male migrants from 1810—Aleck, George, Peter, and Phil—were married to women they had met at Cloverdale; Prince was the father of three children, George had two, and Aleck and Phil each had one. The only recorded union within the group was Old House Prince's marriage to Doctor's Hall Rose. By 1817 Rose and Prince had three children: Betsy, born in 1812; Phyllis, born in 1814; and Patty, born in 1816. Within just seven years the Cloverdale migrants were starting to raise the next generation of furnace workers.[102]

When Tayloe sent slaves to Cloverdale and like places, he was expanding his workforce. But when he sold slaves, he was getting rid of people that he did not want. Of course, he also made money. Between 1808 and 1828 John Tayloe III sold thirty-two females and only twelve males. Tayloe seems to have preferred to sell people whom he did not know very well, because thirty-five of the slaves he got rid of came from the farm quarters and only nine from the home plantation. More than half of these people were sold at one time in 1816, when Tayloe dispatched twenty-six men, women, and children to two professional slave traders, L. Bevan and J. Snide, for $7,210—or $277 apiece—to be auctioned in the Washington slave market.[103] This transaction produced almost as much income as the annual Mount Airy wheat crop. *Appendix 26* combines all of the documentation found in Tayloe's bill of sale to Bevan and Snide with supplementary information gleaned from the Mount Airy inventory book.

It is shocking to find that a bill of sale disposing of twenty-six human beings held by a patrician Virginia gentleman was drawn up so shoddily. Four of the slaves—Celia, Rachel, Jesse, and Cornelius—are misnamed. Milly is incorrectly said to be Patty's daughter. Charlotte is not identified as Patty's daughter, nor is Jenny identified as Nancy's daughter. Reuben and Eve are not identified as husband and wife, and Eve's infant is given no name. Agga, Jesse, and Cornelius are not identified as Rachel's children, and Fanny is given no family connections. Furthermore, the stated ages of the slaves in the bill of sale often conflict with the ages recorded in the inventory book. And several of the prices agreed upon seem to be casually calculated. Milly and Winney, though "diseased in mind and body," are sold at a relatively high price, and Jesse's price is higher than would be expected for a nine-year-old girl.

Much more important, the 1816 bill of sale ignores the fact that seven Mount Airy families are being broken up. Kesiah and her five children were sold collectively to Bevan, but they were almost certainly were not resold collectively by Bevan at the Washington slave market. Patty probably also lost the four children who were sold with her to Bevan, and she definitely lost her other two children who were kept at Mount Airy: thirteen-year-old Mary and ten-year-old Saunders both stayed behind at Hopyard. Likewise Kate lost her four-year-old son Alfred, who was moved from Forkland to the home plantation and trained up as a houseboy, and Rachel lost her six-year-old son Bailor, who stayed at Gwinfield and became a field hand. Nineteen-year-old Fanny was separated from her mother, Gwinfield Kate, her brother James, and her sister Patty. Eight-year-old Agga was separated from her mother, Forkland Peggy, her brother Simon, her sister Nancy, and baby brother Christopher.[104] Most strikingly, Peter, whose textile-working wife Elsy had died in 1815, was sold without any of his five orphaned children. Jenny, a weaver at sixteen, Thomas, a jobber at fifteen, Betsy, a spinner at twelve, Peter, a gardener at nine, and little three-year-old Marcus all stayed on at the home plantation and probably never saw their father again.

Appendix 26 supplies evidence as to why Tayloe chose to sell many of these twenty-six people. Kesiah was "diseased" and past prime labor, her eldest child was likewise "diseased," and three of her other four children would not be working for some years. Patty's case was similar: a marginal nurse in her forties with a "diseased" daughter and two unproductive young children. Nancy was also considered to be weak or sickly, to judge by her low valuation, and her child was very young. Likewise Rachel was forty years old, and two of her children were not yet working. But it is unclear why Tayloe chose Celia, Reuben, Eve, Kate, Tom, and Fanny—all young workers—for sale, except that they fetched good prices. Peter is the biggest surprise. He was a favored house servant (probably a footman or butler) who had worked for many years at Mount Airy during the summer months and at the Octagon during the winter months; he was priced at £120 in 1808 and received a higher valuation (£100) in the 1816 inventory than any of the other twenty-five slaves who were put on sale. Hence it appears that Peter did something undisclosed in the records during the early months of 1816 that greatly irritated Tayloe, who punished him by getting rid of him. Of course, Tayloe also received $500 for Peter.

But Peter's case was exceptional. During the years 1808–1828 the slaves who lived at the home plantation and were known personally by the Tayloes were much less likely than the field hands to be transferred or sold. Of the twenty-eight house servants, stable hands, and gardeners employed at Mount Airy in 1808, only four were transferred to other work sites between 1808 and 1828, and only Peter was sold. The male craft workers were likewise seldom exiled from Mount Airy. Of the twenty-nine shoemakers, blacksmiths, joiners, masons, and carpenters listed in 1808, only four were sent away during John Tayloe III's lifetime—three of them to Cloverdale—and none were sold. Transfers and sales were far more frequent on every farm quarter, but the field hands who lived on the more remote quarters—Gwinfield in Essex County, or Oaken Brow and Hopyard in King George County—had a better chance of escaping exile. At Doctor's Hall, which was close to the home plantation, only twelve of the thirty-four slaves living there in 1808 were still present in 1828; seven had died, and fifteen (44 percent) had been moved or sold. At Gwinfield, by contrast, twenty-three of the fifty-two slaves in 1808 were still present in 1828, while fourteen had died, and fifteen (29 percent) had been moved or sold.

Assessing John Tayloe III's management of his Mount Airy plantation and his Mount Airy slaves between 1791 and 1828, we find a mix of the old and the new. He sustained and solidified the traditional Chesapeake elite social order while participating vigorously in the post-revolutionary economic development of the new republic. As a slave-holder he used his African American field hands to produce large cash crops, while retaining a full third of his slave workers as domestics or artisans who provided him with plenty of creature comfort but little or no added income. Most alarmingly for the African Americans under his control, Tayloe took full advantage of the population growth on his estate by selling over 150 of his surplus Mount Airy slaves (most of them in 1792–1794) and by moving another 150 to new work sites. Meanwhile, he benefited his children by rejecting the traditional Virginia primogeniture inheritance system with a vengeance, and carved up his property among his seven sons.

## *The Slave Division of 1828*

John III's slave sale in 1792–1794 and the division of his estate among his sons at his death triggered the two biggest shake-ups in the history of Mount Airy slavery. Tayloe had started to plan for the future as soon as his first three boys were born in the 1790s, and he gradually acquired enough property so that he could endow all of his sons with land and slaves, and all of his daughters with dowries. As early as 1817 John and Anne decided to stay year-round at the Octagon, and in 1818 John turned the management of Mount Airy over to his oldest son, John Tayloe IV. But John IV died unexpectedly in 1824. At this point an awkward problem developed because neither second son Benjamin Ogle Tayloe nor third son William Henry Tayloe wanted to live at Mount

Airy. Correspondence between Benjamin (known as Ogle) and William in the summer of 1824 indicates that both brothers considered the four Richmond County farms that their father wished to assign to the inheritor of Mount Airy to be worn out and unproductive, and Ogle in particular saw the grand family mansion as too big and old-fashioned for his taste. William was living at Windsor plantation in King George County in 1824, and wanted to stay there, but since Ogle got first pick he reluctantly agreed to move to Mount Airy and assume John IV's inheritance.[105]

Four years later John Tayloe III died at the Octagon on March 23, 1828. In addition to the Mount Airy family seat, he left an imposing estate to his many heirs: twenty-two farms in Virginia, Maryland, and the District of Columbia; four ironworks; schooners on the Rappahannock and the Potomac; the Octagon, the Mansion House Hotel (said to be worth $100,000 with slave attendants and furnishings included), and much other real estate in Washington, D.C.; and three small hotels and a tavern in the Virginia countryside.[106] He also left a huge number of slaves—378 at Mount Airy and somewhere between 700 and 800 altogether.[107] Tayloe's six surviving sons divided up his plantations, farms, and slaves according to the terms of his will. Since John IV had died leaving a young heir, John V, this boy was given two plantations in Maryland. Ogle took Windsor and two other Potomac estates in Virginia and a plantation in Maryland. William received the Mount Airy family seat together with four adjacent Richmond County farm quarters: Old House, Doctor's Hall, Forkland, and Marske. A fifth Richmond County farm, Menokin, was put up for sale, and the more distant Rappahannock farms were distributed to fourth son Edward Thornton Tayloe, who took Hopyard and Doeg; sixth son Henry Augustine Tayloe, who took Gwinfield; and seventh son Charles Tayloe, who took Oaken Brow. Cloverdale went to fifth son George Plater Tayloe, and lifetime use of the Octagon went to their widowed mother, Anne Ogle Tayloe (who lived to 1855), with twelve slaves. Each son received the stock, slaves, and utensils belonging to the properties they inherited. The will specified that the Mount Airy domestic staff should remain with son William, but the slave craftsmen and mechanics at Mount Airy, Neabsco, and Cloverdale were to be divided six ways. And in a final provision the testator freed his favorite slave, Archy, who was fifty-seven years old.

John III declared: "I will that my body servant Archy may be liberated and may be allowed one hundred dollars per annum during his life. My motive for liberating him is his long true fidelity especially since I have been in bad health, and upon one occasion he was the means under the direction of providence of saving my life."[108]

The thorniest problem was how to divide up John III's slave craftsmen and mechanics. The six Tayloe brothers assessed the individual value of fifty-three carpenters, blacksmiths, wheelwrights, joiners, masons, shoemakers, jobbers, and sailors who were attached to Mount Airy, Neabsco ironworks in Prince William County, and Cloverdale ironworks in Botetourt County. At the time of John III's death twenty-six of these men were living at Mount Airy, twenty at Neabsco, and seven at Cloverdale, and the Tayloe brothers reckoned that they had a collective value of $15,100, which meant that each brother should get eight or nine craftsmen valued at about $2,500. The brothers made their selections in order of seniority: Ogle chose eight craftsmen (six from Mount Airy) priced at $2,550; William chose nine (seven from Mount Airy) priced at $2,450; Edward chose seven from Neabsco and two from Mount Airy priced at $2,550; George chose seven from Cloverdale and two from Mount Airy priced at $2,600; Henry chose nine (six from Mount Airy) priced at $2,500; and Charles chose six from Neabsco and three from Mount Airy priced at $2,450.[109] William thus received seven out of the twenty-six Mount Airy craft workers, but he also owned ten younger craft workers who were not included in this division, and he bargained to keep five of the Mount Airy craftsmen chosen by his brothers, sending them seven other slaves in exchange. In the end, six Mount Airy carpenters, a joiner, three blacksmiths, a shoemaker, seven jobbers, and three sailors went to William's brothers.

From the slaves' perspective, the division of Mount Airy in 1828 must have been deeply distressing, especially for the wives and children of the twenty-one craftsmen and mechanics who were removed from the home plantation by William Henry Tayloe's brothers. Though the farm families at Old House, Doctor's Hall, Forkland, and Marske had not been broken up, the numerous children from these families who had been sent to Hopyard, Gwinfield, and Oaken Brow in previous years were now under the control of Edward, Henry, and Charles Tayloe, new young masters with differing agendas. In the Mount Airy inventories, nearly half of the slaves listed in January 1828 suddenly disappeared, and few of these people ever resurfaced in the Mount Airy records.

The new owner of Mount Airy, William Henry Tayloe (1799–1871), did better than any of his brothers except Ogle in the division of 1828. He received the family seat with four farms, some 3,850 acres of Richmond County land, and 193 slaves. But Mount Airy was now a scaled-down agricultural operation. The best Rappahannock farms went to William's brothers in the division of 1828: Gwinfield, Oaken Brow, Hopyard, and Doeg collectively had more than double the acreage of William's farm quarters and supported considerably more slave workers.[110] And twenty-one craft workers left Mount Airy. But William ended up with more than two dozen tradesmen and mechanics plus a cohort of textile workers. After the division of 1828, 82 people were living at the home plantation and 111 on the four farm quarters, so the proportion of domestics and craft workers with their children had swollen to 42 percent of the total population as against 28 percent back in 1809.

By 1829 William Henry Tayloe's establishment had an old-fashioned look. He had taken advantage of his expanded workforce to enlarge his household staff from twelve to seventeen black servants. At age twenty-nine he lived with his wife, Henrietta Ogle Tayloe, and two young children in his ancestral sandstone Palladian mansion, now about seventy years old, and he employed eight house servants, a cook, two dairymaids, a coachman, an ostler, a stable boy, and three gardeners to take care of the family, the house, the horses, and the gardens and grounds. Many of the enslaved domestics had served him at the plantation house ever since he was a child. Coachman Harry, who was fifty-one years old in 1829 and had worked in William's father's stable from 1808 or earlier to 1824, was now the manager of the Mount Airy coach house and stable. The chief housemaid, sixty-three-year-old Franky, who had worked in the big house for over forty years, was at the center of an extensive web of domestic and craft workers. Franky was the widow of weaver Israel and the mother of gardener Dick, spinner Nancy, and house-maid Eliza—all of whom had held these job assignments for many years. William also retained his father's spinning and weaving shop, which in 1829 was staffed by two ginners, six spinners, and one weaver. And in the workshops adjacent to the mansion he employed six carpenters, two sawyers, four joiners, four blacksmiths, three masons, and three shoemakers. His mill was operated by two millers, and his schooner was manned by four sailors.[111]

Lest this description conjure up a romantic image of antebellum moonlight and magnolias, it should be emphasized that William Henry Tayloe was determined to put his slave force to more productive use. Like his father in 1791, he had more field hands and craft workers than he needed to operate his reduced farming system. Furthermore, his slave population in 1828 was exceptionally youthful and expansive. More than half of the people who remained with him at Mount Airy after the division were under the age of eighteen, and the proportion of nonworking young children, which had been 32 percent in 1809, had climbed to 36 percent in 1828. William Henry Tayloe could look forward to an increasing supply of slave workers, with no obvious way of employing these people gainfully in Richmond County, Virginia. His solution to this problem is the topic for Chapter 7.

# 6

# *The Moravian Christian Community at Mesopotamia*

ONE OF THE MAJOR DIFFERENCES between U.S. and British Caribbean slavery is that the North American slaveholders permitted and even encouraged their slaves to join the Christian church from the late eighteenth century through the Civil War, while almost all of the slaveholders in Jamaica and the other British islands strenuously opposed any program of religious instruction for their "heathen" black workers. The Anglican clergy in Jamaica sided with the slave masters, since they supposed that Christianity was for whites only. As late as the 1830s most of the Jamaican planters were outraged because Baptist and Methodist missionaries were daring to spread the Gospel message to the Afro-Caribbeans. Yet, paradoxically, the slaves at Mesopotamia during the ownership of the two Joseph Foster Barhams had access to a Christian church from 1758 onward. This was because the elder Barham was among a small handful of mid-eighteenth-century Jamaican proprietors who believed that the slaves on the island should be converted to Christianity. Barham was an ardent member of the Unitas Fratrum or Moravian Church in England, and in 1753 he and his brother William Foster (another ardent Moravian Jamaican proprietor) urged the Moravians to send missionaries to proselytize among their slaves. The first three missionaries came to Jamaica in 1754, and Mesopotamia soon became one of their principal sites. The Moravians established a station at Mesopotamia in 1758, built a chapel for the slaves, and gathered a congregation of baptized members. With continuous support from the two Barhams, the Moravians sustained an active slave congregation at Mesopotamia for seventy-eight years, until Joseph II's son John Barham cut off all aid to the mission in 1836.

The Moravian creed was highly distinctive. The Unitas Fratrum (with headquarters in Saxony, close to the present German border with the Czech Republic and Poland) practiced a deeply felt religion of the heart: they loved and adored "Our Saviour Jesus Christ" by identifying emotionally and physically with his blood, his wounds, and his sufferings on the cross. And during the eighteenth century they proselytized among non-Christians more vigorously than any other Protestant church. They sent missionaries to the African slaves in the Danish West Indian island of St. Thomas, to the Eskimos in Greenland and Labrador, to the Indians in Pennsylvania, to the African slaves in the English West Indian islands of Antigua, Barbados, and Jamaica, to the Indians and Africans in Dutch Surinam, to the Africans in Guinea, and to the Hottentots in Dutch South Africa.[1]

Aaron Fogleman stresses the radical character of eighteenth-century Moravian pietism: their communal living arrangements and their belief in gender equality, in an androgynous interpretation of the Trinity, and in a sensuous relationship with Jesus, illustrated by Moravian depictions of the Saviour's side wound at his crucifixion in the shape of a vagina. Fogleman also shows how the Moravians in the 1740s and 1750s tried very aggressively to recruit members from the Lutheran and Reformed churches in Germany and British North America (especially Pennsylvania), inciting violent hostility from more orthodox Protestants on both sides of the Atlantic.[2] But the Moravian approach to mission work in Jamaica was very different. Here they did everything possible to placate the masters of the plantations. The missionaries never challenged the slave system or pressed for emancipation. They seldom overtly criticized the management tactics of the attorneys and overseers who routinely flogged the men and raped the women and girls. They preached a doctrine of passive obedience and urged their converts to accept their miserable earthly lot with the promise of future heavenly rewards.

During the seventy-eight years that the Moravians maintained a mission at Mesopotamia, they baptized approximately 290 Mesopotamia people "into the death of Jesus." Thus, they converted about a third of the adult slaves on this plantation to Christianity. The missionaries kept a church catalogue listing the membership of the Mesopotamia congregation, but this catalogue has disappeared. Without the catalogue it is rather difficult to reconstruct the total membership, because the missionaries when baptizing slaves gave them new Christian names to replace the book names in the slave inventories. Sometimes the missionaries referred to the converts only by their book names, and sometimes only by their Christian names. Fortunately, 222 of the baptized Mesopotamians can be identified by both names, and another 41 have inventory names but unrecorded Christian names. These two groups add up to 263 known Mesopotamia church members. In addition, twenty-four people who were baptized between 1770 and 1817 are identifiable only by their Christian names (probably overlapping a lot with those identifiable only by their inventory names), and a further sizable number must have been baptized during the years 1770–1797, 1819–1829, and 1833–1835, when there are big gaps in the Moravian records. This brings the membership total to about 290. A significant number to be sure, but we need to remember that the majority of adults at Mesopotamia did *not* join the Moravian Church. Resistance to Christian conversion is just as instructive as the baptismal rate.

The Moravian missionaries at Mesopotamia—as elsewhere in Jamaica—operated under great handicaps during the slave era. They offered the hope of heavenly salvation, but not what the slaves most wanted, which was earthly improvement. They were constantly hindered and harassed by the attorneys and overseers who operated the estate. And they had serious difficulty in communicating with the slaves. For many of the Mesopotamia people (those born in Africa), English was their third or fourth language, and for most of the missionaries (those born in Germany), English was at best their second language. The slaves who understood what the missionaries were telling them may well have been baffled by the Moravian belief in mystical union with the blood and wounds of Christ, as well as by the eccentric Moravian practice of selecting candidates for baptism and Holy Communion by lot. The missionaries, for their part, were imbued with racial superiority, like almost all European whites, and they disdained the Mesopotamia people's alien beliefs and practices—their "heathen" rituals, their African dancing and drumming, their polygamous marriages, and their sexual intercourse with multiple partners—which the Moravians saw as depraved and uncivilized. The cultural gulf was deep.

J. H. Buchner, a missionary who published the first history of the Moravian evangelizing effort in Jamaica in 1854, considered the Mesopotamia station to be a dismal failure. "This place," he reported, "lies low, in a most unhealthy situation, so that fever and death have made sad havoc in it." Fifteen missionaries and some of their children died there, which was "a waste of valuable life," because the slaves "had no confidence in the missionary, and did not desire his instructions." Buchner claimed that there were never more than forty or fifty baptized Negroes in the Mesopotamia congregation. And he blamed the absentee proprietors, the two Joseph Foster Barhams, for unjustly accusing the missionaries of neglect, while the Barhams' white managers were always in "constant systematic opposition."[3] Many (but by no means all) of Buchner's claims are perfectly true. But if one looks at the situation from the slaves' point of view—as I hope to do in this chapter—then one can see that many of the Mesopotamia people were powerfully attracted by the hope of salvation and deeply affected by their conversion to Christianity.

As they tried to evangelize, the Moravian missionaries left a richly detailed record of their interactions with the Mesopotamia slaves. Here the contrast with Mount Airy is stark, because the Tayloe Papers contain almost no information about the religious beliefs and practices of the slaves the Tayloes held in Virginia or Alabama. We learn that some of the people at Mount Airy attended the Nomini Baptist church, and that William Henry Tayloe, a devout Christian who prided himself on promoting the "religious tendencies" of his slaves, built a church on his Oakland cotton plantation in Alabama, presumably for services conducted by a white minister. But this tells us very little. By contrast, between 1758 and 1832 the missionaries at Mesopotamia sent thirty-eight diaries, most of them in German but some in English, to the church headquarters in Saxony, where they escaped destruction during World War II and are today stored in the Archiv der Brüder-Unität in Herrnhut.[4] Further Moravian diaries and other church records in the Jamaica Archives in Spanish Town contain additional information about the Mesopotamia mission.[5] The Moravian diarists comment extensively on slave life at Mesopotamia, and they sometimes quote the opinions and arguments of individual baptismal candidates and congregants. The Moravians thus provide the best surviving evidence about what the slaves were thinking and doing at Mesopotamia from the 1750s to the 1830s.[6]

## *The Beginnings, 1754–1758*

On October 5, 1754, three members of the Unitas Fratrum—Brother Zacharias George Caries, Brother Gottlieb Haberecht, and Brother Thomas Shallcross—embarked at London on a ship to Jamaica.[7] They were the first Moravians—indeed, the first clergymen of any denomination—to preach the Gospel to the slaves on this island. Brother Caries, who kept a diary, described their elaborate and emotional leave-taking. In the morning at Disciple House they parted with their London brethren, then went to dine with Joseph Foster Barham and his wife Dorothea. Barham, a pious young man of twenty-four, had married Dorothea Vaughan a few months before, and she had brought him into the Moravian Church. Having spent a year in Jamaica in 1750–1751, Joseph Barham had pressed the Moravian brethren in August 1753 to attempt the relief and enlightenment of "those poor souls, the Negroes" in Jamaica.[8] In May 1754 he had offered to give the Brethren 300 acres of land and a house.[9] Now he was sponsoring the Moravians' expedition jointly with his older brother William Foster. The two Foster brothers accompanied the three missionaries to Gravesend, where their ship was waiting. Arriving about 7 P.M., they exchanged tearful exhortations for several more hours. At 2 A.M. the missionaries boarded their ship, and Barham and Foster stayed with them for another hour, helping to put their cabin in order, before finally bidding farewell.

After a two-month voyage across the Atlantic, the three missionaries reached Kingston harbor on December 7, 1754. Here Brother Caries saw a great slave ship filled with Africans from Guinea. Twenty of these Africans were transferred to the Moravians' ship to be taken to Savanna la Mar for sale. They spoke no English, but told an old black woman who understood their language that they expected to be killed. When she said that they were mistaken, they rejoiced, and Caries was pleased to see that the Africans looked healthy and were decently clothed. But on disembarking at Black River he encountered white gentlemen who cursed at him because he wore clerical dress. On December 18 the missionaries reached their destination: the Bogue estate, a sugar plantation in St. Elizabeth parish owned by William Foster, with 400 slaves. The managerial staff (mostly Scotsmen) at the Bogue was far from pleased to see them. At his first service Caries preached to a dozen suspicious whites in one room while fifteen blacks sat segregated in an adjoining room listening through an open window. He tried to preach on Christmas Day, but found that the slaves—who had been given rum and herring to celebrate the holiday—preferred to drink and dance. He tried to preach on Sundays in January 1755, but the slaves told him that, with no free Saturdays during crop time, attending church took precious time away from the cultivation of their provision grounds.

Brother Caries soon found that the Africans had their own strong non-Christian belief system. On January 7, 1755, he attended the burial ceremony for an old man, where the slaves gathered around a great fire, danced to the beat of drums made from a hollow tree, and sang, "Now he is gone out of this World, and has left all sorrows behind! Come and bury me also!" Then they sacrificed a sheep and put brandy and victuals into the dead man's grave to support him until he returned to his own country. Caries also immediately had major difficulty in communicating with the slaves. When a crippled woman told him, "Me pray to God," he asked, "Who is your God?" She replied, "Me no savy Master, you savy." And when he asked, "How do you pray?" she began to bow, singing without words, and often laughing. But another slave told Caries that while he couldn't understand all of his sermon, "what I did understand, did me good here (pointing to his heart)."[10]

When the missionaries set forth for Jamaica, they were cautioned not to baptize the slaves too quickly, because the Moravians held strict standards for conversion. As Bishop Benjamin LaTrobe explained in 1771, "The Missionaries cannot look upon the conversion of a Heathen as real, unless he has felt the want of a Saviour, and his heart is truly directed to Him who . . . as a Man lived, suffered, was wounded, bled and died, to deliver sinful men from their sins and eternal destruction." Thus, says LaTrobe, the Moravians "proceed not to Baptism, until they discover in the hearts of the candidates a real work of the Holy Ghost, and a desire to become obedient to the Gospel."[11] But Brother Caries *did* proceed quickly at the Bogue. Everything seemed to be falling into place for him. The attorney who managed the estate declared that his preaching was less foolish and fanatical than had been expected, and urged the slaves to listen to his sermons. Attendance at church services picked up steadily. The Africans learned to sing Moravian hymns and to sit in separate groups or choirs of men, boys, women, and girls. Caries told them that if they were saved and went to heaven, "You will be no more Blacks, no more Slaves." And they chanted in response, "Yes

Master." By April 1755 the slaves at the Bogue had constructed and whitewashed a simple chapel, and Caries was holding special classes for baptismal candidates. On April 27 (four months after his arrival) he baptized two of these candidates, and by July he had baptized another eleven.[12] Brother Shallcross died in October 1755, the first of many Moravian newcomers to expire in Jamaica. But Caries and Haberecht pressed on. By December 1756 they had brought a total of sixty-nine slaves into the Moravian Church at the Bogue. All of these converts were males. Many of the female slaves also wished to be baptized, but Caries felt that this step could not be taken until some female missionaries were sent to Jamaica to work with the women. As it was, within two years of their arrival in Jamaica, Brothers Caries and Haberecht had incorporated about half of the adult male population of the Bogue estate into their congregation.[13]

Katharine Gerbner, who has studied the initial baptismal pattern at the Bogue, argues that the slaves were the chief motivators for this quick success. "They call me Obea," Caries reported, "which supposedly means Seer, or one who is able to see things in the future." The Bogue people saw this white priest as an obeah man, an African folk figure with magical powers who could not only foretell the future but give them protection. Hence, when Caries and Haberecht baptized the first few men at the Bogue, the other slaves began (as Caries put it) to "overrun us, begging for baptism." Gottlieb Haberecht, worried about keeping the candidates waiting too long, urged Caries to ask the Saviour whether they should think of baptizing more people. Moravians used the lot to discover the Saviour's wishes, so Caries and Haberecht placed two papers in a bowl or a tube, one bearing a message favoring more baptisms, and the other a message calling for delay. After mixing these papers, they drew out the winner, which was positive; the Saviour was in favor of more baptisms.[14] But the rapid gains that were achieved in 1755–1756 suddenly slowed down in 1757 and then ceased. Only one slave was baptized at the Bogue in 1758, none in 1759.

A main reason for this falling off was that the new converts soon discovered that their Moravian "masters" were ineffectual obeah men. When there was a terrible food shortage in 1756 and some slaves died of starvation, the missionaries could do nothing to alleviate the situation, nor could they protect the Bogue people from continuing gross abuse by the white management. Clearly, also, Caries had moved much too fast, baptizing people before they had comprehended the Christian belief system, so that their conversion was superficial. Caries made a further mistake in bringing only men into the Bogue church, thereby distressing the women and failing to establish a Christian family lifestyle among the slaves. A pair of experienced Moravian missionaries, Christian and Anna Rauch, who arrived at the Bogue in December 1756 after working with the Indians in Pennsylvania, criticized Caries's conversion technique. Both Caries and the Rauches were distressed to find that some of the male church members were quickly "backsliding" into heathenish behavior. So the Moravian community on this estate rapidly disintegrated. By 1760–1761 no congregants showed up at many of the scheduled meetings, and it took a decade to revitalize the Christian community at the Bogue.[15]

Meanwhile, in 1755 Joseph Foster Barham and William Foster had set up a house and farm for the Moravians at Carmel, near the Bogue, and supplied them with a cadre of four slaves to grow food and to wait upon them. Caries wrote home that he was passing ownership of the Negroes, horses, and sheep that the Foster brothers had given him to the Moravian leader Count Nicolaus Ludwig von Zinzendorf, because "I want to be poor in this world." But he evidently saw nothing wrong with the Moravians becoming slaveholders.[16] Operating from their base at Carmel, the missionaries began to baptize slave converts at four other large sugar estates in St. Elizabeth parish in the mid-1750s. Three of these estates—Elim, Lancaster, and Two Mile Wood—were owned by William and Thomas Foster, and the fourth (the Island) was owned by Joseph Foster Barham.[17] But Barham also wanted the Moravians to start work at Mesopotamia, which lay some fifty miles to the west of Carmel via very bad roads and rugged terrain. Having joined with William Foster to underwrite the missionaries' expenses, Barham had spent £1,650 (or so he claimed) to set them up at Carmel.[18] And in 1758 the Moravians complied with his wishes by starting an exploratory mission at Mesopotamia.

Zachariah George Caries made the first foray. Back in July 1755, when traveling in Westmoreland, he had encountered five of the Mesopotamia people, who asked him "whether I was not the Parson who was lately arrived from England and whether I should not come soon to them and preach to them."[19] Now, three years later, he was doing so, and trying to avoid the mistakes he had made at the Bogue. Between

October 1758 and October 1759 Caries traveled across the mountains from St. Elizabeth parish ten times, at monthly intervals, to stay at Mesopotamia for a few days. He gathered about a dozen potential candidates and instructed them in small groups, without baptizing anyone.[20] Things looked promising enough so that when Caries went back to Europe, he was replaced by three Moravian missionaries: Christian Heinrich Rauch, Anna Rauch, and Nicolaus Gandrup. This trio moved into the newly built Mesopotamia mission house on November 29, 1759.[21]

So began a long line of Moravian brothers and sisters who ministered to the Mesopotamia slaves between 1758 and 1836. The missionaries who chiefly served at Mesopotamia are all listed in *Appendix 27* together with the diaries that they kept, recording their efforts to reach the slaves. The list shows that resident pairs of missionaries lived on the plantation almost continuously from 1759 to 1796, with short breaks of several months whenever a Moravian couple left or died until their replacements could move in. Four more pairs of missionaries served between 1798 and 1819. Then came a cessation in the 1820s, before the mission was briefly resuscitated in 1830–1835. During the seventy-eight-year history of the station, thirty-nine missionaries took up residence on the estate for months or years, while many others traveled over from St. Elizabeth for short visits of a few days or weeks. Altogether, as many as fifty Moravian brothers and sisters may have come to Mesopotamia. Fortunately for the historian, fifteen of the resident Mesopotamia missionaries kept a total of thirty-eight diaries, which collectively illuminate the history of the mission. The longest and most informative of these diaries were the earliest ones, when the evangelizing effort in Jamaica was just starting, when the missionaries were breaking new ground, and when the Moravians were particularly interested in sizing up the black people they were trying to reach. The later diaries tend to be shorter and more routine, focusing as much or more on the diarists' personal problems as on their converts' spiritual condition. But the final two surviving diaries by Peter Ricksecker provide crucial information about Mesopotamia's role in the 1831–1832 slave revolt. And three pairs of the Mesopotamia missionaries provided valuable documentary evidence about the process by which slaves were chosen for baptism. Between 1798 and 1818 the Browns, the Jacksons, and the Gründers kept a continuous record of their prayer sessions with the Saviour, which they entitled the Minutes of Mesopotamia Conference. This unusual document will be inspected later in this chapter.

### *Matt the Driver: Portrait of a Convert*

The years 1758–1762 were the building-block years of the Mesopotamia mission. Brother Caries made the opening visits in 1758–1759, and the Rauches and Brother Gandrup expanded on his work while they lived among the slaves at Mesopotamia from 1759 to 1762. These initial missionary efforts are well documented, because Caries and Rauch were both excellent diarists. Writing at length in German, they described the opening years of the Mesopotamia mission incisively and in exceptionally informative detail. Caries's diary entries from January to November 1759 are 137 pages in length, and Rauch's diary for November 1759 to December 1761 fills 180 pages. No diary for 1762 has survived. By 1763 Rauch had left Mesopotamia, and his successors kept much briefer records. The three Mesopotamia diaries for 1763, 1764, and 1765 total only fifty-one pages.

Both Caries and Rauch dealt cautiously with overseer Daniel Macfarlane, with Daniel Barnjum, who succeeded Macfarlane in December 1760, and with "Parson Pool," who was the Reverend John Pool, the rector of Westmoreland parish and also Joseph Foster Barham's attorney. This Anglican clergyman must have disturbed the Moravians, because he was certainly no apostle to the Mesopotamia slaves. On the contrary, he was a businessman who profited handsomely from slavery. He received an annual salary of £200 (four times the Moravian missionaries' stipend) for his services to Barham. Pool not only purchased new Africans for Mesopotamia in the 1750s and 1760s but made extra money by hiring some of his own slaves to work at the estate.[22]

During 1759–1761, Caries and Rauch took careful note of the slaves they were trying to convert. They recorded interesting conversations with Frank, a carpenter who was forty-eight years old in 1758; his wife Affraw, a forty-seven-year-old field hand; Ammoe, a forty-one-year-old carpenter; Augustus, an eighteen-year-old distiller; James, a twenty-seven-year-old driver; and Warwick, a twenty-seven-year-old invalid, among others. But by far the most arresting figure to emerge from the early diaries is Matthew or Matt, the Jamaica-born head driver who was about forty-seven years old in 1758. Since his job was to coerce and discipline the first gang field hands, and he was trusted by overseer Daniel Barnjum, Matt was the key black man on the estate.

On a Sunday in October 1758, the only free day of the week for the Mesopotamia slaves, when Caries first visited their huts and sang verses to them, he immediately picked Matt out. "Matheus understood me well," he wrote, "and I felt that most of his words were those of a brother, and he had a welcoming look." Matt told Caries that until the Moravians arrived, the slaves had lived in sin without spending any thoughts on their souls, so he couldn't thank the Saviour enough for sending them. "Matheus kept on saying 'that he wished the Saviour might bless us [the missionaries] for loving their souls, and he wished the same for his master Brother Barham.' "[23] Early on the driver also came to Caries's assistance. The missionary, as he tried to instruct his initial baptismal candidates, became concerned that they were "backsliding" into heathenish practices just as the Bogue slaves had done. So in February 1759 he spoke sharply to his Mesopotamia auditors about their bad behavior. When he returned to the estate on the Sunday after reproving them, Caries found that the candidates for baptism were all waiting for him in Matthew's hut. And it was Matthew who undertook to address him on behalf of the group. Matthew asked forgiveness for their errors and negligence, and protested that they loved their meetings with Caries and that their hearts were really attached to the Saviour. "I increasingly like Matheus," Caries observed three months later. "He is frank and speaks in a language that comes from the heart."[24]

When Christian Heinrich Rauch replaced Caries, he reported on his first Sunday at Mesopotamia in December 1759 that he had observed "one who is called Mathew," and described him as "an understanding Negro." Rauch had been at Mesopotamia since Thursday, so he had witnessed the raucous behavior of the Mesopotamia people on their one holiday night—Saturday night—and noticed that Matthew was among the many slaves who were feasting and dancing. On Sunday he reproved him for this bad behavior. But Rauch also sensed that the head driver was his most likely ally among the slaves, and on the next Saturday he "had a blessed conversation in his one-hour lunchtime break" with Matthew. On the following day (Sunday, December 9) Brother and Sister Rauch designated Matthew as their assistant or "helper" in spreading the Gospel to the slave population. "He organizes people very well," said Rauch.[25]

Matt lost no time in speaking out against the whites ("Weisse" or "blanken") who operated Mesopotamia. In December 1759 he told Rauch that "he had often overheard white people talking about God, yet they always had rejected Him in a rough manner." This led him to wonder about "those people who have learned a lot and who could read in the Bible and find there the will of God, and who yet lived so godless that he thinks that they cannot be real Christians." And Matthew "complained too that the whites talked very bad, and that they were very mad at us, that the Negroes (the depraved seed of Cain) wished to become converts." Matt declared that he had believed in the existence of God since childhood, and he peppered the missionaries with theological questions. Confronted with the Moravians' overwhelming focus on Jesus Christ as the redeeming Saviour, he asked Rauch if the Saviour had a wife and children. Also, having heard that the Saviour had a father, Matthew wanted to know what this father was called in the Bible. And if the Saviour had a father, "he must have been created. And that is not comprehensible for him; how can he then be called Creator if he himself had been created?" On another visit, having learned more about the Holy Trinity, Matt asked "whether God would be willing to take a wife and to sire some children with her, and whether it would be against the will of God to take two and three wives." Here Matt was explicitly raising the troublesome issue of polygamy, which the Moravians particularly objected to. Rauch reports that he answered Matt in kindly language "appropriate for a heathen," as he put it. But the missionary was probably disturbed by Matt's quick interest in multiple wives, and he clearly thought that the driver was asking too many questions. One evening Matthew and the young distiller Augustus visited Rauch and "listened to him with utmost interest; and when Christian wanted to break off, as it was already in the eleventh hour, they still kept on coming up with new questions, so that Christian had to answer them until it was twelve o'clock at night."[26]

During 1760 Matt continued to visit Christian and Anna Rauch often, usually on Sundays, sometimes alone and sometimes in company with carpenter Frank or driver James. One evening Matt and James got Rauch to play some verses on his zither for them. Matt and his wife Love (a forty-nine-year-old field hand) brought presents to the missionaries: grain for their chickens, yams, and coco seeds to plant in their fields.[27] Matt kept hoping that the Saviour would take pity on him and that he could be baptized. When he was told how Philip the Evangelist had baptized an Ethiopian (Acts 8:27–39), Matt said that "he would like to take part in the water-bath just like the Moor." In February 1760 Rauch heard that "the whore of the bookkeeper (the daughter of Mathew) had delivered a dead mulatto boy." This daughter was Hannah, identified in the slave inventories as Matt's Hannah; she was only about seventeen in 1760 and was employed as a house servant, very likely in the bookkeeper's house. The missionaries thought that Matt was embarrassed ("verlegen") by this development, but he was more likely anguished by the white man's fornication with his young daughter. And he had other worries beyond Hannah. In late May 1760 a major slave rebellion spread from the north side of Jamaica into Westmoreland, and the white managerial staff all joined the militia to fight the rebels. Skirmishing continued in the neighborhood of Mesopotamia through June, and Matt had the responsibility of protecting the estate against rebel attack. In July he reported to the Rauches that the slaves were desperately short of food and that the white managers refused to supply rice and flour to keep them from starving. And on a Sunday in August 1760 "Mathew visited us and asked us that when he should die, we bury him in our way and not in the heathen way; and he would also like to lie next to our house."[28]

In December 1760 the Moravians resolved that driver Matthew and boiler Old Sharry were the leading candidates for baptism, but they were disinclined to place Matthew ahead of Sharry. Probably they found him too inquisitive and aggressive, and wanted to make a statement to the other baptismal candidates by preferring Sharry, who was a much more docile person. Also in March 1761, the missionaries "discovered that Mathew's daughter (who is held as a whore for the overseer) had brought a mulatto boy to the world." If this information is correct, Matt's Hannah had been the bookkeeper's mistress in 1759–1760 and the overseer's mistress in 1760–1761.[29] Her second baby fared no better than the first and died when only a week old. The Mesopotamia overseer in 1761 was Daniel Barnjum, who devised the first fully detailed Mesopotamia slave inventory in July 1762. Barnjum was said to be very upset over the loss of his child. The missionaries were inclined to blame Matthew for his daughter's sex life, but Matthew himself was troubled and angry with his white boss. Three days after the baby's death he visited the missionaries and declared that "the whites ["blancken"] were none other than heathen who were completely without God in the world and lived based on their carnal desires, and it astonished him that the whites didn't believe a single word of God, when they should honor him." And in July and August 1761, at subsequent meetings with the missionaries, Matt kept complaining that he had great agitation in his heart.[30]

On August 20, 1761, the missionaries held a conference at Mesopotamia "in which it was agreed: that we should have a baptism, and that the old Jerry should be baptized." Everyone was "joyfully tearful about the prospect of the first [baptism] in Mesopotamia." Old Jerry was known as Sharry in the slave inventories, and was alternately called Sharry and Jerry by the Moravians. In April 1761 he had complained to the missionaries that overseer Barnjum had unjustly hit him on the head and beaten him. Sharry visited the missionaries at least five times between April and August 1761 (once in company with Matthew), professing that "his longing and desire was just for one thing: to belong to the Lord." Born in West Africa, he had come to Mesopotamia sometime between 1744 and 1751 and was described as "old" as early as 1752; he worked in the boiling house during crop time, was a rat catcher the rest of the year, and was probably about sixty at the time of his baptism.

At midday on Monday, August 24, Brother Christian Heinrich told Sharry that he would be baptized that very evening, and the old man "went back to work greatly overjoyed." Dressed in white, Sharry listened to a "penetrating talk" on Acts 22:16: "Stand up and let yourself be baptized and wash off your sins and call upon the name of the Lord." Then he was anointed by Brother Rauch and given the baptismal name Simeon (son of Jacob and Leah). For the missionaries, it was "a great and blessed day for Mesopotamia, the first of its kind." Matthew, however, clearly had more mixed feelings. *"First,"* he told the missionaries, "he had been overcome with a holy shiver and *second* he had to feel ashamed when he saw that the Saviour chose the poor and humble. He [the Saviour] proved Himself so merciful to a simple Guinea-Negro, He would embrace him, while He left an arrogant Creole hollow and behind."[31] This artless comment elicits the triumph of the occasion, but also strongly states that Jamaica-born slaves like Matt considered themselves to be markedly superior to "poor" and "simple" African-born slaves like Sharry.

There were no further baptisms at Mesopotamia in 1762. Missionary activity at Mesopotamia during this year is unrecorded, but it was a time of deep distress for the Moravians. Sister Anna Rauch died at Mesopotamia in August, and Brother Nicolaus Gandrup in November—the first two persons to be interred in the Moravian cemetery on the estate. Brother Christian Rauch moved to Carmel after his wife died, which must have seriously delayed the conversion program. In 1763 John and Susanna Levering succeeded the Rauches. They were familiar with the place, having spent a good deal of time at Mesopotamia during 1760–1761. And in 1763 driver Matthew was baptized. He had waited longingly for four and a half years, and on June 28, 1763, he was "buried in the death of Jesus through the holy baptism." Christian Heinrich Rauch traveled down from Carmel to anoint the "arrogant Creole" with the new name of Jacob (son of Isaac and brother of Esau).[32] Matt's new name was perfect for him, because back in 1758 he had begun to hope for salvation as soon as he learned that Isaac had blessed Jacob—even though Jacob had sinned by deceiving his blind father when he pretended to be his hairy brother Esau. In the Moravian records Matt was henceforth identified only as Jacob, and in 1765 his wife Love (described by the missionaries as "a childlike woman") was baptized by Brother Friederich Schlegel with the name Mary. Also in 1765, Jacob was permitted to observe the missionaries when they celebrated communion, and he told them afterward that he hoped to receive the grace he had just witnessed. His hope came true in December 1766, when Matt was one of the first six black members to receive Holy Communion at the chapel services.[33] He continued to be an active communicant as well as a field gang driver right up until his death at age sixty-five in 1777.

In 1767 overseer Barnjum appointed Jacob's daughter Hannah, whom the missionaries called a "Hure," as the Moravians' chapel assistant. This was a relatively light job assignment for Hannah and an indication that she was still Barnjum's favorite and probably still his mistress. No doubt Matt/Jacob was pleased. The missionaries must have been insulted, but they had to accept her presence in the chapel for the next twelve years, until Barnjum died in 1778. In her last year as chapel attendant and at the age of thirty-five, Hannah gave birth to another mulatto child whom she named Matt in memory of her father, who had died the previous year. Mulatto Matt's white father cannot be identified, but Daniel Barnjum is the most plausible candidate. After Barnjum's death, Hannah was returned to her previous job as a house servant for the white estate managers, and she worked for another twenty-nine years, until she died in 1807. Late in life, in 1802, she was baptized into the Moravian Church (baptismal name unknown) and was described in the Moravian records as a widow (late husband unidentified). Her son Matt Jr. lived to adulthood, although he had a bad case of the yaws as a child and was always in fragile health. His first job was as a cattle boy, then he briefly apprenticed as a mason, but was switched back to stock keeper and died one year before his mother at the young age of twenty-eight. There is no evidence that he ever joined the Moravian Church.

## *The Formation of the Mesopotamia Congregation*

Driver Matthew became the sole member of the Mesopotamia congregation when old Simeon/Sharry died in December 1763, but he soon had company. In February 1764 Brother Peter Paul Bader baptized the first female convert: Mary, an African washerwoman renamed Susanna. Then in December 1764 Bader baptized three more people: carpenter Frank (renamed John) and his invalid wife Affraw (who became Sarah), and another washerwoman named Quasheba (baptized Salome).[34] After 1764 the number of converts quickly accelerated. Brothers Bader and Schlegel baptized three men and six women (including Jacob's wife Love) in 1765; Brothers Bader, Schlegel, and Schnell baptized two men and eighteen women in 1766; Brothers Senseman, Schlegel, and Metcalf baptized twenty men and twenty-one women in 1767; and Brothers Senseman and Schlegel baptized fifteen men and eight women in 1768. By 1769 the missionaries were running out of candidates. Brothers Senseman, Schlegel, and Schnell admitted only five men and two women during that year. In sum, 48 males and 58 females were baptized at Mesopotamia during the 1760s, for a total of 106.[35]

This listing of the many baptisms in 1764–1769 tells us that numerous missionaries participated in the formation of the Mesopotamia congregation. John and Susanna Levering were the resident missionaries in 1763–1764, but John died at Mesopotamia in August 1764. Peter Paul and Mary Bader replaced the Leverings, but Mary Bader also died, in May 1765. Joachim and Christina Senseman replaced the Baders in 1767–1769. Friederich Schlegel, John George Schnell, and John Metcalf also frequently visited, and they baptized many of the Mesopotamia slaves between 1765 and 1770. When Caries, the Rauches, and Gandrup are added, at least thirteen Moravian missionaries worked with the Mesopotamia slaves during the first decade of the mission—and four of them died at Mesopotamia.[36]

Obviously the formation of the Mesopotamia church in 1759–1769 was far different from the formation of the Bogue church in 1755–1756. The missionaries had assembled and indoctrinated their candidates much more slowly and carefully. In particular, they had recruited a very large number of women—nearly three-quarters of the adult females on the estate—whereas female converts had been entirely missing at the Bogue. *Appendix 28* compares the slaves baptized during the 1760s with the slave inventory taken on the estate on December 31, 1769. In addition to the large number of women, this newly formed Mesopotamia congregation had three other notable features. A high percentage of "head people" had joined the church: two of the three male drivers (plus a retired male driver and a future male driver), six sugar boilers and distillers, and the five most senior carpenters, masons, and blacksmiths. Among the women, the sole female driver, the chief female housekeeper, Clarissa (who had attended the Barhams when they lived in Jamaica), and midwife Kickery were also converts. By contrast, relatively few of the more menial field hands or stock keepers were members—although the Moravians did baptize a watchman with a wooden leg, Stump Sharry, whose leg had been amputated as punishment for running away. The missionaries clearly had their greatest success with the older slaves: 91 percent of the baptized were above the age of thirty, and 72 percent were forty or older. Practically everyone over fifty had joined. In 1769 just one man in this age group, Barton, a fifty-three-year-old cooper, was not baptized. Finally, while the Moravians drew both Jamaica-born people like Matt and Africa-born people like Sharry, the African slaves were more strongly represented. Over 63 percent of the converts in the 1760s came from Africa, though Creoles considerably outnumbered Africans at Mesopotamia by 1769.[37]

By happenstance the Moravian missionaries started to recruit converts just when the Mesopotamia overseers and bookkeepers first supplied fully detailed information about the age and health of the slave population. As discussed in Chapter 1, the first fully articulated slave inventory, dated July 10, 1762, shows that the 268 slaves were in generally poor shape. There were extremely few children, not enough young workers, and twenty-three invalids among the adult women. These conditions continued throughout the 1760s. Despite the purchase of thirty-two new Africans, the Mesopotamia population dropped from 268 in July 1762 to 249 in December 1769. And although ten of the old, incapacitated women died during the 1760s, there were still fourteen nonworking invalids among the seventy-eight adult women in 1769. Facing the imminent prospect of death, it is not surprising that these sick and/or elderly women sought to achieve heavenly salvation by baptism into the Moravian Church.

A female field hand named Affraw is representative of this group. She was about forty-seven years old when the Moravians arrived, was married to carpenter Frank, and had a young daughter named Bella. By 1762 Affraw had become an invalid. Sister Levering saw that she was interested in the Moravian message, and kept asking her if she felt a longing to belong fully to the Lord. To the missionaries' delight, Affraw responded that "she knew she was a great sinner, but because she heard joyous things from us, that the poor sinners had a gracious God in the Saviour, she believed with assurance that He will forgive all of her sins, and when she dies, she knew for certain that He will take her soul to Him."[38] Affraw was baptized in December 1764. She continued to be a nonworking invalid for another twelve years, and died at the age of sixty-five in January 1777. But many of the other aged or sickly converts died more quickly. Twenty-one of the 106 people baptized in the 1760s were already dead by the close of 1769; another 36 by 1779; and a further 24 by 1789. Only ten of the youngest converts lived past 1800. Among the men, distiller Augustus (baptized at age twenty-seven) lived to 1812, and driver Hector (baptized at twenty-three) lived to 1816. Among the women, field hand Sucky (baptized at thirty) lived to 1810, and driver Sibby (baptized at twenty-nine) lived to 1814.

About a fifth of the Mesopotamia people who joined the church during the 1760s were described in the Moravian records as "heimgegangen" by the end of the decade. Since death was such a constant factor, the missionaries needed to keep finding new recruits in order to sustain their Christian community at Mesopotamia. And if they wished to build an enduring congregation, they needed to attract more of the younger slaves, particularly the relatively few women in their twenties and thirties who were producing children. So far the Moravians had baptized six couples who can be identified: Franky and Affraw, Matt and Love, Guy and Mary, James and Katy, Parry and Juba, and Primus and Priscilla. Primus was Love's brother, and at least one mother and daughter, Prue and Cooba, were baptized by 1769. More extensive family networks were needed to solidify the church. These were some of the many challenges facing the Moravians as they continued to minister to the slaves at Mesopotamia from 1770 onward.

Nevertheless, the Moravians had achieved a solid initial success. By the close of December 1769, 85 of the 166 adults in the Mesopotamia slave population—or 51 percent—were listed as members of the Moravian Church. In fact, the congregation had already reached its maximum size. The crush of candidates for baptism had ended, so the missionaries strove to engage the members already in the church more fully in the Moravian teachings and rituals. This meant enrolling the baptized slaves as candidates for confirmation, the rite of full membership in the church. The confirmation candidates were given considerable instruction, and then the missionaries asked the Saviour via the lot to accept them as communicants. And whenever the Saviour approved, they were admitted to the Lord's Supper or "Abendmahl," which enabled them to participate in the commemoration of Christ's wounds, blood, and sufferings.

In December 1766 the first six communicants were confirmed: Jacob (book name Matt), the head driver; John (Frank), a retired carpenter; Isaac (Parry), a retired driver; Joseph (James), another driver; Sarah (Affraw), a retired field hand and John's wife; and Susanna (Mary), a washerwoman who was the first baptized female at Mesopotamia. These six had all been greatly interested in the Moravian message in 1758–1760; they were among the first to be baptized (except for Joseph, who did not join until June 1766); and they were clearly pillars of the slave church. Four more communicants were added in 1767 and another sixteen during the next two years. By 1772 there were thirty-eight communicants, constituting more than half of the church membership.[39]

Brother Joachim Senseman and his wife, Sister Christina Senseman, were in charge of the Mesopotamia flock from 1767 to 1769. They achieved a lot during their three years at Mesopotamia. Senseman baptized thirty-five men and women, and welcomed twenty new communicants to the Lord's Supper. In 1769 he closed his diary, written in German, by suddenly bursting out into English with a long communal hymn. His crude but swinging verses vividly express the Moravian obsession with the bloody wounded Christ. At Mesopotamia church services the slaves did a lot of singing, and at Easter in 1769 the twenty-six communicants probably sang this hymn during the Lord's Supper:

Blessed Negro Congregation
Dost thou hear the Gospel Sound?
Be, O be in meditation
Over thy Dear Saviours Wounds
Jesus Cross, Blood, Death and Suffering
and Sin expiating offring
Jesus Cross Devine to thee
shall thy One & all things be.

Come, O Come! with Soul and Spirit
To thy Saviours Table Draw
Come to his Dear Holy Merit
With great Respect, with deep Awe
Eat his holy Corpse so Bloody
Eat the Heav'nly Food, his Body
Drink the Blood, which he did shed
Drink the Blood, the Blood so red,

His Dear Corpse for thee was broken
his Clean Blood for thee was spilt
This is more than Heav'nly Token
Takes away all Sin and Guilt
O! how happy is a poor Soul
When she comes to his dear wounds hole
Eats and drinks, and is renew'd
Thro' this Heavenly Drink & Food. . . .

O Lords Supper-Congregation!
At Mesopotamia
Might thy Saviours sweet Salvation
As he his last Breath Did Draw
As he on the Cross hung dying
As he for thy Sins was Crying
B'fore thy Hearts and Eyes appear
And he always to thee near.[40]

## *The Struggle to Sustain a Christian Community of Slaves, 1770–1819*

From the 1750s into the 1830s, Joseph Foster Barham I and Joseph Foster Barham II ordered their attorneys and overseers to maintain a house and chapel at Mesopotamia, paid the missionaries an annual stipend of £50, provided them each year with £60 to £80 worth of rum, beef, and other provisions, and supplied them with slave servants. These actions were by no means entirely altruistic, because the Barhams' tax assessment was lowered by £50 when they installed an extra pair of white people on the estate.[41] Besides, it was comforting to know that the Moravians "teach no other doctrines to the negroes than what lead to order and submission." Or as the younger Barham put it in 1800: "When men are placed (as is the case of the slaves) in a situation really not free from hardship, those lessons are [not] dangerous which prescribe submission to authority, patience under suffering and resignation under wrong, which teach them to regard this life as a trifle and point their hopes and desires to a future existence."[42]

Joseph Foster Barham II had joined the Church of England by the time he became owner of Mesopotamia, but he sought to expand the Moravian program in Jamaica. In the late 1790s he tried to persuade several slaveholders in western Jamaica to fund Moravian mission work on their estates—negotiating with his characteristic mix of lofty aims and dubious motives. Barham urged these planters to do their duty to the people placed under their power, and also "to get hold of the minds of our negroes" and correct their manners in order to improve the slave birthrate. At the same time he told his attorney that the Mesopotamia mission should be maintained "with all the oeconomy which the object admits." Since the Anglican clergy did not want to convert the slaves, Barham saw the Moravians as the most suitable evangelists for Jamaica, because they avoided speculative reasoning and addressed themselves only to the heart, without infusing a spirit of inquiry among the Negroes. They were thus far different from the Baptists and Methodists—then just beginning their mission work on the island—who in Barham's mind were exceedingly dangerous incendiaries.[43] Bishop Christian Ignatius Latrobe, the Moravian emissary to Barham, pointed out that the attorneys and overseers at Mesopotamia greatly obstructed the missionaries. But Barham did nothing to change their hostile behavior. And while he did persuade seven planters to contribute £643 in 1800 toward the Moravian mission, the newcomers all dropped out by 1802. Financial support for the Moravian evangelizing project in Jamaica from sponsors in England thus remained very limited.[44]

Neither of the Barhams seemed to understand that the high death rate at Mesopotamia for both blacks and whites was tremendously damaging to their Christianization project. Four of the first eight missionaries who worked with the Mesopotamia slaves died at the estate between 1762 and 1765, and a pattern quickly developed whereby the Moravian widows and widowers then left for Carmel or returned to Germany, England, or Pennsylvania in order to recover their health. The turnover was continuous, and Mesopotamia soon had a well-earned reputation among the Moravians as an especially unhealthy mission site. And the Mesopotamia slaves had to adjust to a great many different spiritual guides.

Several of the missionaries lasted only a few months at Mesopotamia. Philip Diemer arrived toward the close of 1781 and died in April 1782. Brother Wagner (first name unknown) arrived in March 1784 and died in September. Thomas Ward arrived in June 1818 and died in February 1819. Brother Wagner's short pastorate was particularly poignant. He and his wife sailed from Europe to Jamaica, and Brother Taylor, who escorted the Wagners to Mesopotamia, noted that the new missionary "is deficient in the English language . . . but we hope that he may improve." In May, two months after his arrival, Wagner was thrown off his horse and damaged his leg, an injury from which he never recovered. For six weeks there were no chapel meetings, and when Taylor came to check up on the Wagners in July, he found that Brother Wagner was too ill to travel by horse to Carmel for a change of air, and could not afford to hire a carriage. Wagner was "very bad in his head; & sometimes not quite master of his senses," but was quiet and bore his affliction patiently. The poor man died two months later, and his widow moved to Carmel. The search for a new Mesopotamia missionary couple had to begin again.[45]

The high death rate among the Mesopotamia slaves also undermined the missionaries' efforts. New people were always entering the church, but many too few to counter the mortality. In 1769 the Mesopotamia congregation had eighty-five members, and then began a steady decline. During the years 1771 to 1777, Brother John Metcalf baptized three adults and two children, while twenty-six members of his congregation died. By 1778 the Mesopotamia congregation was reduced to fifty-one baptized members and twenty-four communicants.

And the number of communicants fell precipitately to fourteen by 1782. Brother and Sister Taylor made a great effort in 1782–1784 to recruit younger people, and listed fifty-five adult members in 1784. Membership then fell to forty-eight in 1788, and to thirty-two a decade later. At this low point membership stabilized. The Browns in 1798–1799, the Jacksons in 1801–1808, and the Gründers in 1811–1818 collectively baptized fifty-nine new members, and baptisms slightly exceeded deaths during these twenty years, so that the active membership in 1818 totaled thirty-five. Brother Gründer also increased the number of communicants from two in 1811 to nine in 1817, which helped to revitalize the slave congregation. But church membership was certainly not keeping pace with the size of the slave gang, which had expanded while the congregation was diminishing. Back in 1769, 51 percent of the adults on the estate were baptized, compared with 16 percent in 1818. The church had become a far less significant factor within the Mesopotamia black community than at its peak period fifty years before.

The missionaries who hung on at Mesopotamia despite death and decline were resolute in the face of danger and generous in their attention to the slaves. John Metcalf reported in January 1773, "We have heard on one of our Neibouring Estates that all the Negroes was for Rising in Rebelion to Kill the overseer, but was shortley prevented thro good care, some [rebel slaves] was secured, others Run in the Bush; thus we are in danger always." But, he added, "in the Evning we had our bible hour as usual, in stillness." Metcalf spent much of his time visiting the sick and elderly members of his flock. Sometimes he went to "Neger-Town"—the cluster of slave huts—and sometimes he trekked around the edges of the estate in order to visit the fifteen or so baptized watchmen, all old or feeble, who rarely attended his meetings because they had to guard the cane fields and slave provision grounds from the livestock and lived in remote huts distant from the chapel. One of these watchmen, a man named Philip (book name Flanders, baptized in 1767), was severely injured because he had tumbled into a deep trench of stones and water and lay unconscious for a long time until he came to and was able to creep out. Poor Philip died shortly after Metcalf came to see him.[46]

Five years later, Peter Jacob Planta gave a similar account of traveling among the seven Moravian congregations in St. Elizabeth parish to visit the sick in their Negro huts. Then he traveled to Mesopotamia.

Lavinia (a baptized slave whom I can't identify) came to see Brother Planta, and they had a loving discussion about the Saviour. Other Mesopotamia blacks visited him also, and on Saturday night Planta celebrated the Lord's Supper with the communicants in the "Chappel house" with "an unspeakable happy feeling, and feasted on our Lords Blessed Corpse and precious Blood." On Sunday, Planta spoke with a number of Mesopotamia people, both Christians and those hoping for baptism, and then "had a happy Communion Liturgy with our Black Brothers and Sisters who had enjoyed the highest good with us in the Holy Sacrament."[47]

One special feature of life in Jamaica that Metcalf, Planta, and the other missionaries encountered was spectacularly destructive weather. On October 3, 1780, one of the most powerful hurricanes in all Jamaican history made a direct hit on Westmoreland with tremendous force. The wind was so ferocious that it tore people's clothes off, smashed every building in Savanna la Mar (including the Barhams' storehouses and wharf), and littered the town with corpses. The diarist Thomas Thistlewood reported that every building on his property, including his dwelling house, was completely destroyed, and his garden—which he had been developing for years—was utterly ruined.[48] Mesopotamia was a little further inland and seems to have suffered less total damage. But two Mesopotamia slaves—a fifty-two-year-old watchman named Hazard and a sixteen-year-old cattle boy named Tony—were killed by the hurricane.[49] The slave provision grounds were badly hit, precipitating a food crisis for the slaves. The Mesopotamia cane fields were shredded, leading to the smallest sugar crop on record in 1781—only fifty-nine hogsheads. The Great House at Mesopotamia was so damaged that it had to be rebuilt. And the bell perched atop the Moravian chapel was flung to the ground and broken.

Four years later, in 1784, when Brother Taylor was holding a meeting for the Mesopotamia congregation, a violent storm broke out during and after the chapel service. "We were afraid that the shingles would be blown from our house every Minute, as the wind came with such violence at times, as if it would tare every thing to pieces before it." There was no sleep that night for Brother Taylor.[50] And in October 1812 another tremendous storm kept Brother and Sister Gründer up all night, trying to close their jalousies to block the sheets of rain. Next morning their hall and bedroom were swimming in water, and the

Gründers discovered that some of the slaves' houses were unroofed. Worse, the slaves' fruit trees were smashed, with all the fruit on the ground, and their main food source—the plantain walk in the mountain provision grounds—was laid flat. A new crop of plantains would not ripen until the following summer, and emergency food shipments from North America were unavailable because of the War of 1812. By the spring of 1813 the Mesopotamia slaves were close to starvation.

Violent weather damaged the flimsily constructed mission chapel and living quarters. The Moravian chapel, built in 1759, was rebuilt in April 1780 at a cost of £252. It was a spacious building, forty-four feet long and twenty-one feet wide, with five doors, ten window shutters, and a back piazza or veranda.[51] When the great hurricane hit in October, the chapel must have received a battering, which would explain why Brother and Sister Zander, arriving at Mesopotamia in 1785, found the chapel to be in ruinous condition. Attorney James Graham, whose son had probably been the overseer when the chapel was rebuilt in 1780, airily explained that the last construction was of "very indifferent materials." So Joseph Foster Barham II had two new rooms built for the Moravians, which Graham thought were "very comfortable." One room was the new chapel, the other an apartment for the missionaries; the cost was £451.[52]

Ten years later, in 1799, attorney Henry Plummer reported that the Mesopotamia chapel (rebuilt in 1780 and 1789) and the missionaries' apartment (rebuilt in 1789) were both "in great want of repair." So the chapel was rebuilt for a third time in 1800 at a cost of £346—and Barham sent the bill to Bishop Latrobe, apparently expecting the Moravian Church to pay. A new house for the missionaries was finished in August 1801. Placed on a stone foundation five or six feet above ground, the house was constructed and shingled entirely with American timber, and contained a large hall, two bedrooms, and a piazza.[53]

Within this framework of death, disease, and storm, the Moravian diarists had much to say about slave life and labor at Mesopotamia. They give many descriptions of the Mesopotamia slaves who begged for baptism, a striking number of whom were in sad physical condition. George, a twenty-six-year-old invalid, was "a sickly person, & like a Lazarus in his Body." Brother Taylor described George as "an object of Pity: and the Tears stood in his Eyes while he spoke to us: & said that he was sickly, & he did not know how long he should live." Taylor, in baptizing him, gave him the appropriate Christian name of Job. And George, who had not worked since he was twenty-one, then survived as an incapacitated church member for another eight years. Patience, a twenty-nine-year-old African field hand, went to the seaside at Christmastime to bathe her swollen legs and body in the salt water, and then had to crawl some of the six miles back on her hands and knees in order to return home. She was baptized on her deathbed as Abigail.[54] Cicily, a thirty-four-year-old Mesopotamia-born field hand, was dying of tuberculosis when she asked to be baptized. In health, according to Brother Brown, she had chased after the vanities of this world rather than attending meeting, but in sickness she begged to be baptized, pleading, "I know of nothing else but to cry Day & Night for Help from Him [our Saviour], and that out of pure Mercy He may have Compassion on my poor soul." And Brown did baptize Cicily just before she died.[55]

Many of the long-standing members of the Mesopotamia congregation were in equally bad shape. Carpenter Frank had worked at Mesopotamia for at least thirty years, was among the first slaves to embrace the missionaries in 1758–1759, and was baptized with his wife Affraw in 1764. Frank and Affraw were the fourth and fifth persons to enter the Mesopotamia congregation, and were also confirmed together in 1766. But by 1773, when Frank was sixty-two and Affraw was sixty-one, they were both invalids, and Frank was blind. When Brother Metcalf came to their hut in 1773, Frank was so eager for this visit that he spoke with the missionary for two hours. Let us hope that Metcalf and his successors visited this couple thereafter, because Affraw lived for another four years and Frank for another seven years. Driver Parry was probably the longest-standing retired church member at Mesopotamia. Parry had been baptized as Isaac in 1765 and became a communicant in 1766. First identified in the earliest Mesopotamia inventory as a field driver in 1727, he had continued as a driver through 1763, then lived in crippled retirement for another twenty years, being the oldest person on the estate when he died at age eighty-two in 1783. Taylor composed Isaac's obituary: he was "a little self-righteous; but yet he knew himself as a Sinner & as such he called upon our Saviour: he was a long time sick, and unable thro old age to come to the Meetings, which loss he often regreted, & mourned over it, particularly since the Hurricane he was not able to attend one meeting."[56] By contrast with Frank and Parry, on February 1, 1785, Catherina (book name Diana), a sixty-three-year-old baptized grass cutter, was feeding the sugar mill with cane all night when she suddenly dropped down in the mill house and died on the spot.[57]

A constant theme in the Moravian diaries was the huge problem of dealing with crop time, which lasted for four or five months, from January to April or May, with the boiling house operating twenty-four hours a day, six days a week. Sunday was the only day the slaves could cultivate their food crops, so the Mesopotamia congregants had great difficulty in coming to church. In 1773, to Brother Metcalf's distress, overseer Daniel Barnjum chose to begin harvesting the crop on a Sunday, so the slaves had no chance to go either to church or to the provision grounds. "This is something which one must believe is not right in the Eyes of God, being contrary to his Law," wrote Brother Taylor in 1782. And Taylor also noted that the people who did come to church were very sleepy, "they being obliged to lose every third nights rest all thro the Crop: so that we hear scarcely any thing but complaining from the Negroes of their being worked so hard, & the Severity of the Overseer, & Drivers." Taylor lived at Mesopotamia during the ten years (1778–1788) when John Graham was the overseer at Mesopotamia, and Graham was a particularly aggressive slave manager. Many of the Mesopotamia congregants "are for running away, & say that they would run away if it was not for us," said Taylor. And he concluded that "they truely have it harder than most Negroes on other Plantations."[58]

The drivers had their own story to tell. Head driver Francisco, who was thirty-one years old in 1782 and not a church member, complained that overseer Graham "wanted more work of him & the Negroes than he was able to make them do, & more indeed than was right, & more than they were able to do: he said that they worked harder than any other Negroes, on any Estates round about." Francisco wished to be baptized, but he knew that baptized Moravians were not allowed to swear, and Graham expected him to "swear & teare at the Negroes, particularly when he came to look at them." The missionaries told Francisco that they could not intermeddle with the overseer, "and that we were here only for his souls good."

Francisco's story was confirmed by former driver Joseph (book name James). Joseph, who was one of the initial communicants in 1766 and a leader in the chapel community, had succeeded Jacob (Matt) as head driver sometime in the 1770s. But he had been demoted in 1779 by overseer Graham, who replaced him with Francisco.[59] When the Moravians first met James in 1758–1761 he had been almost as aggressively inquisitive about the Christian religion as Matt, and after his baptism as Joseph and admission to the Lord's Supper in 1766 he had served as a "helper" to the missionaries in spreading the Moravian message. But Graham destroyed this man. The missionaries recorded Joseph's downfall: he was "the chief Driver for a good while: but an overseer [i.e., Graham] coming to the Estate, who was very sharp, & wanted him to whip the Negroes more than he liked, & more than he could for Conscience sake: he was put down from the Chief Driver to the Second: & geting a lame foot he lost that place too, & at last was put to Rat catching: and when he could catch none, he got severe Punishment, which he took too much to heart: till he almost lost his Confidence to our Saviour, and was cast down with grief." In January 1783 Joseph came to a burial service intoxicated, and had to apologize to Brother Taylor for taking too much liquor before a hall full of his fellow slaves. The poor man was still in deep distress, "seemingly heart broke over his hard treatment," when he died of apoplexy in November 1783.[60]

A quite different issue that greatly bothered the Moravians was how to cope with the "backsliding" that had so undermined the initial efforts of the missionaries at the Bogue. Trouble for the Metcalfs started when Magdalena (book name Priscilla, baptized in 1766) complained about "her Husbands bad things with Susanna." Magdalena's husband was distiller Jonathan (book name Primus, baptized in 1767), who had rejected her and was living in adultery with washerwoman Susanna (book name Mary), the first Mesopotamia woman to be baptized, in 1764. Here the Metcalfs faced a double challenge, because Susanna was married to blacksmith David (book name Guy, baptized in 1767). Furthermore, the Metcalfs discovered that Jonathan also had a second wife, field hand Abigail (book name Lucy, baptized in 1767). This domestic crisis thus involved five members of the Mesopotamia congregation, at least two of whom were among the select band of communicants at chapel services. Abigail persuaded Jonathan and Susanna to meet with Brother Metcalf, but Metcalf found them unrepentant, and so he excluded them from the Lord's Supper.[61]

A decade later the Taylors also found numerous examples of "backsliding" among the baptized. Benjamin (book name Hector, baptized in 1767), a thirty-eight-year-old driver, "was spoken pretty sharply &

closely" concerning his unchristian course. Miriam (book name Maria, baptized in 1767), a seventy-two-year-old invalid, was a communicant who had "deprived herself of this Grace for many years thro dryness & indifferency," becoming "a kind of a dead lifeless Christian, who had no right feeling of our Saviours sweet love in her heart & soul." And Esther (book name also Esther, baptized in 1766) was a sixty-year-old African gardener who had "spent her time very poorly, & was a right Lukewarm indifferent soul." The Taylors, implicitly criticizing the missionaries who had schooled and baptized Esther, were "amazed that such a Stupid, Ignorant Creature as she really was in spiritual things, should ever have been baptized into Jesus death."[62]

In 1798 Brother Brown officiated over a particularly messy family blowup within his small Christian community. A boiler named Daniel (book name Occara) and his field gang wife Sybilla (book name Judy) had been baptized in 1782, and then excluded in 1783 for taking new partners. Despite having split, Daniel and Sybilla continued to have sex together and produced four more children, and when the Browns arrived both of them were also attending church services. Learning of their past record, the Browns cast them out of the meeting, which upset Daniel and Sybilla very much. Brother Nathaniel commented that they "do not know themselves what they shall do. Our Saviour alone must soften their hard hearts. It is not in their power." When the Browns tried to persuade them to resume their marriage and forget and forgive each other's offenses, Sybilla became irreconcilably bitter against the missionaries as well as her former husband. In 1799 Sybilla and Daniel came to blows, and he beat her so brutally that "she could eat very little for a long time, and was even in danger of losing her life. This obliged the Attorney to separate them entirely, and give each of them half of their Children, some of whom had been baptized in their infancy by the brethren." In the estate records, Judy (that is, Sybilla) is credited with six children, one of whom died in infancy; the four youngest, who were apparently divided up in 1799, were Bob (age fourteen), Samuel (age ten), Victor (age seven), and Abigail (age four). There was no resolution to this family fight. Sybilla, age forty-four in 1799, had been a member of the second field gang, but (after her beating) was listed as unable to work in the next inventory and remained an invalid until she died in 1804. Daniel, on the other hand, a fifty-year-old boiler and jobber in 1799, was active for another fifteen years. He became head watchman in 1809 and asked to rejoin the church in 1813, but was not readmitted. He died the next year.[63]

While Nathaniel Brown recorded family fights, he also noted the tender feelings that the Mesopotamia people could have for one another: "They will sit for whole hours holding the head of their sick Relations or Shipmates with incredible Patience and doing all manner of service for them without looking for Payment." Yet Brown could not accept the Mesopotamia slaves' addiction to pagan African death ceremonies. He became very upset when in August 1799 a man on a neighboring estate "made a great Supper and Play" in commemoration of his wife's death, "and to our grief we learned that a great Number from here, and even some of our Christians have been there too." The great supper and play continued all Saturday night into Sunday, spoiling the missionaries' morning meeting. And not just young people went to this feast, but gray-headed men and women who "can hardly crawl about, when at home, will at such Occasions take their Crutches in hand and walk more than an English mile, chiefly to get plenty of Rum to drink and a hearty supper, though they are often disappointed in that Respect and get not what they expect yet another time they will try it again."[64]

Even more distressing slave behavior as far as the Browns were concerned took place every December 24 and 25. Instead of encouraging the Mesopotamia people to celebrate the birth of the Saviour, the white management let the workers have several days of vacation in anticipation of crop time in early January, and supplied them with rum for this holiday. The rum encouraged much drunkenness as well as "heathenish" singing and dancing. In 1799 several congregants at the Christmas Eve service were half drunk and began to talk and laugh until Brother Brown reproved them. And the next morning Mingo, a new church member (Christian name unrecorded) who had been baptized in October, disrupted the Christmas service. He had been celebrating far too much and became so ill during the sermon that he had to walk out and almost fell down in the hall. Mingo, age thirty-nine, had been a field driver in the early 1790s, but by the time the Browns arrived was in such poor physical shape that he could only work as a gardener. His mother, Elizabeth (book name Sucky), had doubtless urged him to join the church. She was a sixty-three-year-old widow and one of the oldest and most respected Mesopotamia communicants, having been baptized way back in 1766 by Brother Schlegel. Elizabeth was said to be very upset at Mingo's conduct.[65]

The Mesopotamia congregation included several slaves who had, or were reputed to have, medical skills. One was a pair of midwives, mother and daughter, known as Old Kickery and Young Kickery. The mother was the plantation midwife for many years until her death at seventy-two in 1779. Old Kickery had been baptized as Deborah back in 1766, and she was admitted to the Lord's Supper in 1767. Young Kickery, born about 1745, when her mother was thirty-eight, worked as a field hand until age thirty-two, then became a midwife in 1778 just before her mother died, and continued in that job until her own death at sixty-four in 1810. Thus, the mother trained the daughter, and they were the only black midwives on the estate for half a century. So it was fitting that when Brother Zander baptized Young Kickery in January 1788, she took her mother's baptismal name. She too became Deborah.[66]

Another church member, a man unfortunately not identified by Brother Brown, was a practitioner of folk medicine who was known as a "Doctor for the Kings Evil" or scrofula (now known as tuberculosis of the lymph nodes in the neck), a horribly disfiguring disease that was common among the Jamaican slaves.[67] This scrofula doctor tried to treat a patient from another estate whom he couldn't cure, and when he sent him home, the slave's owner returned the sick man for more treatment. The patient died, and the scrofula doctor immediately buried him with the help of his other patients to prevent the spread of putrefaction. But a Mesopotamia bookkeeper noticed the burial, and the owner of the dead slave demanded that the coroner and a white jury investigate this death. The coroner required Brother Brown to attend the hearing so as to put pressure on the folk doctor to tell the truth. The body was not disinterred for health reasons. The doctor answered all questions fully and freely, so he was cleared, but he was reproved for not telling a white man what had happened as soon as the patient was dead.[68]

Turning from the Mesopotamia congregants to the missionaries, it appears that at least one of the Mesopotamia missionary couples, the Metcalfs, owned a slave. Brother Metcalf reported in March 1774 that he journeyed to Carmel accompanied by "unser Neger Billy." The only member of the Mesopotamia slave force named Bill or Billy at this date was a two-year-old child, so it seems that "unser Neger Billy" was the

Metcalfs' personal attendant.[69] Another point of interest is that Brother Christian Zander, who lived with his wife at Mesopotamia for five years, was of such marginal visibility to the attorney and overseer that he was listed in the estate records as Frederick Lander. Attorney James Graham knew him (in so far as he *did* know him) by that name. In May 1790 Graham reported to Barham that "good old Mr. Lander the missionary died on the 18th instant."[70] The good old missionary had actually had a quite variegated career. In his last months Christian Zander composed a brief autobiography that was appended to his diary for 1787–1790. He was born in Pomerania, became a Moravian brother as a young man, then moved in his thirties to England, where he married Sarah Wiltshire. This mixed German/English pair spent several years as missionaries in Barbados before they came to Jamaica, so they were experienced at working with slaves. And Brother Christian put in long service for his church. He was sixty-three years old at his death.[71]

Two missionaries in particular, Joseph Jackson (1801–1808) and his successor, Samuel Gründer (1811–1818), stand out for their determined efforts to sustain the Moravian congregation. Joseph Jackson was a second-generation missionary who was actually born in Jamaica. Back in 1769 his parents, Brother Joseph Sr. and Sister Mary Jackson, had come to the island and described their adventurous trip to the Caribbean in a letter they sent back to England.[72] But Brother Joseph abruptly died in November 1769. One month later Sister Jackson gave birth to their son and named him for his father. In June 1770 widow Mary returned to England with young Joseph.[73] Now at age thirty-one Joseph was back in Jamaica, and he and his wife Rachel would operate the Mesopotamia mission for seven years.

We met Joseph Jackson in Chapter 2, where he nominated laundress Affy as a candidate for baptism and befriended Affy's rebellious son mulatto Robert. Though he left no diaries, Brother Jackson wrote candid letters to Bishop Latrobe and to Joseph Foster Barham II. He comes through as a kindly, mild-mannered man with considerable backbone. Barham's attorneys described Jackson as "a very respectable character," and he stood up to them more effectively than most of the Mesopotamia missionaries. He told Bishop Latrobe that he had yet to meet a white person in Jamaica "who would not be ashamed of having a serious thought (much less a word) laid to his charge." He acknowledged that there were few black people at Mesopotamia who wanted to hear the Gospel, but went on to observe that successful evangelism was almost impossible in Jamaica. "The local disadvantages under which the Negroes labor in Jamaica," he told Barham, "and which cannot easily be removed, render such an extensive diffusion of Christianity scarcely ever to be expected, as we hear of in the lesser Islands [i.e., Danish St. Thomas or British Antigua]."[74] And he reported to Barham in 1804 that the Mesopotamia missionaries "chiefly sow in hope." Or as he put it to Latrobe, "We jog on softly—no great things, but however not quite without encouragement."[75]

Barham had been toying with the idea of moving the slave gang at Mesopotamia to Island in order to combine his two units into a single large plantation, and he asked Jackson to sound out the "leading" Negroes at Mesopotamia about moving to Island estate, which lay over fifty miles distant. But Jackson declined to do so, saying that any inquiry would soon become general knowledge, with extremely bad consequences. The Mesopotamia people's "minds would be unhinged" at the dreadful possibility of moving, and they would neglect their provision grounds and let their houses decay. Jackson pointed out that many helpless invalids would be unable to make the long trip. Furthermore, "the parting from their homes, from their grounds, and many of them from husbands, wives and children on neighboring estates, might . . . sink them into a state of inexertion"—or break their hearts. Since the Mesopotamia people don't expect to see their master again, he remarked to Barham, they are more attached to the land than to him, and would prefer being sold with the estate. Jackson proposed that Barham move to Island only those "whole families who have no near connexions on the neighboring estates, or other single Negroes."[76] His opposition helped to dissuade Barham from combining Mesopotamia with Island. But it did not deter Barham from transferring 105 slaves who lived at Springfield estate in Hanover parish to Mesopotamia in 1820—causing the very same family disruption and social damage that Jackson foresaw in 1805.

The Gothic novelist "Monk" Lewis, the owner of Cornwall estate, which lay about three miles by road from the Barhams' place, visited Mesopotamia in 1806 and found the Jacksons at work. He reported that about fifty slaves belonged to their church.[77] We have no figures for church membership at the close of 1808, but the congregation was certainly larger than when the Jacksons arrived. It is telling, however, that the Jacksons did not try to present any of the thirty-six newly baptized slaves to the Saviour as candidates for communion. As Brother Joseph wrote plaintively to Latrobe, "We should rejoice to see some more life" among the new people.

Joseph Jackson died in December 1808, his funeral attended by a large company of blacks and whites. Jackson's death was a great loss for the Mesopotamia congregation, made worse by the fact that no successor could be found for more than two years. When Brother John Lang visited Mesopotamia in November 1809, he found that the furnishings in the mission house were mostly gone, and a year later the cellar and outhouse were also stripped bare.[78] But the chapel and living quarters were put in order so that John Samuel Gründer and Sarah Gründer could take residence in April 1811. The Gründers were experienced missionaries, having served for seven years in St. Kitts and Barbados before coming to Jamaica. Attorney John Blyth immediately dismissed Brother Gründer as "a silly quiet man," and the Moravians who visited Mesopotamia between 1811 and 1818 told Bishop Latrobe that Barham's white managers—attorneys Blyth and Grant and overseer John Patrickson—all harassed the Gründers.[79] The troubles they faced at Mesopotamia are well documented, for Brother John Samuel wrote diaries in English in 1812 and 1815, and diaries in German in 1813 and 1817.

The Gründers found that there were only two communicants in the Mesopotamia congregation. One of them, a woman with the baptismal name Cartarina, I cannot identify; the other was Zepora (book name Sibby), a seventy-three-year-old widow who had been a field driver in her prime but was now elderly, in poor health, and a very marginal member of the slave community. In March 1812 the Gründers tried to celebrate the Lord's Supper with these two women, but only Cartarina could participate. Zepora, who had been too sick to work for some time, had the ill luck to meet overseer Patrickson when she came out of her hut to do her laundry. So, as Brother Gründer commented, she was "put in the stocks to Day, for no other Crime but she went to wash some of her old rags."[80]

The Gründers came at a time of exceptional food shortage in western Jamaica, made much worse by the great storm of October 1812 that smashed the slaves' plantain walk. By the spring of 1813 the Mesopotamia slaves were almost out of food. Brother Lang wrote from Carmel in

April 1813, "Whenever we go, the Negroes cry out Massa hunger kill me. And indeed they look so thin that we cannot behold them without pity." Because of the food shortage, the Gründers were now subject to theft. One night "we were awoke by an alarm, a Thiefe having broke into our Kitchen, he was luckily got hold of and sent to the Bilbow." Another time "we found that our stock house had been broken open, and 3 of our laying Hens taken out In the Night." And a third time "we were alarmed by the watchman, on account the Fowl house having been broken open & seven of our Fowls taken out!" Previous missionaries had not experienced these break-ins, which suggests that the Moravians at Mesopotamia were no longer safeguarded by their godly lifestyle. But in the crisis of 1812–1813 the Mesopotamia people were also stealing from one another. When the Gründers visited the Negro houses, "all complain for want of Provision, and constantly of breaking into others houses and steal what they have for their maintenance!"[81]

With desperate hunger at Mesopotamia and Christian ardor in short supply, Brother Gründer tried to evangelize among the other slaves and free mulattoes in his Westmoreland neighborhood. He was pleased to find that eight slaves from Amity estate (three miles from Mesopotamia) attended Christmas service at Mesopotamia, and that seven of them continued to pray with the Mesopotamia congregation. A cook from Paradise estate was also coming to his evening meetings. But Barham's attorneys disapproved of mixing people from several estates, and so didn't want to see slaves from other properties drawn to Gründer's meetings. Four free mulatto men from the local village of Cross Path had likewise come to Mesopotamia services, but Gründer was leery of bringing free "brown" people (as he called them) into his congregation. A free mulatto named Banjum (perhaps another son of Daniel Barnjum) was dying of tuberculosis and asked to be baptized, but Gründer told him that baptism wouldn't bring salvation; he had to believe in our Lord Jesus Christ in order to be saved. Banjam's response was, "I am not as ignorant as Negroes are for I can read a little." This led Gründer to conclude that "mulattoes especially when free are rather too proud to belong to a Negro Church." Within the Mesopotamia community Gründer himself clearly exhibited color preferences. He told Bishop Latrobe that he was keeping school with some mulatto children at Mesopotamia, and that three of them "can read now pretty well" and were working on the New

Testament. Blacks were apparently not deemed suitable for schooling by Brother Gründer.[82]

Like some of the other missionaries, the Gründers had trouble handling their horses. In February 1812 "we had a severe Fright by our old Mare having kick'd the Boy by trying to catch her, he was immediately rub'd with some Rum and so was enabled to proceed with us on our journey" to Carmel. This "Boy" was mulatto George, ten years old at the time, who served as a chapel attendant from 1811 to 1815. Two months later, Sarah Gründer was feeding corn to the horses and brushing away their flies when their young mare gave her a kick in the thigh that threw her down. Sarah's great crisis came in January 1815 when she became feverish, was blistered twice very painfully by the doctor, tried to get out of bed, and fell to the floor in a faint. When the doctor saw little hope of Sarah's recovery, as we saw in Chapter 2, Affy's twenty-two-year-old mulatto son Robert rode a mule to the Moravian community at the Bogue and managed to bring Brother Becker to Mesopotamia a few hours before she died at age fifty-three.[83]

Sarah's death was a terrible blow to Brother Gründer, but he soldiered on alone for more than a year, noting forlornly that fewer of the baptized came to his classes after his dear wife was gone. One bright moment for Brother Gründer came in March 1815 when a new chapel bell arrived from England. He had written in 1814 to Joseph Foster Barham, asking him to replace the bell that had been broken in the 1780 hurricane, and pointing out that the chapel needed repairs. Attorney James Grant admonished Gründer for complaining to the absentee proprietor, but overseer Patrickson was obliged to hang the bell, and when it rang out for the first time "a tollerable number [of church members], and even some straingers" came to the meeting.[84]

In the spring of 1816 Brother Gründer sailed to England in search of a wife. Unlike most of the other missionaries who had experienced death and discouragement in Jamaica, he returned to Mesopotamia in January 1817 with his bride (Christian name unknown to me) and immediately resumed his mission work. Gründer also tried to instruct the children on Friendship estate and to preach at Paul Island estate until he was blocked by the overseers. Seeking some way to expand his influence, he wrote to Joseph Foster Barham suggesting that the little children at Mesopotamia, who currently "grow up most wild," be required to have an hour of religious instruction every Sunday. But Brother Gründer ominously mentioned in a letter to Latrobe that he had a tumor on his left side that had to be removed. The local doctor cut this tumor, which was the size of an egg, out of his shoulder. The wound was said to be nearly healed when Gründer contracted tetanus and died in May 1818 at age fifty-two. "He was a quiet inoffensive kind of man," concluded attorney Blyth in memoriam.[85]

Gründer's death quickly triggered a new crisis. Gründer's colleague Thomas Ward, the missionary at the Island estate, had been quarreling vigorously with Barham's attorney, J. R. Webb. So when Ward heard of Gründer's death he immediately moved to Mesopotamia with his wife and child. But in February 1819 Thomas Ward also died. This was too much for the Moravians. The Elders' Conference at Hernnhut announced in 1821 that they were very sorry Joseph Foster Barham had reaped so little fruit from mission work on his two estates, but "we must give it up for the present" and cease sending missionaries to the Island or to Mesopotamia.[86]

## *Salvation by Lot at Mesopotamia*

The Browns in 1798–1799, the Jacksons in 1801–1808, and the Gründers in 1811–1818 are linked together by the fact that they collectively maintained a minute book in English, an exceptionally interesting document known as the Minutes of Mesopotamia Conference, which shows how the missionaries arrived at decisions about adding new members and new communicants to the Mesopotamia church.[87] Every month or two these missionaries would hold a spiritual conference with the Saviour, in which they prayed and sang hymns to him and then proposed various slave men and women as applicants for baptism or communion, asking the Saviour to approve or reject each candidate. Every decision was made by lot, which the Moravians interpreted as the voice of God. The minute book kept by the Browns, Jacksons, and Gründers from January 1798 to January 1818 registers 109 slaves whose cases were presented to the Saviour during this twenty-year period. There were three categories of cases: seventy-eight people who wanted to be baptized, identified by their inventory book names; twenty-seven baptized people who wanted to become communicants, identified by their Christian names; and four baptized people who were excluded from or readmitted to the congregation, also identified by their Christian names. There was probably much overlap between the applicants for baptism and the applicants for confirmation, but this cannot be traced. And since many of the slaves with Christian names cannot be identified, I deal here only with the seventy-eight fully identifiable people who were proposed to the Saviour as candidates for baptism.

The first two pages of the minute book show how the missionaries conducted their spiritual sessions. On February 8, 1798, Brother and Sister Brown opened their initial Mesopotamia conference with an ardent prayer to the Saviour "that He Himself might be our Guide and be present with us whenever we meet together to consider our poor Negroes here." They then posed their first question: Does the Saviour approve "that we lay before him today some of the new people who have expressed their desire to be baptized, as Candidates for baptism? and our Saviour approved it with his gracious answer: Yes.(*) We accordingly asked with Yes and No first: if our Saviour approve that the Widow Sarah be now received as a Candidate for Holy Baptism? It was Yes.(*) Secondly, the same question about Cain the widower? It was No.(*) Thirdly, if the Cooper Thomas at present single be received Candidate for Baptism? It was No.(*) Fourthly the same Question about the married man Chelsea? It was No.(*)"[88] The five asterisks in this statement signify that the Browns obtained each of the Saviour's decisions by lot.

The four cases of Sarah, Cain, Thomas, and Chelsea neatly illustrate the Saviour's style of decision making. Widow Sarah was sixty-four in 1798; born at Mesopotamia, she had been a field hand until age thirty-three and a washerwoman until age fifty-seven, and was now an invalid. And though she had a long, hard, unremarkable life, she made a very quick entry into the Mesopotamia church, for she became a candidate for baptism on February 8 and was baptized on March 25, 1798—just six weeks after she was first proposed. Chelsea, it will be recalled, was from Three Mile River, a twenty-six-year-old cooper in 1798 who would become the driver of the Great Gang in 1801. Despite his leadership within the slave community, Chelsea spent nine months trying to gain admission to the church. Between February and August 1798 the Saviour rejected him six times as a candidate for baptism. He was accepted as a candidate in September, rejected for baptism in October, but approved for baptism in November 1798. The widower with the unfortunate name of Cain had still more difficulty. Born at Mesopotamia and a cooper until age forty, he was fifty-six in 1798, had long suffered from a fistula, and had been too ill to work for the past eight years. Between February and October 1798 the Saviour rejected him five times as a candidate for baptism, so the Browns gave up. In February 1802 the Jacksons took up Cain's case again, and he was rejected three more times until the Saviour accepted his candidacy in May 1802. Then he was approved for baptism in July 1802—four and a half years after his first presentation, and ten months before his death. And the fourth applicant, Cooper Thomas, almost didn't make it. He came to Mesopotamia from Three Mile River estate in 1786 and was thirty-seven when presented for admission. He was accepted as a candidate in March 1798 but was turned down for baptism by the Saviour seven times in 1798–1799. At this point the Browns stepped in and baptized him—apparently without holding any further conference with the Saviour—just before he died in December 1799.[89]

These four cases show that when the missionaries really wanted someone baptized they almost always got their way—but usually only after repeated and protracted negotiations with Jesus Christ. The Browns, Jacksons, and Gründers never described how they obtained yes and no answers from the Saviour via the lot. The usual Moravian practice was to place three papers in a tube, one with a favorable message, another with a negative message, and a third that was blank, and after mixing these papers, to draw out one of them, which was the Saviour's choice. However, the Browns, Jacksons, and Gründers may have used a different procedure. Certainly the Saviour's pattern of decision making had its oddities. Why did men have a much harder time gaining his approval than women? There were many fewer potential male church members to start with, since the missionaries in 1798–1818 brought forward twice as many female candidates as male candidates. And 42 percent of the men who were presented received three or more no answers, as against 22 percent of the women. Another curious feature of the lot system at Mesopotamia is that the Saviour became more indulgent over time. The Browns in 1798–1799 obtained yes answers 43 percent of the time, the Jacksons in 1801–1808 got yes answers 48 percent of the time, and the Gründers in 1811–1818 got yes answers 51 percent of the time. The Jacksons and Gründers may have practiced the lot by using only two papers, an affirmative message and a negative message. And all of the missionaries seem to have manipulated the result in emergencies. Most times when a slave was fatally ill and implored the missionaries for baptism, the Saviour immediately approved. From the Moravian perspective this demonstrated Christ's gracious mercy.

In the end, most of the seventy-eight candidates presented for baptism between 1798 and 1818 received the Saviour's approval. Twenty-six men were proposed by the missionaries, of whom twenty-four became candidates, and nineteen were baptized. Fifty-two women were proposed, forty-eight became candidates, and forty were baptized. But despite the admission of fifty-nine new members in 1798–1818, the Mesopotamia congregation was not much larger (thirty-five members in 1818) at the close of the Gründers' ministry than at the beginning of the Browns' ministry (thirty-two members in 1798).

In retrospect, the most significant decision made by the Saviour in 1798–1818 was to baptize cooper Chelsea, one of the four candidates presented by the Browns at their first conference on February 8, 1798. The Saviour was not easily persuaded to accept Chelsea, rejecting him seven times. But this decision really bolstered the Mesopotamia congregation. Chelsea was a son of Betty and Qua, who had moved from Three Mile River estate to Mesopotamia with seven children in 1786. (The fully detailed genealogy of Chelsea's family, previously discussed in Chapter 4, is shown at *www.twoplantations.com*.) Chelsea, who was baptized in 1798 as James Carr, became the driver of the Mesopotamia Great Gang in 1801 at age thirty—a position he would keep for twenty-three years. And he brought eight members of his family into the Mesopotamia congregation. Chelsea's wife Charity was baptized in 1798, his mother Betty was baptized in 1802, his sister Sophy in 1813, his sister Bessy in 1814, his nephew King also in 1814, his sister Nancy in 1815, and his niece Wonder sometime after 1818. Chelsea's brother Charles was rejected by the Saviour in 1805, but Charles's wife Doll was baptized in 1802. This is the largest recorded family network in the Mesopotamia congregation—though several members of this family died soon after baptism. Moravian missionary Samuel Gründer recorded the poignant demise of Bessy at age forty-four. Bessy was Chelsea's twin and the mother of seven children, five of whom outlived her. Gründer described Bessy as an ardent member of the congregation and a devoted mother who "always kept her Children very clean and nice," bringing two or three of them to every meeting. Suddenly in September 1815 she had a violent pain in her head that could not be cured. The sick nurse overheard Bessy, two days before her death, praying "fervently to the Lord to spare her life a little longer for her childrens sake."[90]

### *The Moravian Community on the Eve of Emancipation*

From March 1819 until October 1830 there was no resident missionary at Mesopotamia, and the congregation had to be content with occasional appearances by visiting Moravians. One such visitor was Brother Lewis Stobwasser in May 1824, who noted on arrival that the Mesopotamia people called the place "Barham"—the name it has today. He also observed that mosquitoes "appear to be quite at home here." The church members had evidently been protecting the missionary house, unoccupied for five years, because the furniture, books, and utensils were still in place, though cockroaches and worms had infested the books and manuscripts. Attorney William Ridgard gave orders to sweep the chapel, and "at noon from the belfry near the chapel a sign was given for the meeting and soon all the inhabitants of the plantation white and black, young and old assembled together. . . . The Negroes listened very attentively to the preaching on John 14, 16[91] and order and stillness were prevailing throughout. During the singing we perceived with pleasure, that they were still tolerably acquainted with our hymn tunes. Services being over the Negroes flocked round us to testify their joy at seeing and hearing again teachers from the Brethrens church whom they had been so much used to." Stobwasser then spent several hours conversing with the congregants and bringing the Mesopotamia church catalogue (which unfortunately has disappeared) up to date. He found that there were now only four communicants and twenty-four baptized members, including seven who had been baptized as infants.[92]

While the Mesopotamia church was being neglected during the 1820s, the economic, political, and religious situation within Jamaica was changing very rapidly. Sugar prices and planter profits were falling, abolitionist sentiment in Britain was rising, the slave-owning Caribbean planters were being vilified in Britain for their barbarity, and the possibility of slave emancipation by the home government was in the air. In Jamaica the planters were fuming about new restrictions placed on them by Parliament. Barham's attorney John Blyth was disgusted by the "unconstitutional" parliamentary efforts to ruin the slave owners, such as the prohibition of crop work after 7 P.M. on Saturday and before 5 A.M. on Monday, and he lamented that the slaves, who work barely half as much as formerly, now expect to do still less![93] On the religious front, Baptist and Methodist missionaries had become very active in Jamaica, drawing large numbers of slaves to their evangelical revival meetings—which triggered further rage from the planters. The Moravians carefully distanced themselves from the Baptists and Methodists; they continued to preach passive submission to their congregants and disavowed radical social change. Indeed, they were still slaveholders themselves until 1832, when the Moravian conference in Jamaica grudgingly agreed to manumit the black people they owned at their Fairfield station in St. Elizabeth.

The Jamaica slaves were keenly aware that the system they were trapped in was disintegrating. With the alluring possibility of legal freedom, they saw the advantages of membership in the Christian church, and so enrollment in the Moravian congregations began to expand. This new interest in religious participation benefited the Christian slaves at Mesopotamia. Throughout the 1820s, whenever visiting Moravians showed up at Mesopotamia the church members pleaded for the return of a resident mission. In 1829 Joseph Foster Barham, now old and ailing, applied to the Brethren in England for the same purpose, and he agreed to increase the salary for a pair of Mesopotamia missionaries from £50 to £100 per year. Meanwhile, the Moravian brethren in Jamaica increased their visits to Barham's estate. In 1828–1830 Brothers Scholefield and Zorn took turns coming to Mesopotamia every few weeks to conduct services.[94]

Brother Jacob Zorn wrote a diary in which he described his eight journeys to and from the estate between March and September 1830. On one trip he brought along a boy who rode a second horse and carried a portmanteau containing Zorn's extra clothes, because "in this hot climate . . . the traveller is sure to be drenched with perspiration or a heavy shower." On another trip Zorn was riding carelessly on a borrowed horse and holding an umbrella to shield himself from the sun when the horse jumped and he was thrown to the ground on his back. Remounting his steed, Zorn then rode down to the seacoast and cantered along the beach before turning inland to reach Mesopotamia "much fatigued." The next day he officiated at the marriage of Affy's quadroon granddaughter Jane Ritchie, as described in Chapter 2. Since the congregation had not seen any Moravian sister for ten years, Sister Caroline Zorn accompanied her husband on three of these trips to Mesopotamia. The slaves flocked to her, but Brother Zorn complained that "there is not much of that warm love to the Saviour among the members at Mesopotamia," and he hoped that "a brown man on the Estate" who was baptized in 1826 and can read and write, could help spread the gospel message. This brown man was a twenty-five-year-old mulatto carpenter named Henry, the son of housekeeper Kate (baptized in 1815 as Catherine Patrickson) and overseer John Patrickson. But whatever help Henry gave the Moravians was brief, for he was manumitted in 1831 and left the estate upon payment of £100 to Barham.[95]

On October 29, 1830, a new pair of Moravian missionaries, Peter and Sarah Ricksecker, took up residency at Mesopotamia with their two children. For the first time since 1819 the mission was fully reactivated. Peter Ricksecker kept no record in 1830, but he wrote a short diary in English describing what happened in 1831. Initially he drew large crowds to the chapel meetings. There was full attendance for the Good Friday service in 1831, and on Easter Sunday the chapel was filled to capacity. But most of these churchgoers came from the local neighborhood, especially the village of Cross Path. The Jamaica conference of Moravian missionaries asked Joseph Foster Barham for permission to make Mesopotamia an independent station, "freed from the embarrassments consequent to being restricted to one estate," and when Barham consented to this change, the conference talked in August 1831 of moving the Mesopotamia mission house and chapel to Cross Path. But this would be expensive, and for the time being the Rickseckers stayed at Mesopotamia, trying to build up the congregation on the estate. Brother Peter was vexed that the Mesopotamia baptismal candidates and "new people" were not attending the special meetings he organized for them, supposing that going to church was enough. He also discovered that some of his newly baptized members turned out to be fornicators, and decided that it was very difficult to wipe out such sinful practices on this estate.

Nevertheless, the Rickseckers did enlarge their Mesopotamia congregation—more rapidly than at any time since the 1760s. Capitalizing on the slaves' new interest in establishing their religious credentials, they baptized eighteen adults and three children during 1831, instructed fifty-six "new people" and thirty-five candidates for baptism, and expanded the number of communicants from two to nine, with seventeen candidates hoping to be confirmed. By December 1831 the Mesopotamia congregation had forty-one adult baptized members in good standing, a very notable increase from the seventeen baptized adults Brother Stobwasser had found in 1824. But the Rickseckers were clearly having problems with their flock, because they had suspended or excluded another twelve adults for bad behavior.[96]

The Rickseckers did not compile a list of their new members, but by good fortune the Mesopotamia slave inventory dated January 1, 1832, identifies all of the baptized slaves—the first time that the bookkeepers had bothered to list them since the inventory taken in 1768, sixty-four years before. Twenty-six males and forty females are listed both by their "Original" or book names and by their "Christian" names.[97] Unhappily, this inventory does not match very well with the Rickseckers' figures. It lists sixty-three adult and three child "Christians"—whereas according to Peter Ricksecker there were only fifty-three baptized adults (including the twelve suspended or expelled from the meeting) on the estate in December 1831. Some of the "Christian" people on the inventory list had probably quit the Moravian Church or had been expelled before the Rickseckers arrived, and others may have been baptized into the Church of England. But since this is the only membership list, we must work with it as best we can.

Nineteen of the "Christian" people on the 1832 inventory list had been members of the Mesopotamia church since at least 1818, including Affy (Sarah Affir), admitted in 1814; Chelsea (James Carr), admitted in 1798; his sister Sophy (Catherine Carr), admitted in 1813; and his sister-in-law Doll (Eliza Anderson), admitted in 1802. The other forty-seven were new, most of them presumably baptized by the Rickseckers. A big difference from earlier Moravian lists was the styling of the "Christian" names. Back in the 1760s when the Moravian missionaries had baptized the new converts, they had chosen biblical names for them. But over time the baptized began to choose names for themselves. In 1831 there were still a few Isaacs and Abrahams, but most of the slaves adopted the standard English first names that the white people used. About half of the men on the 1831 list kept their book names, or changed nicknames like Bob and Will into Robert and William, while the women tended to choose fashionable-sounding new names like Isabella in substitution for Hannah, or Felecia for Phillis. And nearly 80 percent of these "Christians" chose to have surnames as well. The mulattoes adopted their white fathers' surnames, while a strikingly high percentage of the blacks adopted their absentee owner's surname. The two most popular surnames by far were Barham (thirteen people) and Foster (seven people). And four slaves chose to have *both* names: blacksmith Edward, imported from Springfield in 1819, became Edward Foster Barham; grass cutter Dorinda, imported from Africa in 1793, became Elizabeth Foster Barham; doctress Janet, born at Mesopotamia, became Jane Foster Barham; and first field gang member Queen, also born at Mesopotamia, became Helen Foster Barham. Thus, the "Christian" names borne by these sixty-six Mesopotamia people reflected both their own individuality and—for many—their personal loyalty to Joseph Foster Barham II.

From Peter Ricksecker's point of view, these "Christian" people had a feeble commitment to their new religion. Ricksecker reports that on Christmas Day in 1831 hardly any adults showed up for divine service because they were all enjoying their holiday rum ration from the overseer. Some of those who did come to church were not in fit condition for worship. And to Ricksecker's annoyance, the drunken holiday celebration continued all afternoon and into the night until 3 A.M.[98] A few days later Brother Peter had a very different problem to worry about. He learned that the largest slave rebellion in the history of Jamaica had broken out in St. James parish on December 27, and that the rebels were moving rapidly into Westmoreland. But that is the subject of another chapter.

Reviewing the history of the Mesopotamia congregation from 1758 to 1831, it is easy to see why the missionary historian J. H. Buchner called the Moravian effort to convert slaves on this estate a dismal failure. The congregation reached its peak strength as early as 1769 and gradually but steadily declined over the next sixty years. The Mesopotamia people never fully embraced the Moravian adoration of the blood and wounds of Jesus Christ, while the missionaries were continually frustrated by their black converts' lack of Christian zeal, their polygamous way of life, and their devotion to "noisy play" at holiday time. The estate managers did their best to obstruct the mission, and the absentee Barhams gave backhanded support. The thirty-nine Moravian evangelists who lived at Mesopotamia from the 1750s to the 1830s were devastated by the hot, wet climate and the toxic disease environment, and fifteen of them died on the estate. The longest tenure at Mesopotamia was seven years, the average tenure was less than two years, and the station was left leaderless for a quarter of the time.

Yet while the Moravians did little or nothing to improve living or working conditions at Mesopotamia, they surely made a difference to their Mesopotamia converts. About 180 people joined the congregation during the "decline" period between 1770 and 1833, an average of three new members per year. The men and women who came into the congregation could readily identify with the sufferings of Jesus, which gave them hope and comfort amid the great cruelties of slave life. Even the Mesopotamia people who did not attend the church services could hardly avoid noticing that the Moravians had moral standards in total contrast to the vicious and licentious conduct of the other local whites. The male missionaries were accompanied by their wives (the only white women on the estate), and unlike the Mesopotamia overseers and bookkeepers, they never fornicated with the young black women. We may wish that the missionaries had been abolitionists, or at least that they had stood up more aggressively to the absentee owners and the estate managers. But the evangelizers who came to Mesopotamia were totally focused on heavenly salvation. They were more otherworldly than any of their Protestant or Catholic rivals, and they made extraordinary sacrifices to reach people in need of spiritual help. A hymn composed in Jamaica in 1768 to memorialize William Foster, the elder brother of Joseph Foster Barham I, conveys in its jogging couplets the essence of the Moravian mission at Mesopotamia:

> We praise the Lord, we praise the Lord
> who through his Grace, & precious Word
> the Gospel, calls so many Souls
> to their sweet rest in his wound-holes.
>
> The Gracious Lord, the Lamb of God
> had by his suffring Death & Blood
> Redeem'd the whole World from the Fall
> and Bought us all, he bought us all. . . .
>
> Let them be unjust and unclean
> So full of Wickedness & Sin
> His Blood, his Blood will Justify
> each Sinner who'll to him apply.

The Heath'n are his Inheritance
They shall enjoy his Love immence
Therefore God sends his Word to them
His Grace & Mercy to proclaim. . . .

Ye Negroes hear, with reverence
Think on your Masters Love intents
What his Kind heart has done for you
be to his Children good & true.

Lord let us Count our Days
as Long we are here Living. . . .
That when thy Message comes
t' leave this World transitory. . . .

That we in Grace might be
Cloathed with thy red Dress
God & the Lamb to see
with Joy & Righteousness.[99]

# 7

## *The Exodus from Mount Airy to Alabama*

William Henry Tayloe (1799–1871), the master of Mount Airy from 1828 to 1865, was a different sort of person from his father, John Tayloe III, and he also lived in a different era, when southern leadership was passing from the old seaboard states to the new cotton states. As a slaveholder, however, William shared much in common with John III. Recognizing that his cohort of enslaved laborers was constantly increasing, he took full advantage of this situation by moving surplus workers to more profitable work sites or selling them to make money. William did not proclaim, as John III repeatedly did, that slaves should be made to do their duty. Living in an era of Victorian sensibility, he felt, I believe, considerably more moral responsibility toward his human property than his father ever evinced. But this moral responsibility did not stop him from routinely breaking up the African American families under his control.

To judge by the nearly 14,000 documents he left to posterity, William Henry Tayloe was a conscientious, hardworking, intelligent, and reflective family man. Reading through his voluminous correspondence, he comes across as far less imperious and much more approachable than his formidable father. As a boy William attended Phillips Exeter Academy in New Hampshire, but he did not go on to college like his brothers Ogle and Edward at Harvard, George at Princeton, and Henry at the University of Virginia. In 1824 William married his first cousin Henrietta Ogle, who bore him nine children, six of whom died young.[1] A Whig in politics, he held statewide office in the Virginia House of Delegates for one uneventful term. Lacking John Tayloe III's extraordinary ambition and entrepreneurial zeal, he avoided operating ironworks or directing banks or building hotels in Washington. Instead he concentrated on becoming a successful planter. Like his father in 1792, William had more African American "servants" (as he called them) in 1828 than he could profitably employ. But instead of selling surplus workers, as John III had done in 1792–1794, William sent 120 Mount Airy slaves on a journey of 800 miles to Alabama between 1833 and 1854, sold 23 of them on arrival, and converted almost all of the other 97 into cotton hands. William increased his grain production in Virginia while he was establishing two cotton plantations in Alabama, and by the 1850s was selling his Alabama cotton and his Virginia wheat and corn for today's equivalent of $1,600,000 a year. During the Civil War another ninety-eight Mount Airy slaves were sent to Alabama. And Tayloe's slave force increased from 193 in 1828 to 457 in 1863. Thus, his ever-expanding Mount Airy slave force was the engine for his success.

Thanks to William Henry Tayloe's detailed record keeping, we can track individually the 218 Mount Airy men, women, and children he transferred to Alabama and find out what happened to them after they arrived. These Mount Airy people were part of an enormous migration in which approximately one million African Americans born on the eastern seaboard were moved to the western and southern slave states between 1790 and 1860—the most important development to take place during the entire history of antebellum slavery.[2] This internal mass movement had two components: about two-thirds of the migrating slaves were sold to professional slave traders, while the other third was moved southwest by their owners (such as the Tayloes).[3] New studies emphasize the wrenching, traumatic character of this deportation and focus on the savage behavior of the slave traders.[4] Ira Berlin, Walter Johnson, and Edward Baptist argue that the internal migration from the mid-Atlantic states to the Gulf states was just as brutal and horrific an experience for nineteenth-century African Americans as the Middle Passage had been for their ancestors. In both cases people were snatched from their families and stolen from their homes, never to see their parents or spouses or children again. From the 1820s onward the interregional traders operated holding pens (equivalent to the barracoons on the West African coast) to house the men, women, and children they were assembling for transfer and resale. They shipped some people directly to New Orleans, the chief slave market on the Gulf.[5] Others were sent overland, locked into coffles—where dozens of men were manacled and sometimes fitted with iron collars attached to a long chain—and made to walk along roads leading south and west at a pace set by the lead slaves. Women, less likely to run away, could be roped together by halters around their necks.[6] These coffled slaves walked for hundreds of miles, stopping at places where the traders could get the best prices, and then they would be auctioned off or sold door to door. Along the way, the traders expanded their operations by kidnapping free blacks, and amused themselves by selecting some of the slave women to serve as their "fancy maids."[7]

Walter Johnson in *River of Dark Dreams* and Edward Baptist in *The Half Has Never Been Told* both argue forcefully that the thousands of new plantations established in the cotton states between 1800 and 1860 differed greatly in character from the eastern seaboard farms and plantations where the migrant slaves had previously lived. Their books differ in style and focus, but both historians see the American slave system erupting into a corrosive apogee of unprecedented wealth, intoxicating power, unbridled speculation, and virulent exploitation of black labor. This abrasive new view of the Cotton Kingdom is in most ways compelling. Johnson and Baptist demonstrate that the internal slave trade was one of most disgraceful episodes in American history, and that the cotton production system enabled reckless slaveholders in the Deep South to acquire dangerously great wealth and political power. I doubt, however, that all of the cotton planters were as grossly abusive as these authors suggest. My examination of the Mount Airy exodus to Alabama is a small-scale case study, but it suggests an alternative way of looking at the antebellum slave migration and at slave life on the cotton plantations. The Mount Airy migrants were not sold to slave traders: they were among the 30–40 percent of the migrating slaves who were moved to the cotton states by their masters. William Henry Tayloe's plan, which he achieved, was to make his slave-based Alabama farming a direct extension of his slave-based Virginia farming. The Mount Airy migrants who came to Alabama between 1833 and 1854 certainly suffered. They had to walk 800 miles in order to reach the Alabama Canebrake; a fifth of them were sold on arrival, another fifth experienced serious abuse in the 1830s, and all of them were forced to work very hard. But slave life for the Mount Airy people in Alabama does not appear to have been hugely different from slave life for the Mount Airy people in Virginia. In particular, slave family formation, which was remarkably robust at Mount Airy, continued to be equally robust at Tayloe's Oakland and Larkin plantations in the Canebrake.

My evidence comes from slaveholder William Henry Tayloe's records, whereas Walter Johnson and Edward Baptist draw upon the narratives of slaves who escaped from the cotton states and wrote accounts of their terrible experiences for northern abolitionist audiences.[8] I do not doubt the validity of these narratives, which vividly express the long-term African American struggle for freedom and equality. But I question whether the accounts by these escaped slaves tell the whole story about what was going on in the cotton states before the Civil War, and I have not used them as proxies for slave life at Oakland and Larkin. This may seem inconsistent, since in Chapter 4 I *do* use Thomas Thistlewood's diary as a proxy for slave discipline at Mesopotamia. But Thistlewood lived in the Mesopotamia neighborhood, whereas none of the freedom narrators that I have read describe conditions in the Alabama Canebrake. On the same grounds I have not utilized the WPA Slave Narrative series to find out more about the Mount Airy people in Alabama, because none of the 2,300 ex-slaves who were interviewed in 1936–1938 had been owned by William Henry Tayloe or had lived at Oakland or Larkin.

Certainly William Tayloe's view of the Mount Airy exodus is provocative. He and his brothers claimed that the Mount Airy slaves *wanted* to go to Alabama and that they migrated voluntarily, even eagerly. He also argued that the Mount Airy migrants were contented on arrival and that they were much better off in Alabama than in Virginia. In this chapter we will test these assertions by tracking the Mount Airy migrants person by person as they traveled to Alabama between 1833 and 1862. And we will try to gauge what life was like for them after they reached the Canebrake.

## *The Initial Move to Alabama in the 1830s*

The Mount Airy slaves were moved to Alabama in five stages: thirty-nine people were sent in 1833–1837, forty-five in 1845, thirty-six in 1854, forty-eight in 1861, and fifty in 1862. The pioneers who came in the 1830s were the groundbreakers, and they are listed in Appendix 29. Eight men led the way in 1833, followed by six males and ten females in 1835, six males and eight females in 1836, and a single man in 1837. It is important to distinguish the members of the three parties sent in 1833, 1835, and 1836, because they had differing experiences in Alabama. Thirteen of these early migrants were sold soon after arrival, while another six died at an early age. Though they were twenty-one years old on average when they migrated, only seventeen of the thirty-nine were living on William Henry Tayloe's two cotton plantations (Larkin and Oakland) by 1863. Clearly the move to the cotton kingdom was a grueling experience for these Mount Airy pioneers.

To understand why William Henry Tayloe sent his first pioneering party to Alabama in 1833, we need to review his situation after the six Tayloe brothers divided up John Tayloe III's estate in 1828–1829.[9] At Mount Airy William took an inventory of his slaves in late 1830, and counted 119 men and boys and 102 women and girls, for a total population of 221.[10] His slave force was extremely youthful. Only 14 percent of the men and women were over the age of fifty; 30 percent were in their twenties, thirties, and forties; 56 percent were under the age of twenty; and 37 percent were under the age of ten. So if William Henry Tayloe was a little short on prime workers in 1830, he was assured of a large future labor supply. The slave workers in 1830 were almost evenly divided into two categories: there were twenty-one domestics and forty-four artisans, most of whom lived with their young children at the home plantation, and sixty-six field hands who lived with their young children and a few old retired slaves on three farm quarters. Tayloe had inherited four farm quarters, and consolidated them into three units named Landsdown (formerly Old House), Forkland, and Doctor's Hall, each supervised by an overseer. The proportion of domestic and artisan workers was considerably larger than in John III's day, and from an economic point of view this was the bad news. The domestic workers produced no income and the artisans only marginal income, so William Tayloe depended on the labors of his field hands for almost all of his earnings. And the farmland bequeathed to him was less productive and profitable than a generation earlier.

Between 1805 and 1820 John Tayloe III had generally sold about 7,000 bushels of Mount Airy wheat and 2,000 barrels of Mount Airy corn (5 bushels to a barrel) for $14,500–$16,500 per annum. But two-thirds of this grain was produced at Gwinfield, Oaken Brow, and Hopyard farm quarters, which were bequeathed to three of William's younger brothers in 1828. Grain prices were falling when William began to operate his Landsdown, Forkland, and Doctor's Hall farms, and in 1836 his wheat crop failed almost completely. Thus, in the 1830s Tayloe was marketing about 1,800 bushels of wheat and 950 barrels of corn and selling this grain for $5,250 per annum.[11] William began his farming career with a strong sense that his Virginia acreage was wearing out and that he needed to find better venues for his slave laborers.

William's first idea—copying his father's scheme of sending surplus Mount Airy workers to Cloverdale—was to join his older brother Benjamin (known as Ogle) and his younger brother George (the owner of Cloverdale furnace) in a partnership to operate a new ironworks, Brunswick Forge, in the Blue Ridge Mountains, 200 miles away. William and Ogle each supplied George with six woodcutters for this project in January 1832. William's woodcutters were six Mount Airy field hands, the first of his migrant workers—and thus important people to focus on.

In 1832 Davy Moore (b. 1798), Tom Moore (b. 1800), and Jim Moore (b. 1801) were three brothers in their early thirties. Their mother, Milly, had died in 1811 when they were young boys aged thirteen, eleven, and ten, and their father cannot be identified. For twenty years the Moore brothers had been working together as farmhands at Old House quarter. Perhaps they were married and had to abandon their unrecorded wives and children when they were sent to Brunswick Forge. But it seems more likely that William Tayloe selected the Moore brothers because they were bachelors and therefore could be removed without breaking up families. The other three migrants—Billy Page (b. 1812) from Old House quarter, Ralph Elms (b. 1811) from Forkland quarter, and James Thurston (b. 1807) from Doctor's Hall quarter—were all young men from very large farm families. Billy's mother, Sucky, bore eleven children; Ralph's mother, Fanny, bore nine children; and James's mother, Sally, bore thirteen children. Here Tayloe's operating principle seemed to be that large families could better absorb the loss of individual members than small families.

But of course the loss was still palpable. Billy Page's parents, Billy and Sucky (aged fifty-four and fifty-five in 1832), had already lost four of their children through early death. A daughter had been sold, and a son had been given to Ogle Tayloe, so only three sons and one daughter remained at Mount Airy when young Billy was sent to the Brunswick Forge. Ralph Elms's mother, Fanny (age fifty-seven in 1832), had similarly lost one boy through early death, and two of her daughters and two sons had been taken by other members of the Tayloe family, so Fanny was left with only two sons and a daughter at Mount Airy when Ralph was dispatched to Brunswick Forge. James Thurston's mother, Sally (age fifty-two in 1832), had the largest of these families, and only one son had died young. Eight of her children remained at Mount Airy after James was sent away. But she too was victimized by the Tayloes' policy, since one of her sons and one of her daughters had been sold, and another daughter had been given to Henry Tayloe.

On January 13, 1832, Davy Moore, Tom Moore, Jim Moore, Billy Page, Ralph Elms, and James Thurston left Mount Airy and started walking toward Cloverdale in the Blue Ridge, approximately 200 miles away. A mason named Adam went with them, but he returned to Mount Airy with a string of horses. Ebenezer Jeffries, the brother or son of one of Tayloe's overseers, led this Mount Airy party (probably on horseback) and brought them to Cloverdale by January 23, so they averaged about twenty miles per day. Jeffries reported that Jim Moore from Old House was very lame on arrival.[12] Did he let these Mount Airy men walk freely or tie them into a coffle? Jeffries doesn't tell us, but the evidence from later journeys indicates that the Mount Airy migrants were not chained up.

It soon turned out, however, that the laborious walk from Mount Airy to the Blue Ridge by the six field hand / woodcutters was all in vain. George Tayloe hired William's and Ogle's twelve slaves at $60 per annum, but the three-brother Tayloe ironworks partnership quickly unraveled. George complained that the new woodcutters showed a disposition to run away—hardly surprising, since ironworks slave labor was extremely hard and demanding. William and Ogle distrusted brother George. Ogle protested to William that George's "schemes are as visionary as we had cause to apprehend," while William complained to Ogle that George's manager at Brunswick Forge was a shady character. In March 1832 the brothers canceled their contract with George.[13] And Ogle was already looking for better opportunities in the South. In February 1832 he announced that he had bought half a share of 3,110 acres in the Pensacola region of Florida, "in the belief we should not long hence have to look for a vent for our negroes; besides thinking the culture of sugar & cotton a few years hence will be much more profitable than that of wheat and corn."[14]

In mid-1833 another brother, Henry Augustine Tayloe, whom William and Ogle liked much better than George, came into the picture. Henry was twenty-four years old in 1833, and the owner of Gwinfield farm quarter with about sixty slaves, a small patrimony that dissatisfied him, and he asked his elder brothers for help so that he could start a Tayloe cotton plantation in Alabama. William and Ogle quickly decided to enter into partnership with him by loaning him some money and contributing the slaves they had sent to the Brunswick Forge ironworks, plus further hands from Deep Hole farm quarter on the Potomac in Prince William County, which they owned jointly.[15] "I shall hope," Ogle wrote to William in November 1833, "gradually to remove to Alabama *all* the negro children capable of picking cotton, and such other hands as may be more valuable there than on our plantations. If satisfied in regard to the healthfulness of Henry's location in Alabama, and that our negroes can be supported as cheaply there, as in Virginia, I should prefer to send *all worth removal.*" And, he added, "I feel assured neither Henry nor yourself will engage in any visionary schemes."[16]

Henry Tayloe was young, adventurous, and enthusiastic: appropriate attributes for the leader of what he called "our Alabama speculation."[17] But he was also impetuous, reckless, and something of a visionary schemer. Having put together nearly $4,000 in loans from his brothers to invest in Alabama land, he asked William to supervise his Gwinfield estate while he was in Alabama, and directed William and Ogle to send the laborers they were contributing to Cloverdale. He reported to William on December 16, 1833, that his Gwinfield hands "were so anxious to commence their journey [to Cloverdale] that I sent them about twelve miles yesterday evening."[18] Henry traveled separately, and arrived at Cloverdale on December 26 amid thirteen inches of snow to take charge of forty-four slaves belonging to the three brothers. Henry, with a half share in the enterprise, contributed twenty people from Gwinfield. This was about a third of his total slave population, including most of his prime-aged workers, and shows the huge commitment he was making. William and Ogle each had a quarter share. They were much larger slaveholders than Henry, but at first they invested cautiously in his Alabama workforce. They each supplied Henry with the woodcutters they had sent to the Blue Ridge two years before, and added ten people from Deep Hole and a few further slaves.[19]

William Tayloe contributed thirteen slaves to Henry's pioneering party: eight from Mount Airy and five from Deep Hole. The eight Mount Airy migrants were the six Brunswick Forge woodcutters, Davy Moore, Tom Moore, Jim Moore, Billy Page, Ralph Elms, and James Thurston, plus a thirty-eight-year-old domestic servant named John Moore and a sixteen-year-old houseboy named Emanuel (surname unknown). It is puzzling that Tayloe selected two domestics on an expedition that clearly called for hard physical labor. Probably John Moore was chosen because he had been a gardener and was an older brother of Davy, Tom, and Jim. Emanuel seems to have been a poor choice, since he was about the only Mount Airy migrant who was sent back to Virginia. While all of William's Mount Airy migrants were males of working age, there were fourteen women and six young children from Gwinfield and Deep Hole in the expedition.[20] Henry told William, "Your hands look well and are perfectly willing to remove South."

A twenty-six-year-old carpenter from Oaken Brow named Travers, who belonged to Charles Tayloe, the youngest of the brothers, had managed to escape en route to Cloverdale, but Henry claimed that the others were "in fine health and spirits, they were all anxious to go."[21] One wonders whether they understood that they would be hiking twenty miles a day for the next month and a half. On December 31, 1833, the forty-four Tayloe slaves started on their long journey, led by an unnamed supervisor who rode on horseback. The expedition was equipped with a wagon led by two mules, which carried supplies (including tents for overnight camping), and the youngest children probably rode in this wagon. The others went on foot. We do not know the route they took, but they probably crossed the Appalachian Mountains from southwestern Virginia into Tennessee and proceeded down into Alabama. In 1839, when Ogle Tayloe was sending another complement of slaves to Alabama, Henry told him to send them via Knoxville, Tennessee. He also said that the journey would take forty-five days and emphasized that the migrants should be well shod and clad.[22]

Henry Tayloe traveled ahead of his slave party, and by early February 1834 he was in Greensboro in west-central Alabama, in the heart of the Canebrake region, so called from the thickets of bamboo cane that grew along the river floodplains. This region was also part of the Black Belt, named for the rich, dark topsoil found in a wide band across

Alabama that was ideal for cotton cultivation. Henry was by no means the first Virginian to explore the Black Belt. By 1834 pioneering cotton planters had brought more than 100,000 slaves into Alabama, mainly from the Chesapeake and the Carolinas. The cotton plantations that Henry saw had larger slave gangs than on most Virginia farms, and the gangs were more youthful.[23] He was delighted with the good canebrake land but disliked the white planters who were settling this region. "They place too little value on human life," he told William. "They will kill a man for the slightest provocation."[24]

Canebrake land was expensive, and Henry announced that if he couldn't afford to buy a tract right away, he would hire out his Virginia slaves at $140 per annum—a bleak prospect for the eight newly arrived Mount Airy migrants.[25] But Henry knew that the Tayloe partners had to have land, and in 1834 he bought two adjacent properties in Marengo County. First, he acquired an 853-acre uncleared tract that he named Adventure plantation, and deeded undivided quarter shares of this property to William and to Ogle. Then he bought a working plantation next door to Adventure named Walnut Grove with cleared fields and a gang of about forty slaves—a transaction that cost far more than he could afford. Henry probably paid $10,808 for Walnut Grove, because in July 1834 he sold an undivided half of this property to William and Ogle for $5,404.[26] Though William thus acquired a one-quarter share in Walnut Grove, very few of his Mount Airy slaves ever lived or worked there. But Walnut Grove became Henry's headquarters for the next decade.

Henry placed all of the slaves he had brought from Virginia on Adventure plantation and gave them the challenging task of clearing and planting the land. Thus, William's eight Mount Airy migrants were put to work cutting down thickets of trees, rolling logs, and grubbing underbrush as well as building cabins and fences. After which they had to learn how to plant and pick cotton. In October 1834 Henry told William, "Our hands pick [cotton] very badly not having been accustomed to it, [but] they improve somewhat however and another year will make good pickers."[27] By early 1835 Henry was much more upbeat, reporting that his partnership with William and Ogle had already garnered $1,000 in sales from Adventure and that he expected to make 200 bales of cotton at Walnut Grove and Adventure. Then in July 1835 he told his partners that he was going to take out a loan to buy more Negroes unless William and Ogle sent him some more slaves from Virginia.[28] Anxious to avoid any further loan payments, the brothers quickly sent Henry a second contingent of laborers in October 1835, with William supplying sixteen people and Ogle ten.[29]

As *Appendix 29* shows, the sixteen Mount Airy slaves who were sent to Alabama in 1835 were more variegated than in 1833: there were five male field hands, eight female field hands, two nonworking children, and an elderly nurse. Eight of the migrants were teenagers. The group included a father and son, two young motherless sisters (Anna Marcus and Jane), and a grandmother who was joined by two of her daughters and two of her granddaughters. The grandmother was fifty-three-year-old Nanny Glascow, who was beyond field labor but could take care of her baby granddaughter Mary Glascow. Ogle was critical of William's selection, asking why he was sending Barneby, who was just seven years old. "I should think it best," he wrote, "to send only such as can walk & be of some service on arrival." This needled William, who scribbled an August 1835 memorandum in which he described the men he was sending as "very good at the plough or any plantation business," the women as "first rate" at anything they were told to do, and the whole group as "a lot of hands equal to any in any Country and capable of making us a full crop of Cotton."[30]

Hearing that William and Ogle were sending a second batch of slaves to Henry, their brother Edward Thornton Tayloe, the owner of Hopyard (renamed Powhatan Hall), reported to William in August 1835 that he was now inclined to send some of his own slaves to Alabama. And he did send seven to Walnut Grove plantation in 1835 and eight more in 1836. "I would not send those who are unwilling to go," Edward told William, "but I have no doubt, that when they find out that some of yours' and Br: Ogle's are to go with them, they would all prefer going."[31] Edward's remark suggests the pulling power, not only for the slave owners but also for the slaves, of a change of scene. We tend to assume that while the white slaveholders were eager to improve themselves by moving south and west, their enslaved workers all wanted to stay put. But was this necessarily so? For young African Americans, the move to an unknown new frontier may have seemed much more exciting and promising than the stifling status quo. And, as Edward Tayloe pointed out, his slaves would not be outcasts; they were joining other young people with whom they had connections from William's and Ogle's and Henry's Virginia plantations. They would be reuniting with friends and relatives they had formerly lived with at Mount Airy on a hopeful new adventure.

When William's and Ogle's second party of Virginians arrived in Alabama, Henry Tayloe immediately sold one of William's Mount Airy people—fifteen-year-old Grace, who had been a field hand at Forkland quarter. The others were put to work clearing land at Adventure, sharing in the same grueling labor with the first party of Virginians. In 1836 Henry hired them out to other planters for the year, and then in 1837 moved them all to a new tract that he acquired for the Tayloe partnership named Oakland. The Mount Airy people in this second party thus had to endure at least four or five consecutive years of transforming forest land into farmland. Oakland plantation, situated a few miles from Adventure and Walnut Grove on the border between Marengo and Perry counties, will feature prominently in our story, because William Tayloe took sole possession in 1847, expanded the property to 1,580 acres, and operated the place through the Civil War. Five members of the Glascow family who came to Oakland in 1837 were still living there twenty-six years later. Tayloe's census of 1863 finds grandmother Nanny Glascow (now age eighty-one), daughters Esther (fifty-nine) and Judy (fifty-five), and granddaughters Esther (forty-three) and Mary (twenty-nine) all still together, joined by grandson James (thirty-three), who came to Alabama in 1836, and three young great-grandchildren born to Mary in Alabama. But Barneby, nine years old in 1837, illustrates a very different side of the Mount Airy story. His mother, Priscilla, and his younger sister Lucinda had been sold at Mount Airy in 1832 when Barneby was four, leaving him in the care of his father, Joe. Barneby accompanied Joe to Alabama in 1835, but this arrangement was short-lived, because Joe was sold in 1836. Orphan Barneby was living at Oakland in 1838, but in that year he too—now a child aged ten—was sold for $850, completing the dispersal of this particular family.

What was life like for the first Tayloe migrants to Alabama? Henry Tayloe asserted repeatedly that his hands were in fine health, but his other scattered references to the slaves indicate that they were deeply discontented with their new environment—and hated working for Henry Tayloe. In 1834 he told Ogle that "some of them are very idle and require considerable severity to make them work."[32] He also reported that seventeen-year-old Mary Flood from Neabsco, whom he had bought from William in 1829, "attempted to fire my house a few weeks ago and I sold her for $600. . . . I was afraid to keep her."[33] Henry had a lot of trouble with runaways, especially at the beginning, when the slaves spent most of their time clearing the land for planting. He complained in January 1835 that "another of my hands is in jail in Tuscaloosa," which was about seventy-five miles to the north. In the same letter Tayloe noted that "those hands from Cloverdale are so very idle that they can't bear pushing and run off if they are the least pressed." This suggests that Davy, Tom, and Jim Moore, plus Billy Page, Ralph Elms, and James Thurston—having walked 200 miles from Mount Airy to Cloverdale, having chopped down trees for a year at Brunswick Forge, and having walked another 800 miles from Cloverdale to Alabama—were rebelling against the repeat experience of chopping down trees at Adventure plantation. In February 1835 Henry Tayloe noted that "one of our negroes (Jess) ran off from me a short time ago and was apprehended in Mobile, his recovery cost us about $70. Three of my Cloverdale hands ran off . . . at the same time, but returned after spending a few days in the woods."[34] The champion runaway seems to have been Gowen, belonging to Ogle Tayloe, who "soon became tired of the Cane Brake" and fled at least three times in 1834–1836.[35]

Things settled down after the land was cleared for farming, but not all of the Virginia hands took to working in the cotton fields. "I sent Massy out to pick cotton," Henry reported in 1834, "and she has deserted." But, he added, "it is impossible for a negroe to escape from this country."[36] Massy was a thirty-five-year-old woman from Oaken Brow belonging to Charles Tayloe. William Tayloe acquired her in 1843, whereupon she married David Moore (one of the eight initial migrants from Mount Airy in 1833), and in 1863 Massy and David were living together at Larkin plantation, aged sixty-four and sixty-five. Another of the early Virginia hands had a nasty accident. In 1835 a fifteen-year-old Gwinfield boy named Israel Brown got caught in the gearing of a cotton gin, and his arm was mutilated.[37] Fortunately, Israel survived. William Tayloe acquired him (at the same time that he acquired Massy) from brother Henry in 1843; Israel was working at Larkin twenty years later (valued at $1,200), became a freedman at age forty-five, and was last reported working as a wage laborer at Larkin in 1867. Both Massy and Israel cannot have had very happy memories of their first years in Alabama.

In December 1835, only two months after William and Ogle Tayloe had sent their second party of slaves to Alabama, they learned that Henry had taken out a $5,000 loan to meet current contingencies. He had also bought additional land for $2,000, and expected his brothers to pay for it. William and Ogle came up with the money Henry needed, but were disheartened to discover that the cotton crop from Adventure and Walnut Grove in 1835 added up to only 90 (rather than the expected 200) bales. Henry was now thinking of selling out unless he made a big crop in 1836. He seems to have been quite ready to dispose of all the fifty slaves William and Ogle had sent him in 1833 and 1835. By selling his plantations and slaves at high prices, Henry thought he could pay off his loans and also form a private banking firm with his brothers capitalized at $30,000, which he claimed would easily return $10,000–$12,000 per annum.[38] Henry was caught up in the speculative bubble of 1835–1837, with easy credit available for the purchase of overpriced land and slaves, a bubble sustained temporarily by the high price of cotton.[39]

Ogle Tayloe, who had built a stylish new Washington residence in Lafayette Square and enjoyed collecting art and traveling abroad, was not primarily interested in farming, and he was strongly attracted by Henry's proposal that he move his slaves out of the Chesapeake and sell them in Alabama at one or two years' credit, bearing 8 percent interest. "If we can make a disposition of our negroes in Alabama," Ogle wrote William, "I think it desirable to send all the active hands from the Potomac—they are too near the Abolitionists, that are daily becoming more dangerous."[40] So Ogle (like Henry) was perfectly willing to send all of his best slaves south, sell them, and collect the cash. This was never William's policy. He did later sell some of his Mount Airy slaves in Alabama to cover the cost of buying land, but he was a planter, and once he became committed to Henry's "Alabama speculation" he wanted to take maximum advantage of the cotton boom by establishing his own Alabama plantations operated by his own Virginia slaves.

By 1836 both William and Ogle doubtless recognized that the Tayloe partnership was in trouble. In the spring of 1836 Ogle went to Alabama in order to find out how Henry was doing. He traveled by way of Louisville, where he found forty steamboats lying in the harbor, and embarked upon "the magnificent & splendid new steamboat *Persian,* upon her first voyage" via the Ohio and Mississippi Rivers to New Orleans. When he got to Alabama, Ogle was pleased with Henry's situation, and told William on his return that "the fertility of his land can't be surpassed. . . . The people were healthy & happy. I never saw more contentment any where." Henry, he predicted, would produce at least 500 bales that year. But their brother was also searching for fresh loans. Henry came to New York in the summer of 1836 to meet with moneylenders, and Ogle reported that he "fears an unsuccessful issue." If his New York visit did not work, he planned to see Nicholas Biddle, the president of the Second Bank of the United States, in Philadelphia. Clearly Henry was desperate for money.[41]

In this critical situation William and Ogle decided to send Henry a third batch of slaves in October 1836, with the aim of selling most or all of them in Alabama in order to provide the Tayloe partnership with cash. Ogle was apparently more in favor of this move than William, because he supplied twenty-nine slaves from his plantations, while William sent fourteen people from Mount Airy, and the brothers jointly contributed fourteen from Deep Hole, for a total of fifty-seven: by far their largest shipment yet. There were eight females and six males in William's Mount Airy contingent (see *Appendix 29*), including two mothers with three children. William apparently saw these people as more expendable than the workers he had sent in 1833 and 1835. Seven of them—Rachel and Leana Dixon, Mary and James Glascow, George Bray, Marcus Page, and Fanny Moore—were related to Mount Airy slaves already in Alabama, so they probably were led to believe that they would be joining family members at the end of their long journey. This would explain why they were willing to go.

Henry Tayloe personally led this large party, leaving Richmond on October 25, 1836, "in fine spirits." He reported to William that "all hands are satisfied and *unconfined.* I don't apprehend any running off." Henry's comment indicates that the Tayloes did not lock their slaves into coffles when marching them to Alabama. But doubtless no one told these "satisfied" migrants that most of them would be sold as soon as they completed their long journey. They took three weeks to reach Knoxville, and Henry reported on November 14 that two people had died en route, one of them being two-year-old Leana Dixon from Mount Airy. "The other negroes," he added, "arc tolerably well excepting lameness," a casual way of describing the rigors of this very long trek. Henry himself had had enough of this "tedious journey" and took a stagecoach from Knoxville to Alabama to prepare for the new slaves' arrival. The Virginians finally reached Walnut Grove on December 3, 1836, after walking for six weeks.[42]

The decision by William, Ogle, and Henry to make fifty-seven Virginia slaves walk 800 miles in October–December 1836 and then sell most of them in Alabama was the cruelest act that I have found recorded in the Tayloe Papers. As their father did when he sold twenty-six of his slaves to professional slave traders in 1816, the Tayloe brothers were tearing many African American families apart—even more decisively than in 1816. I have not been able to trace what happened to all fifty-seven of these people, but two of the fourteen migrants from Mount Airy died in 1836–1837, ten were sold in 1836–1838, and an eleventh was sold in 1843. The two family groups in the Mount Airy party were broken through sale. Mary Glascow was disposed of in 1836 without her son James. And Mary from Landsdown was sold in 1837, her young daughter Catherine in 1838, and her son Cornelius in 1843.

Henry needed to sell these people because he had not yet learned how to grow cotton. His latest crop at Adventure was only 58 bales (instead of the expected 500). Hence he quickly plunged into disposing of the newly arrived slaves. On December 12, 1836—only nine days after the Virginians reached Walnut Grove—he told William that he had sold eleven of the fourteen Deep Hole people for $13,150 on long-term credit—"a little in cash and the balance in one or two years. I have likewise sold five of yours for $6,700 payable in one or two years." By January 1837 Henry had also sold fourteen of Ogle's slaves on similar two-year credit terms for $17,910.[43] In less than two months he sold thirty slaves for $37,760 at very high prices—obtaining $1,200 to $1,300 per slave—but only on the promise of future payment with little direct cash. Everything depended on the buyers' ability to produce enough cotton in the next two years to pay for these Virginia slaves. And it soon turned out that Tayloe had made his sales at exactly the wrong time. Cotton prices dropped from eighteen cents per pound in 1834 to six cents in 1837, and planters were unable to pay for the slaves they had bought on long-term credit. This triggered the panic of 1837, which accelerated when the banks that had loaned money to planters like Henry Tayloe pressed for repayment.[44] Thus, William Tayloe, who should have netted around $20,000 from the sale of his Mount Airy and Deep Hole slaves in 1836–1838, could extract little if any of this money from Henry.[45]

Meanwhile, Henry Tayloe placed twelve-year-old James Glascow—the sole surviving member of William's 1836 Mount Airy slave party who was not sold—in his stable at Walnut Grove to help tend his thoroughbred horses. The Tayloe brothers all shared a family devotion to the turf. Ogle began every letter to William with a discussion of Tayloe horse racing and breeding before turning to Whig politics, family news, and business concerns. Henry, searching for a wife, complained that Alabama women "are too inferior for me to admire, they are not sufficiently erect upon their posterns nor clean limbed for a lover of thorough breds to admire."[46] Henry and William jointly owned twenty-one blooded horses valued at $11,700, fifteen of them stabled at Mount Airy, and six in Alabama.[47] In 1837 William sent Joe Moore, his seventeen-year-old Mount Airy stable boy, to join James Glascow in the Walnut Grove racing stable.[48] From this point on, Henry Tayloe became more interested in horses and racing than in cotton production.

By 1837 William Tayloe must have recognized that Henry was doing a poor job as a cotton planter, and that he had invested more money and more slave labor in the Tayloe partnership than was prudent. He was very fond of Henry, and named a son born in March 1836 after him, but sent him no more slaves to be sold. Indeed, under Henry Tayloe's management or mismanagement the thirty-nine migrants from Mount Airy in the 1830s had received a brutal introduction to Alabama. But the pioneers from the 1830s who remained together after the first decade were resilient people who created strong new bonds.

In 1863 twelve of these early Mount Airy migrants were living on William Henry Tayloe's Oakland plantation and another five on his Larkin plantation, ranging in age from twenty-nine to eighty-one. At least thirteen of the seventeen were married and had Alabama families. Three of the eight initial migrants who arrived in February 1834—Tom Moore, James Thurston, and Ralph Elms—married women from Gwinfield belonging to Henry Tayloe—Elsy, Winney, and Becky—who had traveled with them in the first Alabama party. These three couples had a combined total of nine children, but other couples had bigger families. Becky Carrington, who arrived in December 1835, married a Virginia slave named Tom Flood and bore him nine children. Her travel companion

Anna Marcus married Travers Hilliard, who came from Mount Airy in 1845, and they had eight children. Betsy Dixon, also in the 1835 party, married Arthur Thomas, who came from Ogle Tayloe's Nanjemoy plantation in Maryland, and they had seven children. And Joe Moore, who arrived in 1837, married a woman named Sinah who was not from Mount Airy, and they also had seven children by 1863. Collectively, the early Mount Airy migrants who can be tracked into the 1860s were the parents of at least sixty Alabama-born children. Despite many obstacles, they were building new family networks in the Canebrake.

## *The Crisis of 1843–1844*

The panic of 1837 was followed by the deeper panic of 1839, and Henry Tayloe's situation in Alabama was becoming untenable. He had expanded too rapidly and was spending far more than he could recover from his cotton and slave sales. Back in 1835 he had told Ogle, "My object is to make a fortune here as soon as possible by industry and economy, and then return to enjoy myself" in Virginia.[49] But he had not economized. He had bought over 5,000 acres plus a sizable slave gang at Walnut Grove, and was simultaneously operating three cotton units at Walnut Grove, Adventure, and Oakland while starting to clear a fourth property called Woodland. And once his thoroughbreds arrived, Henry became dangerously absorbed in horse racing. He began to bet on his horses, and when he had gambling losses he paid on credit. William Tayloe reported to Ogle in April 1838 that their brother was now in New Orleans "about his Race Course," and that "the Turf engrosses all his time, and attention."[50] William decided to find out what was going on, and made his first visit to Alabama in the winter of 1838–1839. He inspected Henry's books and observed his cotton fields. The cotton crops were now much larger than in 1835–1836, but William did not like his brother's style of management. As he later observed to Ogle, Henry was always "pushing from plantation to plantation" with little care of implements or stock or workers. It was "Rush Boys, Rush!" and "How much Cotton did you weigh?"[51]

In 1840 Henry Tayloe announced that he was tired of the turf and had stopped gambling. But the next year he was sued for a bad debt, and in 1842 Ogle discovered that the Alabama accounts contained egregious errors and hugely inflated running expenses.[52] In 1843 Henry probably went back to gambling at the New Orleans racecourse, and was sued for $12,600, a sum that he couldn't possibly pay. Henry remained as cheerful and full of hope as ever. But his lawyers warned the other five Tayloe brothers that Henry's liabilities far exceeded his assets, and that they should buy his property and not allow Henry to repossess any part of it in order to shield the estate from his creditors. Accordingly, the brothers formed a new partnership called Tayloe & Company, claimed control over all of Henry's assets, and began to settle his most pressing debts.[53] Edward Tayloe went to Walnut Grove in October 1843 to figure out the dimensions of his brother's losses. "I can obtain no satisfactory information from Henry," Edward told William. "He deals only in generalities and is profoundly ignorant of his situation. He does not, poor fellow, seem to be aware, that his is a case of hopeless insolvency." Edward concluded that Henry owed $52,602 to his five brothers and his sister Virginia, and that his largest family debt was the $20,940 he owed to William. Altogether, Henry's assets amounted to $74,271, and his liabilities amounted to $163,781![54]

Henry Tayloe's bankruptcy and his brothers' quick formation of Tayloe & Company to shield the family assets took place amid scenes of wild fiscal chaos in Alabama. Merchants went bankrupt, banks were insolvent, court dockets were filled with debt cases, creditors and debtors were accosting each other on the streets, and some planters were sneaking their mortgaged slaves out of the state. A northern bank agent reported that in Alabama "the people are getting all most desperate—More shooting and killing each other here than you have any idea of."[55] In this environment the Tayloe brothers did better than might be expected. While Tayloe & Company could not recover the slaves Henry had sold or obtain payment for them, the partners did retain possession of the land and most of the slaves on Henry's four plantations. In September 1843 Henry complained to Ogle that he was being treated ungenerously, but his brothers had to employ devious tactics in order to evade his creditors. Finally, in January 1847 they hit upon the expedient of deeding one-quarter of Adventure plantation together with eighteen slaves to Henry's wife Narcissa, a settlement that equaled the value of her dowry. This was all that Tayloe & Company felt able to salvage for Henry from his Alabama speculation.[56]

William had been badly damaged by Henry's mismanagement. He had lost twenty slaves—thirteen from Mount Airy and half of the fourteen from Deep Hole who were sold by Henry in 1835–1838—with little or no compensation. He had gained little income from Henry's cotton crops, and now was saddled with his gambling debts. In the wake of Henry's bankruptcy William shared the ownership of four plantations in 1844–1845 with four of his brothers in a cumbersome partnership where no one was in charge. He was a partner in Tayloe & Company, which held title to Henry Tayloe's estate: an undivided half share of Adventure plantation with sixty-nine slaves in 1843, an undivided half share of Walnut Grove plantation with sixty-two slaves, an undivided third share of Oakland plantation with forty slaves, and an undivided half share of Woodland plantation (not yet planted), plus a warehouse and other undeveloped property. In his own right William held an undivided quarter share in Adventure, Walnut Grove, and Woodland, and a third share in Oakland. Seven of his Mount Airy slaves were attached to Adventure plantation, and fifteen to Oakland plantation. So William Henry Tayloe could not walk away from his Alabama investment; far too much was at stake.

Besides, by playing a more active role in the development of the Tayloe cotton plantations, William could expect to make a good deal of money in Alabama. In 1843–1845 Adventure averaged 242 bales of cotton per year, Walnut Grove 161 bales, and Oakland 223 bales. William's share of the cotton proceeds in these years was about $6,000 per annum, which was more than his corn and wheat sales in Virginia.[57] He visited Alabama twice in 1844, a sign of increasingly active interest. These visits confirmed his belief that Henry's management style must be changed. He told Ogle that a slave of his named Giles had an injured spine caused by abusive mistreatment, and that he was palsied as a result. And William added a telling phrase: "There shall be no more cruelty."[58]

The year 1844 was a time of great personal tragedy for William Tayloe. His wife Henrietta (known as Etta) and his eldest son, Benjamin (named for his brother and also known as Ogle), had both been sickly for several years. Etta was distressed by the deaths of her last three children in infancy, and both parents worried about the failing health of Ogle, their firstborn and a promising youth of eighteen. There is no diagnosis of Etta's and Ogle's maladies in the family correspondence, but thc symptoms suggest tuberculosis. William wrote to his brother in May 1844 that "Ogle & Etta are declining gradually—she can scarcely get down stairs to dinner, & Ogle is only little better. Thank God, they continue free from pain. . . . My prospects are truly gloomy." Ogle died on July 3, and William found some measure of consolation by telling his brothers that his son had "a spirit so pure" that there was no cause for fear and everything to hope for his heavenly salvation.[59] Etta followed two weeks later on July 18 at the age of forty-four, leaving her husband a grief-stricken widower with three girls and a boy ranging in age from fifteen to eight. William never remarried and never fully recovered from the loss of Etta and Ogle. For the rest of his life he kept recalling the memory of his saintly wife to his children, and he always particularly extolled Etta's skill at slave management. William's grief was sharpened in the late 1850s and 1860s, as we shall see, by his disappointment in his one surviving son, Henry, whom he saw as a very inadequate replacement for eldest son Ogle. And young Henry had to bear the heavy burden of fulfilling all of his father's large expectations.

From July 1844 onward, William Henry Tayloe had to leave his Mount Airy overseers in charge whenever he absented himself for visits to Washington or Alabama. In September 1844 John Jeffries reported cheerfully that at Mount Airy "the servants were all well and not one out in the woods"[60]—a remark suggesting that temporary escape from work was fairly common among the Mount Airy slaves. And three months later, there was big trouble. On a Sunday morning, December 22, 1844, when William was in Alabama, the governess in charge of his children at Mount Airy took them to church. Shortly after their carriage left, the house servants discovered two fires, one under the governess's bed on the second floor and the other in a pile of old lumber in the garret. They threw the governess's bed out the window and put out the bedroom fire, but they could not extinguish the garret fire, which soon spread out of control into the lower floors of the house. The eight slaves who were in the house managed to rescue the ground floor furniture and most of the family portraits before the roof and lower floors fell in, leaving the Tayloes' stately mansion a gutted shell.

How did the two fires start? It was obvious that they had been deliberately set, and the house slaves had a candidate: they accused the laundry maid Lizzie Flood of arson. Lizzie was a twenty-two-year-old who had come to Mount Airy in 1828 with her family from the Neabsco ironworks when she was a young child. Her father was a carpenter and her mother was a spinner, and Lizzie herself was employed as a housemaid at Mount Airy from about 1836 to 1842. Then she was shifted to laundry maid. Young Lizzie had no recorded husband or children. She may well have resented her demotion to the laundry job, and she was certainly on bad terms with the other black domestics, who suspected her of setting fire to a curtain in October. Her family had a reputation for arson, because in 1834 Henry Tayloe had accused Lizzie's older sister Mary Flood of trying to burn down his house at Walnut Grove in Alabama. On the morning of December 22, at about breakfast time, Lizzie brought the weekly wash from the laundry and delivered the clothes to the big house. When the fires were discovered, she was nowhere to be seen. And Lizzie had an alibi: she claimed that she was miles away en route to the Nomini Baptist church when the fire broke out.

Edward and Charles Tayloe, both living nearby, arrived quickly to take care of William's children and to help interrogate all of the house slaves. On December 30 Edward wrote a long report to brother William that gives a fascinating view of relationships among the Mount Airy black domestics. Sixteen-year-old Kitty Wormley, a housemaid who previously had Lizzie's job as laundry maid, testified that Lizzie usually delivered the laundry on Saturday evening, not Sunday morning, and stated that when she heard someone going upstairs on the morning of December 22 she followed to find out who it was and met Lizzie at the top of the stairs on the bedroom floor. Kitty went on to say that ten minutes after the Tayloe party left for church she went upstairs again and saw a fire in the governess's bedroom, and she also found the garret door (which was always locked, with a hidden key) to be unlocked. Jenny Moore, another housemaid who was on particularly bad terms with Lizzie, testified that she went upstairs with Kitty after the Tayloes left for church and found a third small fire in the young ladies' room, which she quickly put out.

When Lizzie was questioned, she said that she suspected somebody of setting the fire but was afraid to say who. She countered Kitty's testimony by claiming that she didn't go into the house with the laundry; instead, Kitty took the clean clothes from her. When asked what she did after delivering the laundry, Lizzie said that she had breakfast with her father, Tom Flood, got dressed for church, and started walking to Nomini Meeting House. But it started to rain when she was partway there, so she returned home. Lizzie's father, Tom, testified that he had breakfasted that morning with his wife. And shoemaker James Dixon, who was on the road Lizzie would have taken, did not see her, and pointed out that the house had been burning for more than an hour before the rain started, and that anyone wishing to attend the Baptist service who started walking at the usual time would have got to the Nomini church before the rain started.[61]

From this testimony it appeared that the fire could have been started by Lizzie Flood or by Kitty Wormley or by Jenny Moore. But Lizzie's account was challenged by her own father and by James Dixon. Edward Tayloe was convinced that Lizzie had set the fires, and he had her put in jail until William returned home. But the evidence against her was too circumstantial to stand up in court, and she was handed back to her master. On December 30 Edward had written to William, "In my opinion, you should get rid of all that family of Floods." And William did so. In 1845 he exiled Lizzie's parents and her two brothers to Deep Hole farm quarter on the Potomac, and he sold Lizzie herself in Richmond for $602.[62]

The other issue was what to do about the ruined building. William's brother Ogle had always considered the ancestral plantation house to be unsuitably large and old-fashioned, and when he heard of the conflagration he remarked to William that Mount Airy House was out of character with the family's means—"a sort of anchor that can not be hauled up." So he assumed that William would not rebuild. But William thought otherwise. In Alabama, after learning about the fire, William penned the following note on January 1, 1845: "directed the walls of the burnt building to be protected, to be roofed at a convenient season."[63] When he returned to Virginia he employed his five carpenters (led by Bill Grimshaw, who ran away a few months later)[64] together with Edward's and Charles's carpenters to construct a temporary roof for the building. For nearly three years William Tayloe's family lived in a service wing while the interior of the mansion was rebuilt. This construction job cost nearly $4,000 and was finally completed in October 1847.[65]

## *1845 to 1860*

In January 1845, right after the house fire, 211 slaves were inventoried at Mount Airy, which was 10 fewer than in 1830. This was not due to natural decrease: births were continuing to exceed deaths. But William Henry Tayloe had sent thirty-nine people from Mount Airy to Alabama during the years 1833–1837, which fully accounts for the decline in

Virginia. The Mount Airy population in 1845 remained almost as youthful as in 1830, with a few more elderly people and a few less children under the age of ten, but a very large number of teenagers. More than a quarter of all the Mount Airy slaves—fifty-nine boys and girls—were between the ages of ten and nineteen, and most of these adolescents were employed at the three farm quarters. So William had an ample supply of young Virginia laborers who could be switched to cultivating cotton in Alabama.

Starting in 1845, William began to separate himself from Tayloe & Company so that he could manage his own cotton operation independently—a long and difficult process, not completed until 1855. During these ten years William focused on Oakland plantation, where most of his Mount Airy slaves were living and working. In April 1845 he acquired brother Ogle's one-third undivided share in Oakland land and stock, and in October 1845 he sent forty-five slaves down from Mount Airy, placing almost all of them at Oakland. Also in 1845 he hired an able and efficient overseer, Richard H. Donnahan, who worked for him until 1858. And in February 1847 he bought the remaining third of Oakland from Tayloe & Company and thereby acquired fifteen slaves already working on the place (four of whom had been sent from Mount Airy in the 1830s). Now William Henry Tayloe was the sole owner of a promising cotton plantation.[66]

Ogle, on hearing in 1845 that William was assembling a new party of slaves for Oakland, offered the following advice: "Rebellious spirits should be sent, in order to be put under sensible government, and as an example to those left behind who might be disposed to be disorderly."[67] But this was not William's policy. Instead he mainly chose young people who looked to be promising cotton laborers (see *Appendix 30*). Whereas William had started cautiously in the 1830s, he was now sending a large party of forty-five from Mount Airy. The group as a whole was very young: twenty-two were teenagers. Excluding the eight nonworking children under ten, there were twenty-five field hands, eight craft workers, and four domestics, almost all of whom would become cotton workers. Also, unlike in his previous shipments, William Tayloe was now sending most of the people in family groups: Alfred and Sinah Lewis from Landsdown farm quarter with their six children, Lucy Dudley from Doctor's Hall with five of her six living children (but without her sixty-eight-year-old husband Jarrett), shoemaker James Dixon with six of his siblings and a nephew, three young members of the Page family, two Thomas sisters, two Carrington brothers, and Winney Grimshaw, whose history is chronicled in Chapter 3, with her son John and her brother James. Thirteen members of the party were singles, and all of the migrants left close relatives in Virginia. The Mount Airy slaves were accompanied by a pair of men from Deep Hole and seven people sent by Virginia neighbors.[68] This large group traveled with five horses and mules and a heavy wagon. They took a slightly faster route than previously used, going from Richmond and Lynchburg across the Blue Ridge at Salem, through western Virginia to Rogersville and Knoxville, Tennessee, then south by way of Chattanooga to Summerville, Georgia, and into Alabama via the small towns of Gaylesville, Montevallo, and Marion. This route required about thirty-eight days; they left Mount Airy on October 14 and reached Oakland by the end of November.[69]

The Virginians must have seen immediately that they were in for hard times. They probably did not have to cut down trees and clear land for planting, because Oakland plantation had been in operation for eight years. But the place was poorly equipped. There were not enough cabins to accommodate the migrants, and there was no gin house for cotton production and no stable for the horses and mules. Unfortunately for the new people, extra cabins and a stable weren't finished until early 1847, and a gin house still hadn't been built. Oakland had produced 223 bales of cotton in 1845, but under the supervision of Richard Donnahan (managing his first cotton crop) the inexperienced newcomers could only make 109 bales in 1846. Donnahan claimed that the Oakland crop was decimated by ball worms and army worms. During 1846 William Henry Tayloe came down to confer with Donnahan and to inspect the new house that was being built for him. The Oakland slaves constructed a portico in front of this building and a piazza in the rear, but this was no plantation great house à la Mount Airy.[70] An inventory taken in February 1847 listed sixty-seven slaves and fifty-four working hands at Oakland. Ten of the workers had come from Mount Airy in the 1830s, and thirty-two in 1845; the remaining twelve were mainly Tayloe slaves from Virginia, though not from Mount Airy. Three young children died in 1846–1848 (including two who had journeyed down from Virginia in 1845), but there were three new births by 1847, a hopeful sign that the population was not going to shrink.[71]

There is no description in the Tayloe Papers of cotton production at Oakland, but in one of his notebooks William Tayloe reckoned on ten acres of cotton cultivation per hand, and hoped for seven bales of cotton per hand, which gives us a start. In his 1847 Oakland inventory Tayloe meticulously listed his thirty-nine working hands, only nine of whom were thirty or older, while twelve were in their twenties, and eighteen in their teens.[72] Overseer Donnahan probably made these slaves plant about 400 acres in cotton that year. In March (when the temperature was warming up to 60 degrees) they prepared each field by plowing a series of long, parallel rows, three to five feet apart, then seeded the rows around the first of April. After the tender seedlings sprouted, the slaves had to cull the weaker plants carefully and chop out the invasive grass in order to nurture rows of evenly spaced healthy plants. As the cotton shrubs slowly grew, they were hoed repeatedly into July, when they flowered and developed bolls that started to open in late August. Then the three-month picking season began. The plants continued to bloom into early December, so the slaves picked each row at least three times and often more. In early October 1848, for example, Donnahan reported that the slaves had picked 140 bales and the fields were still white; they finished picking that year on December 10, producing a total of 280 bales.[73] Even this brief synopsis shows the demanding character of cotton cultivation. The slaves had to work nearly year-round on the crop, and every step from plowing, seeding, and hoeing to picking required speed and care. During the long months when the cotton plants were small, Donnahan could readily survey his rows of workers to catch laggards, and during picking season it was easy to impose production quotas on all of the individual hands. In 1847 the Oakland hands—most of them newly arrived from Virginia—could produce only 230 bales, or 5.9 bales per hand, but in 1848 they reached Tayloe's goal of 7 bales per hand.

In the Mississippi valley, the cotton planters imported much of the wheat, corn, beef, and pork needed to feed the population, and the slaves had meager rations.[74] This was not true at Oakland, where the inventories show large numbers of fattening hogs, in addition to ample stores of corn to feed the slaves as well as the cattle. In late 1848 Donnahan planned to kill one hundred hogs and put another one hundred into the woods "for another year's fattening."[75] And Richard Donnahan never complained, as Henry Tayloe had in the 1830s, that his workers were constantly running away. But there was evident unrest in the neighborhood. In November 1849 Donnahan reported that Dr. Reese's gin house was burned the previous week, apparently fired by his slaves. Mr. Harris's gin house had been burned in September, and like Reese he lost twenty-five to thirty bales of cotton. Judge Ormond's gin house had also recently been burned, and again one of his black hands was named the arsonist.[76] And in June 1847 there was a virulent epidemic of typhoid fever at Oakland caused by poor sanitation. "I fear all the new hands will take it," the local doctor commented. The slaves who caught this highly infectious disease had severe head and neck pains, followed by diarrhea, high fever, and general prostration. At least fifteen of the people from Mount Airy (eleven of them newcomers in 1845) became so ill that they couldn't work for a month, which seriously delayed the cotton crop. Worst of all, four of the Mount Airy people died: eighteen-year-old Scipio Denny, sixteen-year-old Alfred Lewis, twenty-three-year-old Letty Dixon, and nineteen-year-old Lucinda—all of whom had come to Oakland in 1845.[77]

The Oakland slaves, while well enough fed, were inadequately clothed. Donnahan reported in 1850 and again in 1851 that they were "nearly naked for shirts," and he told Tayloe that they needed more than two shirts per year.[78] But living conditions at Oakland seem to have gradually improved. Certainly the cotton crops were larger and more profitable. In 1847–1848 the Oakland slaves produced a two-year total of 510 bales that fetched $14,302, and in 1850–1851 the two-year total was 500 bales selling for $20,813.[79] By the early 1850s Tayloe was receiving $250 per year from the labor of each of his hands. Donnahan reported in December 1848 that "our hands have worked very well this fall"—by which he meant throughout the picking season. By 1851 he could boast that he had built two corn cribs, a smokehouse, a blacksmith's shop, and a carpenter shop, plus sheds for the plows and wagons, and that he expected to construct two more rooms for shelling corn and storing oats. Oakland produced 289 bales of cotton and nearly 20,000 pounds of pork in 1851. "Can you beat that on your Mount Airy Estate?" he asked William Henry Tayloe. Furthermore, "the Negroes have worked better this year than they have ever since I have been on the farm and with less trouble. There has not been one of them whipped since we commenced working the crop. . . . There has been 10 Children Born here in less than 12 months All doing very well except one little Sinah that is small and puny."[80] By 1853, twelve Mount Airy people who had migrated in the 1830s and thirty-two who had come in 1845 were living at Oakland, and the inventory listed twenty children who had been born since 1845. Despite the typhoid deaths in 1847, the Oakland population had increased in six years from sixty-seven to seventy-nine.[81]

William Henry Tayloe did not move any more of his people from Virginia to Alabama between 1846 and 1853, but he did expand his shareholding significantly in two of the Alabama properties held by the Tayloe family partnership. In 1847 he bought from Tayloe & Company an undivided half share in Walnut Grove, which lay next door to Adventure, and an undivided half share in Adventure, where some of his Mount Airy slaves had been working since 1834.[82] By 1850, according to the U.S. Census taken that year, William Tayloe owned 71 slaves at Oakland and a half share of 124 slaves at Adventure and Walnut Grove, for a total of 133 slaves in Alabama, which was not far from his total of 166 at Mount Airy. He was grossing close to $20,000 in annual cotton sales from Oakland, Adventure, and Walnut Grove. But he was operating only one Alabama plantation, Oakland, on his own.

Then in 1854–1855 the brothers decided to liquidate their Tayloe & Company partnership. William sold his share of the land he held at Adventure and Walnut Grove to his brothers, while keeping possession of his slaves who worked on these two plantations.[83] Also in 1854, Ogle and William sold Deep Hole plantation in Virginia, which they jointly owned, and divided up the slaves who lived there.[84] Now William was in a position to establish a second cotton farm staffed entirely with his own slaves. In 1854 he bought a 1,203-acre tract named Woodlawn in Perry County, not very far from Oakland, for $24,060. He transferred overseer Richard Donnahan from Oakland to Woodlawn and directed him to start cotton production on the new site in 1855, paying him a salary of $800. During 1854–1855 Tayloe moved some of his Oakland, Adventure, and Walnut Grove hands to Woodlawn. But he needed more workers. So he assembled thirty-six slaves from Mount Airy and eighteen from Deep Hole, and sent this party of fifty-four Virginians down to Alabama in September 1854.

This 1854 party was the last contingent that Tayloe sent from Virginia before the Civil War. The Deep Hole migrants came as families; the husbands and wives were mainly in their twenties, and they brought seven small children who were far too young to work. The Mount Airy migrants, on the other hand, were almost all selected for immediate field labor. They are listed in Appendix 31. William Tayloe sent one blacksmith, a jobber, two domestics, a spinner, twenty-nine field hands, and only two nonworking children in this party of thirty-six. As in 1845, a good many of the Mount Airy people were unattached singles. There was only one complete family: Mark and Franky Bray with their nine year-old son Reuben and their young daughters Rose and Rosetta. The oldest Mount Airy migrant, forty-two-year-old Bridget, came with her adult daughter Sarah. Nineteen of the Mount Airy workers were under the age of twenty, and most of these youngsters traveled with a brother or sister: there were four young Browns, three young Yeatmans, three young Wormleys, two young Dudleys, two young Thomases, and two young Wards. Gabriel Dudley, at age eleven, was making his third 800-mile trip between Mount Airy and Alabama. He had been sent to Oakland with his mother and four siblings in 1845 when he was only two years old and had soon been returned to Mount Airy. Now he was rejoining his family. Looking ahead, in 1870 Gabriel was still living at Oakland as a twenty-seven-year-old freedman, with his wife and baby.

The most distressing feature of the 1854 migration is that six of the Mount Airy newcomers abruptly disappeared from sight. After being placed at Woodlawn, they were apparently sold to help pay for William Henry Tayloe's extensive new land purchases.[85] As usual, Tayloe was breaking up families. Bridget Thurston was separated from her daughter Sarah, and Becky Lewis, Isaiah Wormley, Michael Thomas, and Anne Yeatman were all separated from their brothers or sisters. Isaiah and Anne were only fourteen and ten years old, respectively, when they were sold. One of the youngest migrants, Rose Bray, died immediately after arrival. The remaining twenty-nine were divided between Woodlawn and Oakland. Slave inventories taken at Woodlawn and Oakland in 1855 show the central role of the Mount Airy slaves on both plantations. Oakland had a population of eighty-two, with forty-four of the fifty-six workers coming from Mount Airy. Woodlawn had a population of seventy-five, with thirty-one of the sixty workers coming from Mount Airy.[86]

William Tayloe quickly became dissatisfied with Woodlawn, probably because it was not close enough to the new line of tracks being laid out by the Alabama and Mississippi Rivers Railroad. In 1857 he bought a better sited tract of 808 acres from John Larkin for $20,200, sold Woodlawn in 1858, and moved the Woodlawn slaves to Larkin. Tayloe's slaves, having spent three years clearing land at Woodlawn, now had to repeat this arduous work at Larkin. Between 1857 and 1860 Tayloe expanded Larkin plantation to 2,085 acres with three additional land purchases that cost him $39,801.[87] He also transferred all of his remaining slaves from Adventure and Walnut Grove. Finally William Tayloe had achieved his goal. From 1858 to 1865 he was the sole owner of two big cotton plantations less than ten miles apart. Larkin was situated east of Uniontown in Perry County, and Oakland was northwest of Uniontown on the eastern border of Marengo County (in Hale County after 1867 when county lines were redrawn).

Ever since they first arrived in the 1830s, the Mount Airy migrants who labored for Tayloe in the Alabama Canebrake had been victimized by a very poor transportation system. No railroad line had been built in this region, and Tayloe's slaves had to cart his cotton bales thirty miles east to the Alabama River or twenty miles west to the Tombigbee River, to be shipped downstream to Mobile. The Canebrake roads were notoriously muddy and sometimes impassible. Finally, in 1857 the Alabama and Mississippi Rivers Railroad began to operate close enough to the neighborhood of Oakland and Larkin so that William Tayloe could ship his cotton by rail east to Selma on the Alabama River. The following year the Alabama and Mississippi Rivers Railroad completed a line running from Selma west to Uniontown. "Our neighbors are lively, looking for the Iron Horse to enter Uniontown in April," Tayloe wrote in March 1858. "Only think of cars from Uniontown!" The Selma-Uniontown line ran through the edge of Larkin plantation, and in 1859 the railroad directors built a platform at Larkin to accommodate Tayloe. At about this same time the Newbern Railroad ran a line to Marion (the Perry County seat) that passed through Tayloe's Crossing in the Oakland plantation woods, enabling Tayloe to ship his Oakland cotton via the Newbern and Marion Railroads to Selma.[88]

As yet there was no rail line running east from Selma to Montgomery. When Tayloe traveled from Alabama to Virginia around 1854, he took a stagecoach from Marion to Montgomery, then went by rail to

Richmond via Atlanta and Augusta, Georgia; Branchville, South Carolina; Wilmington and Weldon, North Carolina; and Petersburg, Virginia, changing cars at all six of these stations. This was certainly better than walking the whole distance, as five parties of his slaves did between 1833 and 1854. And on the eve of the Civil War, William Tayloe was able to make his journey in only three days: he left Montgomery on May 7, 1859, stayed overnight in Atlanta, and reached Richmond via Knoxville by May 9.[89]

Tayloe was living in Alabama much more now that he was operating two cotton plantations. And he was making a lot of money. His cotton receipts for 1858–1860, which may not be complete, far exceeded any of his earlier earnings. In seven months of 1858 he sold 572 bales for $34,502. In nine months of 1859 he was credited with 860 bales fetching $47,965. In eight months of 1860 his brokers sold another 801 bales for $34,659.[90] Meanwhile, William had not neglected his Mount Airy grain fields. In fact, by practicing crop rotation, draining and ditching wet land, and applying a new "plaister" form of fertilizer, he was marketing 1,000 more bushels of Virginia wheat in the mid-1850s than he had in the mid-1830s, and was earning about $9,500 per annum for his Mount Airy corn and wheat—nearly double what he had made twenty years earlier.[91] Brother Ogle noted enviously that William was out-cropping him by a lot in both Alabama and Virginia, and could now provide ample inheritances for his daughters Sophia and Emma and his son Henry. "You are indeed in worldly matters wonderfully blessed," he told William in 1859, "& with your last years crops in Va. & Ala. now on hand, will I suppose be wholly free of debt, & prepared to spend your money any way you wish."[92]

William, as we have seen, had invested heavily in Alabama property. In 1857–1859 he paid almost $40,000 for the land at Larkin, and built a new house for himself at Oakland at a cost of more than $21,000.[93] Having spent his money so freely, Tayloe complained that "the Alabama bills are enormous and swallow up our Cotton." He pictured himself as always having to scrimp and save: "I am sick of screwing, screwing, screwing," he cried out in 1858. And he complained also about the burdens of plantation management. The overseers he hired to operate Oakland and Larkin did not satisfy him. In early 1858 he parted company with Richard Donnahan, who had done him very good service, mainly because he disapproved of Donnahan's sexual relations with Winney Grimshaw, as described in Chapter 3. A few months later he fired the overseer at Larkin who had run off to Selma without notice. And he reported that E. W. Holcroft, the overseer who had replaced Donnahan at Oakland, was "too heavy, and self-conceited," though he continued to employ Holcroft throughout the Civil War.[94]

Tayloe especially wanted more help from his son, Henry Augustine Tayloe, who turned twenty-one in 1857 and was an immature and irresponsible youth. In William's opinion Henry kept bad company and drank too much, and William warned him of the family history: his uncle John Tayloe IV and many of his Tayloe cousins "would not be warned in time, and died sots." The father kept pressing the son to get married and take full charge at Mount Airy, as he himself had done as a young man back in 1824. He told Henry that as soon as he saw evidence that he could manage overseers and slaves judiciously, he would hand over all of his Richmond County, Virginia, property to him. "You are my only son," he wrote in another plaintive letter from Oakland, "and if you do not qualify yourself to manage our property with prudence, what will become of the Human Beings under us. Owned by us. Humanity demands their care."[95]

We can catch a few glimpses of Tayloe's relations with the Alabama human beings under his care in the late 1850s. Several of the migrants from Mount Airy in their twenties were getting married: Dangerfield Perkins (arrived 1845) to Eliza Lewis (also 1845), Paul Myers (1854) to Eve Brown (1854), and James Grimshaw (1845) to Arabella Ward (1854). William Tayloe was pleased when Amphy Wheeler (1854) "asked my consent to his marriage with Isabella from Deep Hole. . . . A nice girl. A capital match for him. Amphys friends will be glad to hear of it."[96] But we also learn that John Smith from Gwinfield got a local white man to forge a pass for him and took off one night from Oakland on a mule. And Tayloe had to confess that the cabins he built for his slaves at Oakland needed improvement. "I find out," he told his son Henry, "Negro Houses should be on sills about two feet from the ground, and with floors. 96 rats were killed in a House which I built in a hurry."[97]

With such poor sanitary conditions, it is not surprising that there was much illness among the slaves in 1857–1858. Edward Holcroft, the Oakland overseer, reported "a heap of sicknes" but no deaths in October 1857, and listed seventeen ailing people, eleven of whom had come from Mount Airy. The next year another outbreak of typhoid at Oakland struck the three young sons of Joe Yeatman, the foreman at Landsdown farm quarter. "I am sorry to tell Little Joe," William Tayloe wrote on July 16, 1858, "all his boys had the Typhoid fever. Lemuel and Lewis are well but poor little Billy died. We did our best to save him." Billy had come to Alabama with his older brothers in 1854, and was sixteen when he died. Actually, Billy Yeatman had died in February, more than four months previously, so notifying father Joe about his son's death was not William Tayloe's top priority.[98] But Tayloe did make some effort to help the Virginia slaves keep in touch with their loved ones in Alabama. In 1857, writing from Oakland, Tayloe wanted Jane and Amelia in Mount Airy to know that their sons John Smith and David Saunders were both doing well and sent their love. The next year he sent a message to Martha, Eliza, Delphy, Sally, Mildred, Letty, Georgina, and Kesiah that their children were all very well. These eight Mount Airy mothers had eighteen sons and daughters living at Oakland or Woodlawn in 1858. Spinster Eliza Ward was also informed that her daughter Arabella had received the letter and the gold dollar she sent from Mount Airy. And Tayloe's sixteen-year-old body servant William Henry Harwood, coming into the Oakland kitchen with a load of wood while his master was writing, sent a cheery message to his mother, Martha: "Give her my love, give my love to all of them."[99]

During the late 1850s the populations on William Tayloe's two Alabama plantations rose rapidly. An informative list of the 123 slaves at Oakland in 1859, grouped into twenty-seven households, illustrates the expansive character of this community.[100] The cabins were small, probably two rooms, and eight of the cabins had six to eight inhabitants, with the youngest children having to share a bed. There were forty-eight migrants from Mount Airy and another fifteen men and women acquired by Tayloe in Alabama. Among these sixty-three people were twenty-four married couples and two single mothers. The couples and single mothers were the parents of sixty children born at Oakland, about half of them under the age of five. Appendix 32 looks at eleven of the twenty-seven Oakland households.

Tom and Becky Flood in cabin 18 and Michael and Patty Smith in cabin 3 were older couples who had been in Alabama since the 1830s, and they had teenage children. Becky Flood, whose husband Tom was not from Mount Airy, bore nine children in Alabama, six of whom were living with her in 1859 in one of the most crowded cabins. Michael and

Patty Smith, a Gwinfield couple acquired from Henry Tayloe, had at least six children in Alabama, three of whom were living with them in 1859. The married couples who had come to Oakland more recently, in 1845 or 1854, typified by Jim and Arabella Grimshaw (described in Chapter 3) in cabin 2, had younger families. A number of the males who arrived in 1845 or 1854 were out of luck on the marriage front because there were not enough young women at Oakland. Four of them, headed by Albert Ward and his brother Sydney, shared cabin 5. There was a different solution in cabin 24, where Lucy Dudley and her second husband, Peter Smith (not from Mount Airy), lived with four of Lucy's Mount Airy bachelor sons. A fifth son, Thadeus, was married and lived in cabin 20 with his wife and two children. And in cabin 26 Alfred Lewis was starting a second family. His first wife, Sinah, had died of typhoid fever in 1858 after giving birth to nine children, seven in Virginia and two in Alabama. Two of Alfred's daughters, Anne and Sally, were married, respectively, to David Carrington in cabin 9 and to Thadeus Dudley in cabin 20. One of his older sons, General Lewis, was living with his wife and children in cabin 22, while young Daniel, born at Oakland, remained with his father. Now Alfred was cohabiting with Sophia Dixon, whose sister Nancy was General's wife, and the two young children Sophia brought into the household were very likely Alfred's as well. With very limited marriage choices, there was a great deal of interconnection among the Oakland slaves.

By 1860, according to the U.S. Census taken in that year, Tayloe had 152 slaves at Larkin and 125 slaves at Oakland, for an Alabama total of 277. There were many other large slaveholders in the Canebrake. Perry County, where Larkin was situated, had a slave population of 10,106 in 1860, and fifteen of the planters held more than a hundred slaves. Marengo County had a slave population of 13,166, and in the eastern division where Oakland was located there were thirty planters with more than a hundred slaves.[101] The concentration of slaveholding in large units was much different from the Northern Neck of Virginia, and more closely resembled Westmoreland parish in Jamaica. And the Tayloes were leading members of the Canebrake plantocracy. In 1860 William was the largest Alabama slaveholder in the family, but three of his brothers and two of his nephews—Ogle Tayloe, George Tayloe, Henry Tayloe, Ogle's son Edward Thornton Tayloe, and George's son John Tayloe—were operating cotton plantations in either Marengo

County or Perry County, and collectively the Tayloes held a total of 776 slaves.[102]

The 1860 census enables us to compare the status of William Henry Tayloe with his younger brother Henry, who had led him into the Canebrake in the 1830s. Ironically, Henry Tayloe was the least successful planter in the family, having never recovered from his bankruptcy in 1843. By 1860 he had moved to a plantation close to Demopolis, where he had fifty-four slaves—fifteen fewer than the slave force he inherited at Gwinfield in Virginia, thirty-two years before. Henry remained on friendly terms with his brother, and in 1852 he named his third (and only surviving) son William Henry Tayloe, returning William's gesture in 1836 when he had named his third (and only surviving) son Henry Augustine Tayloe. In 1860 Henry's William was a young boy, who would grow up to become an Alabama lawyer in the late 1870s, while William's Henry was a newly married man and in charge of Mount Airy estate. There were 169 slaves at Mount Airy in 1860—3 more than in 1850, which is impressive when one considers that 36 people had been taken from Mount Airy to Alabama in 1854. But William Henry Tayloe now controlled a considerably larger and stronger slave force at Larkin and Oakland. The center of gravity for the Mount Airy Tayloes had shifted to the Deep South.

## *Motherhood and Family Life*

We have seen in Chapter 4 that the Mount Airy women had many more children and much larger families than the Mesopotamia women. Among the 191 Mount Airy women of childbearing age there were 147 mothers who produced 636 children between 1809 and 1863, for an average of 3.3 births per woman, 4.3 babies per mother, and 11.6 newborns per year. At Mesopotamia, about 139 of the 299 women of childbearing age produced 420 children between 1762 and 1833, for an average of 1.4 births per woman, 3.0 babies per mother, and 5.8 newborns per year. Less than a quarter of the Mount Airy women aged sixteen to forty were childless during these years, while more than half of the equivalent Mesopotamia women bore no children. The most striking feature of Mount Airy motherhood is that forty-six of the women (31 percent of the mothers) had six or more babies. These women bore 359 children, or 56 percent of the total, as shown in *Appendix 19*. Spinner Eliza Ward was the leader, with thirteen children, followed by field hand Georgina Page, spinner Nancy Carter, and field hand Sally Yeatman with twelve, seamstress Amelia Carrington and spinner Fork Eve with eleven, and field hands Crissy Carrington and Lucy Dudley with ten.

To be sure, Mount Airy's robust birthrate was tempered by very high early childhood mortality: 19.5 percent of the boys and girls born on the estate died before the age of four—a statistic almost identical to the 19.4 percent of children under age four who died at Mesopotamia. The death rate among the older children and adults at Mount Airy cannot be tracked, because nearly half of these people were transferred to other Tayloe properties or sold. But there were 216 more recorded births than deaths at Mount Airy between 1809 and 1863, so despite the transfers and sales the population was increasing naturally.

The growth rate at Mount Airy propelled the exodus of slaves to Alabama, but was it sustained among the Mount Airy migrants who lived in the Canebrake? Some commentators talk about the slave breeding states (for example, Virginia) versus the slave consuming states (for example, Alabama), and it is generally agreed that cotton labor in Alabama and Mississippi was considerably more stressful and demanding than the mixed agriculture practiced in nineteenth-century Virginia and Maryland. So it is pertinent to ask whether cotton labor impeded slave motherhood on William Henry Tayloe's Alabama plantations. Unfortunately, Tayloe's reportage on slave vital statistics in Alabama, 1834–1865, is seriously flawed. He frequently moved his Alabama slaves from one plantation to another, from Adventure, Walnut Grove, and Oakland in the 1830s to Oakland, Woodlawn, and Larkin in the 1840s and 1850s, and he failed to list the slaves that he held on these five Alabama plantations in the years 1839, 1844, 1846, 1848–1852, 1857–1858, and 1864—a total of eleven missing years.[103] Tayloe never kept a birth register, and he often neglected to list young children until they were several years old. Hence there were probably a good many unrecorded births and infant deaths in Alabama.

What we do know is this: A total of 105 Mount Airy women bore all of their children in Virginia during the years 1809–1863, and they had 455 recorded births for a family size of 4.3. Eleven Mount Airy women had children in Virginia and also in Alabama, and they produced 68 babies for a family size of 6.2. And thirty-one Mount Airy women bore all of their children in Alabama during the years 1834–1863, recording 113 births for a family size of 3.6.

These totals probably understate the fecundity of the Mount Airy women in Alabama. Two-thirds of the Mount Airy women and girls sent to Alabama before the Civil War came in 1845 or 1854; most were teenagers when they arrived, and many were in their twenties when freed in 1865. These women did not have the time during enslavement to produce exceptionally large numbers of children. The eighteen females who came to Alabama in 1835–1836 had the longest recorded time span for childbearing, but only seven of them were candidates for motherhood on William Henry Tayloe's Alabama plantations. Nine were sold soon after arrival, one girl died on arrival, and the oldest woman (Nanny Glascow) was past childbearing when she was sent down. Six of the remaining seven who came in 1835–1836 bore children on Tayloe's Alabama plantations, and collectively they had 4.8 children each—which is higher than the Mount Airy average. The two Alabama mothers with the largest recorded slave families, Becky Carrington with nine children and Anna Marcus with eight, both came in 1835. Becky, who was sent to Adventure plantation in 1835 at age seventeen, married Tom Flood and bore her children in Alabama between age nineteen and thirty-eight. Anna was thirteen on arrival at Oakland, married Travers Hilliard (who arrived in 1845) about ten years later, and had her children between age twenty-three and thirty-nine. If the female migrants in 1845 and 1854 continued after 1863 to produce children at the same rate as the 1833–1835 female migrants, then the Alabama mothers probably had as many children as the Virginia mothers.

*Appendix 33* examines the relationship between motherhood and occupation, categorizing the 191 Mount Airy women of childbearing age by the jobs they held as adults in Virginia or in Alabama between 1809 and 1863. As we saw in Chapter 5, the Mount Airy field hands who worked in John Tayloe's Virginia farm quarters performed arduous yearlong physical labor, were assigned to drudge agricultural tasks such as ditching, grubbing, weeding, and hoeing, and were pushed by Tayloe's overseers to produce unusually large crops of wheat and corn. The Mount Airy field hands almost certainly worked harder than the Mount Airy female textile workers or the female domestics, and they had smaller families, suggesting that occupation had some impact on childbearing, as at Mesopotamia. Yet 81 percent of the Mount Airy field hands were mothers as against 72 percent of the domestics and 75 percent of the spinners, and eleven of the eighteen Mount Airy women with eight or more children were field hands. So the difference between the occupational groups of mothers was fairly slight. Turning to Alabama, the occupational impact is almost impossible to measure, because practically all of the women were cotton pickers. Tayloe operated a no-fringe system in Alabama, with only a housekeeper, a seamstress, an elderly cook, and an elderly nurse on each of his plantations. For what it is worth, the two Mount Airy domestics in Alabama *did* have larger recorded families than the fifty-three Mount Airy field hands in Alabama.

Since the Mount Airy migrants to Alabama had been selected for their youth, it is not surprising that there were many more births than deaths within this population in 1834–1863: a total of 134 recorded births and fifty-seven recorded deaths, as shown in *Appendix 5*. It would be interesting to know whether this was the standard pattern on other Alabama cotton plantations or whether the slaves who had been brought to Alabama collectively by their masters (like the Mount Airy slaves) had more favorable vital statistics than the slaves who were sold in Alabama by slave traders and flung together among strangers on new cotton farms. Very possibly the Tayloe case was exceptional. Edward Baptist offers another measurement. He argues that the cotton planters coerced their slaves into increased productivity per laborer year after year, and that there was a 400 percent increase in the amount of cotton picked per hand between 1800 and 1860.[104] The slim documentation for William Henry Tayloe's cotton production shows no such dramatic change in output per hand. At Oakland plantation in 1843, thirty-five hands produced 210 bales of cotton, or 6.0 bales per hand. In 1845, thirty-five Oakland hands produced 233 bales, or 6.7 bales per hand. In 1847, thirty-nine Oakland hands produced 230 bales, or 5.9 bales per worker. In 1848, forty Oakland hands produced 280 bales, or 7 bales per hand. And in 1859, sixty-eight Oakland hands produced about 430 bales of cotton, or 6.3 bales per worker.[105] So while cotton production more than doubled at Oakland between 1843 and 1859, this increase appears to have been mainly due to the expanded size of the workforce. Thus, it is certainly possible that William Tayloe treated his Alabama "servants" more humanely and pushed them less hard than most of the other cotton slaveholders.

Whether the Mount Airy slaves were living in Virginia or in Alabama, they clearly lived in nuclear families and—despite constant violation of these families by the Tayloes—they managed to sustain strong kinship ties. The Mount Airy evidence strongly supports Herbert Gutman's argument that the family was the key organizational unit by which African American slaves were able to preserve their cultural values and traditions.[106] The Tayloes recognized slave marriages, and their slave inventories demonstrate the vitality of Mount Airy family life by identifying fathers as well as mothers. Some families can be traced back to the 1780s, making it possible to construct genealogical trees that link parents, children, and their descendants through four or even five generations. The Mount Airy families all bore distinguishing surnames that William Tayloe accepted and recorded from the 1840s onward. How these surnames originated is a puzzle I have not solved. The roll call of Mount Airy slave surnames—Carrington, Dixon, Dudley, Glascow, Hilliard, Lewis, Moore, Page, Thomas, Thurston, Ward, Wormley, Yeatman, and so on—may indicate past connections with white families in the Northern Neck. But if so, the connection is difficult to establish. Census returns identify eighty-six householders in Richmond County who held twenty-five or more slaves between 1810 and 1860; only two of these slaveholders' surnames—Carter and Smith—overlap with any of the Mount Airy slaves' surnames.[107] The 1850 census also identifies forty-nine free black households in Richmond County with thirty-five different surnames, none of them overlapping with the Mount Airy slave surnames.[108] What we do know is that the Mount Airy slave families valued their surnames, and usually retained them after they were freed in 1865. Compared with the broken Mesopotamia families discussed in Chapter 4, the Mount Airy people are more fully identified, their families are larger in size, and they can be traced over a longer time period.

A prime example of four-generation motherhood is Sally Thurston (1780–1843), a field hand who lived all of her recorded life at Doctor's Hall farm quarter and bore thirteen children—the first five of them before the initial Mount Airy inventory of 1808, and the last eight between 1810 and 1826. Her very extensive family is fully charted on the website accompanying this book, *www.twoplantations.com.* Sally's first husband was a Doctor's Hall field hand named Amphy Thurston (1782–1815). I have not been able to identify her second husband, whose surname was Hilliard. The first eight of Sally's children took the surname

Thurston, and the last five Hilliard. Six of Sally's eight daughters can be tracked throughout their careers, and five of them bore children: Winney had seven, Sarah six, Kesiah five, Anna two, and Bridget one—for a total of twenty-one. At least three of her five sons were also parents. Anderson married twice and had eleven children by his wives Lizza and Rose; Travers's wife Ann Marcus had eight children; and James's wife Winney had two, giving Sally forty-two traceable grandchildren. By 1863 sixteen of Sally's grandchildren were dead or unaccounted for, and another seventeen were childless, most of them too young to marry or to have children. But by 1863 five of Sally's granddaughters and four of her grandsons had produced twenty-four children. Four of these great-grandchildren had died, and the other twenty were all twelve or younger in 1863. The nine grandchildren with young families and the seventeen grandchildren without families in 1863 doubtless produced a great many additional progeny after the Mount Airy records terminated, so that Sally probably had well over fifty great-grandchildren. And her brood would look even larger if we could trace the careers of three of her children—John, Mary, and Jane—who were sold before they married or had children.[109]

Another woman with a great many descendants was Franky Yeatman (1766–1852), a housemaid at Mount Airy whose husband was a weaver named Israel Yeatman (1766–1821). Her family is also charted in full detail at *www.twoplantations.com*. Franky had six children, many fewer than Sally, but because they were born a few years earlier, we get a better look at the fourth generation and a peek at the fifth generation of the Yeatman family. Franky and Israel's two sons and four daughters were born between 1786 and 1807. Two of them disappeared early from the records: Tom was transferred to another Tayloe property at age twenty-eight, and Fanny was given to Ogle Tayloe at age twenty-three. But their four siblings had large recorded families. Spinner Nancy and spinner Eliza both had thirteen children, weaver Jane had four, and gardener Dick's wife Betsy had five, for a total of thirty-five grandchildren. In the next generation, two of Dick's children died or were sold when young, but the other three produced twenty-nine children. Four of Eliza's children died early, and another three were childless in 1865, but the remaining six had twenty recorded children. Jane's four children were all transferred away from Mount Airy before they were old enough to have children, and eleven of Nancy's brood either died or were transferred at a young age, but two of Nancy's girls bore ten children, so that by 1865 Franky had fifty-nine recorded great-grandchildren. And by this date five of her great-granddaughters had given birth to eleven great-great-grandchildren—a total that probably reached over one hundred during the late 1860s, 1870s, and 1880s.

Both Sally Thurston's family and Franky Yeatman's family were considerably thinned out by early death. One of Sally's sons, twelve of her forty-two grandchildren, and four of the twenty-four great-grandchildren charted at *www.twoplantations.com* died young. In Franky's case, none of her six children, but eleven of her thirty-five grandchildren, fifteen of her fifty-nine great-grandchildren, and six of the eleven great-great-grandchildren charted at *www.twoplantations.com* died young. Several members of both families were sold, and others were transferred to William Tayloe's brothers in the division of 1828. But despite these losses, both families increased and multiplied in biblical fashion. And of course William Henry Tayloe was the beneficiary. Sixty-six members of Sally's family and sixty-eight members of Franky's family appear on Tayloe's Alabama and Virginia slave inventories taken in 1863.

It is clear from this Tayloe evidence that the living and working conditions impeding slave births and family development at Mesopotamia—the brutal labor regime, the toxic disease environment, the semistarvation diet, and the rejection of motherhood by some women—were minor factors or nonfactors in the Mount Airy situation. Most critically, perhaps, the Mount Airy people were generally in sound health and always had enough to eat. Some commentators on antebellum southern slavery want to believe that the masters forced their slaves to couple and breed. There is absolutely no reason to suppose that William Tayloe ever engaged in such a practice. He had no need to. His slaves were young and vigorous, and—like most people—they enjoyed sex. So the move from Virginia to Alabama, despite its many negative features, failed to deter the continuous growth of this African American community.

## *The Civil War in Alabama*

William Henry Tayloe was never a rabid secessionist, but the eve of the Civil War found him in an unusually aggressive mood. In December 1859, two months after John Brown's raid on Harpers Ferry, he wrote to brother Ogle from Alabama, declaring that since the radical abolitionists led by "Old Brown" have hurried the crisis, "we are making ready to fight, for fight we must, if the Union is dissolved." He added: "All classes cry out 'War is preferable to our present position.'" For William, the question—who is to be our next President?—had become irrelevant, since the North was plundering the southern slaves. "The time has come at last for War, or the Yankees must let us alone."[110] William Tayloe, being sixty-two years old and in frail health, was not in a position to join the fight. But he assumed that the South could defeat any military effort by the North to prevent secession. In December 1860, a month after Lincoln's election, he proposed to his son Henry "a union of the Tayloes in case of Disunion of the U.S. and buying a large tract of unimproved first rate Cotton lands, to be gradually settled, by detachments of our Negroes—converting the Virginia farms into grazing fields and introducing a better system of tenantry."[111] To William, the break with the North thus offered an opportunity to expand and strengthen his family's slave-based plantation system.

Benjamin Ogle Tayloe cannot have shared William's aggressive stance, because he lived in the Federal capital and continued to believe in the Union, and during the crisis of 1860–1861 tried as best he could to help avert the Civil War. In February 1860 Ogle visited South Carolina and found to his alarm that everyone was in favor of secession. As the presidential election approached, he fervidly supported John Bell of the Constitutional Union Party (who carried Virginia and two other border states), believing that Douglas of the Northern Democrats was "impossible," that Breckenridge of the Southern Democrats (who carried Alabama and all of the Deep South) was "a man of straw," and that Lincoln and the Republicans would incite civil war. In August 1860 Ogle met with President Buchanan and tried to persuade him to form a fusion ticket that could block Lincoln's election. Immediately after the November election he wrote to President-elect Lincoln urging him to support the constitutional rights of the South, and he also sent a letter from "A Southern Cotton Planter" to a Washington newspaper, the *National Intelligencer,* asking his southern colleagues to give Lincoln a fair trial. In April 1861, on the eve of Virginia's secession, Ogle was given an interview by the president, who told him that he had decided to stand firmly on the Republican platform, which stated that "the Union of the States must and shall be preserved." So Ogle's peacekeeping efforts came to nothing.[112] The other southern cotton planters in the Tayloe family all supported the Confederacy, and several of William's and Ogle's nephews became high-ranking officers in the Confederate Army.

While William Henry Tayloe was buoyant about southern prospects in the secession crisis, he also realized that if war came, Mount Airy would be temptingly close to the Union lines. His Virginia "servants" might try to escape to the Yankees. So in December 1860, while proposing to Henry at Mount Airy "a union of the Tayloes in case of Disunion," he also advised his son to send "all your active Negroes South before long."[113] Son Henry delayed for a year, which proved to be a mistake, but in early December 1861 he made preparations to bring fifty of the Mount Airy people to Alabama. On December 16, when Henry set forth with this party, two men who had been selected for membership—Elias Harrod and Isaac Bray—were "not in place to go and were left." The forty-eight African Americans who made this trip are listed in Appendix 34. Their journey to Alabama was much quicker than previous ones. We have no details, but the migrants must have traveled by railroad, probably all the way from Fredericksburg or Richmond, Virginia, to Montgomery, Alabama. There was as yet no rail connection in 1861 between Montgomery and Selma, so the slaves probably walked to Selma and then traveled by train to William Henry Tayloe's junction at Larkin plantation. They reached Larkin on December 24 after a journey of only eight days.[114]

When Henry Tayloe returned to Mount Airy in early 1862, he discovered that six of his most valuable people—carpenter Elias Harrod or Harwood (age twenty-seven), field hand Isaac Bray (age forty-two), miller Jacob Carrington (age thirty-eight), field hand Paul Page (age thirty), carpenter Richard Thomas (age thirty), and carpenter Jerry Glascow (age twenty-five)—had fled to the Yankees on December 29. Henry Tayloe valued these six men collectively at $7,500, a major loss.[115] Another missing man was Ralph Ward (age fifty-five), the Mount Airy coachman and groom. These desertions prompted son Henry Tayloe to send more of the Mount Airy people as fast as he could to his father in Alabama. While he was assembling a second party, two further men—field hand Urias and mason Edward—absconded on March 18, 1862. Urias seems to have crossed into Maryland successfully, but Edward drowned while attempting to cross the Potomac River.[116] Henry Tayloe quickly dispatched another fifty slaves to Alabama under the care of Conway Reynolds, who was the overseer at Doctor's Hall quarter. The members of this second party are listed in Appendix 35. The only information we have about this second journey is that the travelers arrived at Larkin plantation on April 1, 1862, so they too must have come by railroad. Overseer Reynolds, for his part, was one of many Tayloe white employees who was barely literate. After he returned to Virginia, he sent a letter to William Tayloe that began as follows: "eye roat you sum time a goe."[117]

Between December 16, 1861, and April 1, 1862, the Tayloes moved ninety-eight slaves—forty-one males and fifty-seven females—from Mount Airy to Larkin and Oakland plantations. This was by far the largest exodus from Mount Airy, and also the last. These migrants came almost entirely in family groups. There were only three singles in December 1861 and four in March 1862. Harrington and Fanny Smith in 1861 and Richard and Sally Yeatman in 1862 traveled with seven children. Six other married couples and four single parents were accompanied by young sons and daughters. Many of the migrants in these two groups were interrelated: twenty-two members of the Yeatman family came in five groups from three different branches of this large family. Altogether, there were many more young children than in any of the previous parties from Mount Airy: seventy of the ninety-eight migrants were under the age of twenty. Twenty-six boys and girls were in their teens, and thirty-nine of the youngest migrants were not yet of working age. The Tayloes' aim was no longer to move young laborers who were especially suited for cotton production to Alabama, but to hide as many people as possible—both current workers and future workers—from the Union Army.[118]

The most senior Mount Airy couple sent south in 1861–1862 was Gowen and Georgina Page, both field hands in their fifties. Georgina was a prolific mother, with four recorded children by her first husband, Fork Joe, and eight by her second husband, Gowen. Five of her brood had died young; two had been sent to Oakland in 1845, one to Oakland in 1854, and one (Catherine Yeatman) to Larkin in 1861; and one (Paul Page) had deserted to the Yankees in 1861. Georgina's two remaining children, Susan (age twenty-eight and married with a young child) and John (age sixteen), accompanied their parents to Alabama in March 1862. Young John Page had broken his leg in December 1861, so he was probably not fully recovered on arrival in Alabama. But when he complained that his ankle hurt, William Tayloe put him down as an idler.[119]

Three years later, in February 1865, at the age of nineteen, he was impressed into Confederate Army service with three other Tayloe slaves. The military appraisers valued him at $4,500. According to the enrolling officer, John stood five feet three inches, weighed 147 pounds, and had a "copper" complexion rather than being "black" like his three companions. He was given two suits of clothes, a blanket, a bushel of meal, and twenty pounds of bacon.[120] How long John served is not known, but he survived the war, worked at Larkin in 1866 as a free man, and in 1870 was described in the U.S. Census as a field hand who was still working at Larkin, unmarried and living by himself, but close to his aged father, Gowen Page, who lived in an adjacent Larkin cabin.

The Mount Airy slaves congregated at Larkin and Oakland had absolutely no chance of escaping to the Yankees, because the Alabama Canebrake was in the center of the Confederacy and was one of the very last areas to be reached by the Union Army. The wartime atmosphere in this region was rabid, and the whites (being heavily outnumbered by the blacks) were paranoid about the possibilities of African American insurrection. In the summer of 1861 a local vigilante committee tried to keep the slaves from communicating with one another or moving around at all, not even allowing those men whose wives lived on neighboring plantations the liberty of visiting their mates. In March 1862 William Tayloe reported to his son, "Mob law is rampant among us." Five slaves who had attempted to rob an overseer were taken out of jail and nearly lynched, and in August 1862 a slave who killed a tyrannical overseer was hanged "in double quick time." The Alabama whites talked savagely, Tayloe reported, offering no possibility of compromise with the North.[121]

Despite white hysteria in the Canebrake, William Tayloe reported to his son that the ninety-eight black newcomers from Mount Airy were "wonderfully contented," and he insisted that they were far better off than they had been in Virginia. There was apparently much more room for them at Larkin than at Oakland, because ninety-four of the people who arrived in 1861–1862 were placed at Larkin, and only four at Oakland. Tayloe himself lived at Oakland but made frequent visits to Larkin. "All the Mount Airy set," he remarked in March 1862, "seem determined to be dutiful obedient and faithful. They are contented."[122] And he may well have been right, because in 1861–1862 many of the families that Tayloe had been pulling apart since the 1830s were suddenly reunited.

To illustrate the situation, let us return to the extensive progeny of Sally Thurston, whose family is charted at *www.twoplantations.com*. Anderson Hilliard, who was Sally's youngest son and had been the foreman of the Doctor's Hall field gang until he came to Alabama in 1861, was now living at Larkin in company with his brother James, who had been moved to Alabama in 1833, his sister Anna, who had been moved to Alabama in 1835, and his sister Becky, who had been moved to Alabama in 1854. Anderson's brother Travers, who had come to Alabama in 1845, lived close by at Oakland. In the next generation, Anderson's niece Isadora Page, who had been a Mount Airy field hand until 1861, was now reunited with her brothers William and Chapman and her sister Grace, who had all been moved to Alabama in 1845. Anderson's nephews Ralph and Jonathan Wheeler, arriving in 1861–1862, were similarly reunited with their brother Amphy, who had come to Alabama in 1854. And Anderson's own children Anna, Travers, Martha, Jane, and Sallie, who came with him to Larkin in 1861, rejoined their older sister Mary, who had been in Alabama since 1854. In the fourth generation, Isadora's children William and Josephine, having been brought from Mount Airy to Larkin in 1862 when they were eight and four years old, now found seventeen young cousins at Larkin and Oakland who had all been born in Alabama.

Another extended family similarly reunited at Larkin and Oakland was the Carringtons: the ten children, thirty-one grandchildren, and two recorded great-grandchildren of David and Criss Carrington. This family is also fully charted at *www.twoplantations.com*. David Carrington was a free mulatto man who evidently lived in the Mount Airy neighborhood; he surfaced in the Virginia census of 1830 (the first census to itemize free black and colored people) but not in the census of 1840, and is otherwise unidentified.[123] His wife Criss or Crissy (1793–1836) was a field hand at Forkland who bore ten children, all of whom followed her into slavery. Two of her progeny died in infancy, and Mary died at twenty-one in 1838, but the other seven were all alive at the outbreak of the Civil War, six of them married with a total of twenty-seven living children. Criss's daughter Becky (eight children) had been sent to Alabama in 1835; her sons Austin and David (two children) went to Alabama in 1845; while Godfrey (eleven children), Israel (three children), Jacob (one child), and Kitty (five children) stayed at Mount Airy. Godfrey and Israel were Forkland field hands until they moved to

Larkin in 1861–1862. Jacob (whose wife, Winney Grimshaw, was taken away from him in 1845) operated the Mount Airy mill until he deserted to the Yankees in 1861. And Kitty (who accused Lizzy Flood of setting fire to the Mount Airy mansion in 1844) was a housemaid who remained at the big house until 1865. By 1862 five of Criss Carrington's seven living children were at Larkin or Oakland, together with twenty-four of her twenty-seven living grandchildren. But not everyone in this large group was satisfied. Seamstress Amelia Carrington, the wife of Godfrey, arrived from Mount Airy in April 1862 with her six youngest children and was reunited with her four oldest children who had been sent to Alabama in 1854 and 1861. But Amelia bluntly opened up to her master, "saying 'she won't tell a story about it—that she wants to go back.' "[124]

By April 1862 William Henry Tayloe seemed to be in great shape. He had nearly 400 slaves securely hidden from the Yankees, and about 225 working hands (aged ten to fifty) available for employment. But already the war was causing enormous problems. The Union blockade of Mobile cut off all exports, so Tayloe's Alabama cotton could no longer be shipped to Liverpool. In 1861 he had sold his crop to the Confederate government at 15 cents per pound; in 1862 he was storing his bales in pens covered with boards until they could be sold; and during the rest of the war he grew cotton only for homespun textile production. In 1864 Larkin produced a total of eight bales, and Oakland one bale, and Tayloe thought that his workers were becoming lazy "from want of the excitement of the cotton crop."[125] In place of cotton Tayloe set his hands to producing food crops for human consumption and fodder for horses and cattle, and bartered these goods to local merchants. In September 1862, for example, he sent a shipment of butter, corn, and bales of fodder valued at $781 to McClure & Thomas at Selma, and received in return eleven barrels of molasses and seven sacks of flour valued at $674.[126] In 1863 Tayloe acquired additional agricultural land by renting a tract adjacent to Larkin, and he placed 101 of his Larkin people on it.[127] From January 1864 to March 1865 overseer J. W. Ramey sold Larkin produce valued at $9,723—mainly corn, but also some fodder, potatoes, and butter. This matched Tayloe's average wheat and corn sales at Mount Airy in the 1850s, but amounted to only a fifth of his Alabama cotton sales in 1859.[128]

In 1862 Tayloe predicted that the Civil War was going to be a long struggle. The Federal capture of New Orleans in April 1862 was a severe shock, and early southern victories on other fronts were not being followed through with sufficient force "to conquer a peace." The war immediately cut off all manufactured goods from the North, so Tayloe warned his son that it was impossible to buy anything in Alabama, and urged him when he visited in 1862 to come well supplied, especially with calico for the field women and house servants. A number of Tayloe's hands (not identified in the records) were working for the Confederate government in 1862, laying down a new railroad line from Demopolis, Alabama, to Meridian, Mississippi. Meanwhile, forced to establish a self-sufficient economy, Tayloe fixed up looms to make cloth, and set his Mount Airy shoemakers to practice their craft. Seamstresses Amelia Carrington and Ailse Carter, who had both just come down from Virginia, were put to work making hats, and Richard Yeatman, who had been a gardener at Mount Airy for thirty-three years, starting when he was eleven, laid out a new garden at Larkin. In May 1864 William Tayloe was dressed in a homespun "sack" made by Amelia, "who is one of the very best of our servants," and he also noted that the Larkin garden, cultivated by Richard, "is the best filled of any I have seen."[129]

In 1863, when William Henry Tayloe took his last complete Alabama census, 250 African Americans were living at Larkin and 141 at Oakland, for a total of 391 on the two plantations. And 76 percent of these people were directly or indirectly from Mount Airy. Ninety-five males and ninety-two females had lived at Mount Airy and migrated to Alabama, and they had produced 110 Alabama-born children—60 boys and 50 girls. The other 24 percent of the slave population consisted of twelve people who came from William's Deep Hole plantation in Virginia and Ogle's Nanjemoy plantation in Maryland, forty-one Alabama hands (most of whom had previously worked for William's brother Henry Tayloe at Adventure or at Walnut Grove), and their forty-one Alabama-born children.[130] By 1863 only sixty-six slaves still lived at Mount Airy, under the management of William's son Henry. Together, William and Henry Tayloe held a total of 457 slaves, which was more than double the number that William had started out with in 1828. And 86 percent of these people were living in Alabama.

But the war was going badly. By May 1863 William was very anxious about the fate of Vicksburg, and after the garrison surrendered he wrote on July 22, 1863, "I know of no safe place since Vicksburg fell."[131] Alabama was now open to possible invasion, and the Confederate government was particularly anxious to protect Selma, the chief military manufacturing center in the Deep South, producing rifles, cannon, and much of the ammunition for the Confederate Army. During 1863 a huge three-mile fortified earthen wall was built in a semicircle around the city, fifteen feet wide at the base and eight to twelve feet in height. Thirteen of William Tayloe's hands can be identified as working on the Selma fortifications, and there may have been others not listed in the surviving documents.

Confederate military impressment records for the year 1863 in the Tayloe Papers give us a mixture of information about the African Americans from Oakland and Larkin who worked at Selma. On February 10 an officer in Selma issued the following pass: "The boy Kit, impressed for Government service, has permission to return to his home in Perry County. He is authorized to travel on the Rail Road." The "boy" in question was thirty-eight-year-old Kit (or Christopher) Lee, who had been born at Mount Airy, was formerly owned by Henry Augustine Tayloe, and was now living at Larkin. On March 18, General Lewis and Dangerfield Perkins, both of whom had come to Oakland in 1845, were impressed and issued one shovel and one spade, which apparently they had to share, and sixty-seven pounds of bacon. On September 1, Jim Moore was released after working for fifteen days on the Selma fortifications, for which William Henry Tayloe received $15. The next day Joe Moore, Thadeus Dudley, Jim Glascow, and Dangerfield Perkins were given passes to return to Oakland plantation with their tools (a shovel and spade each) after working at Selma for thirteen days. Since these four men put in a total of fifty-two days of labor, they had earned Tayloe another $52. Also on September 2, Bill Wormley, John Dudley, Cornelius Elms, and John Wright from Oakland, and Ralph Wheeler and Peter Yeatman from Larkin, were given passes to travel home to Tayloe Station on the Selma-Uniontown line after working at Selma for thirteen days each. Their labor totaled seventy-eight days, bringing their master an additional $78. And three months later Peter Yeatman was released on December 1, after working for ten days at $1 per day. Peter Yeatman and Dangerfield Perkins were impressed twice during 1863. All thirteen of Tayloe's men who are recorded as working on the Selma fortifications in 1863 were field hands, and they all had come to Alabama from Mount Airy.[132]

William Tayloe himself was still hoping to keep his slave force. In July 1864, while lecturing his son as usual on plantation management, he presented himself as a model patriarchal master. "I have not done as much good as was in my power when in prosperity," he wrote, "but my best acts are my examples in managing Negroes. The effort to promote cleanliness, domestic comforts and religious tendencies on my plantations influenced my [Alabama] neighbours and within a few years the position of the Negroes in this County is much ameliorated." Tayloe was expecting to resettle Mount Airy after the war with "a party of good hands."[133] But he faced increasingly challenging conditions at Larkin and Oakland. In the spring of 1864 a diphtheria epidemic broke out at Oakland, and despite his efforts to nurse the sick he recorded eight slave deaths. Tayloe became the treasurer of the Canebrake Soldiers' Relief Association, and paid $1,553 in 1864 (compared with $169 in 1862) in a combination of taxes and contributions toward clothing and medical care for the Confederate soldiers. Between July and November 1864 another fifteen Tayloe hands were impressed to work for the Confederate government; five of these men had been previously impressed in 1863.[134]

By the close of 1864 William Henry Tayloe was becoming despondent. "My Neighbours are whipped, and will do anything to try to save property," he wrote. "The Railroad is near worn out, and *a smash* is a common thing." Three more slaves were impressed from Oakland in January 1865, and four more from Larkin in February. Tayloe thought too many of his people were being taken, but he couldn't get to Marion to plead with the enrolling officer, because the road was impassable for a buggy and he was too frail to ride eighteen miles on horseback. "The truth is I am becoming unfit for service," he admitted. "Anxiety and care is wearing me out." He predicted that the Canebrake would soon be as open to Yankee raiding as the Northern Neck, and that Selma would be attacked. "Slavery is doomed any how," he concluded. "Shermans success and Hoods misfortunes make me think secesh is whipped and I am ruined."[135]

The end came very suddenly and decisively. On April 2, 1865, a Federal army of 9,000 men attacked Selma, which was defended by less than 4,000 men and boys, far too few to man the three miles of fortifications effectively. In only two hours the Yankees broke through the earthen wall that the Mount Airy hands had worked on for months in 1863. Having captured the city, the soldiers spent a week destroying the Selma arsenal, and moved on to occupy Montgomery on April 12. Meanwhile, Lee had surrendered to Grant at Appomattox Courthouse on

April 9. The war was over, nearly 400 African Americans at Oakland and Larkin were now freedmen, and William Henry Tayloe had lost property worth more than $250,000 by his valuation.[136]

Looking back at the exodus from Mount Airy to Alabama between 1833 and 1862, it is evident that most of the Mount Airy people escaped the worst features of the internal slave migration. Those who came to the Canebrake before the Civil War walked the entire distance, but they were not chained into coffles, and it is indeed quite possible that they looked forward to the adventure and came willingly. On arrival, twenty-three of the Mount Airy migrants were sold, and the others worked on William Henry Tayloe's cotton plantations. Their living and working conditions appear to have been very harsh, but less brutal than on the plantations described by Charles Ball, Solomon Northup, and the other escaped slaves who described their captivity for abolitionist readers. Of course, our only reports are from Tayloe and his overseers. It is worth pondering what Richard Donnahan, the Oakland overseer, wrote to Tayloe in May 1851: "The Negroes have worked better this year than they have ever since I have been on the farm and with less trouble. There has not been one of them whipped since we commenced working the crop."[137] Reading between the lines, the Oakland slaves had not been working to Donnahan's full satisfaction between 1845 and 1850, and there was still some degree of unspecified "trouble" in 1851. There had been no whippings as yet in 1851, but the cotton crop had only been in the ground for two months, and the picking season was three months into the future. The whip was always there, whether in use or as a threat, the ultimate instrument of white mastery and black degradation.

What seems most important about the Mount Airy migration is that these people all came to Alabama in the company of relatives and friends, which is why they wanted to come. From the outset they were bonded together. A New Orleans slave auctioneer decreed to one of his buyers: "It is better to buy none in families, but to select only choice, first rate, young hands from 16 to 25 years of age (buying no children or aged negroes)."[138] By this auctioneer's criteria, almost all of the Mount Airy migrants were choice, first rate, and young, and *all* of them were in families, even those youngsters who traveled to Alabama as singles, because they had grown up with the other Virginia migrants and knew them intimately. Once in Alabama, the migrants quickly established nuclear households similar to the family units they were taken from in

Virginia. Here we have a rooted community, with shared endurance for the cruelties of forced labor and the whip—a community very different from Mesopotamia, where a variegated mix of Creoles and Africans, blacks and colored, with widely differing backgrounds and experiences, had been forced to live together by their white masters.

We have watched this new Alabama black family network take gradual shape. Those pioneers in the 1830s who weren't sold or quickly died began to marry and to have children. The migrants who came in 1845 found friends, relatives, and marriage partners who had been in Alabama since the 1830s. The migrants who came in 1854 found a larger number of friends, relatives, and marriage partners who had preceded them in the 1830s and 1845. And the Civil War newcomers in 1861–1862 were reunited with the migrants from 1833–1854. The Tayloes did not plan this reassemblage of the Mount Airy slave community in Alabama. It was supposed to be a temporary method of hiding the slaves from the Yankees. But it became permanent. Once the war ended and the Mount Airy people were freed, the great majority continued to live close together in the Canebrake. In 1865 most of these Virginia slaves became Alabama freedmen and freedwomen. And most of their children, grandchildren, and great-grandchildren were still living and working in the cotton South until at least the 1930s. This was the lasting consequence of the exodus from Mount Airy.

# 8

# *Mesopotamia versus Mount Airy: The Social Contrast*

WHILE THE DEMOGRAPHIC CONTRAST between our two plantations is strong and clear, the social contrast, though equally strong, is much harder to synthesize. At Mesopotamia there was a constant influx of new people; at Mount Airy there was a constant exodus of young people. At Mesopotamia a great many of the slaves were born in West Africa, bringing with them their vital native culture; at Mount Airy none of the slaves were born in Africa. At Mesopotamia interracial sex was openly practiced and well documented; at Mount Airy interracial sex was secretly practiced and undocumented. At Mesopotamia corporal punishment was generally unrecorded but seems to have been frequent and severe; at Mount Airy corporal punishment (again mainly unrecorded) was apparently far less ferocious. At Mesopotamia the people were desperately hungry much of the time; at Mount Airy there was never semistarvation. At Mesopotamia the field hands spent most of their time yearlong on the sugar crop; at Mount Airy the field hands cultivated half a dozen crops and were switched to new tasks every few days. One of the chief features of life at Mesopotamia was a Christian mission organized by the Moravian Church. One of the chief features of life at Mount Airy, utterly different in character, was the removal of over 200 people to far-distant Alabama. A kaleidoscope of multiple refractions makes it difficult to integrate these contrasts in a meaningful way. The most obvious conclusion—and also the most easily overemphasized—is that physical conditions were a great deal harsher in Mesopotamia.

It is helpful to look instead at the commonalities on the two plantations. At both Mesopotamia and Mount Airy there was a significant internal split between two categories of workers: the craftsmen and domestics versus the field hands. In both places there was also a fault line between the people who tended to cooperate with their white owners and the people who tended to rebel. The testing of this fault line came in two major events that form twin climaxes to our story: the great slave revolt in western Jamaica in 1831–1832 and the Civil War in Virginia and Alabama in 1861–1865. And it is striking to find that the Barhams' cooperators and rebels during the Jamaican revolt turned out to be quite different from the Tayloes' cooperators and rebels during the Civil War. Examination of this social contrast forms the subject matter of this chapter.

## *Head People at Mesopotamia*

Reading Thomas Thistlewood's Jamaica diary opened up for me the significance of rebels versus cooperators within a slave community. At first I was most struck by Thistlewood's frequent flogging of the people who worked for him and his obsessive sexual assaults upon slave women—as discussed in Chapter 4. Then I saw that a good many slaves were always challenging Thistlewood's management system by repeated thievery or by running away. The diary can be construed as a perpetual war between a tyrannical master and his rebellious slaves.[1] But the Caribbean slave system—like the antebellum U.S. slave system—was based on the premise that a majority of the slaves would cooperate. And I soon found that a sufficient number of slaves were always coming to Thomas Thistlewood's assistance. Certainly he was at war with his most rebellious subjects, but he remained the master because he received essential support from the cooperators.

At Thistlewood's first post, when he single-handedly managed some forty slaves at Vineyard pen in 1750–1751, his diary identifies seven males and one female who ran away during the course of the year. Five of these runaways had to be caught, and two youngsters who were food thieves had to be looked for again and again. Seven of the Vineyard people provided crucial help to Thistlewood by catching these runaways and thieves.[2] Similarly in 1765, when Thistlewood was the overseer at Egypt sugar estate, ten of the slaves ran away during the year, several of them repeatedly. But there were other slaves—fifteen men and two women—who caught these malefactors and ran important errands, for which they were generally rewarded with bottles of rum. The line between rebellion and compliance was often blurred. A jobber named Quamina was flogged during the year for being insolent to his white supervisors, but he was essentially a cooperator, and received at least three bottles of rum during 1765 for his prowess in catching runaways.[3] The picture that emerges shows a significant division within the Egypt slave community between the people in compliance and the rebels. Was this also true at Mesopotamia?

Mesopotamia was a much bigger operation than Thistlewood's Egypt, and the workforce was more formally organized. Thistlewood frequently shifted some of his slaves from one occupation to another, from crab catching to field work, or from field hand to seamstress, whereas the Mesopotamia slaves held the same jobs year after year and changed assignments only because of age or illness. And Mesopotamia was also more hierarchical, with many more head people who were the leaders of specialized work groups. Driver Johnnie and Thistlewood's mistress Phibbah were the only head people at Egypt in the 1760s, whereas there were always a dozen to fifteen head people at Mesopotamia: at least three drivers in charge of the principal field gangs; the head carpenter, who led the other carpenters; and similarly the head cooper, head mason, head blacksmith, head boiler, head distiller, head pen keeper, head carter, and head watchman. Among the females, there would be one or two drivers for the grass gang and one or two head housekeepers. The head people at Mesopotamia were generally listed among the highest-priced slaves, but their intangible leadership skills could not be measured in pounds, shillings, and pence.[4] The strong presence of the head people at Mesopotamia exacerbated the sharp division on this estate between the skilled specialists and the subordinate field hands.

The most prominent set of specialists at Mesopotamia were the eighteen male field drivers who led the field gangs between 1762 and 1833. By definition these people complied with the slave system, and they were given an extraordinarily challenging assignment, with no real equivalent at Mount Airy. Their main job was to force all of the men and women who planted, weeded, and harvested the cane to keep in pace with their fellows, day after day, and in order to satisfy their white bosses they needed to press and harass the gang members continually—or to seem to be doing so. Their secondary job was to flog any slave identified by the overseer as needing correction for misbehavior. And these men had to live in the same tight village community with the people they were yelling at and whipping during working hours.

In 1762 there were four male field drivers with symmetrically stated ages: Parry (age sixty), Matt (fifty), Ralph (forty), and James (thirty). Parry had been a driver at Mesopotamia since at least 1727, when he was probably in his mid-twenties; he was also listed as a driver in 1736, 1743, 1744, 1758, and 1762, and in the past had probably been the head driver. But he was well past his prime and retired from his driver's job in 1764.[5] Matt was the head driver in 1762. We have met Matt in Chapter 6: an intelligent and articulate man, with long experience at Mesopotamia, having worked as a mason from 1736 into the 1750s, when he became a driver. Ralph, about a decade younger, had a similar career trajectory; he had been a child of about five in 1727 and a teenage field hand in 1736, and was probably promoted to driver in the 1750s. And James seems to have been chosen around 1762 as a successor to Parry; he first appeared on a Mesopotamia inventory in 1743 as a young boy, and like Ralph had previous experience as a field hand. Overseer Daniel Barnjum (1760–1778) was evidently satisfied with Matt, Ralph, and James because he continued to employ this trio as his three field drivers through 1776. During these fifteen years the field gangs produced modest crops averaging 157 hogsheads of sugar per year (see Appendix 14). Matt died in 1777, Ralph retired the next year, and James became head driver in 1778.[6]

When Barnjum died, the new overseer, John Graham (1778–1788), wanted greater production and stricter discipline over the field gangs, so he assembled a new team of more aggressive drivers: Francisco, Hector, and Cuffee Tippo. Francisco, who replaced James as the head driver in 1779, complained to the Moravians in 1782 that Graham made him swear and tear at the Negroes more than he liked, but he continued as head driver for twenty years until he died in 1798.[7] Hector, a Mesopotamia-born field hand, became a driver in 1775 at age thirty-one, and in 1797 killed a cook named Cudjoe in a scuffle. Acquitted of murder, Hector succeeded Francisco as head driver in 1798, and was still employed as the head driver in 1816 when he died at age seventy-one. Cuffee Tippo also had a lengthy stint, serving from 1778 to 1797. These three men were thus in charge during the difficult years of the American Revolutionary War and well into the boom period that followed, when the

Mesopotamia field gangs were enlarged and the slaves began a thirty-year peak period of sugar production, with the crop averaging 250 hogsheads during the years 1788–1817. Two men who had been drivers at other Jamaican plantations became drivers at Mesopotamia in this period: Qua from Three Mile River estate in 1786, and Warwick from Southfield estate in 1791. Two field hands, Mingo and Dennis, were also tried out as drivers in the 1790s but were quickly demoted. And then there was Bristol, who became a driver in 1798, was promoted from managing the second gang to joining Hector on the first gang in 1802, and died horribly when he fell into a copper in the boiling house in January 1809.

At the turn of the century, Chelsea, who was born at Three Mile River and worked as a house servant and cooper at Mesopotamia, joined Hector as a driver of the first gang in 1801 at age thirty. Chelsea was subordinate to Hector, but when Hector died he took over as head driver from 1816 to 1823, after which he became the head watchman. In 1814 an African named Smart, having been a driver at Cairncurran estate, became a driver at Mesopotamia. Similarly, in 1820 Quaco, a driver at Springfield estate, joined the Mesopotamia drivers. In 1826, when sugar production was declining, Mesopotamia-born Bernard became the head driver at age thirty-one. Eight years previously, in 1818, when he was only twenty-three, he had been appointed driver of the third gang and was soon switched to pen keeper. Bernard was the head driver from 1826 to 1832, and very likely continued to the end of slavery in 1834. The other field drivers in the final years of slavery were also born on the estate: Edinburgh, an experienced sugar boiler, and Sammy, an experienced rum distiller. Edinburgh and Sammy, as we shall see, played a significant role when the great slave revolt of 1831–1832 broke out in western Jamaica.

As this roster indicates, there was a lot of turnover among the Mesopotamia field drivers in 1762–1832, and some of them had short stints. In twenty-five years (mainly between 1775 and 1804) the inventories identified four men as drivers; in thirty-three years three drivers were listed; and in thirteen years (all after 1808) only two were listed—so in those years there were probably unidentified men who drove the minor field gangs. Four of the drivers had previously managed field gangs at Three Mile River, Southfield, Cairncurran, and Springfield, and they probably continued in this role because they knew how to push the field hands who came with them to Mesopotamia. Most important, seven of the eighteen drivers served for fifteen years or more, with Hector holding this important job for an amazing forty-one years.

Next to the drivers, the chief specialists at Mesopotamia were the boilers who at crop time converted the cane juice into crystallized sugar, and the distillers who processed molasses into rum. Like the drivers, these men were by definition cooperators, and they were crucial players because they possessed the necessary skills for making a successful sugar crop. The boilers knew how to temper the cane juice with just the right amount of lime so as to clarify it by separating the dirt from the juice, and they also knew, once the boiling juice had been ladled from copper to copper, the exact moment to "strike" when the liquid was at the point of crystallization and ready to granulate. The distillers had to mix the right proportions of water, yeast, and sugar skim with the molasses, and then ferment this mixture into alcohol. This was unpleasant and dangerous work. The boilers had to endure terrible heat, and both the boilers and the distillers had to cope with nauseous stench. Two Mesopotamia men during our period were fatally scalded in the boiling house, and another was fatally burned in the distillery. During crop time a pair of boilers and a pair of distillers were needed at Mesopotamia, because the sugar factory operated in double shifts round the clock, with a small army of underlings who stoked the boiling house fires, ladled, stirred, and skimmed the boiling cane juice, and assisted with the distilling. There were white supervisors stationed at the boiling house and distillery to prevent theft of sugar and rum, but it was the enslaved boilers and distillers who actually made the sugar and distilled the rum.[8]

Boiler and distiller were seasonal jobs, and out of crop time these men worked as field hands, masons, stock keepers, jobbers, or drivers, while the older sugar workers became watchmen. Unfortunately, the inventories identify only twenty-four sugar makers, when there were probably double that number. No distillers are listed in thirteen of the years after 1803, and no boilers in six of those years. The occupational information that we do have shows that six boilers and three distillers worked for ten years or more as sugar makers between 1762 and 1832, and some spent their entire careers at these jobs. Caesar was a teenage boy in 1727 and a boiler in his early twenties by 1736, and he continued to manage the boiling house until age fifty-five in 1766. A distiller named Cudjoe was about the same age as Caesar. He also was a teenager in 1727 and working in the distillery by 1736, where he continued until 1764. Another distiller, Augustus, who was twenty-two in 1762, worked at this craft for over forty years, until 1804, when he became the head watchman at age sixty-five. Most of the sugar makers toiled for many years as field hands until they were chosen in their thirties or forties to be boilers or distillers, and if they were good at their new jobs they continued into old age. And some were unsuccessful. Davy was made a distiller in 1768 but lasted only one year before he was put back into field work, and in 1773 three distillers—Frederick, London, and Grafton—were dismissed after two years as rum makers and sent back to the field gangs.

A striking fact about the sugar makers is that most of them came from West Africa: six of the ten identifiable boilers were Africa-born, as were eight of the thirteen distillers. The Mesopotamia-born sugar makers mainly got their jobs after Barham bought his last new Africans in 1793, so the proportion of Africans would probably have been higher had Joseph Foster Barham II kept buying from the slave ships until 1807. As noted in Chapter 4, most of the slaves imported directly from Africa were excluded from skilled jobs at Mesopotamia. Only eleven of the seventy-one carpenters, coopers, masons, and smiths came from the slave ships, and only eight of the fifty carters, cattle boys, pen keepers, and stock workers, while 63 percent of the 164 male field hands who lived at Mesopotamia between 1762 and 1793 came straight from Africa. It may be that African slaves were placed in the boiling house and distillery because nobody else wanted this hot and dangerous work. But of course the overseers had to find men who had the requisite skill. Perhaps they discovered that the African slaves worked well with fellow Africans when given a responsible assignment.

The Mesopotamia field drivers, boilers, and distillers—as well as the other head people—played a very prominent role in the formation and maintenance of the Moravian church. As shown in Chapter 6, Moravian missionaries came to Mesopotamia in 1758 offering the promise of eternal salvation to slaves who were baptized in the blood and wounds of Jesus Christ, but they had no program for social change or physical improvement here and now. Instead, they urged the slaves to obey their worldly masters. When they arrived bearing this message the missionaries were warmly greeted by driver Matt and his fellow head people, and also by large numbers of (mostly female) old and sickly slaves. It is hardly surprising that elderly enslaved people, close to death, would identify with the sufferings and resurrection of Jesus, giving them hope and comfort amid the great cruelties of slave life. But why were the head people, the most skilled and responsible members of the slave community, also immediately attracted to the Moravians?

Brother Christian Heinrich Rauch in his diary for 1759–1761 didn't have much to say about the elderly candidates for baptism, but he took an active interest in drivers Matt, Parry, and James, boiler Sharry, distiller Augustus, and carpenters Frank and Ammoe, who visited him frequently in the evenings after work, talked freely, and asked many questions. These men all had enough self-confidence to deal very directly with these strange newcomers. They could see right away that the devout brothers and sisters dressed in black garb were totally different from any white people they had met before, and they were clearly intrigued by the missionaries' efforts to reach them. When the head people learned that the Moravians had been sent to Mesopotamia by their absentee master, Joseph Foster Barham, they probably figured that by converting to Christianity they might gain some leverage against the overseers and bookkeepers who mistreated them. And of course they saw an opportunity to engage with the white man's religion, a set of beliefs that until now had been largely concealed from them.

The Moravians, for their part, were eager to work with the head people, whom they saw as the leaders of the slave community, and Brother and Sister Rauch quickly made head driver Matt their helper in reaching out to his fellows. Matt obviously enjoyed this role, and he seized the opportunity to complain about his irreligious white bosses to the Moravians while stressing that he badly wanted to become a Christian. Matt visited Brother Rauch in the evening after work at least eight times during Rauch's first month at Mesopotamia. He told the missionary that he had believed in the existence of God since childhood, but had been worrying about his sinful state for the past five years—probably ever since he became a driver and spent his days harassing and whipping the field hands. In May 1760 he complained to the Moravians that his job as a driver was giving him a lot of trouble and blocking his path to baptism, but the missionaries—anxious not to clash with the estate managers—persuaded Matt to accept his worldly lot.[9] Matt's fellow driver James put the missionaries in a more perplexing dilemma. James was feeling very anxious because he had two wives and wanted to know whether this was acceptable to the Lord. Rauch answered ambiguously that since he had two wives he should keep them both, but a believer in the Saviour should have only one wife.[10] Distiller Augustus presented the missionaries with a different sort of insoluble problem by complaining to them that Daniel Barnjum was having sex with his wife at the Great House. Once again the missionaries had no satisfactory remedy.[11]

As they ventured into a minefield of unacceptable "heathen" custom (such as polygamy), and unacceptable white abuse (such as rape), the Moravians moved slowly and carefully in forming a Christian congregation at Mesopotamia. Boiler Sharry was the initial Mesopotamian to be baptized in 1761, followed by driver Matt in 1763. But the next baptismal candidates were largely drawn from the bottom of the Mesopotamia social hierarchy. Thirty-three slaves were brought into the congregation in 1764–1766, an odd assemblage of seven male drivers, sugar makers, and craftsmen and twenty-six women, most of whom were elderly and/or invalided. And when the Moravians confirmed six of the baptized members as communicants in 1766, enabling them to participate with the missionaries in the "Abendmahl," the same pattern held: the first communicants were three drivers (Matt, Parry, and James), a carpenter, and two old women. Back in 1758, the Moravians had told Matt that the Hebrew patriarch Isaac had blessed his son Jacob even though Jacob had deceived the blind Isaac by pretending to be his brother Esau, and this had given Matt hope that God might be merciful to him as well. So it was fitting that when the Moravians baptized Matt (aged about fifty-two), his new name was Jacob. Later they baptized old driver Parry (aged about sixty-four) as Isaac (Jacob's father), and driver James (aged about thirty-five) as Joseph (Jacob's son), and young Hector (aged twenty-three), who would soon become a driver, as Benjamin—another of Jacob's sons. Thus, by design or accident, the Moravians branded these four Mesopotamia drivers as an Old Testament dynasty.

From the beginning the Moravians had difficulty in reaching the main body of Mesopotamia slaves, the field hands, especially the male field hands. The missionaries may well have alienated most of the first gang members by quickly enlisting Matt, the person who knocked them around during working hours, as their Moravian helper and recruiter. Philander, the first active male field hand to be baptized, in 1767, was the thirty-sixth person to join the church. And only six other male field hands joined Philander by 1769, when 106 Mesopotamia slaves had been baptized. Half of the Mesopotamia adults belonged to the Moravian Church by 1769, but the field hands and livestock workers who constituted 55 percent of the population were underrepresented. Only about a tenth of the male field hands were church members. Since the Mesopotamia women had scarcely any option except field work, there were twenty-two female field hands in the church by 1769—but this was just a third of the women in the field gangs. Furthermore, 72 percent of the converts were beyond the age of forty when they were baptized. The Moravians had conspicuously failed to attract the young men and women of prime working age who could best sustain the congregation in future years. They had formed a church led by a few head people and mainly populated by senior slaves.

Over the years, the field hands and the young people continued to stay away from the Mesopotamia church. *Appendix 36* compares the structure of the Mesopotamia slave congregation at three stages in the history of the Moravian mission: in 1761–1769, when 106 slaves were baptized to form the congregation; in 1798–1818, when 77 candidates applied for baptism; and in 1831, when 66 people were identified as "Christians" on the slave inventory.[12] The comparison shows striking continuity in the composition of the slave church. There were always more women than men, especially among the baptismal candidates in 1798–1818. More than half the members were always over age forty, with many elderly invalids and very few young people in their teens and twenties. Not surprisingly, the membership shifted from an African majority in the 1760s to a heavily Creole majority by 1831. In occupational terms, new head people and craft workers kept joining the congregation, while the field hands and livestock keepers remained underrepresented. Only three male field hands were among the baptismal candidates in 1798–1818, and one of them was rejected.

Overall, ten of the eighteen men who served as field drivers between 1758 and 1833 were members of the church. Between 1763 and 1823 there was always at least one Moravian field driver, and in most years there were two, and sometimes three. Eight of the ten identifiable boilers were baptized, and five of the thirteen distillers. There were also ten carpenters, ten coopers, six masons, and five blacksmiths—or 44 percent of the Mesopotamia craft workers—including most of the head carpenters, head coopers, and head masons. By 1831 half of the male churchmen were drivers or craftsmen, a much higher percentage than on earlier lists. The head carpenter, the head cooper, and the head mason were all churchmen, as well as two of the three male drivers on the estate. Through seven decades, while the congregation kept shrinking in size and successive pairs of missionaries kept struggling to find new recruits, the head people at Mesopotamia remained loyal to the Moravian Church.

## *The Jamaican Slave Revolt of 1760*

At the beginning and at the end of the Moravian mission at Mesopotamia, two major black rebellions took place in Westmoreland parish: the slave revolt of 1760 and the slave revolt of 1831–1832. These were the two largest slave rebellions in all of Jamaican history, and the Mesopotamia slaves played an active and interesting role in both rebellions.

The slave revolt of 1760 was devised as a coordinated island-wide conspiracy, led by a secret network of Coromantee (Akan) slaves from Ghana. It broke out at a time when Britain and France were fighting the Seven Years War in the Caribbean, with well-trained army and navy troops stationed in Jamaica. Just what the rebels hoped to achieve remains unclear; possibly they intended to kill all the whites and divide Jamaica into African ethnic states. These rebels excelled at striking with sudden surprise. The revolt started, apparently prematurely, in St. Mary parish on the north side of the island on April 7 under the leadership of a Coromantee chieftain named Tacky, who attacked several plantations, seized arms, and butchered a number of whites. Tacky was quickly defeated and killed by a counterforce of British soldiers, local militia, and Maroon warriors, and the revolt in St. Mary was suppressed by the end of April. But then it spread. The rebels, now numbering up to 1,200 and drawing many slaves from local plantations, stealthily regrouped in an unsettled mountainous area of Westmoreland near the Hanover border.[13] Suddenly, on May 25, they attacked Masemure estate, about six miles west of Mesopotamia, assaulting the Great House and killing four whites.

On May 26 the Moravian missionaries at Mesopotamia reported that overseer Macfarlane "came to us in a great hurry and gave us a distressing message: that Negroes had broken out in rebellion against the white people not far from here, and they had gruesomely murdered many, and that all the white people in this parish were going to rifle practice."[14] Though the missionaries don't say so, it appears that Daniel Macfarlane remained on duty at Mesopotamia while Daniel Barnjum and the other members of the white managerial staff immediately departed to serve in the militia, taking with them many of the Mesopotamia slaves they most trusted to serve as baggage carriers.[15] Head driver Matt was not among the baggage carriers; his job was to keep the field gangs operating with diminished manpower. The sugar crop was completed by this date, or nearly so; Mesopotamia shipped 185 hogsheads of sugar and forty-two puncheons of rum in 1760. But out of crop the Mesopotamia field hands were required to work nearly twelve hours a day, from 5 A.M. to 6:45 P.M., with two hours allowed for breakfast and dinner.[16]

During May 26–28 the rebels attacked at least eight plantations in western Westmoreland and two in Hanover, causing much damage and loss of life. Thomas Thistlewood's Egypt estate was closer to the action than Mesopotamia, on the main road between Savanna la Mar and the rebel headquarters, and he reports vividly on the white and black refugees fleeing to the port town, and on the army, navy, and militia detachments marching to confront the rebels. Thistlewood kept guard at Egypt day and night, extremely worried that the Egypt slaves would join the rebels.[17] On May 29 the Westmoreland militia attempted to attack the rebels' barricaded encampment and was soundly defeated, sending Thistlewood into a near panic. Three days later, on June 1 at Mesopotamia, driver Matt told the Moravian missionaries that he could not come to their Sunday church service because "he had to send four loads of mules with Provisions from this estate to the Soldiers" who were assembling for a second and much larger assault. The next day, June 2, British soldiers and sailors stormed the rebels' barricade, supported by militia units from three parishes and Maroon warriors, and drove them out in a two-hour battle, with scores of rebels killed and captured.[18] This was the turning point in the revolt. The remaining rebels now had no hope of beating the British army in a pitched fight, so they turned to guerrilla warfare, scattering in small bands, moving through the mountainous backcountry, and staging raids on plantations within their reach.

Eight days after the storming of the rebel barricade, on the evening of June 10, the Moravians reported "a great noise" at Mesopotamia, when a detachment of regulars and militia flushed out a band of rebels on the mountainous northern edge of the estate. The Moravians were told that in this skirmish "40 of the rebels were killed and 50 were taken captive." But Mesopotamia was now on the firing line, and since the rebels in previous attacks had targeted the residences of white planters, the Mesopotamia Great House needed protection. So later that evening driver Matt "brought his money and his best things [to the missionaries] so we could watch them in the house because he had to guard the big house all night." On June 16 the Moravians observed that "these past few days there were several of prisoners from the rebellion led past us toward the Bay." These prisoners would be burned to death by slow fire, or they would be strung up on gibbets for public display until they died of thirst.[19]

The climax of the rebellion for the Mesopotamia slaves came on June 22–23, 1760. According to the Moravian missionaries' somewhat convoluted account, on Sunday, June 22, "our negroes caught one of the rebels on their plantation; he was even busy making his killing knife sharp. They kept him in the floor until they could give him over to Mr. Macfarlane. Our negroes have now legitimated themselves in the neighborhood." Who were "our negroes"? They were the Mesopotamia slaves who had been attending the Moravian missionaries since 1758 as candidates for baptism, and the missionaries were obviously pleased that they had "legitimated themselves in the neighborhood" by seizing a rebel and delivering him to overseer Macfarlane, thereby demonstrating that the Moravian doctrine of passive obedience posed no danger to the social order. Unhappily, the Moravians didn't bother to identify the men who captured the rebel. He was caught "on their plantation" (the slave provision grounds) and then probably taken to one of the slave huts, where they "kept him in the floor." On June 23, "the Rebel, who was caught by our negroes yesterday, was taken to prison today, and they held a [hasty trial] over him and burned him alive. He was found guilty of murdering 2 children two days before."[20] The Moravians were seemingly untroubled by the rebel's grisly fate. For them, the main point was that the Mesopotamia baptismal candidates had demonstrated their stout opposition to the slave revolt.

As sporadic fighting continued, militia and army units kept trying to track down the remaining rebels. Numerous Mesopotamia field gang workers had been serving as military baggage carriers since late May, and on June 25 a further contingent of slaves from the estate was ordered to bring provisions to Daniel Barnjum and the other Mesopotamia white militiamen. The next day driver Matt complained to the missionaries that "he has so few negroes at home to help him work his plantation; a few of them hadn't been home for four weeks etc."[21] And by early July a new problem had developed at Mesopotamia. The slaves' provision grounds, almost certainly heavily raided by the rebels, were producing no food, and the slaves were desperately hungry. Matt told the Moravians on July 4 that "several had not even had a bite to eat," and that attorney Pool had refused to supply emergency rice and flour. A week later, the Moravians learned that the Reverend Mr. Pool had placed twenty slaves "in the Stock[s], because they had complained about their hunger etc. and the Attorney was afraid that if they went out at night, they would steal."[22] Such was the reward for dutiful behavior during the slave revolt.

By mid-July the rebellion in Westmoreland was winding down, but the whites continued on active militia duty to the end of the year. At Mesopotamia, Daniel Macfarlane fell ill and died in December 1760, and attorney Pool replaced him with Daniel Barnjum. When Barnjum wrote Joseph Foster Barham to give him this news, he reported that he was now the only white man on the estate—indicating that the other four or five Mesopotamia staffers remained on militia duty, and that the two male missionaries, Christian Heinrich Rauch and Nicolaus Gandrup, didn't count as white men. Barnjum told Barham that he was also still out on patrol three days a week until after Christmas.[23]

Meanwhile, most of the remaining rebels had moved east to the uninhabited Nassau Mountain area of St. Elizabeth parish. This gave the Moravian missionaries fresh alarm, because all of their missions except Mesopotamia were on sugar estates in St. Elizabeth, close to the rebel hideout. There were Moravian missions at the Island, which was owned by Joseph Foster Barham, and at the Bogue, Elim, Lancaster, and Two Mile Wood, owned by Barham's relatives. And it is interesting to note that Barham's Island slaves, doubtless told by the Moravian missionaries that rebellion was wicked, resisted the rebels with exceptional vigor. On Sunday, November 2, a Moravian diarist wrote, "In Island there was a big commotion because of the rebelling Negroes. They killed some of the rebels, cut their heads off and stuck them on poles. They captured one of them alive and then hung him the next day."[24] In short, the Island slaves were combating the rebels in just the same fashion as the white planters and soldiers. And they were acting far more aggressively than the Mesopotamia slaves did in 1760.

## *Runaways at Mesopotamia*

By 1761 the slave revolt was over. But from 1764 onward the Mesopotamia records repeatedly document another form of turbulence on this estate by identifying the numerous slaves who ran away. The slave inventories provide most of the data, stating, for example, that twenty-five-year-old Dublin was a "run away" in 1764, that twenty-nine-year-old Rory was a "great runaway" in 1792, that thirty-eight-year-old Austin was a "notorious runaway" in 1811, and (the most common notation) that eighteen-year-old Springfield Joe "runs" in 1824. Since evasion of work was endemic on all the Jamaican sugar estates, I believe that the bookkeepers who listed Dublin, Rory, Austin, and Joe as runners were not describing single disappearances of a day or two. These Mesopotamia absconders were like Thistlewood's Adam, Derby, and Plato at Egypt, slaves who tried repeatedly to escape, or who succeeded in getting away for weeks or months at a time. Since Mesopotamia was bordered to the north by undeveloped mountain land, it was relatively easy to hide in the bush. And there is supplementary evidence in the estate expense accounts to show that many of the Mesopotamia runaways trekked for long distances before they were picked up and put in jail. On more than fifty occasions the estate was billed for charges incurred by Mesopotamia slaves who were incarcerated in the Westmoreland, Hanover, St. James, or St. Elizabeth parish workhouses, sometimes for several months.

Altogether, 140 Mesopotamia slaves were identified as runaways—104 males and 36 females. They ranged in age from eleven to sixty-four, and ninety-two of them were repeat absconders. Most of them started young: thirty-three were in their teens when they first ran away, and fifty-seven were in their twenties. Five of the men and two of the women managed to escape permanently, and another five men were transported off the island for bad behavior. As a group they differed markedly from the head people we have been discussing. Most significantly, 115 of them (82 percent) were field hands, proof, if any were needed, that toiling in the cane fields was to be evaded if at all possible. Only one of the female runaways was *not* a field worker. Among the males, nine were craft workers, but four of these men were mulattoes who were probably very bitter about being kept in slave status when they were half white.

Another seven were stock keepers, and the remaining eight had marginal jobs: five watchmen, two field cooks, and a jobber. None of them were head people, and just 3 of the 140 identifiable runaways were members of the Moravian Church. Carpenter Primus, who was baptized in 1773, fled in 1777. Watchman Austin, who was baptized in 1814, ran in 1828 and escaped permanently in 1829. And chapel attendant Prudence (the sole female absconder who was not a field hand) was on the run in 1831 because she didn't get along with the Moravian missionaries, Brother and Sister Ricksecker. Two other Mesopotamia runaways *did* become church members—but only after they stopped trying to escape. Jeremiah, a field hand, ran in 1805, 1807, and 1809, then became a candidate for baptism in 1811 and was baptized in 1817. And Sambo, another field hand who was listed as a runner in fifteen of the years between 1780 and 1825, was baptized in 1831. Overall, there was a clear, sharp separation between the Mesopotamians who joined the Moravian congregation and the Mesopotamians who ran away.

Over the years the runaway pattern at Mesopotamia kept changing. In 1764 ten slaves—seven men, two women, and a young girl—disappeared. Seven of the ten were born at Mesopotamia. They seem to have slipped off separately, because Portland was caught near the Hanover border, and Jack spent thirty-two days imprisoned near the St. Elizabeth border, while fourteen-year-old Abba got as far as Montego Bay.[25] Where the others went is unknown, but all ten were back at Mesopotamia by the end of 1764. Three of the culprits, Fidelia, Portland, and Charles, were threatened with transportation off the island, but instead they remained on the estate until they died in 1766, 1781, and 1799, respectively. Among these ten people, only an African field hand named Frederick was recorded as a repeat absconder.

Overseer Barnjum made no mention of the 1764 runaways in his letters to Barham, but he did report in September 1765 that he was pleased with the eight new African men and two African women who had just been brought to Mesopotamia. It soon became evident that the new Africans were not very pleased with Barnjum. In January or February 1766, at the start of their first crop time, three of them fled, probably collectively. Cato and Polydore were picked up by the Maroons and spent twenty-six days in the Montego Bay jail during March, while Othello was nabbed and jailed in April. Othello was probably weakened and injured during his flight, because he died the next month in a "sullen" state. By October 1766 Cato and Polydore were back in jail together with two of the other new Africans, Richmond and Fabio, and the four men stood trial on the charge of rebellious conspiracy. Unfortunately, Barnjum doesn't supply any details about this alleged conspiracy. Fabio was released and Richmond was acquitted, but Cato and Polydore were found guilty and sentenced to transportation.[26] Barnjum reported in August 1767 that Richmond probably knew about Cato and Polydore's conspiracy, and regretted their departure, for "from being a very good Negro, he became a very worthless one." Richmond, according to Barnjum, ran away frequently after he was released from jail, plotted rebellion with some neighboring slaves, "took to eating dirt and neglected himself to so great a degree that he seemed perfectly Indifferent what became of himself and at last dyed."[27] Murdock, another of the new Africans and another dirt eater, also died in September 1767. Thus, five of the eight African men who had come to Mesopotamia in 1765 were dead or gone within two years.

In the late 1760s there are no further references to slave runaways or rebels in the Mesopotamia records, but from 1772 onward the annual inventories listed at least one absconder every year through 1831. From 1772 to 1782 there were only two or three runaways reported per annum, but some of them were determined repeaters. Mesopotamia-born Ned, for example, was advertised for in 1773, fled again in 1774 and 1776, disappearing this time for at least two years, and was finally sold to the Reverend Mr. Pool for £60 in 1779. And an African named Levant fled in 1779, was jailed for nine months at Montego Bay in 1779–1780, and escaped permanently in 1784. In the years between 1783 and 1795, a period when many new slaves were brought onto the estate, the average number of runaways suddenly jumped from three to eight per annum. Mesopotamia-born Sam started running in 1780 and was sold for £40 in 1787 and transported. An African named Rory, bought in 1785, was called "a great runaway" in 1789, and disappeared so much of the time during the next six years that he was also sold and transported in 1795. And the slaves brought to Mesopotamia from Southfield in 1791 seem to have been particularly dissatisfied with their new home. Attorney James Graham told Barham in 1791 that "there are three or four Negroes on your estate that no management can reclaim from Breaking Negroe houses, plundering the grounds and running away."[28] Ten of the thirty-two Southfield men ran away often, sometimes in groups of three or four, and two of these Southfield men—Anthony and Garrick—managed to get away permanently.

Things settled down somewhat during the next two decades. Between 1796 and 1818 only three men and one woman on average were identified as runaways per annum. Again, some of the men were very persistent. Mesopotamia-born Poorman began running in 1787, continued in 1789–1790, 1792, and 1801–1802, was placed in a workhouse in 1804, was gone for at least eight months in 1807, fled again in 1809, and was taken up for the last time in 1810, shortly before he died at the age of forty-one in 1811. African Ralph began running in 1802, continued in 1813–1814, and made a successful escape in 1816.

Bigger trouble started in 1819 when Joseph Foster Barham II, having bought Springfield estate in Hanover parish, directed his attorneys to move the 112 Springfield slaves to Mesopotamia. Springfield was six miles from Mesopotamia as the crow flies, but ten miles away by foot. There was no road connecting the two places. In June 1820, 105 of the Springfield people spent a week carrying their beds, tables, and other household goods along a difficult mountain track to a halfway point where carts could transport their goods the rest of the way. Many of the Springfield people were very upset by this move, because they were separated from their mates, children, and friends back in Hanover. At first Barham's attorneys allowed them to spend weekends at Springfield until their new provision grounds at Mesopotamia produced enough food for them to live on. They were supposed to arrive at Mesopotamia on Mondays by noon and leave on Friday afternoons. But by August 1820 up to a dozen Springfield people were staying away all week, a habit that the attorneys tried to correct by dismantling their old huts so that they would have no places to live in at Springfield. This did not stop the absenteeism. In 1824 fifteen Springfield people never showed up during the entire crop, and when attorney Ridgard distributed cloth to the slaves in June he found that twenty Springfield people were missing.[29] Seven of the Springfield people never came to Mesopotamia. Another seventeen Springfield men and eight women were recorded as runaways, at least five of whom spent time in the Hanover workhouse after they were caught. Congo Phillis disappeared permanently in 1823. Congo Margaret ran off in 1824, 1825, and 1827, and after a month's disappearance in 1830 was found dead in Hanover. Springfield Mingo absconded in 1819 and was caught by the Maroons, fled again and was imprisoned in the Hanover workhouse in 1825–1826, and was transported in 1827. More than a dozen of the Springfield slaves were still on the run on the eve of emancipation.

When the Springfield slaves ran away, they were going back home to Hanover parish, but most of the other absconders in the late 1820s did not have a destination. Here they differed greatly from runaway slaves in the antebellum South who aimed to reach the northern states, where they could hope to be protected by abolitionists, or to get all the way to Canada. For Jamaican runaways there was no Canada. Hence most of the Mesopotamia absconders seem to have disappeared for relatively short periods to get out of work, not to live permanently in the bush. Finding food and shelter in the wild was a huge problem, and the runaways who stole from the Mesopotamia provision grounds incited anger from their fellow slaves who had been robbed. Joining the Maroons was not an option, because the Maroons received monetary rewards when they delivered escaped slaves to the white planters. The Mesopotamia accounts show payments to the Maroons ranging from £1 to £2 12s. for catching runaways. At least four Mesopotamia runaways died from exposure during the last years of slavery, including a twelve-year-old boy named Ben who ran away in 1831. The only person who escaped successfully in the 1820s was an elderly watchman and Moravian Church member named Austin, who was sixty-three years old when he disappeared in 1829 and unlikely to last very long on his own.

From 1824 onward nearly twenty runaways were listed per annum at Mesopotamia. A rising expectation that slavery would soon be terminated made coerced plantation labor more burdensome than ever, and bad health conditions on the estate must have accelerated the discontent. In August 1825 exceptionally large numbers of the slaves came down with fevers, bone pains, and general debility; there were fifty people daily in the hospital. In May 1826 a second epidemic forced the attorney to halt the sugar harvest, with between 115 and 135 people (including half of the first gang) in the hospital, this time suffering from influenza, pleurisy, and violent bowel pains.[30] During the years 1826–1831, when on average sixty-nine Mesopotamia men and boys and ninety-four Mesopotamia women and girls were working in the first and second field gangs, twenty-seven of the males and eighteen of the females were described as running away at least once during these six years, and twenty-eight disappeared repeatedly. Thus, more than a quarter of the field workers ran away during these closing years of slavery. And the number of absconders kept increasing year by year, with the highest total in 1831, when eighteen males and ten females were listed as runaways. Such was the situation when a massive slave rebellion broke out on December 27, 1831, in St. James parish and spread rapidly toward Mesopotamia.

### *The Jamaican Slave Rebellion of 1831–1832*

This rebellion, the largest slave revolt in the history of the British West Indies, was quite different in character from the Coromantee-led revolt of 1760. The slaves were not trying to kill all the whites and establish African-style rule. They were demanding the immediate abolition of slavery. By 1831 there was massive agitation in Britain for abolition and much pressure on the home government to liberate the slaves. In Jamaica Baptist and Methodist missionaries, who were openly critical of the slave system, were attracting large numbers of slaves to their evangelical meetings. The revolt is often called the Baptist War because the initial slave leaders were converts to the Baptist Church who were inspired by the Bible to strike for freedom. They were emboldened by a rumor—allegedly spread by the Baptist preachers—that the king of England was going to free the slaves on January 1, 1832. Ironically, this slave action was deeply distressing to the pious members of the Unitas Fratrum who had been trying for more than seventy years to bring Christianity to the Jamaica slaves. In 1831, as in 1760, the Moravians were totally opposed to slave rebellion.

For several years preceding the rebellion the Jamaican planters were in a state of helpless outrage. They hated the impending termination of slavery but were in no position to declare independence from Britain. Their attitude is well captured in the pugnacious articles published in the *Jamaica Courant and Public Advertiser,* a daily Kingston newspaper (known as "the beastly *Courant*" by its critics) that spoke out for the defenders of the old order. During the course of 1831 the *Courant* continually referred to the enslaved Jamaican population as "our peasantry" and characterized the blacks as "idle, sensual, but cheerful, merry and contented," and far better off than the half-starved mechanics in England. The *Courant* printed a letter from a free mulatto slave owner named T. F. who declared that "the more slaves we have, the more are we interested in the rights of property." The editor objected to a bill in the Jamaica Assembly prohibiting the flogging of female slaves, since women are harder to govern than men, and always the ringleaders of trouble. Throughout 1831 the *Courant* continually railed against the "saintcraft" of "Willy" Wilberforce and the "idle and hypocritical vagabonds" who operated the Anti-Slavery Society in Britain. It noted that Nat Turner's rebellion in Virginia (August 1831) was led by a fanatical black preacher. When discussing religion in Jamaica there was no mention of the Moravians, but the *Courant* attacked the "humbug" Society for the Propagation of Christian Knowledge, as well as two "reverend ruffian" Anglican clergyman from the island who had joined the abolitionists in Britain, and especially the "base and sordid" Methodists and "their greasy brethren" the Baptists. Yet the *Courant* also printed a Methodist sermon with a message it approved: telling the slaves that Christianity does not meddle with the civil relations between masters and slaves, and urging them to accept their present condition.[31]

In July 1831 the *Courant* cited with approval the resolutions of a meeting in Westmoreland that denounced attacks on the Jamaican system of slavery—property held by necessity, not choice—and called for committees of correspondence in each parish to organize a petition to the king. In an interesting inversion of American revolutionary ideology, the *Courant* was also alarmed by the Reform Bill in Parliament that was rejected in 1831 but passed in 1832, because this measure would end the rotten borough electoral system and thereby destroy "every opportunity which the Colonies possessed, of preserving virtual representation in the Nation's Parliament." The editor was well aware that the sugar islands *did* benefit from the fact that absentee planters like Joseph Foster Barham II became virtual representatives by buying seats in the House of Commons, and the *Courant*'s worst fears were realized in 1833 when the reformed Parliament quickly passed the Slavery Abolition Act, which brought about emancipation in Jamaica in 1834.[32]

Against this background, a large-scale slave rebellion exploded on Tuesday, December 27 in St. James parish, directly northeast of Westmoreland, just as the slaves' rum-infused Christmas holiday celebration was drawing to a close. In 1760 the rebels had operated by stealth and aimed to kill white people. In 1831 the rebels aimed to destroy property, and quickly set spectacular fires that lighted the night sky. In St. James, thousands of rebels fired at least 103 properties, destroying over 40 sugar works and the houses of nearly 100 planters. On most properties the rebel leaders appear to have been the head people, or (as Barry Higman puts it) men "who had achieved some degree of status within slave society and saw the possibilities of complete liberty."[33] Very quickly the destruction spread west to Hanover, east to Trelawney, south to St. Elizabeth, and southwest to Westmoreland. The timing was crucial, for all of the white estate supervisors were forced into militia duty just before crop time in January.

The diary entries of Moravian brother Jacob Zorn, who lived in St. Elizabeth, give a dramatic day-by-day sense of the spread of the rebellion. Zorn first heard of "serious disturbances" in St. James on the second day of the rebellion, December 28, 1831, and "in the evening our fears were augmented by the reflection of fires in the north west, extending to a great distance." On December 29 he met a trooper who told him that many estates had been burned and that the St. Elizabeth militia was under arms. On December 30 Zorn reported that in Black River, the coastal town in St. Elizabeth, the troopers were flying about in all directions, the great guns were pointed up the principal street, and the drums were beating to arms. "In the evening we saw the flames of Ipswich works [in northern St. Elizabeth] rising high into the air." The next day the Zorns started for Spring Vale, where they were ministers to a large Moravian congregation, in order to counsel the slaves to obedience. But when they heard that every white person from the St. Elizabeth mountains had fled to Black River, they likewise retreated to the port town. Martial law was proclaimed on December 31, and an order issued that on every estate where the master's house was burned, the Negro houses should be immediately fired and their pigs and poultry destroyed. Jacob Zorn closed his diary for the year with reflections on the horrors of the rebellion, with 30,000 slaves in open revolt in St James and Hanover.[34]

On January 1, 1832, at Black River, Zorn heard still more alarming reports of spreading destruction. "The revolted slaves have united in large bodies, & being well-armed, threaten to carry all before them." There was fear that the town of Montego Bay in St. James would be fired, and the vessels in the harbor were filled with ladies who had fled to safety. "The fires seen at night, & especially one observed last night, are very alarming as they seem to advance nearer." Nevertheless, January 1 being a Sunday, the Zorns did go to Spring Vale to meet with their Moravian congregation. They found the Spring Vale slaves very glad to see them and quite fearful of reports that the rebels gathered in the woods were enchanted, "so that no sword or ball could do them any injury." Back at Black River on January 2, the Zorns heard that "a number of the rebels have Baptist tickets, & that some are even Leaders in the Church." And on January 3 Zorn reported for the first time that the white population was blaming the Christian missionaries, Moravians included, accusing them of preaching sedition and of telling the slaves that the king had made them free. According to Zorn, the public press, led by the diehard Tory daily *Jamaica Courant and Public Advertiser,* had added fuel to the flames by announcing that now was the time to get rid of *all* the missionaries.[35]

What was the scene like at Mesopotamia? Back on December 30, 1831, the *Jamaica Courant* had printed a letter from Westmoreland: "Don't believe a word they will tell you of the negro insurrection here—there is as much likelihood of a rebellion among the steers." But reality turned out to be a little different. Brother Peter and Sister Sarah Ricksecker had been living at Mesopotamia for a little over a year, and they first learned about the insurrection on December 30—the day the *Courant* letter appeared, and two days after Brother Zorn had heard of "serious disturbances" in St. James. The Rickseckers were immediately alarmed about the rebels, because "we heard of their fast approaching this parish, burning all properties they came to." Like Brother and Sister Zorn, they could now see new fires to the north every night. Overseer William Turner and all the other members of the Mesopotamia managerial staff immediately left on December 30 to join the Westmoreland militia and fight the rebels, and for the next month the Rickseckers were the only white people in the neighborhood. And though the Mesopotamia slaves remained quietly at work, the missionaries prayed to the Saviour that "none of them might become infected with the evil spirit of rebellion."[36]

On Sunday, January 1, 1832, the Rickseckers "entered the year filled with great anxiety & fear, not knowing, how the dreadful insurrection of the negroes might end." The text Brother Peter chose for his sermon that morning was Psalm 102, a cry to the Lord by an afflicted people overwhelmed with trouble, and Ricksecker reports that he focused on the psalmist's closing affirmation (verses 25–28) that the Saviour is eternal and will protect his servants. Ricksecker also noted that an unusual number of strangers came to the service, and the chapel was so jammed that some people could find no room and left.[37] At the church service,

Richard Gilpin, the head mason at Mesopotamia, noticed a drunk and disorderly African stranger named Billy Grant and wondered what he was doing there. Gilpin was thirty-five years old and had come to Mesopotamia from Springfield estate a dozen years before. He had worked his way up from cooper to mason to head mason at Mesopotamia, but took his surname from his old master William Gilpin, the former owner of Springfield. When Richard Gilpin found out that Billy Grant was a field hand from Prospect estate, a property a few miles to the east that had just been torched, he sensed trouble.

After the church service Billy Grant remained on the plantation, and Gilpin heard him asking the Mesopotamia slaves for gunpowder and urging them to join the fight for freedom. "You are a young man," Gilpin retorted, "and you know nothing" about the rebellion. Then Gilpin got blacksmith Edward Foster Barham (also from Springfield) and William Prince (the mulatto head cooper) to help him seize Grant. Gilpin, Barham, and Prince were all church members. Having tied Grant up, they asked a fourth church member, possibly field driver William Samuels, to join them in telling Brother Ricksecker what they had done. Ricksecker reports in his diary that toward evening on January 1 "four of our baptized members came & informed me that a negroe from a neighbouring Estate inquired for gunpowder etc. saying 'Now was the time to fight for freedom, or else they would never get it.' This negroe was then secured by our people." Brother Ricksecker gave them a pass to take Billy Grant the next morning to the militia headquarters at Savanna la Mar.[38]

During the night of January 1, after Billy's capture, the Rickseckers were awakened by musketry from the militia, and they saw a blazing fire hardly three miles distant—probably at Deans Valley Dry Works estate, where the rebels set fire to the trash house but were prevented by the militia from destroying the sugar works. The Rickseckers packed their best clothes in preparation for flight, but the crisis passed. On the next night, January 2, driver William Samuels—a Mesopotamia-born church member, aged thirty-four—was standing sentry at the Cabarita River bridge with Beresford, a young non-Moravian field hand, when a man named George Watson from Hanover parish tried to cross. Watson claimed to be heading for Fort William estate, but he had no pass, so Samuels and Beresford brought him to the overseer's house, where cooper William Prince and blacksmith Edward Foster Barham searched Watson and found that he had gunpowder and flint. They tied him up and brought him to Brother Ricksecker the next morning. The missionary reported in his diary that "a negroe of this property who is a member of our church took up a Negroe who possessed gunpowder, bullets, &c." Watson confessed to Ricksecker that he had helped to attack the Dean's Valley Dry Works the previous night, "& being chased by the militia, he threw away his gun, & had lost his way." Ricksecker then gave Samuels a pass, and he took his prisoner to Savanna la Mar.[39]

Three weeks later, both Billy Grant and George Watson were tried by court-martial. At Grant's trial on January 20, three Mesopotamia slaves—Richard Gilpin, Edward Foster Barham, and William Prince—appeared and charged Billy with instigating rebellion. Gilpin testified about his argument with Grant. Barham testified that he didn't hear Gilpin's argument but helped to seize Grant, and Prince testified that Grant was drunk and disorderly at church. The seven white military officers hearing the case found the charge by the three Mesopotamian slaves well proven, but sentenced Billy to a comparatively light punishment: a whipping of one hundred lashes.[40] At Watson's court-martial on January 19, William Samuels (identified in the court records as "a churchman") and William Prince (identified as "a Christian") charged the prisoner with setting fire. The officers found Watson guilty, and though he seems to have been a less dangerous rebel than Billy Grant, he was sentenced to transportation.[41]

Through the first week of January Brother Ricksecker went out daily to check up on the Mesopotamia slaves, and found to his satisfaction that they were quietly at work. On about January 3, Mesopotamia overseer William Turner made a quick visit to the estate and told Ricksecker that if the rebels had succeeded in burning down the Deans Valley Dry Works estate on the night of January 1, many of the Westmoreland Lowlands slaves would have joined the rebellion. So the capture of Billy Grant and George Watson by Gilpin, Samuels, Barham, and Prince not only saved Mesopotamia from attack but clearly helped to stop the spread of the revolt in their neighborhood.

As it turned out, the rebellion in Westmoreland quickly fizzled out. Throughout the first week of January the Rickseckers continued to see fresh fires lighting the sky at night, but they came from places in the mountains bordering St. James and Hanover. The last significant Westmoreland firing was at Darliston pen on January 6 in the mountains eight miles east of Mesopotamia, though the Westmoreland militia spent another month hunting down rebel fugitives in their mountain retreats. The Mesopotamia attorneys William Ridgard and Duncan Robertson did not find time to write to Joseph Foster Barham until February 10, 1832, and when they did write they were still on duty at military headquarters in St. Elizabeth.

Westmoreland saw less damage than the other four western parishes, and only about fifty properties—most of them small places in the mountains—were burned. Deans Valley Dry Works and Roaring River estate were the two big lowland sugar estates that were attacked. The Jamaica newspapers, when reporting on the rebellion, paid little attention to events in Westmoreland, mainly stressing that most slaves were quietly back to work by mid-January 1832.[42] Only six slaves were killed in this parish during the rebellion. As elsewhere, the authorities responded savagely to the insurrection, instructing the militia to destroy the slave villages, livestock, and provision grounds on properties that had been fired by the rebels. In Westmoreland, rebels captured during the fighting were tried either by military court-martial or in civil slave courts. Most of the thirty-two military trials were held during the week of January 16–21, and twenty-six of the accused slaves were convicted. Another fifty-two slaves were convicted in the Westmoreland special slave courts. The military judges executed almost half of the slaves they convicted, whereas the civil judges used the lash much more. None of the seventy-eight convicted slaves came from Mesopotamia. These seventy-eight "guilty" rebels were all males; only one was a driver, seven were craftsmen, and 86 percent were field hands. So the head people, who led the rebellion in St. James, did not do so in Westmoreland. Thirty-two of the convicted were executed, twelve were transported, thirty were whipped, two were imprisoned, and two escaped or were liberated. In St. James, by contrast, everything was on a much bigger scale: 124 slaves were killed during the rebellion, and 113 captured rebels were brought to trial and executed.[43]

The great western slave revolt lasted only about two weeks at high intensity, but the rebels forced most if not all of the white managers of the sugar estates in the six western parishes into a full month of militia duty. On the great majority of properties, the sugar works and cane fields had not been destroyed, but the slaves on these properties were suddenly freed from white supervision just at the beginning of crop time. In this situation, the people on Joseph Foster Barham's two plantations—

Mesopotamia in Westmoreland and the Island in St. Elizabeth—responded very differently.

It will be remembered that during the Jamaica slave revolt of 1760 the Island slaves had vigorously combated the rebels, killing some of them and then cutting off their heads. They did not repeat this in 1832. When the Baptist War broke out, the Island slaves no longer had an active Moravian congregation, and the missionaries' message urging obedience to their absentee master was forgotten. On January 1 the Island slaves stopped working, joining a labor strike that spread through nearly all estates in St. Elizabeth. Interestingly, the one estate in the parish where the slaves didn't strike was the Bogue, which had the largest Moravian congregation in Jamaica. The Island slaves not only went on strike, but plotted to destroy the place on January 4; six Island slaves were the ringleaders, led by a mulatto carpenter named Campbell. When head driver John Benjamin learned of this plan, he alerted the militia. The rebels burned down Benjamin's house in reprisal, but the militia killed Campbell, cut off his ears, and gave them to Benjamin to show the Island people. A few slaves then went to work, and others gradually followed. Meanwhile, the authorities rounded up the five remaining ringleaders. One was hanged, with his head cut off and nailed to a pole. The other four were whipped, the punishment ranging from 300 to 500 lashes with a cat-o'-nine-tails.[44] Thus, the revolt achieved nothing at the Island except two deaths, four humiliations, and general residual smoldering discontent.

Meanwhile, as William Ridgard and Duncan Robertson reported to Barham in their letter of February 10, "at Mesopotamia the Negroes have behaved remarkably well. Having been left entirely to themselves they took up two Rebels who came to entice them from their duty and took them to the Guard." But this was not all that they did. "The Mill being about before Christmas," the attorneys added, "they took off the liquor in the distilling house themselves and since the first [of January] have had the Mill about and have made 55 hhds of sugar and a proportion of rum"—which was one-quarter of the annual crop. Starting up the sugar mill and harvesting the cane required the participation of the three field drivers. William Samuels, the third driver, had already demonstrated his loyalty to the estate by seizing George Watson; he had worked for three years as the head distiller, and thus knew how to manage the distillery without white supervision. Edinburgh, the second driver, was a baptized member of the church with the "Christian" name of James Barham, and he had served for three years as the head boiler, so was qualified to operate the boiling house. William Samuels and James Barham were probably the key players. Bernard, the head driver, was not listed as a church member on January 1, 1832, but he acquired the "Christian" name of William Ellison by the final inventory of August 1833, and seems to have joined with the chapel slaves during the slave revolt. The Ricksecker's reported that "from the 30th of December we saw ourselves in imminent danger . . . yet it was a satisfaction to us to perceive that all the negroes belonging to this property continued quietly at work, and conducted themselves in the most exemplary manner." Barham's attorneys, while praising the initiative shown by the Mesopotamia people, could not resist adding that the rum produced during the rebellion "will be found cloudy it being taken off under management of the Negroes when all the white people were on militia duty."[45]

Joseph Foster Barham II was too old and ill to respond to the news of his slaves' actions during the rebellion of 1831–1832, and he died on September 28, 1832. His son John Foster Barham (1799–1838), who inherited the Jamaica properties, sent a watch to Richard Gilpin as a reward for his meritorious behavior at Mesopotamia. But attorneys Ridgard and Robertson didn't want Gilpin to get this watch, because it would provoke discontent among the other deserving Mesopotamia head people, especially drivers William Samuels and Bernard and head cooper William Prince. Instead they persuaded Barham to give the watch to the Island head driver, John Benjamin, who had risked his life to preserve the Island sugar works in January 1832. This wouldn't create jealousy, since Benjamin was about the only meritorious person on that estate. Ridgard and Robertson got the Jamaica Assembly to give Benjamin a reward of £40, and they also rebuilt Benjamin's house and supplied him with extra clothing. At Mesopotamia, Gilpin and Samuels each received £10 from the Jamaica Assembly, plus extra allowances of fish at Christmas. And in the next crop the attorneys reserved two hogsheads of inferior sugar and two puncheons of weak rum to give the Mesopotamia people at Christmas "in consequence of their good behavior."[46]

Ironically, the Moravian missionaries, who had tried their best to encourage this good behavior, were accused by many white Jamaicans of instigating the insurrection—along with the Baptists and Methodists. Brother Pfeiffer was arrested and brought to trial on the trumped-up charge that he had proclaimed freedom to the slaves on New Year's Day. The missionaries feared that several of their chapels in St. Elizabeth would be destroyed by white mobs. And when they held their first conference after the revolt in March 1832, the missionaries lamented that some of their congregants had joined the rebels ("though without committing any acts of outrage"), while others had joined the labor strike.[47]

At Mesopotamia, the Rickseckers had an especially bad time. They were criticized by attorney William Ridgard and the new overseer, William Johns, for conducting "private speakings" with the candidates for communion, though Brother Peter protested in his diary that private instruction is essential for candidates who are so ignorant. In March 1832 Ricksecker felt forced to discontinue evening meetings "on account of the agitated mind of the public against evening services." And attendance at the Sunday morning meetings was depressingly low. But the greatest damage came from an unexpected source: Sarah Affir's eighteen-year-old quadroon granddaughter, Jane Ritchie. As we saw in Chapter 2, Jane was the Rickseckers' house servant, and she greatly irritated the missionaries with what they considered to be insolent behavior. When the Rickseckers made the big mistake of dismissing her in September 1832, they discovered too late that she had been gossiping to the other "brown" domestics for two years about the polygamists and adulterers within the Mesopotamia congregation. Jane and her "brown" friends gave overseer Johns the names of congregants who had been excluded by the missionaries from church meetings, and Johns confronted the Rickseckers with complaints from these excluded people. This greatly undermined the Rickseckers' status among the slaves, and during the last months of 1832 attendance at the church services sagged still further.

"Even the little negroes who came to us every morning were now also staying away," Ricksecker wrote in December 1832. On Christmas morning only twenty people from the estate, plus twelve free brown people, accepted the Rickseckers' invitation to attend morning prayers, "whilst the Estate negroes were all waiting for their [rum] allowance." The final days of December 1832 became "days of the deepest sorrow to us, concerning our flock; the greater part as we could observe have taken share of the noisy & most savage plays, which continues nights & days on this Estate; from the neighbouring estates many negroes came here with their idols, & they went from house to house; & opposite the Chapel on the public road." On Sunday, December 30—exactly one year after he first learned of the slave revolt in St. James—Ricksecker lamented that "the noisy play of the people drove away our sleep during the past night, which continued also nearly all day long, notwithstanding my sending a note to the Overseer's house praying to stop the nuisance."[48]

The Rickseckers withdrew from Mesopotamia in February 1833, their ministry an obvious failure. But on paper, at least, the Moravian congregation looked to be in somewhat better shape by mid-1833 than at the time of the slave revolt. The total number of slaves identified as "Christian" fell from sixty-six in January 1832 to sixty-three in the final Mesopotamia inventory of August 1833, but the newly baptized people clearly added strength, and some of the women who left the church were probably not missed. Among the newcomers were head driver Bernard (listed as William Ellison), mulatto carpenter George (George Slight), mulatto mason James (James Brownfield), mason Davy (David Reid, who was Sarah Affir's son), field hand Ruthy (Henrietta Roberts), and three house servants: mulattoes Eleanor (Jane Cothman) and Fanny (Fanny Fisher) and sixteen-year-old quadroon John (John Bell), the younger brother of Jane Ritchie. Among the departed were carpenter Dundee, quadroon Jane Ritchie, and mulattoes Elizabeth and Becky, both of whom worked at the Great House, together with Becky's two young quadroon daughters. The withdrawal of Jane and her house servant friends must have eased tensions within the Moravian congregation, and clearly not all of the "brown" people on the estate were in Jane's camp, since four mulattoes and one quadroon joined the church to replace the five who left. And the principal community organizers during the slave revolt—Richard Gilpin, William Samuels, Edward Foster Barham, and William Prince—were all still identified in August 1833 with their "Christian" baptismal names.[49]

Reviewing the events of January 1832, the big question is why the Mesopotamia head people acted as they did when the white managers left them alone for a full month. Being church members, the head people very likely believed what the Moravians had long been preaching: that they had a Christian duty to obey their worldly masters. Peter and Sarah Ricksecker, by remaining on the estate, certainly helped to deter violence. Furthermore, the head people had a personal stake in protecting Mesopotamia: they had the most responsible jobs, they had family and friends, and they were the leaders in building a self-contained black community out of very disparate elements. Difficult as it may be for us to understand, the head people probably also remained loyal to their absentee master because they felt comparatively well treated by Joseph Foster Barham. And most basically, they could easily figure out that burning the place down was not an answer to their problems. It would merely bring fresh misery and loss of life.

The great majority of the Mesopotamia slaves, however, were not members of the Moravian chapel, and many of the field hands had been openly resisting the sugar labor regimen. When the rebels approached from Hanover, there were surely plenty of young Mesopotamia workers who were eager to join. So in January 1832, how did a handful of head people take command? There is a parallel here with the situation at Mesopotamia in 1758–1759 when the missionaries first came, and in 1760 when the slave revolt broke out. Back then the head people, led by driver Matt, had aggressively courted the missionaries and had worked hard to protect the estate from rebel attack. By 1831–1832 religious ardor at Mesopotamia was in short supply, but the head people still had plenty of self-confidence and shrewd judgment. So they seized the moment and assumed command of the Mesopotamia slave community in order to block outside agitators like Billy Grant and to persuade the field workers to start the sugar harvest. It cannot have been easy, but the head people prevailed, and did far more service for their absentee masters than the Barhams deserved.

## *Migration and Rebelliousness at Mount Airy*

To move from Mesopotamia in 1760–1832 to Mount Airy in 1800–1860 is to enter a vastly different environment. It was not possible in the Northern Neck of Virginia, nor in the Alabama Canebrake, to stage a slave uprising in any way equivalent to the slave revolts of 1760 and 1831–1832 in Westmoreland. Slaveholders in the antebellum South were expanding in wealth, power, and ambition, not struggling like the Jamaican slaveholders to keep their system in operation. Planters like the Tayloes did not depend on enslaved collaborative head people like driver Matt of Mesopotamia to make their labor system work. And for the enslaved, any form of overt resistance was much riskier in antebellum Virginia and Alabama than in Jamaica. The white population was far larger, white surveillance was far more effective, and there were fewer wilderness areas in which to hide. Despite these immense obstacles, there was plenty of slave rebelliousness in the antebellum South, but it generally took a more hidden form.

As we have seen, Mount Airy—like Mesopotamia—had two basic categories of workers: the domestics and craftsmen versus the field hands (who at Mount Airy were also livestock keepers). But the division worked out differently on the two plantations. At Mesopotamia, all the slaves were grouped together in one slave village, so that the drivers and the runaways lived side by side. At Mount Airy, the field hands lived on scattered farm quarters, miles away from the domestics and craft workers who lived adjacent to the Tayloes' plantation house, so that the two sets of slaves were socially separated. And as the Tayloes moved their surplus slaves to Alabama, the occupational division became much less significant, because on a cotton farm there was really only one category of worker: the cotton hand.

Most of the Mount Airy slaves who were sold or moved by John Tayloe III and William Henry Tayloe came from the farm quarters. Of the thirty-six people dispatched in 1810 to work at the Cloverdale iron furnace in the Blue Ridge Mountains, thirty-one were field hands.[50] And when twenty-six slaves were sold to a pair of slave traders in 1816, all but two came from the farm quarters.[51] The thirty-nine migrants to Alabama in the 1830s included thirty field hands.[52] Among the forty-five migrants in 1845, thirty-three came from the farm quarters.[53] And thirty-three of the thirty-six migrants in 1854 were drawn from the farm quarters.[54] In these five parties, the Tayloes sent a total of 182 people to new work sites or new owners. Among the workers (those who were age nine or older), 121 were field hands, 6 were jobbers or ditchers, 11 were craftsmen and millers, and 13 were domestics or nurses. Thus, 80 percent of the workers who were moved or sold were field hands.

When sending people to new work sites, both John Tayloe III and William Henry Tayloe chose young people in their teens and twenties who could most readily adjust to new jobs or new crops. They particularly opted for boys and girls just entering the workforce. Seventeen of the 36 Mount Airy people John Tayloe sent to Cloverdale in 1810 and 65 of the 120 Mount Airy people William Tayloe sent to Alabama before the Civil War were aged ten to nineteen. From the Tayloes' point of view, the migration process worked out with hardly a hitch. None of the 49 people John Tayloe sent from Mount Airy to Cloverdale between

1810 and 1827 and none of the 120 people William Tayloe sent from Mount Airy to Alabama between 1833 and 1854 disappeared en route. In 1861–1862 William Tayloe's son Henry completed the transfer by sending ninety-eight Mount Airy people down to his father in Alabama. These two large parties, sent by railroad in December 1861 and March 1862, included many more young nonworking children than in 1845 and 1854, and a higher percentage of domestics and craft workers than in earlier groups. But most of the job holders—80 percent—were field hands.[55]

Field hand Sally Thurston (1780–1843), whose very large family is charted at *www.twoplantations.com,* was directly affected by the Tayloe policy of moving and selling people from the farm quarters. Three of her thirteen children were sold, and five were moved to Alabama before the Civil War. Sally had one son, Travers Hilliard, who became a blacksmith, and daughter Winney Myers became a spinner, but all of her other children were farmhands like their mother, and this pattern continued into the next two generations. Being agricultural workers, they were prime candidates for movement to Alabama. Twelve of Sally's grandchildren went to Alabama in 1833–1854, and ten in 1861–1862, and another eleven were born in Alabama. By 1863 two of Sally's daughters, two of her sons-in-laws, and a granddaughter were still living at Mount Airy, but the other sixty-one living members of her family who can be tracked were down in Alabama. Except for blacksmith Travers, the forty-three Alabama members of this clan who were old enough to work in 1863 were all field hands at Larkin or Oakland.

By contrast, housemaid Franky Yeatman (1766–1852), whose large family of domestics and craft workers is also charted at *www.twoplantations.com,* was much less affected by the exodus to Alabama. Franky and her husband, weaver Israel (1766–1821), had six children, all of whom became craft or domestic workers like their parents. Dick Yeatman was a gardener, Tom Yeatman a blacksmith, and the four daughters—Jane, Nancy, Fanny, and Eliza—all joined their father in the textile shop as spinners or weavers. In the next generation a majority of Franky's grandchildren were also craft or domestic workers. Of the twenty-one traceable grandchildren who survived childhood, fifteen became artisans or domestics. Because they were craft workers, two of Franky's children and five of her grandchildren were appropriated by William Henry Tayloe's brothers in 1828 when their father's slave force was divided up. But none of Franky's children moved to Alabama, and only ten of her numerous grandchildren and great-grandchildren went to the Canebrake in the 1830s, 1840s, or 1850s. Another thirty of her grandchildren and great-grandchildren were sent down during the Civil War to keep them away from the Yankees. When slavery was abolished in 1865, Franky's son Dick and her daughters Nancy and Eliza still lived at Mount Airy, as did six of Franky Yeatman's grandchildren and eleven of her great-grandchildren.

When the final ninety-eight migrants from Mount Airy reached Alabama in 1861–1862, four of them were placed at Oakland and all the rest at Larkin. They matched up well with the workforce William Tayloe had already assembled at Oakland and Larkin, because his cotton plantations employed very few domestics or artisans, and a very large number of field hands. William's last full Alabama inventory lists a population of 391 slaves with minimal indication of occupation, but it appears that there were only about seven domestics and nine carpenters, blacksmiths, and spinners at Larkin and Oakland. Some 87 percent of the job holders on the two cotton plantations were field hands.[56] By contrast, the sixty-six slaves left at Mount Airy, when inventoried by Henry Tayloe in 1863, included more house servants and artisans than in Alabama: twelve domestics, ten craft workers, sixteen elderly retired cooks, gardeners, spinners, and weavers, and just ten field hands.[57]

As we saw in Chapter 7, families that had been torn apart by the 1833–1854 migration were now reunited in Alabama. Unlike most of the million black people forcibly moved from the upper South to the lower South, the Mount Airy newcomers were now living with the sons and daughters and brothers and sisters and uncles and aunts and cousins they had lost in the 1830s, 1840s, and 1850s. William told his son Henry in March 1862 that the people at Larkin were "wonderfully contented," and he insisted that "the Negroes in Alabama are far better off than in Va." In a second letter he reaffirmed this view. "All the Mount Airy set seem determined to be dutiful obedient and faithful. They are contented."[58]

But not everyone was contented. There is a long (though poorly recorded) history of resistance to the Tayloe style of management. I have found reference to about forty slaves who ran away or otherwise rebelled against John Tayloe III and William Henry Tayloe between 1792 and 1865, and there were doubtless many more resisters whose actions are unrecorded in the Tayloe Papers. While the identifiable Mount Airy insurgents are fewer in number than the 140 identifiable Mesopotamia slaves who ran away and/or rebelled between 1762 and 1833, some of them did escape successfully, and at least one Mount Airy slave managed to inflict a great deal of property damage.

Back in 1792 when John Tayloe III took charge of Mount Airy, he soon encountered opposition. Some slaves were unhappy when Tayloe sold them, others when he bought them. After he held his big slave sale in 1792–1793, newspaper advertisements announced that two of the Mount Airy men who were sold had run away from their new owners. George had fled from his new home in the Shenandoah valley and was thought to be "lurking" in the neighborhood of Mount Airy, his old home. George had heavy whiplash scars on his back, though whether these scars were inflicted by one of Tayloe's overseers or by the new owner is not revealed. A second fugitive, postilion Jerry, had fled from the city of Richmond, and the new owner feared that he would sell his clothes in exchange for liquor.[59] Another runaway named Sambo, now owned by Tayloe, was thought to have returned to the neighborhood of his former owner in Richmond County.[60] And two other Tayloe slaves named Bob and Duke ran away three times between 1792 and 1794 because they wanted a new master or a new owner. John Tayloe stated in his ads that Bob and Duke were trying "to compel me to hire or sell them." But Tayloe added, in characteristic style, "I will do neither till they return to a sense of their duty."[61]

John Tayloe had an efficient estate superintendent, William Holburne, at Mount Airy and overseers on each of his many farm quarters, who generally kept tight control over his huge slave force. In the late 1790s four slaves—Garrett, Isaac, James, and Harry—ran away from Spring Hill quarter in King George County, a Northern Neck farm Tayloe no longer operated by 1808. The first three of these men were "taken up" at a cost of £3 15s., and were all living at Doctor's Hall quarter in 1808. But Harry, who had been gone for six months when he was advertised for in 1798, very likely escaped permanently or was sold, because he was not living at Mount Airy when the surviving slave inventories start in 1808–1809.[62]

We have a bit more information about two slaves, documented via the Mount Airy slave inventories, who came into conflict with John Tayloe in the 1810s. House servant Peter we have met in Chapter 5. He was among Tayloe's favored domestics, one of the eight servants who attended the family at Mount Airy during the summer months and accompanied them to the Octagon during the winter months. Peter's wife, spinner Elsy, died in April 1815, leaving him with five motherless children. Not long afterward, in September 1816, Peter was sold to a slave trader for $500—not only dismissed from Mount Airy but permanently separated from all of his orphaned children, aged sixteen to three. Peter was incongruously placed with a group of farm quarter people Tayloe wanted to get rid of who were mostly weak or sickly. Hence it appears that Tayloe was punishing Peter for some undisclosed misconduct.[63] The case of miller Reuben is somewhat clearer. In 1808 Reuben was a thirty-one-year-old field hand valued at £90 who lived at Hopyard quarter in King George County. In 1815 he was promoted to miller, and four years later he ran away and was caught by a neighboring overseer who wanted a reward of $50. "I carried him to Fredericksburg Jail," the overseer wrote to Tayloe, "and I assure you I took no small pains to hold of him."[64] After his jailing Reuben disappeared from the Hopyard inventory list without comment; presumably he was sold.

William Henry Tayloe took command of Mount Airy in 1824, and in 1827 he was billed $4.75 for runaway Nancy's two-week stay in the Lancaster jail, twenty miles down the Northern Neck. Nancy was a seventeen-year-old field hand at Doeg farm quarter, and after William recovered her she had a new baby in 1828. Nancy and the baby disappeared from William's inventory lists after 1828, not because they were sold but because William's brother Edward Thornton Tayloe took possession of Doeg quarter in the family settlement of 1828.

A much more notable flight occurred in December 1833, when Henry Tayloe dispatched a party of slaves to Cloverdale, where they would join with other slaves sent by William and Ogle Tayloe and head down to Alabama. One of the members of this party was a twenty-seven-year-old carpenter named Travers who belonged to Charles Tayloe, and he escaped during the march to Cloverdale. Travers was born and raised at Mount Airy, where he had worked as a joiner and carpenter from the age of seven. He had spent a year or two at Cloverdale and was then acquired by Charles Tayloe in the family settlement of 1828. By 1833 Travers was married. He may have figured that Alabama would be as bad as Cloverdale, and he certainly didn't want to be separated from his wife. Charles Tayloe, being the youngest of the brothers, and having the smallest slave force, was very upset by Travers's escape. "I am afraid that we shall lose him," Charles wrote to brother William. "He has written to his wife. I have sent his letter to you . . . [and] he has promised to write his wife again. I want you to write brother Ogle so that he may take proper steps to secure them. I am willing to spend Travers' value to get him for an example." Meanwhile, Henry Tayloe had his own plans for Travers. "I hope," he told William, "that Charles will recover him and ship him to New Orleans, thence to Mobile, from whence I may get him."[65] Travers's letter to his wife is unfortunately not in William Tayloe's file of correspondence, and whether Charles Tayloe managed to recover him is unknown. But here we have an absconder who not only knew how to write, but was probably helping his wife to escape as well.

We have seen that a number of the first Mount Airy migrants to Alabama in the mid-1830s rebelled against Henry Tayloe's management at Adventure plantation by running away. And one of them, Mary Flood from Neabsco, tried to burn his house down. Jess, who had probably been a sailor at Mount Airy in the 1820s, got all the way to Mobile, some 150 miles to the south, and Henry had to spend $70 to recover him. Gowen, belonging to Ogle Tayloe, ran away at least three times in 1834–1836. But Jess and Gowen and all the other Alabama runaways were rounded up. And from the 1840s onward the newcomers to William Tayloe's Alabama plantations were no longer trying to escape.

Meanwhile, as we have seen in earlier chapters, there was bigger trouble back at Mount Airy. In December 1844, when William Tayloe was in Alabama, someone—probably laundry maid Lizzie Flood (the younger sister of Mary Flood)—started a fire in the garret of the plantation house that destroyed the interior of the mansion. It took nearly three years, at a cost of $4,000, to restore the building. Tayloe dealt with Lizzie Flood as his brother Henry had dealt with Mary Flood, by selling her. He also exiled Lizzie's parents and brothers to Deep Hole farm quarter. In 1845, about six months after the Mount Airy fire, Tayloe's head carpenter, Bill Grimshaw, was whipped by a white supervisor and ran away, getting all the way to Canada. He was the only Mount Airy slave who managed to escape from William Tayloe between 1828 and 1861. And Tayloe reacted to Bill Grimshaw's flight in the same way he had reacted to Lizzie Flood's arson. He sold Bill's wife, two daughters, and youngest son to separate buyers, and sent Bill's two remaining children, Winney and James, to Alabama. There were no further major instances of rebelliousness among the Tayloe slaves in either Virginia or Alabama until the 1860s.

### *The Civil War in Virginia*

When war broke out in April 1861, many thousands of slaves in the border states were able to flee to the Yankees, whereas slaves in the heart of the Confederacy could make no move until the Union Army reached them. The Tayloe slaves at Oakland and Larkin were immobilized until April 1865, when the Yankees finally reached the Alabama Canebrake, whereas Mount Airy was only fifteen miles from the border with the Union at the Potomac River. It was by no means easy to cross the Potomac to Maryland from the Northern Neck. The lower reaches of the Potomac are very broad. But if a runaway slave could procure a skiff, he could cross at night, and on the northern shore in Charles County, Maryland, he would find thousands of other black people encamped with the African American soldiers who had enrolled in the colored troops of the Union Army.

As early as December 1860, before fighting began, William Tayloe urged his son Henry to send "all your active Negroes South before long."[66] But Henry stalled for a full year while indecisive opening battles were being fought some eighty miles to the north close to Washington, D.C. At some point during these twelve months Ralph Ward, the fifty-five-year-old Mount Airy coachman, disappeared. Ralph had served William Henry Tayloe since 1825—initially as his houseboy, then as his ostler, and for the past twenty years as his coachman and groom—so his desertion must have been particularly galling. No details of his flight are known, but presumably he had access to a horse, which made his escape much easier. When he took off, Ralph left behind at Mount Airy his fifty-three-year-old wife, spinner Eliza, his daughters Mildred, Eliza, Julia, and Ellen, and his son Cornelius. His sons Albert and Sydney and daughter Arabella were in Alabama, and another daughter, Betsy, was in Georgetown working for William Tayloe's daughter Sophia. And it is worth noting that—unlike Bill Grimshaw—Ralph reconnected with his family after the war. In 1870, at age sixty-three, freedman Ralph Ward was back at Mount Airy, living with his wife, daughter, and two grandchildren.

The key moment for the Mount Airy slaves came in December 1861, when it became evident that Henry Tayloe was assembling a party to bring to his father in Alabama. There were at that date nearly eighty slaves living on the plantation who were old enough (and not too old) to be candidates for desertion to the Yankees. Twelve of these people were domestics: house servants, nurses, gardeners, or stable hands. Another seventeen were artisans: carpenters, blacksmiths, masons, sailors, or millers. The great majority—fifty males and females—were field hands. How did they react to the news that Henry was about to take a large number of them to Alabama?

Henry Tayloe selected fifty people to move by train to Alabama, but when he left Mount Airy on December 16, 1861, two of the intended migrants—Elias Harrod or Harwood and Isaac Bray—were "not in place to go" and were left behind. Tayloe returned to Mount Airy as quickly as he could in early 1862 to find that Elias and Isaac had deserted to the Yankees on December 29. Elias, age twenty-seven, had started work as a scullion at Mount Airy, became an apprentice carpenter at fourteen, and served the Tayloes as a carpenter for a dozen years. William Tayloe valued Elias at $1,000.[67] Isaac Bray, age forty-one, was a field hand, and had worked all his life at Forkland quarter. William Tayloe valued Isaac at $800. But Harrod and Bray were not the only missing ones. Jacob Carrington (age thirty-eight), Paul Page (age thirty), Richard Thomas (age thirty), and his brother Jerry Glascow (age twenty-five) had also joined the flight on December 29, 1861. Jacob was the Mount Airy miller and had worked at the mill for twenty-six years; he had been married to Winney Grimshaw in 1844–1845 before she was taken from him and sent to Alabama. William Tayloe valued him at $1,500. Paul, like Isaac Bray, was a field hand at Forkland quarter, and was valued at $1,200. Richard and Jerry were both joiners and carpenters. Richard had worked at Mount Airy for twenty years, and was valued at $1,500, while Jerry had put in fifteen years and was also valued at $1,500. Unlike Ralph Ward, none of these six men left recorded wives and children behind at Mount Airy when they fled.

Hastening to prevent further defections, Henry Tayloe assembled a second large Mount Airy party to send to his father in Alabama. While he was doing so, two further men—Urias Myers and Edward Carter—absconded on March 18, 1862. Urias, aged thirty-four, was a field hand who had worked at Landsdown and Doctor's Hall quarters for more than twenty years. He had no recorded wife, but left his mother and sister behind at Mount Airy when he fled. Edward, aged forty-seven, was a mason who had worked at this craft since he was a boy in 1826, and his flight ended in disaster: he drowned while attempting to cross the Potomac River. Probably scared by what happened to Edward, no further Mount Airy people attempted to get away in March 1862, and Henry Tayloe sent fifty more migrants by train to Alabama. Not wishing to leave Mount Airy unattended, he engaged overseer Conway Reynolds to take the second party to Larkin. Henry had now sent south nearly all the Mount Airy slaves who were suited for employment on a cotton plantation.

With the removal of ninety-eight slaves to Alabama in 1861–1862, and the escape of nine others to the Yankees, there were less than seventy people left at Mount Airy. The sixteen active adults who were retained in Virginia could keep the old place functioning at a minimal rate: house servant Armistead Carter, blacksmith Bill Moore, field foreman George Yeatman, sailors George Wormley and Peter Richardson, coachman Ruffin Moore, carpenter Tom Thomas, weaver Alice Richardson (married to Peter Richardson), field hand China Myers (who was too pregnant in 1862 to move to Alabama), field hands Delphy Wright and Ibby (whose husbands were held by the Carters of Sabine Hall), cooks Kesiah Brown and Martha Harrod, along with her daughter Nancy, and spinners Nancy Thomas (married to Tom Thomas) and Eliza Ward Junior. All of the other slaves left at Mount Airy were too old, too sick, or too young for productive labor. A quarter of them were over the age of sixty in 1862, and a third were under the age of ten. The home plantation had become a refuge for the elderly and a child care center for the young.

The Northern Neck, though close to the Union lines, was one of the few regions in Virginia to escape heavy fighting during the war. But in late April 1862, a month after Henry Tayloe sent his second party of Mount Airy "servants" to Alabama, a Federal regiment from New Jersey raided King George County just to the north of Mount Airy and occupied Edward Thornton Tayloe's Powhatan plantation, which before 1828 was called Hopyard quarter and had been part of Mount Airy. When Ogle Tayloe in Washington learned of this raid, he obtained a pass and traveled down the Potomac on a U.S. government steamer to see what damage had been done to his brother's property. He found that Edward Tayloe had fled from Powhatan ahead of the Union troops, and that the New Jersey soldiers were also gone, having robbed the place of everything they could carry off. Thirteen of Edward's slaves had left with them. Ogle then rode from Powhatan to Mount Airy, and saw en route a good many black wanderers who would normally be confined to their plantations. He was relieved to discover that the Federal raiders had not reached Mount Airy, but the slave population had little or no white supervision, because Henry Tayloe, like his uncle Edward, had taken flight. During his short visit to Mount Airy, Ogle wrote a letter to brother William in Alabama on May 8, 1862—the only wartime letter between them that has survived—in which he criticized Henry and Edward for running away and exposing their slaves to the Yankee raiders. And he argued that Henry ought to maintain residence at his plantation house during the war. The Northern Neck Negroes, Ogle said, are demoralized, and "it looks as if all the large plantations of this Neck are broken up."[68]

Ogle Tayloe was exaggerating as usual; large Northern Neck plantations such as Mount Airy were not yet broken up. But the white masters were not very good at defending their human property. In August 1862 William Tayloe urged son Henry to reside at Mount Airy or at least to install a tenant. "Armistead [Carter] and Ruffin [Moore] and other faithful servants," he argued, "may become unfaithful, if made to believe we have deserted."[69] Henry Tayloe and his wife, Courtenay Chinn Tayloe, did return to Mount Airy and lived there most of the time from mid-1862 to 1865, joined by Henry's father-in-law, B. C. Chinn. And Armistead and Ruffin remained faithful. But a few more Tayloe slaves defected.

At some point in 1862 three Mount Airy girls—Betsy Ward, Caroline Moore, and Betsy Harrod or Harwood—all working for William Henry Tayloe's married daughter Sophia in Georgetown in the District of Columbia, took flight. Betsy Ward (age twenty-three) was the daughter of deserter Ralph Ward; Caroline Moore (age eighteen) was the daughter of coachman Ruffin Moore; and Betsy Harrod (age fourteen) was the daughter of cook Martha Harrod and the sister of deserter Elias Harrod. Since they were already behind the Union lines in Federal territory, these girls were in an excellent position to slip away. And they did not resurface after the war. Once they were freed they very likely all married and acquired new surnames, because they cannot be tracked in the census of 1870.

There was one further desertion from Mount Airy during the Civil War. In May 1863 Peter Richardson (age twenty-nine) took off, leaving behind his mother, spinner Nancy Richardson, his wife, weaver Alice

Richardson, and their three young daughters, Harriet, Ella, and Becky. The son of a sailor, and a sailor himself since the age of nine, Peter obviously had the water skills to cross the Potomac River. Interestingly, his fellow sailor at Mount Airy, George Wormley (also the son of a sailor), did not join him in 1863. George was married to chambermaid Kitty Wormley, who as a girl in 1844 had accused Lizzie Flood of setting fire to the Tayloes' mansion, and perhaps he was more interested than Peter in sticking with his wife and children. In 1870 George and Kitty were living at Mount Airy together with two of their children, whereas Alice Richardson was living in Richmond County with her daughter in a white farmer's household, both employed as the farmer's domestic servants. And Peter Richardson—like virtually all of the Tayloe absconders—was not to be found in the 1870 census. If he survived the war, he seems to have changed his name.

Compared with Larkin and Oakland, Mount Airy was now dominated by the very young and the very old, with only eight people out of sixty-six who were between the ages of twenty and forty. When Henry Tayloe took a census of the remaining Mount Airy people on February 18, 1863, he enumerated twenty-eight males and thirty-nine females (including runaway Peter Richardson). *Appendix 37* arranges these people in family groups. The census shows that forty-one of these remaining Mount Airy slaves had parents, or sons and daughters, or brothers and sisters, who were living in Alabama. And twenty-eight of them were closely related to the thirteen slaves who had fled to the Yankees in 1861–1863. Most of the adults were too elderly to attempt flight themselves, and people like Eliza Ward, whose husband and daughter had deserted, or Alice Richardson, whose husband and two brothers had deserted, must have been dreadfully apprehensive about what had happened to their loved ones.

Henry Tayloe noted that one of the Mount Airy slaves was blind, eight were "chargable," twelve were "worthless," and one was "crazy."[70] Chambermaid Kitty Wormley was both "crazy" and "cracked." Eliza Ward and Nancy Carter, who had both produced thirteen children, were seen as "worthless." Obviously Henry Tayloe felt burdened by having to provide food and shelter for the many retirees who lived at Mount Airy during the war. But his disdain was not, I think, shared by his father. When William Henry Tayloe appraised all of his Alabama slaves in 1863, he gave a dollar value of zero to the five oldest women, but he didn't characterize these women as worthless. In the Tayloe Papers there are a great many slave lists in William Tayloe's handwriting composed over a period of forty years, and none of these lists contains demeaning or insulting commentary.

Altogether, ten Mount Airy males and three females fled to the Yankees in 1861–1863. Six of them were craft workers, four were domestics, and only three were field hands. Four of the deserters—coachman Ralph Ward, his daughter housemaid Betsy Ward, carpenter Elias Harrod, and his sister housemaid Betsy Harrod—were members of Franky Yeatman's extended family of craft workers and domestics, charted at *www.twoplantations.com*. There is irony in the fact that the slaves who personally served the Tayloes were much more likely to escape when they had the chance than the agricultural workers. Of the seventy-nine slaves living at Mount Airy and the three at Georgetown in 1861 who were old enough (and not too old) to be candidates for desertion to the Yankees, 37 percent of the craft workers and 27 percent of the domestics who might have escaped actually did so, as against just 6 percent of the field hands. Probably most of the field hands actually *wanted* to go to Alabama in 1861–1862—particularly when they learned that they would be traveling in railroad cars instead of on foot—because they wanted to reconnect with their lost relatives.

The relatively small number of identifiable Mount Airy slaves who ran away or rebelled between 1808 and 1845 fit the same pattern as in 1861–1863. Most of them were craftsmen or domestics: house servant Peter in 1816, miller Reuben in 1819, carpenter Travers in 1833, housemaid Mary Flood in 1834, sailor Jess in 1835, laundry maid Lizzie Flood in 1844, and carpenter Bill Grimshaw in 1845. There were field hand rebels as well, of course, especially when they were mistreated at Adventure plantation in the 1830s. But the Mount Airy craft workers and domestics were clearly the chief risk takers.

Thus, the internal split between skilled workers and field hands at Mesopotamia and Mount Airy played out quite differently during the last two generations of slavery on the two plantations. At Mesopotamia the head people led the way in joining the Moravian congregation that the Barhams sponsored, and they also led the way in defending their absentee owners' estate during the slave revolts of 1760 and 1831–1832. At Mount Airy the opposite happened. Here the slaves with the greatest skills and the house servants who personally served the Tayloes were more likely to rebel or escape when they had the chance. The one Mesopotamia slave who clearly fits the Mount Airy pattern was the quadroon domestic Jane Ritchie, who dared to outmaneuver Moravian brother and sister Ricksecker.

What is the larger significance of this contrast? I certainly do not claim that head people on the Caribbean sugar estates always collaborated with their white masters. On the contrary, the head people on most sugar estates in western Jamaica appear to have been the chief rebels during the slave revolt of 1831–1832. And in the antebellum South it is certain that craft workers and domestics were seldom the chief rebels; most of the slaves who escaped and wrote about their adventures for a northern audience were field hands. Setting aside for a moment the motivations of the Mesopotamia collaborators and the Mount Airy rebels, both sets of (mostly male) slaves stand out because—in a time of crisis, and in an environment of suffocating oppression—they acted boldly and purposefully against the grain. At Mesopotamia these self-appointed leaders took control of their fellow workers in the revolt of 1831–1832. At Mount Airy during the Civil War they broke from their comrades and acted independently in a quest for freedom. In both cases they exhibited special qualities of individuality, independence, and leadership.

It may seem surprising that the head people at Mesopotamia, where conditions were really bad, defended the place during two slave revolts, whereas at Mount Airy some of the most skilled slaves, who were far better treated, rebelled against the Tayloes when they saw the chance. But both patterns of behavior make sense. The Mesopotamia head people lived in a black world with a small cadre of abusive whites ordering them around. As long as slavery was the only option, the slaves with the most responsible jobs wanted to protect the stake they had achieved in that black world, and they could reckon that the rebellions of 1760 and 1831–1832 were going to fizzle out. The wisest course was to follow the Moravians' teachings: to oppose the revolt, to preserve their community, and to protect their small leadership stake.

At Mount Airy, the skilled workers and domestics lived in a mixed black-white world with many more controlling white people close at hand. Rebels like Mary and Lizzie Flood, sick of dealing with the whites on a daily basis, were clearly revenging themselves on their owners. Two domestics who did not rebel, both of whom were characterized in 1862 by William Tayloe as "faithful" servants—Mount Airy house servant Armistead Carter, whose brother deserted, and coachman Ruffin Moore, whose daughter deserted—must have been strongly tempted to flee, but the risks were great. Three of the men who fled in 1861–1863 were carpenters, reflecting the fact that these workmen operated much more independently than the field hands and didn't take kindly to discipline, as exemplified by Bill Grimshaw. And all of the thirteen Tayloe slaves who deserted during the Civil War had a broader view of the world and its possibilities than most of their colleagues. Working in or near to the big house, they were in the best position to learn about the course of the war and the possibility of freedom through flight. So the Mount Airy rebels, like the Mesopotamia cooperators, were exercising their intelligence, courage, and good judgment to achieve their goals.

# 9

## *Emancipation*

THE SLAVES in the British West Indies were freed thirty years before the slaves in the United States, but there were broad similarities in the emancipation process in both regions. In the British Caribbean there was a four-year preparatory stage, from 1834 to 1838. Slavery was officially terminated, but the ex-bondsmen were thought to need training for their new role as free wage workers, so they were designated as apprenticed laborers in continued (though somewhat lightened) service to their former owners. In the United States the preparatory stage, totally different in character, was a four-year war between the slave states and free states from 1861 to 1865, during which a great many slaves fled to the Yankees, Lincoln issued his Emancipation Proclamation, and a high percentage of the remaining bondsmen were released from slavery by the Union Army in 1863–1864. When all of the British Caribbean slaves were fully liberated in 1838, and all of the U.S. slaves in 1865, the freed people in both places wanted land of their own so they could operate independently, while their ex-owners wanted to retain them as wage laborers. And newly created intermediaries appeared: special magistrates in the British West Indies and agents of the Freedmen's Bureau in the former Confederate states—both authorized to settle disputes between the freed people and their former owners.

There were also huge differences in the two situations. The British Caribbean legislatures had approved the abolition of slavery on condition that the slaveholders on the islands were paid £20 million for the loss of their property, whereas the Confederate States of America had fought a tremendously destructive four-year war before being forced to accept military defeat and the loss of their slaves. The abolitionists who secured black emancipation in the Caribbean lived thousands of miles distant in Britain, while the abolitionists who secured black emancipation in the United States controlled the victorious Federal army and government. Having achieved emancipation, the British abolitionists lost interest in the Caribbean blacks, and the British government acted to sustain the political and economic status quo in the West Indies. In the United States, the Radical Republicans in Congress sought to reconstruct the defeated South from top to bottom. Outraged whites in both places responded to black freedom with virulent racism, but the Confederate whites were far more embittered and far more numerous. Black and colored people constituted over 90 percent of the population in the British islands, but only 40 percent in the southern states, and this population differential turned out to be very significant.

Emancipation at Mesopotamia and at William Henry Tayloe's three plantations in Virginia and Alabama illustrates how these broad similarities and huge differences played out.

## *Emancipation at Mesopotamia*

The surviving documentation for Mesopotamia during apprenticeship and emancipation is distressingly thin. John Barham, who became the owner of the estate in 1832, immediately terminated the slave inventories as an unnecessary expense. The Moravian missionaries in 1833–1835 kept no diaries, and in 1836 John Barham shut down the Moravian mission at Mesopotamia. John Barham became mentally deranged in early 1836, and ceased all correspondence.[1] After he died in 1838, his widow took the Barham family papers with her when she married the fourth Earl of Clarendon in 1839. Ownership of Mesopotamia briefly passed to John Barham's brother William Foster Barham, and at William's death in 1840 to his next brother, the Reverend Charles Foster Barham (1808–1878), but neither William nor Charles has left a paper trail that I have found. Hence all plantation records for Mesopotamia end in early 1836. Nevertheless, enough evidence survives among the Colonial Office records and in the Jamaica Archives to piece together what was happening on this estate in 1834–1844.

The abolition of slavery in Jamaica (and throughout the British Caribbean) was initiated from across the Atlantic. Propelled by a tidal wave of antislavery agitation in Britain, and by the Great Reform Act of

1832, which reshaped the House of Commons, Parliament passed the Slavery Abolition Act on August 28, 1833. This legislation, drafted in the Colonial Office, decreed that all of the slaves in the British Caribbean colonies were to be emancipated in a two-stage process. On August 1, 1834, slavery was officially ended, and slave children under six were declared free, while every slave over the age of six became an "apprenticed laborer." Agricultural slaves were apprenticed to their former owners for six years, until August 1, 1840, while domestic slaves were apprenticed for four years, until August 1, 1838. All agricultural apprentices were to work without pay for a maximum of forty-five hours per week, which was construed as three-quarters of their previous ten-hour, six-day work week, and they were to be paid for any work done beyond forty-five hours—in the hope that this would encourage them to become disciplined full-time wage laborers after 1840. The apprentices were to receive customary allowances of food, clothing, shelter, and medical attention from their masters. Special magistrates were appointed by the crown to adjudicate disputes between the apprentices and their former owners, and these magistrates had the sole authority to require extra labor from balky apprentices or to whip or imprison them for misbehavior. Finally, every colonial assembly was required to confirm the Slavery Abolition Act in order to share in a parliamentary grant of £20 million in compensation for the slaveholders' lost property—a prodigious gift in the eyes of the antislavery party, but far less than the slaveholders wanted.

The home government was well aware that apprenticeship would be especially problematic in Jamaica, where half of the British Caribbean slaves lived, where a major slave rebellion had taken place in 1831, and where the legislative assembly—dominated by the attorneys and overseers of the absentee planters—was implacably opposed to emancipation.[2] The Colonial Office wished to preserve sugar production in Jamaica by ensuring that the planters had an adequate labor supply, and saw that a conciliatory policy was needed in order to persuade the apprentices to continue working for the planters after 1840. It was obvious that in small, high-density islands such as Barbados and Antigua, where all the land was under cultivation, the apprentices, once freed, had little choice except to continue on the sugar estates as wage laborers. However, in Jamaica—and also in Trinidad and British Guiana—there was a great deal of unoccupied, uncultivated land, so the ex-slaves would be able to desert the sugar estates and set up independent peasant settlements in the bush. But the Jamaica Assembly was in no mood to conciliate the former slaves. In order to get its share of the £20 million compensation fund, the assembly was compelled to accept Parliament's Slavery Abolition Act, which it adopted in December 1833.[3] But it inserted a number of revisions to the parliamentary statute that were bound to irritate the apprentices, imposing draconian punishments on those who failed to work or were "insubordinate," while the maximum penalty for a misbehaving master was only a £5 fine. In an odd twist, the Jamaican Assembly reduced the required weekly hours of unpaid labor by the apprentices from forty-five hours to forty and a half hours, apparently in order to make them work for five eight-hour days, whereas the Colonial Office expected the apprentices to work four and a half ten-hour days, giving them half of Friday as well as Saturday and Sunday off.[4]

At Mesopotamia, sugar production had dropped from 233 hogsheads in 1832 to 125 hogsheads in 1833, which was the lowest total in a decade, but the attorneys told John Barham that his slaves—despite the clamor over emancipation—were behaving in an "unexceptionable manner," and the sugar crop rose to 175 hogsheads in 1834.[5] While the attorneys were apparently not anticipating any trouble on August 1, 1834, the new governor of Jamaica, the Marquis of Sligo, was not so sanguine. He had arrived in Jamaica in April 1834, was immediately at war with the assembly, and by August had only twenty-eight special magistrates on hand, less than half the number he wanted and needed. Sligo issued a proclamation in April urging the slaves to show gratitude for the great sacrifices made to them by their masters and by the British government, and he followed this up with a second proclamation in May, urging them to work diligently when they became apprentices. Everyone works, the Marquis explained to the Jamaica slaves: some with their heads, and Negroes with their hands.[6]

Emancipation Day on Friday, August 1, 1834, was a public holiday in Jamaica, and except for some resistance on the north coast, the newly styled apprentices went back to work the next week. But in September attorneys William Ridgard and Duncan Robertson told John Barham, "We dread the consequences of so few hours of continuous labour" at crop time, and they warned that wages would have to be paid to the Mesopotamia apprentices for extra labor. In December the attorneys went further and recommended that Barham also pay the apprentices for extra labor after crop time, which "would be an inducement to their becoming industrious, and working for hire."[7] These recommendations were not to John Barham's liking. Since Mesopotamia was no longer earning him a handsome income, he was looking for every possible way to cut expenses. On the advice of his sugar merchants, in October 1834 Barham reduced the annual supply of clothing and food sent from London for the apprentices from £608 in value to £318.[8] At the end of the year he drew up a set of instructions for his brother William, who sailed to Jamaica in early 1835 to inspect conditions on the Barham estates. John Barham was pleased with the good brown sugar and the extremely well-flavored rum produced at Mesopotamia, but reviewing the records sent by Ridgard and Robertson for 1831–1834, he found many other things to complain about. He claimed that the hiring of jobbing gangs for extra labor at Mesopotamia never worked well under the old (slavery) system, and he worried that paying wages to the apprentices under the new system would become an unbearable expense. He thought that Ridgard's and Robertson's salaries should be reduced and that all aid to the Moravian missionaries should be stopped. Barham also fretted about the high cost of imported lumber for making sugar casks and rum puncheons, the continuous shrinkage of the Negro population at Mesopotamia, and the deaths of eighty cattle at Mesopotamia in 1833. Finally, he told brother William that Mesopotamia needed some dogs—"good English Rat Catchers"—to deal with the annual rat infestation, because every spring the rodents came down the Cabarita River and ate the young cane.[9]

At Christmas 1834, Ridgard and Robertson gave the Mesopotamia apprentices their customary allowances of beef, fish, rum, sugar, and salt, and told Barham that many proprietors had not done so, causing resentment among their workers. To deal with the 1835 sugar crop, the attorneys formed two "spells" of people to operate the Mesopotamia sugar works, the first shift working from 4 A.M. to noon, and the second from noon to 8 P.M. from Monday through Friday. The field hands also worked eight-hour shifts five days a week, and the Mesopotamia apprentices collectively received £70 for extra labor during crop time. The crop was finished by late April, and totaled only 155 hogsheads of sugar, a drop of 20 hogsheads from 1834. But the attorneys blamed dry weather rather than labor problems for this shortfall, and noted that Mesopotamia produced a better crop than Friendship and Black Heath, the closest neighboring estates. And they seemed to be pleased with the terriers Barham had sent to attack the rats, since 12,000 rats had been killed at Mesopotamia in 1834.[10]

In March 1835 William Barham and his wife arrived in Jamaica for their inspection tour; they first visited Island and then came to Mesopotamia in May. They returned to England in July. There were labor problems at Island, where the apprentices (according to attorney Robertson) were very insolent and insubordinate. Throughout the apprenticeship period, as during the slave revolt of 1831–1832, the Island people appear to have been much more overtly rebellious than the Mesopotamia people. In June or early July 1835, with William Barham in attendance, a special magistrate heard Robertson's complaints and ruled that the Island Great Gang must work four full days beyond the required work week in order to make up for recent short days, with threats of severe punishment for future transgressions.[11] What William Barham thought of conditions at Island and Mesopotamia is unknown. In August he was back in London, and in December he complained that his brother owed him money for his trip to Jamaica. On December 26, 1835, John Barham responded indignantly. Protesting that he had paid William's Jamaican bills totaling £300, he declared, "The less we talk of Jamaica the better but I am ready of course to pay for your voyage & &—the expence of the Woman [Mrs. William Barham] however does not concern me."[12] The next month John Barham became incapacitated by mental illness, and his correspondence file shut down. The final communiqué from Jamaica in the Barham Papers is a letter from Ridgard and Robertson, dated December 29, 1835, announcing that the Mesopotamia apprentices had returned to work cheerfully after Christmas, were busy with the crop for 1836, and had already made twenty-six hogsheads of sugar.[13]

Dispatches from Jamaica to the Colonial Office indicate that there was very significant tension between the apprentices and their former owners throughout the apprenticeship period. Both sides jockeyed for position in the fall of 1834, with labor-management quarrels easing a bit in 1835–1836 and getting worse in 1837–1838. During these four years the planters were generally trying to extract the maximum amount of labor from the apprentices while paying them low wages for extra work, while the apprentices were concentrating on their provision grounds in order to supply their families with more adequate food than in slavery days and to raise surplus produce for sale or barter in the local weekly markets. It was reported from Westmoreland in 1835 that the head people, cartmen, mill hands, and boiling house workers were readier to work for wages than the field hands.[14] The apprentices also exhibited several other tendencies that pointed toward the near future when they would be fully freed. When offered wage work, they preferred individual task work to gang labor, and sugar factory work to cane planting or harvesting, and they sometimes rejected wage work at their home plantations and accepted lower wage offers at neighboring plantations. Most strikingly, women workers were often the most militant and vocal critics of the apprenticeship system, which posed a challenge to the planters, since the Abolition Act decreed that women could not be disciplined by flogging.[15]

During apprenticeship the special magistrates were the key players. They were given the daunting challenge of reshaping relations between the ex-slaves and the ex-slaveholders—a challenge they failed to meet. The magistrates who served in Jamaica in 1834–1838 were overworked and underpaid, and whenever they ruled in favor of the apprentices they were vilified by the planters and the assembly and sometimes prosecuted in the local courts and assessed high fines. Usually they ruled against the apprentices, who soon saw them as a new breed of slave driver. Most of these magistrates were outsiders from Britain with liberal beliefs and a detestation of slavery, but a great many were also deeply imbued with racial prejudice. For example, Special Magistrate John Anderson, who wrote a detailed journal describing his service in St. Vincent from 1836 to 1838, immediately saw the population of this island ranked into four immutable categories: at the bottom the heathen savage African blacks; next the childish, lazy, dishonest Creole blacks; above them the free colored, who possessed some ability because of their white blood but were absurdly foppish, irresponsible, and improvident; and at the top a small cadre of whites, the only inhabitants with positive characteristics.[16]

The special magistrate adjudicating labor relations at Mesopotamia was an Englishman named Thomas M. Oliver, who was probably among the many half-pay retired army and navy officers chosen by the Colonial Office. He was appointed on June 10, 1834, and served for about three and a half years.[17] As one of the three special magistrates in Westmoreland Parish, Oliver was expected to visit some twenty-five properties every two weeks, including Mesopotamia and seventeen other sugar estates. This required him to travel sixty to eighty miles per week and to hold court sessions on Saturdays. Governor Sligo described Oliver in 1834 as "steady, active, good tempered, and quiet but determined," and likely to perform his business well. In fact, however, Oliver performed his business poorly. He was often ill and failed to make his weekly rounds. He was a lazy record keeper who wrote brief bland monthly reports and never supplied concrete information about what was going on at Mesopotamia or any of his other plantations. He was also a harsh disciplinarian. In May 1836, for example, Oliver heard thirty-seven complaints on nine of the properties he visited, and administered 175 lashes to offending apprentices, while the other two Westmoreland magistrates heard twenty-seven complaints and administered 20 lashes. In the three months from April to June 1836, Special Magistrate Oliver ordered a total of 515 lashes, while magistrate Philp ordered 210 and magistrate Kelly ordered 45.[18] To be sure, Oliver used the whip much more moderately than a typical Westmoreland overseer would have done during slavery. And by 1838 he was no longer in a position to whip people. Oliver became so mired in debt that he spent several months in a Jamaica prison until he applied for release as an insolvent debtor, and divided his assets among his creditors. He was dismissed as a "Bad Magistrate" probably by the close of 1837.[19]

During the four years of apprenticeship in Jamaica, the new constraints on labor had an immediate effect on sugar production. Sugar exports from the island in 1835–1838 were about 25 percent lower than in 1831–1834, and nearly 50 percent lower than during the peak production years of 1799–1817.[20] At Mesopotamia, the decrease was less dramatic. This estate had produced 257 hogsheads annually in 1799–1817, dropping to 200 hogsheads in 1831–1834, and in 1835–1837 production fell again, but not by much: to 183 hogsheads annually. The 1838 total at Mesopotamia is unknown, but it probably matched the 1835–1837 average.[21] Occasional references in the Colonial Office files to Mesopotamia indicate that conditions were less volatile here than on many plantations. The Mesopotamia apprentices worked a five-day week in crop time and a four-and-a-half-day week out of crop, and received wages for extra labor. Labor-management disputes in 1836–1837 appear to have been less frequent than in 1834–1835. Most significantly, Special Magistrate George Gordon (who replaced Thomas Oliver) reported in January 1838 that Mesopotamia was one of four Westmoreland sugar estates where schools had been set up for the instruction of the apprentices' children.[22] This was an important initiative, because there were about 5,000 black children under the age of ten in Westmoreland parish, and during the first three years of apprenticeship little or nothing had been done to educate them.

Meanwhile, it was becoming increasingly obvious that the halfway character of apprenticeship suited neither the apprentices nor the planters in Jamaica. In 1834–1835 Governor Sligo had tried to employ the special magistrates as buffers between the planters and the workers, but when the Colonial Office overruled his dismissal of a disruptive magistrate in 1836 he resigned. His successor, Sir Lionel Smith, set out to win over the assembly and the planters by reducing the authority of the magistrates, and discovered too late that the assembly was unwilling to make any accommodations that might encourage the ex-slaves to continue working on the sugar estates after 1840. Reports reached Britain that female apprentices who complained about their mistreatment were being brutally flogged in Jamaican houses of correction, and Parliament was again besieged with petitions for the immediate termination of apprenticeship. The Jamaica Assembly independently opted for immediate termination. Hoping to get better leverage over their black workers in a free-market environment, and also wanting to get rid of the special magistrates and interference from the Colonial Office, in June 1838 the assembly passed legislation abolishing apprenticeship for all ex-slaves two years ahead of schedule, on August 1, 1838.[23]

Governor Smith recognized that the end of apprenticeship was likely to bring major change. The island was governed to suit the interests of fewer than 20,000 whites—5 percent of the population—with about 60,000 colored people pressing for a share in power, and some 310,000 blacks now totally free. The planters wanted economic control over the blacks, but they were not interested in social control. Seeking to reduce expenses by eliminating marginal and part-time workers, they planned to eject all freed slaves not willing to work as regular wage laborers from their estates. This precipitated conflict, because many of the women in newly freed families wished to withdraw from field labor. Besides, the freed slaves believed that the houses they had built and inhabited, as well as the provision grounds they had been tending, in many cases for several generations, belonged to them. In July 1838 Smith issued a proclamation warning the new "FREE LABOURERS" that after a three-month grace period they would have to start paying rent on their houses and provision grounds, but he was undermined by the Jamaican attorney general, who ruled that the planters could immediately start charging rent. In September a magistrate at Montego Bay reported that hundreds of "the late apprentices" had visited him to consult about wages and to have the legal status of their houses and grounds explained to them. Some planters, eager to make money, were charging higher weekly rents than they paid out in weekly wages.[24]

What was the impact of these new policies? On October 29, 1838, three months after the end of apprenticeship, the two special magistrates in Westmoreland compiled a highly informative survey of the labor situation on thirty-two sugar estates, including Mesopotamia. The magistrates claimed that the employment pattern gives "every satisfaction to the labourers as well as to their employers," especially since no rent had yet been charged on any of these properties.[25] But their survey actually shows that huge changes had already taken place. Eighteen of the thirty-two sugar estates they surveyed were among the twenty-seven Westmoreland estates examined in Chapter 1 at the termination of slavery on August 1, 1834. *Appendix 38* compares the number of enslaved field laborers on these eighteen sugar estates on Emancipation Day 1834 with the number of resident field laborers on October 29, 1838. On every estate the number of resident field hands was drastically reduced, overall to about a quarter of the 1834 enslaved workforce. Mesopotamia saw the least change. It had the smallest reduction since 1834, and listed far more resident field hands in 1838 than the other seventeen sugar estates. This was probably because Mesopotamia (along with Fort William) offered the highest wages in Westmoreland: two shillings sixpence per day for first-class laborers; one shilling eightpence for second-class laborers; one shilling threepence for third-class laborers; and twenty to thirty shillings per acre for job work. The other sugar estates offered a top daily wage of one shilling eightpence, and only thirteen to twenty-two shillings per acre for job work.

Another very obvious difference between the field labor pattern in 1834 and that of 1838 was the dramatic change in sex ratio. In 1834, 60 percent of the enslaved field hands were female; in 1838, 59 percent of the resident field hands were male. Not only had large numbers of women withdrawn from regular field work, but many marginal workers—old people, and young boys and girls—must also have quit or been rejected by the planters as resident agricultural laborers. On ten of the Westmoreland sugar estates there were fewer than fifty resident field hands in October 1838, and during crop time these estates would have to hire additional hands on daily wages. Clearly many of the freed slaves were no longer working regularly on their old estates. And it is likely that some of them had quickly moved to new work sites in search of better job opportunities. It would be interesting to know whether the 110 resident field hands listed at Mesopotamia in *Appendix 38* had all previously been slaves on this estate, or whether some of them were newcomers from neighboring Black Heath, Friendship, or Blue Castle estates—all of which were undermanned in 1838, and offered poorer wages. The situation was fluid, and the emancipated blacks were already breaking loose from their past imprisonment.

In their survey of October 1838, the Westmoreland magistrates reported that at Mesopotamia "the labourers are behaving very well, and [the] estate [is] forward." But in December 1838 overseer James Kirby began to charge rent, and there was an immediate adverse reaction. During December Kirby ejected six Mesopotamia residents—Thomas Morning, Thomas Allen, Marianne Dixon, Marianne Rodney, Sarah Barker, and Molly Noble—for demanding excessively high wages. All six came to terms with Kirby and remained on the estate. But in January 1839 Kirby ejected eleven other Mesopotamia residents—Mary Ewart, Sarah Cunningham, John Williams, Robert Morning, Mary Ann Hill, Joseph Brown, Mary Thompson, Mary Ann Myrie, Rachel Howell, Margaret Brown, and Daniel Bramwell—nine of them for refusing to work on the estate, and the other two for working badly.[26] It is significant that eleven of the seventeen people ejected by overseer Kirby were women; clearly he was trying to get rid of all the women at Mesopotamia who were withdrawing from field labor. Another interesting point is that only one of these seventeen people can be positively identified as having lived at Mesopotamia during slavery. Mary Thompson had been brought from Cairncurran coffee plantation as a six-year-old child in 1814, and was last reported in 1833 as employed in the first field gang at age twenty-five. Eleven of the other ejectees may also have been Mesopotamia slaves, because they shared first names with people who were listed without surnames on the 1833 slave inventory. But at least five seem to have been newcomers, because they cannot be found on the

Mesopotamia slave inventories.[27] This would corroborate magistrate George Gordon's complaint in March 1839 that the Westmoreland freemen were often leaving their home estates to find work at the same wages on estates a great distance away.[28]

The 1839 sugar crop at Mesopotamia was in trouble from the start. It was reported in early February 1839 that about 90 people were working on the crop at Mesopotamia, including 35 harvesters (compared with 110 field hands in October 1838). The harvesters worked short hours, starting at 8 A.M. and quitting at 3 P.M., in which time they cut twelve or thirteen cartloads of cane, which made one hogshead of sugar per day instead of three hogsheads as in prior years. By late February, production at Mesopotamia had increased to ten hogsheads per week. "A great number have been ejected on this Estate," magistrate Thomas Abbot observed, "and [they] have all subsequently resumed their work and are working better."[29] But Mesopotamia shipped only ninety-four hogsheads of sugar and fifty-six puncheons of rum in 1839, the smallest total since 1787, and barely half the average annual shipment during apprenticeship. Throughout Jamaica, sugar production in 1839 was the lowest in many years, but the precipitate drop at Mesopotamia—where the crops had been larger than on most estates during apprenticeship—was striking. William Johns, who had been the Mesopotamia overseer in 1832–1834, was now the attorney. He became the attorney/overseer in 1840 and personally managed this estate for absentee owner Charles Barham into the 1850s. Johns did not report the wages he paid to his workers in 1839, but he collected £326 in rent for housing and provision grounds.[30]

In 1840 the situation became more critical. Mesopotamia produced the smallest crop since 1781, after the great hurricane of 1780 had decimated the cane fields. The estate shipped only 65 hogsheads of sugar and 30 puncheons of rum, compared with an annual average of 257 hogsheads of sugar and 109 puncheons of rum in 1799–1817—a drop of 75 percent. We have no information about the labor pattern at Mesopotamia in 1840, but considering the small crop size, there was probably a further reduction in the workforce. William Johns stopped collecting rent in 1840, apparently belatedly recognizing that he was driving his workers away. And he charged the parish of Westmoreland £20 for repairs to the Mesopotamia chapel, indicating that he was holding church services for the workers, or perhaps operating a school for their children.

These adjustments helped the estate to survive. During the years 1841–1844 Mesopotamia must have drawn a more dependable labor force, because it shipped significantly larger crops, averaging 142 hogsheads of sugar and 71 puncheons of rum per annum, and thereafter production increased a little, averaging 157 hogsheads of sugar and 111 puncheons of rum from 1846 to 1853.[31] While William Johns kept Mesopotamia going, a third of the Jamaican sugar estates closed down by the mid-1840s. Overall sugar production in Jamaica fell to a new low in 1840 of about 32,000 hogsheads—which was half of the 1835–1838 average and a quarter of the 1799–1817 average—and stayed at this level for the next thirty years. The Jamaica sugar planters had failed to convert effectively from slave labor to wage labor, and most of the freedmen and freedwomen had left the sugar estates. By midcentury the small island of Barbados was shipping more sugar than Jamaica.

Regrettably, I have almost no information about how many of the Mesopotamia slaves stayed on the estate after emancipation. We may speculate that Jane Ritchie, who bore William Johns an octoroon daughter in 1834, could have continued to live with Johns at Mesopotamia. Jane would have been thirty years old in 1844. We know that Mary Thompson lived on the estate briefly until she was ejected in January 1839. And two former Mesopotamia slaves named Joseph Foster and James Barham surfaced in 1843 when they bought livestock from William Johns. I cannot identify Joseph Foster, because the only Mesopotamia slave with this name in 1833 was two years old, but with the surname Foster he was almost certainly an ex-Mesopotamia slave who adopted a new name after 1833. James Barham *can* be identified, and it is pleasing to be able to trace him ten years after the final slave inventory. He was born at Mesopotamia in 1793 and named Edinburgh. His mother, Patty, had come from Southfield estate in 1791, and Edinburgh was the fourth of her five children. He was a cattle boy and carter in his youth, worked in the first field gang from 1815 to 1826, then became the head boiler from 1827 to 1829, and the second driver from 1830 to 1833. By 1831 he had joined the Moravian Church and displayed his loyalty to his absentee master by choosing James Barham as his baptized name, and he seems to have played a central role in starting the Mesopotamia harvest of 1832 without white supervision as described in Chapter 8. At age fifty in 1843 he was still working on the estate or living nearby. And if James Barham was connected to Mesopotamia, it is certainly possible that some of the other head people who had shared his leadership role in managing the estate during the rebellion of 1831–1832 were also still connected—such as head mason Richard Gilpin, driver William Samuels, blacksmith Edward Foster Barham, or cooper William Prince, four collaborators from 1833 who would have been in their mid-forties by 1843.

Most of the men and women who had formerly been slaves at Mesopotamia, however, were almost certainly gone by the mid-1840s. In 1842, if not earlier, William Johns was beginning to replace native black labor with imported labor. In the Mesopotamia crop report of 1842, payments were listed from "Emigrants" who had received tools and cash advances from Johns. These people were probably white workers. John Barham had talked of sending English laborers to his Jamaican estates in 1835, though he went mad before he had time to do so.[32] John's younger brother Charles Barham (who became owner in 1840) seems to have followed up on this plan. Some 3,700 immigrant laborers were brought to Jamaica from Britain and Germany in 1834–1842, so white workers were available.[33] The 1842 Mesopotamia crop report does not reveal whether the "Emigrants" worked for daily wages or on yearly contracts, but Johns collected £448 in rent in 1842 and a further £689 for rent and pasturage during 1843–1844. If these imported workers were British or German, they probably soon left or died. The Jamaica census returns indicate that the number of British-born people living in Westmoreland parish fell from 689 in 1844 to 204 in 1861, and the number of Germans fell from 196 to 58.[34]

By 1847 William Johns had turned to another source of imported labor. No Asian contract workers were resident in Jamaica in 1844, but 4,289 Indian immigrants were imported in 1846–1847, and Mesopotamia was one of the first estates to employ these people. In 1847 Johns paid a "Coolie tax" on contract workers imported from India.[35] As far back as 1806 Joseph Foster Barham II had looked to East Asia for potential laborers. He wanted to form a company to import Chinese workers to Jamaica on fourteen-year indentures, and his papers include a large boxful of proposals documenting his continued advocacy of the Chinese, desirable because of their "industry, ingenuity, frugality and orderly conduct."[36] When the "coolies" (now Indian rather than Chinese) came to Mesopotamia, they established a long-term trend. From 1847 onward they constituted the core workforce at Mesopotamia, supplemented in busy times by local black workers. The Indians labored under long-term contracts that defined their pay and working conditions, and restricted their movements. And they seem to have lived separately from any black workers on the estate. "Coolies" were still working at Mesopotamia in 1864 after Charles Barham sold the place to new owners.[37]

It remains to ask where the Mesopotamia freedmen and freedwomen went when they quit their old estate. Individual movement is unrecorded, but the general pattern is clear. Most of them acquired small tracts of uncultivated land, often on the fringes of sugar estates, at a cost of £2 to £10 per acre, and gathered with other freed people to form peasant farming settlements. Probably some of the Mesopotamia people remained in their old neighborhood, helping to create the numerous settlements that today lie within several miles of what is now called Barham Farm.[38] The rapid increase in small peasant holdings is suggested by the following estimates: in 1838 there were about 150 small freeholds in Westmoreland parish; by 1840 the number had increased to nearly 600, by 1844 to 1,700, and by 1860 to 3,750.[39]

This Jamaican peasant system evolved naturally out of Jamaican slave life.[40] The Mesopotamia slaves had been required to grow their own food, and were allocated land for that purpose. As they struggled to survive, they came to regard the kitchen gardens and fruit trees adjoining their slave huts and their provision grounds three to six miles into the interior of the estate as their personal property, and they devoted as much time and energy as possible to growing extra plantains and root vegetables that they could sell in the weekly market. They also raised poultry, pigs, and goats, and in 1809 Joseph Foster Barham's attorneys told him that the Mesopotamia slaves possessed upwards of one hundred cattle.[41] On weekends and holidays the slaves utilized these assets to participate in an internal economy that operated independently of the planters' economy. Having harvested whatever surplus produce they could spare, they would walk to Savanna la Mar, carrying their sale goods on their heads, and return home with clothing, trinkets, tobacco, alcohol, or small amounts of cash.

Once freed from bondage, the Mesopotamians who became independent peasants turned the slaves' internal economy into a full-scale way of life. They could now build houses adjacent to their provision grounds, spend as much time as necessary to grow food to feed their families, and take surplus produce to the weekly markets in exchange for a wide range of consumer goods. The more ambitious farmers could add to their incomes by growing export crops such as ginger and arrowroot. The Mesopotamia women could retire from sugar labor and spend their time gardening, marketing, and raising their children. Unless their farm plots were devastated by storms or drought, the freedmen and freedwomen could escape from one of the cruelest features of slave life—the hunger season every summer, when their provision grounds, neglected during crop time, had produced too little food. Best of all, perhaps, they could now live in a black world where it was possible to pass throughout an entire year without encountering a single white man.

Many of the peasant farmers, however, chose to work part time on nearby sugar estates in order to supplement their incomes. It appears that those who signed up as field hands only showed up for work when it suited them. The detailed daily wage records at Worthy Park sugar estate in 1842–1846 show many freelance workers on the pay list for short periods during the spring crop and fall planting seasons, and only a small core of residential workers during the slack times. Overall, there was a huge variation in daily employment at Worthy Park. Michael Craton, who has analyzed this data, found that the Worthy Park women rapidly withdrew from field work, changing the gender ratio from female dominant in 1838 to male dominant by 1842—echoing what we have seen at Mesopotamia. And Craton also found a great deal of turnover, with almost none of the people employed in 1838 still working at Worthy Park by 1846.[42]

Everyone in command—the Colonial Office, the royal governors, the assembly, the plantation owners—deplored the black flight from the Jamaican sugar estates. During the 1840s the Colonial Office stopped trying to protect the ex-slaves from the ex-slaveholders, and the British abolitionists who had campaigned so hard to end slavery lost interest in the Jamaican blacks once they were freed. Increasingly, the proponents of West Indian sugar production expressed their vexation with the blacks in virulently racist language. Echoing past slaveholders like Joseph Foster Barham, who called his Mesopotamia workers "dreadful idlers" because they objected to working without pay for sixty to ninety hours a week, commentators in the 1840s pilloried the Jamaica freedmen as apelike savages because they worked for themselves rather than for the sugar planters. Thomas Carlyle was the most famous exponent of this view. "Our beautiful black darlings are at last happy," he declared in

1849, "with little labor except to the teeth, *which,* surely, in those excellent horse-jaws of theirs, will not fail! . . . Sitting yonder, with their beautiful muzzles up to the ears in pumpkins, imbibing sweet pulps and juices; the grinder and incisor teeth ready for every new work, and the pumpkins cheap as grass in those rich climates; while the sugar crops rot round them, uncut, because labor cannot be hired, so cheap are the pumpkins." Carlyle went on to propose that the island governors and legislatures force the blacks back to the sugar estates and make them servants for life.[43] But the white Jamaicans had neither the legal authority nor the military power to make this happen.

It is a striking fact that the Jamaican blacks, who were documented in exceptionally full detail during the closing years of slavery, were so poorly observed after emancipation that they seem to have disappeared into historical oblivion. In the slave registration of 1817, the 321 men, women, and children living at Mesopotamia, and 344,931 other slaves living elsewhere in Jamaica, were all individually identified by name, sex, age, color, origin, and location. Additional tallies taken during 1807–1834 provided a great deal of further information about the slave communities on every plantation and the overall composition of the labor force, as we saw when examining the leading Westmoreland sugar estates in Chapter 1. After the slaves were freed, an island-wide census was taken in 1844, followed by another census in 1861, but both of these tabulations were mere summary head counts. There was no attempt to identify individual householders, whether white, brown, or black. There was no information about the new settlements where many of the ex-slaves were living. The feebleness of the Jamaica censuses in 1844 and 1861 reflects the declining political vitality of this colony as sugar production was collapsing. The record keepers, with scant interest in the freed blacks, ignored the fact that a very large social and economic transformation was taking place on the island. The 1844 census counted 132,252 people (constituting 52 percent of the total population above age ten and below age sixty) as "agricultural laborers" without attempting to find out how many of these people were still sugar laborers and how many had abandoned their old jobs. Judging by the collapse in Jamaican sugar exports, more than half of the resident workers in 1838 had left the sugar estates by 1844. And by 1861, when the census found 147,468 "laborers," probably two-thirds of this huge undifferentiated group had become independent peasant farmers.

It is tempting to romanticize the birth of the Jamaican peasantry, but life was very hard for them, whatever Thomas Carlyle might suppose. From Monday through Thursday they labored long hours on market gardening or on sugar estate work, on Friday they prepared for market, on Saturday they walked many miles to and from market, and on Sunday many attended church. As Thomas Holt observes, "They valued most the very things persistently denied them as slaves—a home and control of their family's labor."[44] Clearly they were living better than at any time during slavery. From 1655 to 1834 there had always been more black deaths than births, and now the overall Jamaican black population was actually *increasing:* from 293,128 in 1844 to 346,374 in 1861. In Westmoreland, black population growth was especially vigorous: from 18,880 in 1844 to 26,693 in 1861.[45] Yet these people were still living close to subsistence, with no protection against hurricanes and drought. And they faced continuous white oppression. They were politically powerless, without the right to vote. They were regularly abused in land disputes and in the local courts. They paid extortionate taxes on their land and on imported goods. They had access to churches but to very few schools. And most of these privations would continue far into the twentieth century. Jamaicans did not become citizens until 1962.

## *Emancipation at Larkin, Oakland, and Mount Airy*

On June 8, 1865, two months after the Confederate surrender at Appomattox, William Henry Tayloe on Oakland plantation in Alabama wrote a long letter to his daughter-in-law Courtenay Tayloe at Mount Airy, Virginia, telling her that his neighborhood in the Canebrake region of Alabama was now fully controlled by Union troops. "The blue coats are in Selma, Demopolis, Uniontown and other towns. Negroes have crowded to them. They are advised (not ordered) to return home, and to work on the plantations. Many did so, but told they are free, they are very much demoralized and idle." Thirty-six of Tayloe's former slaves had left Oakland and Larkin plantations to meet with the Yankee soldiers, and all of them had come back at least temporarily. "What they intend doing after the crops are made, God only knows," the old master lamented. "They meet in consultation, but probably have not arrived at any conclusions, or keep well their secrets. Poor creatures! Their castles tumble, subjecting them to great disappointment." Tayloe then moved on to consider the future. "Mount Airy negroes," he told Courtenay, "(with few exceptions) have been faithful, and I believe will serve you well anywhere—they deserve indulgences. I cannot turn off the old who have worked for me, nor starve the children. We need not expect to live as heretofore." And he had some further questions to ask his Mount Airy daughter-in-law. "How many will you take back to Virginia? How should they travel? How many can you shelter and feed? For what number can you find teams, and implements? Do you want them?"[46]

The contrast between William Tayloe's real personal concern for the Mount Airy freedmen and John Barham's total indifference to the Mesopotamia freedmen is of course very striking. Tayloe could identify with all of the people he had owned at Larkin, Oakland, and Mount Airy, Barham with none of the people he had owned at Mesopotamia. It is also evident that Tayloe considered his former slaves to be unfit for freedom. The blacks were now "demoralized and idle" because they were suddenly set loose. They had built castles in the air, supposing that emancipation would bring big changes in their lives, and were now disappointed. So it was Tayloe's paternal duty to continue to take care of them. This intrusive stance by Tayloe, and by thousands of other southern ex-slaveholders, was a big part of the emancipation problem facing the freedmen at Larkin, Oakland, Mount Airy, and everywhere else. How could they untangle themselves from the white planters who had habitually controlled them and were determined to continue to do so?

By June 8, 1865, when William Henry Tayloe wrote his letter to daughter-in-law Courtenay, great changes had taken place in the two months since Lee surrendered to Grant on April 9. The assassination of President Lincoln on April 14 produced a violent shift in executive policy toward the defeated South. In January Lincoln had pushed the Thirteenth Amendment through Congress, abolishing slavery, and the Freedmen's Bureau was created in March, so he was clearly aiming for deep social change, but the president was still feeling his way on the crucial issues of black suffrage and black access to confiscated land when he was murdered. His successor, Andrew Johnson, quickly acted to restrict the freed slaves. On May 29, 1865, he restored the landed property rights of all Confederate participants who pledged loyalty to the Union, and called for elections in the southern states based on an all-white franchise. Here Johnson was taking a position quite similar to the British government's in the 1830s: he saw the freed slaves as permanently second-class semicitizens, workers for their former owners, with minimal civil rights or economic opportunity. But in other respects the emancipation issue in mid-1865 was far more volatile than it had been in 1838. Radical Republicans in Congress, finally victorious after a brutal war, were pursuing an agenda almost opposite to Johnson's. Committed to reconstructing the slave South, they demanded black suffrage, partly to help the freedmen achieve meaningful emancipation, and partly to build a durable Republican political party in the South. The southern planters, totally resistant to such fundamental change, were far more numerous and vigorous than the Caribbean planters, and despite military defeat they were determined to keep the ex-slaves as their dependent laborers. And the ex-slaves for their part were trying as hard as they could to exercise their new freedom.[47]

William Henry Tayloe was now sixty-six years old and in frail health. He was directly affected by Andrew Johnson's reconstruction policy, because the president blamed the large slaveholding planters for starting the Civil War, and his proclamation of May 29 specified that all southern owners of property valued at more than $20,000 had to apply personally for presidential pardons. Tayloe was thus at risk of losing his Virginia and Alabama plantations. He wrote to son Henry at Mount Airy on June 22, 1865, that he was about to travel from Alabama to Washington in order to try to salvage his confiscated property. He had started cotton crops again at Larkin and Oakland for the first time in three years. The great majority of his former slaves were still living and working on his two plantations; he was supplying housing, food, and clothing to them as in the old days, and he bought five horses and three mules to make their work easier.[48] His hands stayed with him in the expectation that they would receive a share of the cotton crop at the end of the year, but Tayloe reported that many were discontented. The hope of getting back to Virginia was helping the Mount Airy–born people "to behave well," but five of his former slaves who were *not* from Mount Airy—Archy and Charles Williams (from Deep Hole, Virginia) and John Peter Coleman, Paris Coleman, and John Mingo (all acquired in Alabama)—were making trouble at Larkin.[49] Tayloe still had his overseers from the war years, Ramey at Larkin and Holcroft at Oakland, and when he departed for Washington he put them in charge, promising his black workers that he would return.

Tayloe absented himself from Alabama for four months. Accompanied by his black body servant, twenty-four-year-old William Henry Harwood, he obtained a pardon from the president and visited his family in Washington and Virginia for the first time since 1861. On October 9, 1865, Tayloe and Harwood started a leisurely return to Alabama, traveling by train to Cincinnati and Cairo, Illinois, in order to enjoy the scenery in the northern states of Ohio and Illinois, then down the Mississippi River by steamer to Memphis, and by train through war-damaged Mississippi. Tayloe was back in Alabama by October 21, and saw that everyone at Larkin and Oakland was on edge. Lewis Yeatman, a twenty-eight-year-old Larkin worker who had come down from Mount Airy in 1854, was stabbed (not seriously) in a scuffle with poor whites at the Uniontown railroad station. When Tayloe found out that the Freedmen's Bureau refused to transport any of the Mount Airy people to Virginia, he offered to loan son Henry enough money to bring Horace Yeatman, Charles Carrington, and eighteen to twenty other youngsters back to Mount Airy: "They are mighty anxious to return to you." And when he heard that his Oakland hands refused to enter into work contracts for next year, expecting that the government would provide them with land, he tried to persuade them that they would not get free land, and that it was better to stay on his plantation than to migrate to the local towns.[50]

In November 1865 Tayloe counted 224 freedmen as resident at Larkin, which indicates that about 30 people had left this plantation since April 1865. But he still had seventy-one male and fifty-five female working hands, all but three of whom were former Larkin slaves.[51] This favorable situation was about to change. In late December, when Tayloe was trying to negotiate contracts with "my free Negroes" for their labor in 1866, a crisis erupted at Larkin—as doubtless on many other southern plantations. Tayloe had always gotten along poorly with overseer J. W. Ramey, and when he learned that Ramey had been electioneering with the best Larkin hands to move with him to a new work site in 1866, he fired the overseer. There was a stormy scene in which Tayloe accused Ramey of feasting the freedmen at his expense and selling his cotton when he was in Virginia, confiscating the gold that Tayloe had entrusted him with, and stealing a mule. Ramey called him "a damned old rascal," threatened to kill him, and apparently did hit him.[52] And this sparked a general black exodus from Larkin. Thirty-eight adults and working children left with Ramey, another fifteen moved to Pool's plantation, thirteen to Booth's plantation, and smaller numbers to various other work sites. By January 1866 only thirty-nine of Tayloe's former slaves were still living at Larkin. Those who left had good reason to stay away, because Tayloe seems to have paid none of the deserters for the work they had done for him in 1865.[53]

Up to this point the Mount Airy story sounds like a repeat of the Mesopotamia story: the U.S. freedmen in 1865–1866, like the Jamaica freedmen in 1838–1839, were quickly abandoning the plantations where they had been enslaved. But turning from Larkin to Oakland, we find less dramatic upheaval. On this plantation about 150 men, women, and children had been emancipated in April 1865. Tom and Becky Flood quickly departed with their seven children, as did six young single men and women. And in December 1865 when Tayloe fired overseer E. W. Holcroft, the aggrieved overseer persuaded another fifteen people to leave with him. But around 120 former Tayloe slaves stayed on at Oakland into the next cotton year. In early 1866 Tayloe listed seventy-four Oakland men, women, and teenagers of working age, which was not a huge reduction from the workforce he had on this plantation back in 1863.[54] The resident freedmen he listed in 1866 were all deeply rooted at Oakland. Nine of them (including eighty-one-year-old Nanny Glascow) had come from Mount Airy to Oakland way back in the 1830s, another seventeen in 1845, and seven in 1854, while most of the younger people had lived on this plantation all their lives. Oakland was a more homogeneous place than Larkin, especially since only four of the ninety-eight people who had been brought down from Virginia in 1861–1862 were placed at Oakland, and all the others at Larkin. Furthermore, William Henry Tayloe probably exercised more direct influence at Oakland because he lived there, and mostly stayed away from Larkin because of his antipathy to overseer Ramey.

By early 1866, despite relatively stable conditions at Oakland, William Henry Tayloe was in a gloomy mood. "I am sick at heart," he lamented. "Our country is ruined. Our position changed, and we should submit quietly to our fate." But Tayloe did not submit quietly. "The rascality of my Agents and folly of the Negroes is enough to disgust any good Christian," he protested. A good example of rascality and folly from Tayloe's point of view was the exit in December 1865 of Nannie Key Wormley, the daughter of sailor George and chambermaid Kitty

Wormley at Mount Airy. In 1862 Henry Tayloe had sent Nannie (age twelve) and her older brother Austin (age seventeen) down to Larkin, leaving the other five members of the Wormley family in Virginia. Once the Wormleys were freed, George and Kitty asked to have their children returned to Mount Airy, but Nannie (now sixteen) decided to join Ramey's party instead, leaving brother Austin behind at Larkin. Young Austin (now twenty-one) promised Tayloe that he would take Nannie back to her parents at Christmas 1866, but Ramey refused to give her up. Tayloe applied to the Freedmen's Bureau "for a release of Nannie's unlawful detention," but was turned down, so Nannie stayed with Ramey.[55] By 1870, according to the U.S. Census, George and Kitty Wormley with two of their sons were living at Mount Airy; Austin Wormley was at Oakland plantation; and Nannie had disappeared from sight.

Between December 1865, when Nannie Key Wormley left Larkin, and June 1870, when her parents and three brothers were identified in the federal census, Alabama and the other former Confederate states experienced Reconstruction with a capital R. President Andrew Johnson's quick restoration of southern state governments based on white supremacy encouraged the legislatures in those states to pass Black Codes in 1865–1866 that returned the newly freed people to semislave status. The Republican-controlled Congress retorted by issuing the Fourteenth Amendment in 1866 (ratified by the states in 1868), which declared equality before the law for all black citizens, in defiance of Johnson. In 1867, again over Johnson's veto, Congress passed Reconstruction Acts that divided the South into military districts, and established conditions for new state governments: black suffrage, adoption of the Fourteenth Amendment, and disfranchisement of former Confederates for five years. Alabama, after holding a constitutional convention with black participation, was readmitted to Congress in 1868. And the election of Grant as president in 1868 assured that Reconstruction would continue, though in modified form.[56] Thus, much greater political change took place in the U.S. South than in Jamaica following emancipation. Black men in eleven rebel states received civil rights and voting rights in 1866–1867 that black men in Jamaica did not attain until 1944. But the economic and social change was smaller than in Jamaica. The U.S. freedmen did *not* receive free access to land. Nor were they protected from the Ku Klux Klan and other forms of white racist terrorism that would soon crush black political activism in the South until the 1950s.

In March 1866, while the president and Congress were bitterly quarreling over black civil rights, William Henry Tayloe retired from Alabama to Georgetown in the District of Columbia, to live with his elder daughter Sophia. He installed W. W. Gwathmey (the husband of his niece Mary Tayloe Gwathmey) as his tenant at Oakland and Larkin.[57] During the Civil War, Gwathmey had been an undercover agent for the Confederate government; in 1861 he bought guns manufactured in England for the Confederate Army. During 1866–1868 Gwathmey lived at Oakland, but supervised both of Tayloe's Alabama plantations, and received half of the net income from the two cotton farms. T. T. Woolley was the new overseer at Oakland, and Joseph Taylor at Larkin. In 1869 Tayloe turned to a new arrangement. He placed Confederate general Thomas Munford (the husband of his younger daughter Emma), who had served as a Virginia cavalry officer throughout the war, as his tenant at Oakland.[58] This appointment was designed to be permanent. In May 1869 Tayloe divided up his estate among his three children, giving Oakland plantation to Emma, forgiving Thomas Munford the $12,000 he owed him, and giving Larkin plantation to Henry, in addition to Mount Airy, which Henry had been operating since 1858. Having divested himself of his three plantations, William Tayloe continued to live with Sophie at Georgetown until his death on April 9, 1871.[59]

Under the management of Gwathmey and Munford, the workforce at Tayloe's two Alabama plantations was in constant flux between 1866 and 1870. Neighboring planters and overseers were always trying to entice Tayloe's hands away, and Gwathmey and Munford were always looking for new recruits. Peter Kolchin observes that in these years "the pressure was on the planter to find laborers rather than on the Negro to find employment."[60] But the black laborers could be cheated. The Gwathmeys reported in December 1866 that many of the freed people who left Larkin with Ramey a year ago were very dissatisfied with him, but they "remain with Ramey because he refuses to pay them anything unless they contract for next year [1867] when he promises payment in full."[61] Tayloe's former slaves generally moved in family groups. Richard and Sally Yeatman switched from Larkin to Pool's plantation in 1866 together with all nine of their children, who ranged in age from twenty-six to eight.[62] The Yeatmans didn't stay at Pool's for long, and the census of 1870 finds Richard and Sally back at Larkin with six of their children. Nancy and Etta Yeatman (both missing in 1870) had probably married, and young Richard had probably died. To give another example, Henry and Grace Dixon joined the ousted overseer E. W. Holcroft and left Oakland in December 1865 with their five children (ages fourteen to four).[63] Henry Dixon continued to work for Holcroft in 1867–1868, and in 1870 all seven of the Dixons (the children now ages nineteen to nine) were living close to Holcroft in Hale County, Alabama, probably on his cotton farm. Dixon was a bit unusual in staying with Holcroft; the tendency during the late 1860s was to return to Oakland and Larkin. In December 1869 Thomas Munford reported that Thadeus Dudley, Jacob Smith, and Harrington Smith all applied to return "to the old land" with their families.[64] Harrington Smith did bring his family to Oakland, but Thadeus Dudley and Jacob Smith must have gone elsewhere, because they were not living at Oakland in 1870.

Cotton production at Oakland and Larkin plummeted in 1866–1868, just as sugar production had plummeted at Mesopotamia in 1839–1840. During the peak years 1859–1860 on the eve of the war, Tayloe had sold 830 bales per annum from his two plantations. The figures for 1865 are conflicted. At Oakland, Tayloe brought 170 bales to his railroad crossing in December 1865, and was paying a guard $5 per night until the bales were picked up, but was this his entire crop? At Larkin, 456 bales were marketed in 1865, probably including cotton that had been stored during the war. In addition, some bales may have been stolen. In October 1865 Tayloe reported, "My stolen cotton was recovered," but in 1867 he told Gwathmey, "My waggoners had received $30 a night to haul off cotton," and that at least sixty-four bales were unaccounted for.[65] Whatever the truth about the missing bales, the total for 1865 looks pretty good when one considers the disruption at the end of the war. But with reduced manpower in 1866, production fell to only 58 bales at Larkin and 62 bales at Oakland according to one report, and to a combined 166 bales by another report. Gwathmey blamed a blight of cotton worms for the low yield of one bale for every four acres. During the next four years the only figures are from Oakland: 77 bales in 1867, about 70 bales in 1868, climbing to 129 bales in 1869, and 200 bales in 1870.[66] If we presume that the Larkin totals are similar, Tayloe's plantations were producing about half as much cotton in 1870 as in 1860.

Meanwhile, labor relations were rapidly changing. The whip could no longer be used, and the workers had to be paid. In 1866, seventy-four Oakland hands received $5,583.20, or $75 apiece on average. The top earner at $143 was blacksmith Travers Hilliard (age forty-seven), who had come to Oakland from Mount Airy in 1845. The two leading cotton hands, earning $120, were Joe Moore (age forty-six) and Mike Smith (age fifty-five), both of whom had been sent to Alabama in the 1830s and had worked at Oakland since 1840. Another twenty-two men were paid more than $100, while the best-paid woman, Lucy Rice (age forty), received $96. There were only thirty women and girls on the list, suggesting that women were beginning to withdraw from field work. Several of the adults, including Jim Grimshaw and his wife Arabella, were paid very little, apparently because they left Oakland in mid-crop. Among the eleven boys and girls, Solomon Glascow (age eleven) received the lowest pay: $13.75.[67]

From 1867 onward, Gwathmey and Munford stopped paying wages and instead negotiated annual contracts in which the workers received one-quarter of the cotton crop, which became the standard formula in Alabama.[68] Both the laborers and the planters seem to have preferred sharecropping to wage work; it gave the cropper an incentive to grow more cotton and to feel more independent. Gwathmey soon recognized that his Oakland hands resented traditional gang labor, and in 1868 he divided them into three smaller squads led by Mike Smith, John Lawson, and Arthur Thomas, a system that worked better.[69] In 1869 Joseph Taylor, the manager at Larkin, observed that freedmen were renting land whenever they could and setting up for themselves, so it was important to fix up the dilapidated cabins at Oakland and Larkin as an enticement to stay. By 1869 a further tendency was observable: the Oakland and Larkin workers were forming into tight communities, wanting to choose who else to work with and to reject intruders they didn't like—turning a white owner's plantation into the equivalent of a Jamaican peasant community. Joseph Taylor noted that "a man by the name of Sidney (who belonged to Pool) has returned to his old home driven away by these that are here" at Larkin. And Thomas Munford told Tayloe, "The negroes who formerly belonged to you have taken up the idea, that the place will be more or less under their control, they object to others coming in and do all in their power to keep them off."[70]

A chance comment by William Gwathmey reveals that William Henry Tayloe had built a church on his Oakland plantation—a fact not mentioned elsewhere in the Tayloe correspondence that I have read. Presumably Tayloe had engaged a white clergyman to hold religious services for the slaves in this building, and attended services himself when he was resident at Oakland. According to Gwathmey, the freed people abandoned this reminder of slavery days. "As the Negroes will not use the church," he wrote William Tayloe in October 1866, "it is falling to pieces. I shall next week move it into the yard as a kitchen for Mary [Gwathmey]." But he soon found a different use for the church. Gwathmey converted the building into a double cabin for two black families, placing John Lawson, his wife Becky, and sons Jesse and Ogle in one cabin, and Albert Ward and his wife Ginny in the other.[71] The 1870 census finds the Lawsons in Beat 8 dwelling house #543 and the Wards in Beat 8 dwelling house #544 on Oakland plantation, still sharing Tayloe's slave church.

Perhaps the most illuminating feature of William Henry Tayloe's correspondence with his family members in Virginia and managers in Alabama, 1865–1870, is a new sense of vexation and frustration whenever they talked about the freed people. As in slavery days, there were the old familiar charges: the blacks are lazy and irresponsible, and steal all the time. Added to this litany were new complaints: the children are going to school and the women are having babies instead of working in the fields; the men are always riding their horses and brandishing their guns; they don't trust any southern whites; and they are being duped by "the worst style of Yankee the 'Carpet Bagger.'"[72] There was no mention in this correspondence of white vigilantes or the Ku Klux Klan. But there was obvious irritation at having to negotiate directly, person to person, with individual freedmen and freedwomen. Tayloe himself urged patience. Advising his son to feign cheerfulness, he observed that "freedmen have to learn their position and will improve after suffering some hardships." And when he heard that Jim Glascow, who had come to Oakland in 1836 at age eleven, was taking his wife and son back to Virginia in 1866, he exclaimed: "Poor weak creatures! None can tell one day what they will do next."[73]

The Gwathmeys were more censorious. Mary Gwathmey told her uncle that "free negroes talk of work, but they dont or wont do it," and when some Oakland hands reported sick on hot days she declared that "nothing in my opinion will make some of them work except the lash." Mary Gwathmey was particularly indignant at young women who withdrew from field work. When Judy Glascow (age fifty-eight), her daughter Mary Glascow (thirty-one), and her sister Esther Barnes (sixty-two)—who had all come to Oakland in 1835 when Mary was a baby—complained that they were being worked too hard, Mary Gwathmey decided to dismiss Mary Glascow. But Judy's daughter outmaneuvered her; first she moved in with Robert Pen as his housekeeper, and then she married John Peter Coleman "and is happy doing nothing."[74] In 1870 Mary Glascow Coleman, Judy Glascow, and Esther Barnes were all still living at Oakland. Similarly, William Gwathmey in 1867 took a particular dislike to two "infernal scoundrels" in his workforce, Dangerfield Perkins (thirty-three) and General Lewis (thirty-four)—both migrants from Mount Airy in 1845—because they were agitating for better working conditions. Gwathmey was determined to get rid of General Lewis, "as I caught him last year with a pig in his house," and he became more exercised when Perkins and Lewis, with twelve other Oakland men, absented themselves without leave in August 1867 to attend a political barbecue, and spread the word that land was soon to be parceled out to them.[75] But Gwathmey admitted that Perkins and Lewis were both good workers, which overrode other considerations, and both men were still living at Oakland with their families in 1870.

During 1867, when Reconstruction was taking hold in Alabama, Gwathmey became increasingly pessimistic. "As the Negroe advances in freedom," he argued, "he becomes more difficult to manage and most of them more worthless than ever." Arthur Thomas, who had come to Oakland from Ogle Tayloes' Nanjemoy plantation in Maryland, was "looking forward confidently to Confiscation, to heal all their wounds and set them up in the world."[76] Gwathmey repeatedly complained that there were too many old people past productive labor at Oakland, and in 1868 he told Tayloe, "This place needs weeding out of all the worthless Negroes and the sooner it is done the better."[77] In January 1869, when Thomas and Emma Munford replaced the Gwathmeys, the atmosphere at Oakland temporarily reverted nostalgically to old slavery times. No one in the Oakland black community had seen Emma for many years, and she told her father that they "pay me the compliment of telling me I am not much broken. They all inquire after you and 'Miss Sophie.'" General Munford, adopting black dialect, reported that "the old ones all said to me, Old Master did this, and that, and so, but now things is 'mighty' changed everywhere, all the plantations are going down, and nobody cares, as all the 'Figguring' people is dun stop coming out here." If this was how the old people felt, a younger generation was taking charge within the plantation workforce. After a year at Oakland, Munford told his father-in-law that he had to exercise all of his patience in dealing with the freedmen. "There are some six or eight youngsters on the place who would torment a saint."[78] And the general was perhaps not very saintly.

This was the situation in Alabama, where 86 percent of William and Henry Tayloe's slaves were liberated at the close of the Civil War. But there were also more than sixty newly freed people at the home plantation in Virginia in 1865. Here emancipation worked out rather differently. During the war Henry Tayloe had sent ninety-eight slaves to Alabama, and another ten slaves had escaped to the Yankees. Mount Airy thus lost most of its workforce and became a retirement center for the Tayloes' elderly retainers. In 1863 Henry Tayloe had characterized twenty-two of the men and women who lived there as "chargable" or "worthless" or "crazy."[79] Many of Tayloe's slaves, once freed, seem to have felt as little affection for him as he had for them, because at least twenty-five of the Mount Airy people quit the plantation in 1865. Or perhaps they were dismissed. Many of those who left headed for Washington, D.C. By 1866 only about thirty-five freed people remained with Henry Tayloe.

It is hard to determine what sort of freedom the infirm and elderly people at Mount Airy experienced. In June 1865 William Tayloe had remarked to Henry's wife, "I cannot turn off the old who have worked for me, nor starve the children." But he soon changed his tune. In February 1866 he told Courtenay Tayloe, "You should send the old people [at Mount Airy] to the poor house or Freedmans Bureau," arguing that the Tayloes were taxed to support the poor house, so they should make use of it.[80] By 1865 five of Henry Tayloe's "chargable" or "worthless" men and women had died. Of the remaining seventeen, eight left the plantation shortly after they were emancipated. Nancy Carter (age seventy-three) went to Washington with her daughter Martha Harwood; the others had no stated destination. I have found only two of the eight who moved away from Mount Airy in the 1870 census. Tom Lawson, an invalid in 1863, was a field hand in 1870 at age fifty-six, living near Mount Airy with a young woman who kept house for him. And Tom Page, a field hand in 1863, was an invalid in 1870 at age seventy, living with a non-Tayloe black farmer's family. The remaining nine people on Henry Tayloe's negative list stayed on at Mount Airy after 1865, and four of them were still there in 1870: "crazy" Kitty Wormley (age forty-one) and "worthless" Winney Yeatman (forty-eight), Eliza Ward (sixty-two), and Fanny Bray (seventy-eight), all four living at home and no longer employed according to the census return.

After emancipation, very few of the freed people at Mount Airy continued to work for Henry Tayloe. In 1866 he had twenty-one employees, but only ten were his former slaves. Of the thirteen men hired as field hands, just two—George Yeatman (age fifty-six) and John Moore (age forty-one)—came from Mount Airy. Tayloe paid these thirteen men a total of $900, or $69 apiece, for a year's work, wages that were somewhat lower than the $75 average yearly pay received by the Alabama freedmen at Oakland in 1866. George Yeatman seems to have been the foreman and was the highest paid, at $5 a month, with a supplement of about $40 in December, for a total of $100. To operate his household Tayloe employed eight domestics (all former Mount Airy slaves) at still lower rates. George Wormley, a former sailor and "crazy" Kitty's husband, probably took charge of the stable and horses, and received about the same wages as George Yeatman, but the five women were paid only $1 or $2 per month, plus food and lodging.[81]

One might expect Henry Tayloe to solve his labor problems by inviting some of the Mount Airy freedmen at Larkin and Oakland to work for him, since he knew from his father that they were very anxious to return to Virginia. But he chose a different path. In 1867 he divided the three farm quarters at Mount Airy into nine much smaller farms, and rented eight of them to white tenants. To operate the home farm, his sole agricultural venture, he contracted with four of his former field hands—George Yeatman, John Moore, Jim Moore (John's brother from Larkin), and Cornelius Ward—in a sharecropping arrangement. Tayloe promised to furnish the four farmers with teams, implements, good housing, and rations (meal, meat, and fish), and in return they were to receive one-third of the corn they raised after repayment for fertilizer. The four farmers pledged to work for a year and were subject to dismissal without wages for misbehavior or improper conduct to Tayloe or his agents. At the same time Tayloe engaged four Mount Airy former slaves and a Mount Airy former slave's wife as his domestic staff: handyman George Wormley, coachman and ostler Ruffin Moore, and house servants Eliza Ward, her daughter Ellen Ward, and the wife of Cornelius Ward (who was Eliza's brother). This nine-person retinue was quite a slim-down from 1809, when Henry's grandfather had employed 29 enslaved domestics at Mount Airy and 153 enslaved field hands on his eight farms.[82] And Henry Tayloe told his father in 1871 that he had had a series of crop failures and was so deeply in debt that he needed to reduce the number of laborers still further. Henry's retrenchment policy was in strong contrast to the situation in Alabama, where William Tayloe's agents were always trying to enlarge the workforce at Larkin and Oakland.[83]

Between 1866 and 1870, a number of former Mount Airy slaves in Alabama tried to return to their home plantation. In February 1866 Jim Moore (sent to Larkin in 1862) left for Virginia with Chapman Page (sent to Oakland in 1845) and Chapman's wife and children. Jim was hired to work with his brother at Mount Airy, but Chapman was urged to move on to Washington. In March 1866 Jim Glascow (sent to Oakland in 1836) came back to Virginia with his wife Anne and son Solomon; Jim was hired by Henry Tayloe as soon as he arrived in April, but quit in disgust in December and wrote to Travers Hilliard, advising him to stay in Alabama.[84] In December 1866 Jim Flood, a carpenter from Deep Hole (sent to Alabama in 1854), left Larkin to visit his blind mother, Annie Flood (age seventy-eight), at Mount Airy, but didn't stay. And in March 1870 a sizable Alabama contingent arrived unexpectedly: Joe Yeatman (George's brother, sent to Larkin in 1862) with his wife Letty and family, together with Edmund Wright (sent to Alabama in 1854) with his younger brother Ben (sent to Oakland in 1861). Henry Tayloe found room for Joe Yeatman, but told his father that "we have no use for" Edmund and Ben Wright.[85] In 1870, when Henry Tayloe was now the owner of Larkin as well as Mount Airy, he contemplated visiting his Alabama plantation, "though I do not see what I could do after I got there, but the negroes would be better satisfied." Instead, he asked Joseph Taylor to tell the Larkin freed people that "they can get neither labor nor houses on this estate [Mount Airy] nor any help from me."[86]

The severe economic downturn in Alabama and Virginia in 1865–1870 was similar to the severe economic downturn in Jamaica in 1838–1844. The newly emancipated black people in the South, as in the British Caribbean, had to find their way into a new life during an era of agricultural depression when cotton/sugar production was falling and wages were low. But the challenges facing the freed people in the United States are far easier to track, because black life in the South during Reconstruction is so much more fully documented than black life in Jamaica after 1838. When comparing Mount Airy with Mesopotamia during emancipation, the unfortunate termination of the Barham Papers in 1836 means that only a tiny number of Mesopotamia freed people surface after 1838, whereas the Tayloe Papers identify hundreds of Mount Airy freed people moving to new locations or continuing to live and work at Larkin, Oakland, and Mount Airy. And there is an equally sharp contrast between the Jamaica census of 1844, which identifies no individuals by name, and the U.S. Census of 1870 (far more comprehensive than any earlier federal censuses), which provides detailed information about the total population, listing 38,555,983 individuals by name, age, sex, color, location, occupation, and place of birth.

Starting on June 1, 1870, the census takers—armed with printed sheets of census schedules, inkstands, ink, and pens—traveled systematically from house to house, numbering each dwelling sequentially. They were required to write down the name, age, sex, color, and birth place of each member of every household.[87] The digitized U.S. 1870 manuscript census schedules in the National Archives, accessed through Ancestry.com, make it possible to search for the men, women, and children who had been enslaved by William Henry Tayloe. In 1863 (their last complete inventory) William and his son Henry had recorded 457 slaves at Larkin, Oakland, and Mount Airy, and 452 of these people had surnames and were listed in family groups. I have searched for all of them in the 1870 manuscript census schedules. I was pleased to find 251 men, women, and children in the census whom I feel confident had been inventoried by the Tayloes in 1863, including 141 members of the Thurston, Yeatman, and Carrington families, described in Chapter 7 and charted at *www.twoplantations.com*.

Linkage between 1863 and 1870 is not easy. For tracking purposes, the most vital information that the census takers obtained was family name and first name, and in 1870 almost all of the formerly enslaved Tayloe householders were illiterate and unable to verify the correct spelling of their names. So the enumerators wrote down what they heard—which was often far removed from the spellings recorded in 1863. Godfrey Carrington was transmuted into Godfrey Kasing in the census, and Paul Myers became Paul Mills, and I never could have identified either family except that Godfrey's wife and six children and Paul's wife and three children were listed correctly by name and age.

Only about half of the people I found in 1870 knew their correct ages within a year or two, so the census age statements often match very poorly with the birth data in the Tayloe slave inventories.[88] Some of the elderly ex-slaves who are missing had died during the seven years between 1863 and 1870, and many of the young single women had married and adopted their husbands' names. But the chief reason, I think, why nearly half of the Tayloe freed people couldn't be tracked in 1870 was that they had changed their surnames or the census takers had radically respelled their surnames. I had especially poor luck searching for the thirteen Tayloe slaves who fled to the Yankees in 1861–1863. The only deserter who surfaced was coachman Ralph Ward, who returned to his family and was allowed back at Mount Airy. On the plus side I *did* locate 55 percent of William Henry Tayloe's former slaves, and I hope that some student of African American genealogy may be interested in tracking these people, clustered in ninety-one family groups, through the censuses of 1880–1940 to find out what happened in subsequent generations after 1870.

The 251 people I searched out in the census were all living in just five locations: 103 were in Hale County, Alabama; 96 were in Perry County, Alabama; 6 were in neighboring Marengo County, Alabama; 44 were in Richmond County, Virginia; and 2 were in Washington, D.C. Thus, most of William Tayloe's former slaves were still congregated in or near Larkin, Oakland, and Mount Airy five years after emancipation. The residential division between Alabama and Virginia was also little changed. In 1863, 86 percent of Tayloe's slaves lived in Alabama, and 82 percent of the freed people I tracked in 1870 were still in Alabama.

Hale County, Alabama, the seat of Oakland plantation, shows a particularly interesting residential pattern.[89] While Oakland plantation is not named in the census, twenty-seven of the dwelling houses in the Hale County manuscript schedule for Beat 8 can be identified as black workers' cabins at Oakland, and in 1870 twenty-two of these cabins were occupied by former Oakland slave families, with former Larkin slave families living in another four.[90] Only one non-Tayloe black family was working at Oakland in 1870. So the Tayloe freed people not merely continued to live at Oakland; they were actively keeping non-Tayloe workers out. Some of the Tayloe people who broke away from Oakland and Larkin in December 1865 were also living in Hale County in 1870. Former Larkin overseer J. W. Ramey had moved from Perry to

Hale County, and was farming in Beat 7. Three of the Tayloe freedmen who had gone with him from Larkin in 1865—Archy Williams, Amphy Wheeler, and his brother Jonathan Wheeler—lived with their families adjacent to Ramey in 1870, so were probably working on his farm. Frances Carter, who had also left Larkin with Ramey in 1865, was his domestic servant in 1870. Another former Tayloe employee, E. W. Holcroft, who had been the overseer at Oakland during the war, was likewise farming in Hale County, in Beat 8 close to Oakland. Four Tayloe freedmen—Abner Dudley, his brother Carter Dudley, Henry Dixon, and Ruffin Page—lived with their families near Holcroft in 1870 and seem to have been laborers on his farm. Abner Dudley and Henry Dixon had been working for Holcroft since 1866; the other two joined them between 1867 and 1870.

In Perry County, where there had been a major bust-up at Larkin plantation in 1865, the census shows a significant migration by Tayloe freed people back to their former home. Thirty-six workers' cabins can be identified at Larkin in 1870, and twenty-five of them were occupied by former Tayloe slaves.[91] Six families, led by Godfrey Carrington, Henry Thomas, James Pen Glascow, Gowen Page, Ralph Elms, and Cornelius Elms, had lived regularly at Larkin since 1865, and gradually some of their old companions had come back to join them. Six of the Larkin families that left with Ramey had returned by 1870, as had four that moved to Booth's plantation in 1866, six that moved to Pool's plantation in 1866, two that moved to Walnut Grove in 1866, and one that moved to Tate's plantation in 1866. As at Oakland, these people were finding strength in numbers by returning to Larkin and consolidating with their old mates. But the bonding was less exclusionary than at Oakland, since eleven non-Tayloe families were also living and working at Larkin in 1870. At some distance removed from Larkin, Charles and James Williams, brothers from William Tayloe's Deep Hole plantation in Virginia, lived in adjoining dwellings close to their elderly mother, Milly. And another six Tayloe families lived scattered in other parts of Perry County. At least three of them—headed by Bill Wormley, Joe Saunders, and Reuben Winters—seem to have rented land and were farming on their own.

The two families found living in Marengo County in 1870, physically separated from the other Tayloe freedmen who had congregated in or near Oakland and Larkin, had always been outliers. Tom Flood (age fifty-one in 1870) was born at Neabsco in Virginia, and was sent to Adventure plantation in Alabama as a youth in the 1830s. His family became stigmatized as arsonists, because Tom's elder sister Mary was accused of trying to burn down Henry Tayloe's house in Alabama in 1834, and Tom's younger sister Lizzie was accused of setting fire to the Mount Airy mansion in 1844. Tom became a carpenter, married Becky Carrington from Mount Airy (sent to Adventure in 1835), and they had eight children. By 1856 the Floods were living at Oakland, and in 1865 Tom was among the first freed people to leave Tayloe's plantation. Becky had died, and three of their older children were living at Oakland in 1870, but Tom himself was in Marengo County with two of the youngest children and his new wife Martha. The other Marengo settlers were Gowen and Anna Davis (age fifty-five and sixty-four, respectively, in 1870). In the Tayloe slave division of 1828, Gowen had been given to Edward Tayloe, and Anna to Henry Tayloe. Like Tom Flood, they were both dispatched to Adventure plantation in the 1830s and were acquired by William Tayloe in the 1850s. The Davises were among the many freed people who left Larkin in 1865 with ex-overseer Ramey, but they didn't remain with him. Anna had two children, a married son who was living at Oakland in 1870, and a married daughter who left Larkin in 1865 and then disappeared. In 1870 Gowen was working as a field hand, and Anna Davis was living at home with her daughter's child, eleven-year-old grandson Ben.

The pattern at Richmond County, Virginia, is also significant, because it shows that the Mount Airy freed people in Alabama had little opportunity to return home. Some forty slave families had lived at Mount Airy in 1860, but after the war Henry Tayloe greatly reduced his workforce, so only eleven former Tayloe black families were living on the home plantation in 1870. Several of the Tayloe slaves sent to Alabama had managed to return to Richmond County. John Moore had moved back to Virginia during the war, and his brother Jim Moore had come up from Larkin in 1866; both were employed by Henry Tayloe in 1870. Joe Yeatman and Billy Shears had arrived from Larkin in the spring of 1870 with their families and found work with Tayloe's tenants. So had Thomas Lewis, who was living with his wife and daughter on the old plantation in 1870. But when Jim Glascow brought his family from Alabama to Mount Airy in 1866, he quit working for Henry Tayloe after a few months. In 1870 Glascow was still in Richmond

County, but living some distance from Mount Airy. And Ben Wright, who had showed up from Larkin in 1870, found no work at Mount Airy, so he too was living away from the home plantation.

The census shows that Tayloe's former slaves had very narrow occupational choices in 1870. All but six of the eighty-four male householders were listed as laborers. The six exceptions were three blacksmiths, a carpenter, a teamster, and a minister. Tom Thomas, who had worked as a carpenter for forty years at Mount Airy, was the minister. One wonders whether Tom Thomas had been preaching at Mount Airy when he was a slave carpenter. He continued to live on Henry Tayloe's plantation, and was one of the very few Tayloe ex-slaves who told the census takers that he was able to read and write. He was also credited in 1870 with personal property valued at $100. Unfortunately, the census supplies contradictory information about the occupational status of the Tayloe women. The census taker in Hale County listed almost all of the Tayloe wives as keeping house, while the census taker in Perry County listed all but two of the wives as laborers along with their husbands. Carpenter Charles Carrington and his brother, blacksmith David Saunders, lived next door to each other in Perry County, a short distance from Larkin plantation, and their wives Caroline Carrington and Mary Saunders were said to be the only Tayloe women in Perry County who avoided field work, which is hard to believe. Just three old men and two old women among the 243 former slaves had no listed occupation. Some of the cotton laborers were far from young: Ralph Ward was sixty-three, Gowen Page was sixty-one, George Yeatman was sixty, and fifteen field hands were in their fifties. On the positive side, the Tayloe families found in the census were expanding rapidly, as the young people married and started having children. The ninety-one listed households contained nearly 400 people, including some 80 children born since 1863. As in slavery days, almost every boy or girl over the age of ten was working in the cotton fields. Five girls, but no boys, were attending school, and only ten members of this large cohort—eight of them children—were described as literate. The Reverend Tom Thomas and nineteen-year-old Ruffin Page were apparently the only adults who could both read and write.

Let us take leave of the Mount Airy people with vignettes of three slaves turned freedmen, linking their experiences before 1865 with their experiences after emancipation. The first sketch is of William Henry

Harwood (also spelled Harrod), who was William Tayloe's body servant. This lad was a member of Franky Yeatman's very prolific family, charted at *www.twoplantations.com*. His grandmother Nancy Carter (a Mount Airy dairymaid and textile spinner) had thirteen children, but William Henry knew only one of his uncles and aunts, because six of them died young, two were sold, and three others were moved from Mount Airy before he was born at Mount Airy in 1841. His mother, Martha Harwood or Harrod (a Mount Airy housemaid), gave birth to nine children, five of whom became Mount Airy domestics or craftsmen. William Henry, very likely named for his master, started working in the Mount Airy dining room at age ten, accompanied Tayloe in 1857 at age sixteen to Oakland, where he was installed in the overseer's house, and waited on Tayloe when he was in Alabama. We heard from him in Chapter 7 when he came into the Oakland kitchen with a load of wood while his master was writing. When Tayloe asked for his message to his mother, he said, "Give her my love, give my love to all of them."[92]

During the war William Henry's eldest brother, Elias, a carpenter, deserted to the Yankees in 1861, as did his sister Betsy in 1862. William Henry had no opportunity to escape. He continued to attend Tayloe at Oakland throughout the war, and in June 1865 (now a freeman, aged twenty-four) he accompanied Tayloe on his four-month visit to Virginia and Washington, and traveled back with him on a roundabout sightseeing tour via Cincinnati, the Mississippi River, and Memphis. Tayloe, when he retired from Alabama in March 1866, wanted to take William Henry to Washington with him. But young Harwood elected to stay at Oakland, to be close to his "lady love" Frances Carter (as Tayloe described her), who had recently departed from Larkin with Ramey. So Tayloe instructed William Gwathmey to keep him on the Oakland domestic staff. But in July 1866 William Gwathmey fired Harwood for "his trifling and troublesome conduct," claiming that he was supplying his mistress Anne Carrington (the wife of David Carrington) with food from Gwathmey's table.[93] At this point William Henry disappeared from the Tayloe records. His mother, Martha, had moved to Washington in 1865 from Mount Airy with his grandmother Nancy and his youngest siblings, Julius and Meta; his sister Nancy had also departed from Mount Airy with two young children; and his brother Beverley (the only field hand in the family) was working at Oakland in 1868, though he was no longer there in 1870. The census tells us that in

1870 William Henry's former lady love Frances Carter had a two-year-old boy and was still working for Ramey as his domestic servant in Hale County, Alabama. Harwood's alleged mistress Anne Carrington was still living at Oakland plantation, where she shared a cabin with her husband David and five children, three of them born since 1863. I have not found William Henry's mother or grandmother or any of his siblings in the census, but he was very likely in touch with several of them, because William Henry Harrod (now twenty-nine, but stated to be thirty-nine) was listed as a teamster in Washington, D.C., married to a woman named Jane who kept house for him. Jane's ten-year-old daughter Laura went to school, where she had learned to read and to write. And William Henry was also at least semiliterate; he could read, although he couldn't write.

The second sketch is of field hand Joe Yeatman, a distant relative of William Henry Harwood, so his family is also charted at *www.twoplantations.com*. Joe, born at Mount Airy in 1815, was a generation older than William Henry; his father, Dick Yeatman, was the elder brother of William Henry's grandmother Nancy Carter. Joe's parents were domestic and craft slaves: Dick Yeatman was the chief gardener at Mount Airy, and Joe's mother, Betsy, was a textile spinner. Joe started to work at age seven as a miller's boy, but at age seventeen, after ten years of milling, he became a field hand like his older brother George. For the next thirty years Joe farmed at Landsdown quarter, and eventually became the foreman of the Landsdown farmhands in parallel with his brother George, who became the foreman of the Forkland quarter farmhands. William Tayloe labeled him "Little Joe"—probably because he was not very tall. Around 1835 Joe married Fanny Smith, daughter of a Mount Airy carpenter, and by 1842 they had six children. But in 1843 Fanny did something unexplained in the records that must have angered William Tayloe, and she was sold together with her brother Joe Smith and her little son George and little daughter Jane. This left poor Joe Yeatman with four small children—Lemuel, Lewis, Grace (who died in 1844), and Billy. He soon married Letty Richardson, who bore him another three boys and three girls. In 1854 Joe's three remaining children from his first marriage, Lemuel, Lewis, and Billy Yeatman (aged eighteen to thirteen), were sent to Alabama. A typhoid epidemic broke out at Oakland in February 1858, and as we learned in Chapter 7, William Tayloe sent a terse message to Joe Yeatman four months later. The news was bad. "I am sorry to tell L. Joe all his boys had the Typhoid fever," Tayloe wrote. "Lemuel and Lewis are well but poor little Billy died. We did our best to save him."[94]

When the war broke out, Joe and Letty Yeatman were sent from Mount Airy to Alabama in 1861–1862 together with four of their five surviving children. Joe was now reunited at Larkin with Lemuel and Lewis, but seven-year-old Dickerson Yeatman was left behind at Mount Airy. When slavery ended in 1865, Joe Yeatman was fifty years old. His oldest sons, Lemuel and Lewis, soon quit Larkin, but Joe and Letty and their four children stayed on. Daughter Betsy died in 1866, and the next year Joe asked the Larkin manager, Joseph Taylor, to find out how his son Dickerson (now twelve), his brother George (now fifty-seven), and Dickerson's grandmother Nancy Richardson (now sixty-seven) were getting on at Mount Airy. One hopes that he got an answer. Joe and Letty apparently continued to work at Larkin until 1870, when they showed up at Mount Airy in March with two of their youngest children, Jesse (seventeen) and Nettie (nine), together with Joe's son, probably Lemuel (thirty-four), and his son-in-law, who was Billy Shears, the new husband of Joe's daughter Airy (twenty-one). Henry Tayloe was probably not very pleased to see them, but his Mount Airy farm tenants hired Joe and his sons Jesse and Dickerson and his son-in-law Billy Shears.[95] So now Joe was back with son Dickerson (fifteen) and brother George (sixty). I have not found Lemuel in the 1870 census, and Lewis (thirty-three) was living in Alabama. But Joe (age fifty-five) and his son-in-law Billy Shears (thirty-seven) had adjoining cabins at Mount Airy. Wife Letty (forty-nine) was keeping house for Joe, Jesse, Dickerson, and Nettie, while Airy was keeping house for Billy Shears, their two young children, and an elderly woman who was probably Billy's mother.

The last sketch is of Amelia Carrington, whose family is also charted at *www.twoplantations.com*. She was the daughter of Peter and Nancy Richardson, a Mount Airy sailor and spinner couple, and was born in 1824, so was a bit younger than Joe Yeatman but older than William Henry Harwood. As a child Amelia lived at Mount Airy with her parents and started work as a dairymaid, but soon became a seamstress and textile spinner, and developed into an accomplished needlewoman. She had three brothers and four sisters, and three of the Richardson girls married farmers. Amelia's sister Letty married Joe Yeatman, and Amelia married Godfrey Carrington, who was a field hand at Forkland quarter for over thirty years. Godfrey identified himself (and his children) as mulatto, because his father David Carrington was a free mulatto. Godfrey had nine siblings, all of whom were born at Forkland quarter and most of whom worked there. Amelia got to know Godfrey's brother Jacob Carrington, briefly the husband of Winney Grimshaw, and Godfrey's sister Kitty Carrington, who accused Lizzy Flood of setting fire to the Mount Airy mansion. And Amelia produced eleven children between 1839 and 1860. She had her first baby, a girl who died in infancy, when she was fifteen. Her first son, David, seems not to have been Godfrey's, because he took the surname Saunders, but he grew up in the Carrington household. David became a blacksmith, the third son, Charles, became a carpenter, and all of Amelia's other children became farmhands. David was sent to Alabama in 1854, but the rest of the family remained at Mount Airy until the war, when they were also sent to Alabama. Philip, Charles, and James (aged eighteen to fourteen) went in 1861, and Amelia and Godfrey followed with the six youngest (aged twelve to one) in 1862. And Amelia didn't like the move to Alabama, saying to Tayloe that "she won't tell a story about it—that she wants to go back."[96]

During the war, Amelia lived at Larkin with Godfrey and all of her children except David, who lived at Oakland. She earned Tayloe's gratitude with her energy and efficiency in increasingly difficult conditions. Since clothing from the northern manufacturers was no longer available, Amelia constructed hats for her fellow slaves and stitched a homespun suit for her master, who called her "one of the very best of our servants."[97] In 1865, when the Carringtons were free to move, it might be supposed that Amelia would urge her husband to return to Virginia. Instead, they were among the few former Tayloe slaves at Larkin who didn't abandon that plantation in December 1865. Amelia's oldest son, David, was at Oakland, and remained there after 1865, while her second son, Philip, and her oldest daughter, Milly (who were both married), joined with ex-overseer Ramey. But Charles, James, Brooke, Chris, Jacob, Mary, and Arthur Carrington (ages eighteen to five) stayed with their parents at Larkin. The 1870 census finds Amelia (now forty-six) and Godfrey (fifty-two) living at Larkin in a very overcrowded cabin. Daughter Milly had returned with two small children, so Amelia and Godfrey were living with seven children and two grandchildren. Every member of this family was listed as mulatto in the census. Amelia's oldest son, David Saunders (thirty), had left Oakland and was living next door to Charles (twenty-three) in Perry County, quite close to Larkin. These sons were still craftsmen, David a blacksmith and Charles a carpenter. The rest of the family—all except nine-year-old Arthur—were listed as farm laborers. But Amelia actually was the housekeeper at Larkin house, where the white manager lived. When William Tayloe's nephew E. Thornton Tayloe (son of Benjamin Ogle) visited Larkin in January 1870, he reported that Amelia made him feel very comfortable, and that Larkin house was scrupulously neat and clean. Thornton Tayloe also told his uncle that when he went out into the field to talk to the hands, trying to discourage the homesick Virginians from wanting to return, "they asked many questions after 'Old Master,' how he was, how he looked." And they didn't believe that Henry Tayloe would come to see them.[98]

It is ironic that William Henry Harwood, who was far more decently treated by William Tayloe than Joe Yeatman or Amelia Carrington, was the only one of the three people presented here to break with his old master. And Joe Yeatman, who was far the worst treated, was the only one of them who returned to Mount Airy. And Amelia Carrington, who said she wanted to go back to Virginia, stayed in Alabama. All three vignettes suggest that these people after 1865 had a freedom that was truly liberating, and that they were bonding with their families in ways that had never been possible under slavery. Yet opportunities for economic advancement remained almost as restricted as during slavery.

Unfortunately, there is no way of presenting parallel vignettes of Mesopotamia slaves turned freedmen. But it is evident that during 1838–1844 the Mesopotamia ex-slaves moved in an alternative direction from the Mount Airy ex-slaves during 1865–1870. Many of them immediately began to separate themselves from Mesopotamia, from sugar work, and from white control. They opted instead to buy small plots of land and farm independently in autonomous peasant settlements. This gave them the ability to reject coercive labor, to escape the brutal exploitation they had suffered under slavery, and to achieve a full measure of economic and social independence.[99] The Jamaica peasants survived in obscurity for many years after 1838, but in retrospect they were fortunate to be largely left alone by a British government with rising imperial ambitions in Africa and Asia, and with white man's burden assumptions about how to manage dark-skinned people.

Freedom obviously turned out very differently for the Mount Airy people. In 1866–1867 they were granted civil rights and political rights that the Jamaican peasants didn't receive, but in the 1870s most of these new perquisites of citizenship were forcibly taken from them. The Mount Airy people that we have been able to trace were tremendously hampered by their inability to secure land of their own. They wanted to move back to Virginia, but economic conditions at Mount Airy were even more depressed than in Alabama, so most of them stayed in the Deep South. A very sizable number continued to work for the Tayloes at Oakland and Larkin, still growing cotton, but now as sharecroppers. This was a major improvement, to be sure, and the Tayloe Papers present compelling evidence that the Mount Airy ex-slaves acted more aggressively and adventurously in 1865–1870 than at any time during slavery. The 1870 census gives proof that the Oakland and Larkin people banded together for community solidarity, confirming Thomas Munford's statement in 1869 that they wanted the cotton farm they worked on to be "more or less under their control, they object to others coming in and do all in their power to keep them off." What we don't hear about in the Tayloe Papers is the mob violence that the freed blacks were subjected to. But we know that it happened, and that tortured race relations in the postwar South gravely damaged the individuality, dignity, and selfhood of the newly freed African American people.

*A Tale of Two Plantations* has now reached its terminus, without supplying either a proper opening or a proper conclusion. Historians rightly love to write biographies, in part because a well-told account of an interesting life has a natural evolutionary shape from birth to death. But in this book I have aimed for something different. I have tried to show a Jamaican slave community in action during the final seventy years of British Caribbean slavery (1762–1834) and a Virginia-Alabama slave community in parallel action during the final seventy years of United States slavery (1792–1865). My story has no proper beginning because, by virtue of the sources I employ, it starts in midcourse. Edmund Stevenson assembled the first group of slaves to work his Mesopotamia sugar plantation around 1680, but the makeup of this slave community is hidden past recovery until the 1760s, when Joseph Foster Barham ordered his overseer to supply documentation that suddenly brought the men and women on this plantation to life. Similarly, William Tayloe started to buy the slaves who operated his Mount Airy tobacco farm in the 1680s, but the characteristics of this slave community are obscured for an even longer period, until the 1790s, when John Tayloe III began his methodical record keeping. And my story has no proper conclusion because it stops where my records run out. I part company with the Mesopotamia ex-slaves in 1838–1844 and with the Mount Airy ex-slaves in 1865–1870 just as they were entering into a new life, with the death of slavery and the birth of freedom.

Appendixes

Notes

Acknowledgments

Index

# *Appendixes*

## *Appendix 1* Slave Population Changes at Mesopotamia, 1762–1833

A. During the ownership of Joseph Foster Barham I, 1762–1789

| | Male | Female | Total |
|---|---|---|---|
| Population in July 1762 | 152 | 116 | 268 |
| Increase: | | | |
| Born at Mesopotamia | 53 | 56 | 109 |
| Transferred from other Barham estates | 0 | 0 | 0 |
| Bought from Africa | 86 | 21 | 107 |
| Bought locally in Jamaica | 31 | 22 | 53 |
| Total | 170 | 99 | 269 |
| Decrease: | | | |
| Died at Mesopotamia | 145 | 85 | 230 |
| Transferred to other Barham estates | 0 | 0 | 0 |
| Sold for transportation | 2 | 0 | 2 |
| Escaped | 1 | 0 | 1 |
| Manumitted | 2 | 2 | 4 |
| Total | 150 | 87 | 237 |
| Population in December 1789 | 172 | 128 | 300 |

B. During the ownership of Joseph Foster Barham II, 1790–1833

| | Male | Female | Total |
|---|---|---|---|
| Population in December 1789 | 172 | 128 | 300 |
| Increase: | | | |
| Born at Mesopotamia | 155 | 156 | 311 |
| Transferred from other Barham estates | 0 | 0 | 0 |
| Bought from Africa | 11 | 19 | 30 |
| Bought locally in Jamaica | 114 | 111 | 225 |
| Total | 280 | 286 | 566 |
| Decrease: | | | |
| Died at Mesopotamia | 287 | 234 | 521 |
| Transferred to other Barham estates | 0 | 0 | 0 |
| Sold for transportation | 2 | 0 | 2 |
| Escaped | 4 | 2 | 6 |
| Manumitted | 2 | 6 | 8 |
| Total | 295 | 242 | 537 |
| Population in August 1833 | 157 | 172 | 329 |

*Appendix 2* Mesopotamia Population Pyramid, 1762

*Appendix 3* Mesopotamia Population Pyramid, 1789

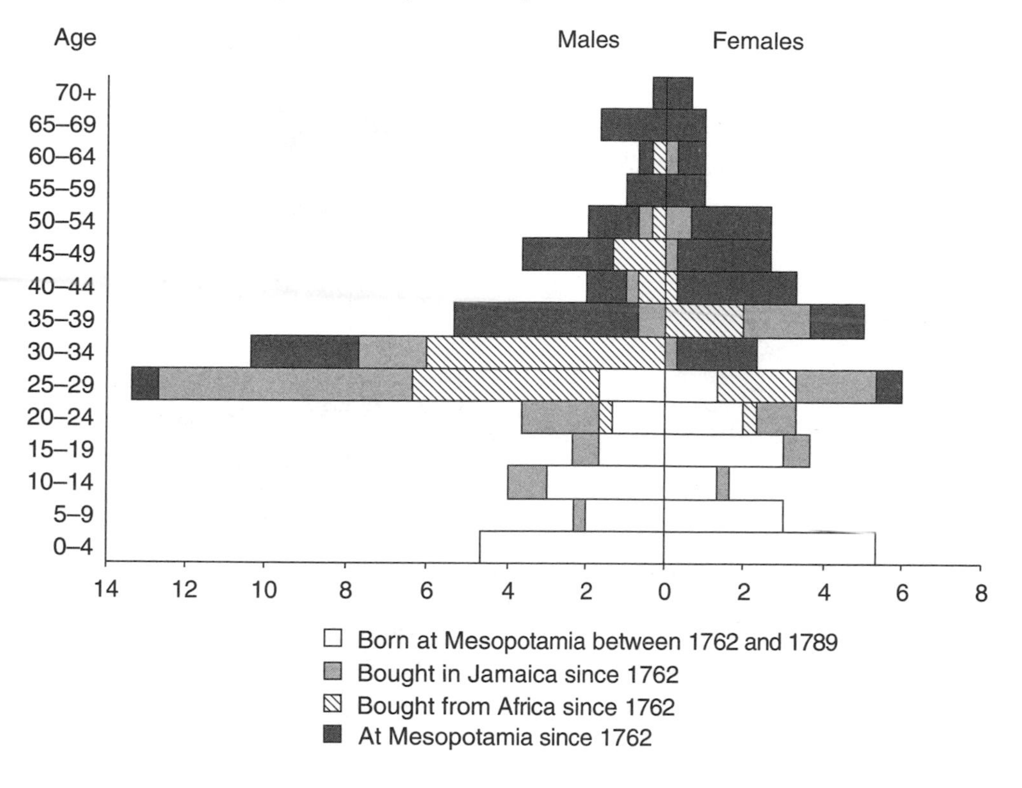

*Appendix 4* Mesopotamia Population Pyramid, 1833

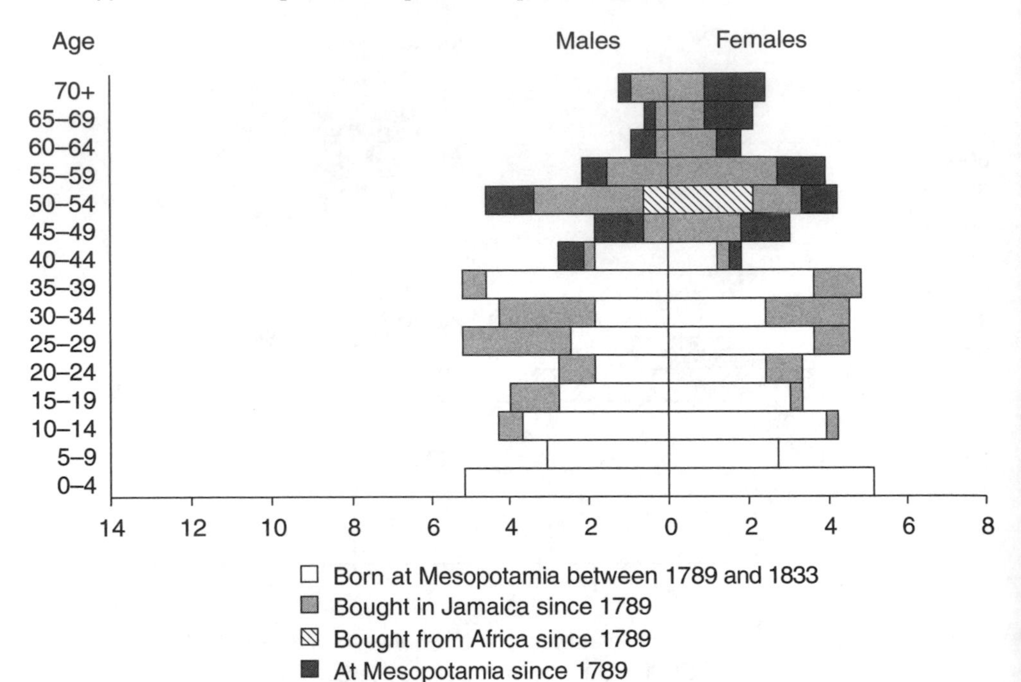

*Appendix 5* Slave Population Changes at Mount Airy, 1809–1863

A. During the ownership of John Tayloe III, 1809–1828

| | Male | Female | Total |
|---|---|---|---|
| Population in January 1809 | 218 | 164 | 382 |
| Increase: | | | |
| Born at Mount Airy | 136 | 116 | 252 |
| Purchased | 4 | 0 | 4 |
| Transferred in from JT3's other Tayloe Chesapeake properties | 20 | 16 | 36 |
| Total | 160 | 132 | 292 |
| Decrease: | | | |
| Died at Mount Airy | 80 | 62 | 142 |
| Transferred out to JT3's other Tayloe Chesapeake properties | 72 | 37 | 109 |
| Sold | 12 | 32 | 44 |
| Total | 164 | 131 | 295 |
| Population in January 1828 | 214 | 166 | 380 |
| Transferred from Mount Airy to WHT's brothers, 1828–1829 | 104 | 78 | 182 |
| Transferred to Mount Airy from WHT's brothers, 1828–1829 | 6 | 5 | 11 |
| Total loss | 98 | 73 | 171 |

(*continued*)

*Appendix 5* (continued)

B. During the ownership of William Henry Tayloe, 1828–1863

| | Male | Female | Total |
|---|---|---|---|
| Mount Airy population after settlement of JT3's estate | 116 | 93 | 209 |
| Increase in Virginia: | | | |
| Born at Mount Airy | 122 | 128 | 250 |
| Transferred in from WHT's other Tayloe Chesapeake properties | 8 | 12 | 20 |
| Purchased in Virginia | 9 | 11 | 20 |
| Total | 139 | 151 | 290 |
| Decrease in Virginia: | | | |
| Died at Mount Airy | 86 | 58 | 144 |
| Sold in Virginia | 8 | 17 | 25 |
| Escaped in Virginia and District of Columbia | 11 | 3 | 14 |
| Total | 105 | 78 | 183 |
| In Alabama: | | | |
| Moved from Mount Airy to Alabama | 107 | 111 | 218 |
| Born to Mount Airy mothers in Alabama | 74 | 60 | 134 |
| Deaths of migrants and their children in Alabama | 28 | 29 | 57 |
| Sold in Alabama | 10 | 13 | 23 |
| Migrants from Mount Airy in Alabama, 1863 | 95 | 92 | 187 |
| Children born to Mount Airy migrants in Alabama, 1863 | 60 | 50 | 110 |
| WHT's non–Mount Airy slaves in Alabama, 1863 | 57 | 37 | 94 |
| Total population in Alabama in 1863 | 212 | 179 | 391 |
| Population of Mount Airy in 1863 | 27 | 39 | 66 |
| Total Mount Airy population in Virginia and Alabama, 1863 | 182 | 181 | 363 |
| WHT's total population in 1863 | 239 | 218 | 457 |

*Appendix 6* Mount Airy Population Pyramid, 1809

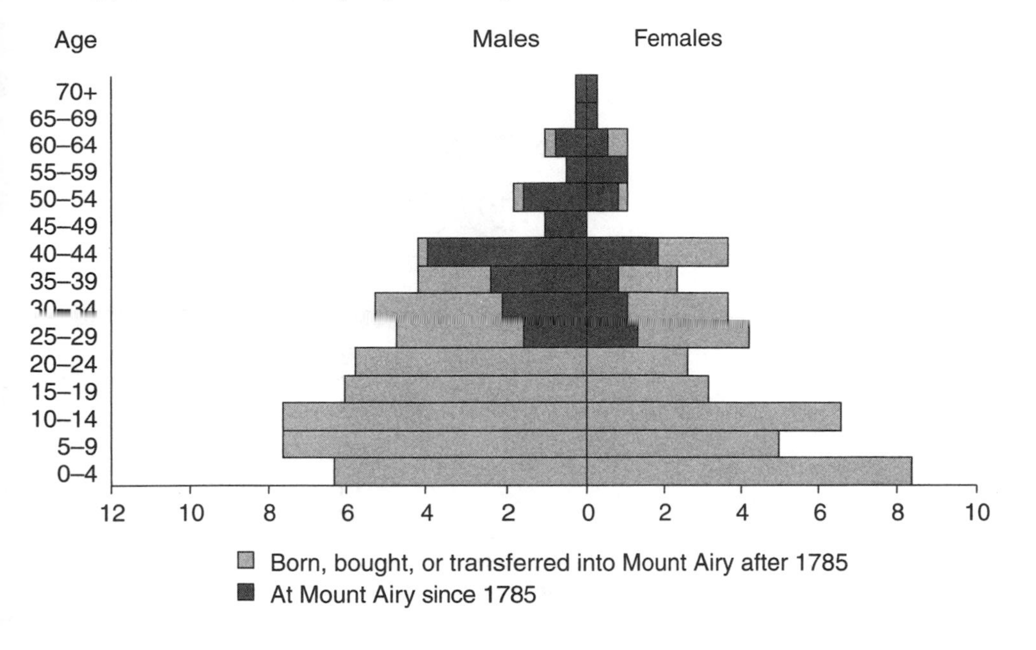

*Appendix 7* Mount Airy Population Pyramid, 1828

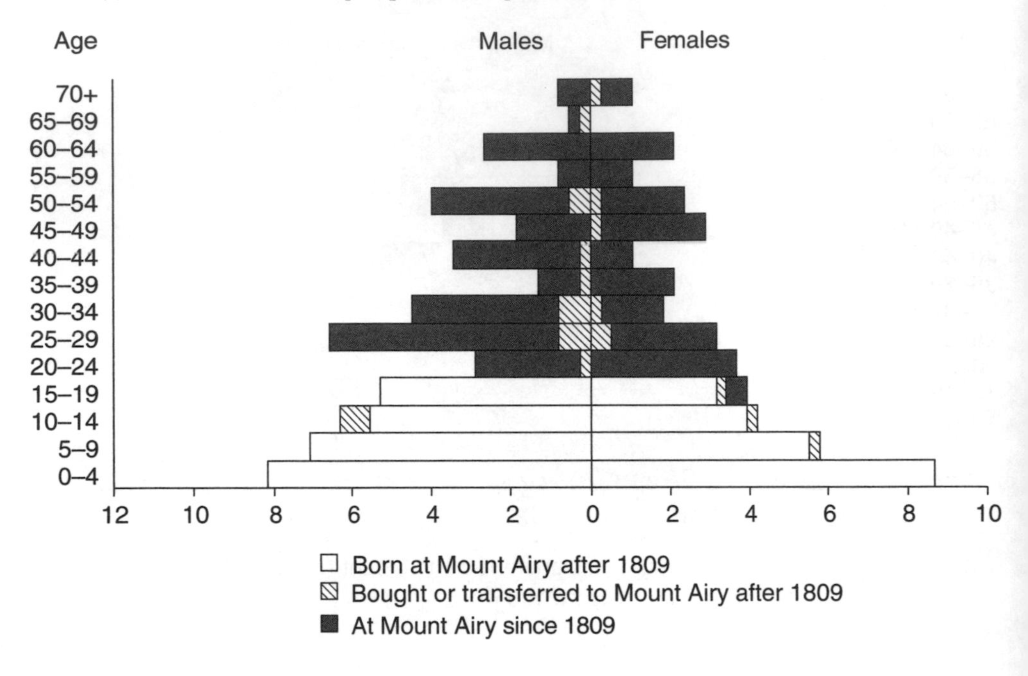

*Appendix 8* Mount Airy and Alabama Population Pyramid, 1863

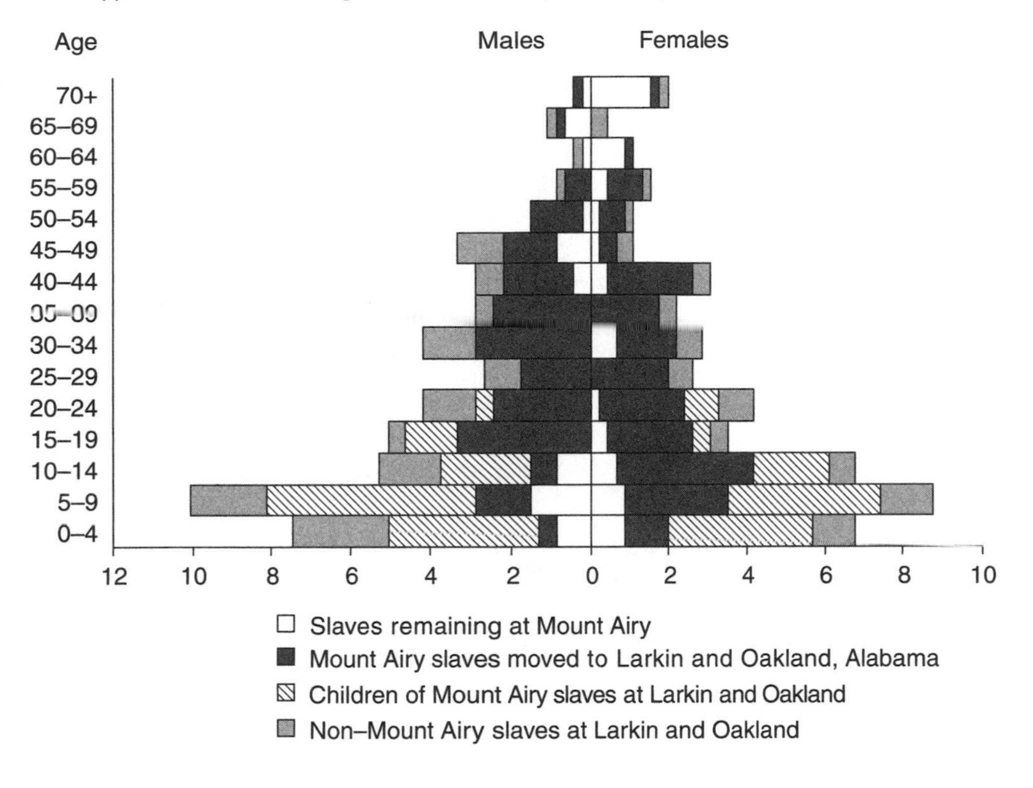

*Appendix 9* Slave Population Changes on Twenty-Seven Westmoreland Sugar Estates, 1807–1834

| Name | Pop. 1807 | Pop. 1814 | Pop. 1817 | Surplus births 1817–32 | Surplus deaths 1817–32 | Pop. 1834 | M/F sex ratio 1817 | M/F sex ratio 1834 |
|---|---|---|---|---|---|---|---|---|
| Bath | 236 | 213 | 194 | | 49 | 210 | 113/100 | 88/100 |
| Black Heath | 350 | 381 | 328 | | 61 | 353 | 80/100 | 77/100 |
| Blackness | 307 | 318 | 310 | | 45 | 263 | 75/100 | 73/100 |
| Blue Castle | 275 | 310 | 286 | | 47 | 204 | 86/100 | 79/100 |
| Bog | 241 | 310 | 241 | 24 | 0 | 260 | 93/100 | 98/100 |
| Canaan | 267 | 217 | 247 | | 19 | 200 | 89/100 | 79/100 |
| Carawina + pen | 229 | 265 | 339 | | 32 | 318 | 98/100 | 85/100 |
| Cornwall | 353 | 312 | 285 | | 52 | 243 | 89/100 | 77/100 |
| Delve | 242 | 184 | 203 | | 17 | 184 | 85/100 | 80/100 |
| Fontabelle etc. | 432 | 505 | 829 | | 223 | 246 | 90/100 | 80/100 |
| Fort William etc. | 559 | 461 | 433 | | 86 | 344 | 78/100 | 66/100 |
| Friendship etc. | 409 | 390 | 367 | | 29 | 262 | 90/100 | 78/100 |
| Georges Plain | 282 | 302 | 331 | | 63 | 255 | 107/100 | 95/100 |

| Kings Valley | 299 | 264 | 326 | | 55 | 324 | 87/100 | 85/100 |
|---|---|---|---|---|---|---|---|---|
| Lenox + pen | 279 | 291 | 291 | 9 | 0 | 309 | 99/100 | 97/100 |
| **Mesopotamia** | **332** | **355** | **321** | | **92** | **316** | **90/100** | **88/100** |
| Meylersfield + pen | 521 | 504 | 501 | | 32 | 439 | 73/100 | 74/100 |
| Midgham + pen | 321 | 303 | 270 | | 28 | 232 | 93/100 | 98/100 |
| Mount Eagle | 225 | 300 | 304 | | 64 | 255 | 100/100 | 81/100 |
| New Hope | 216 | 212 | 192 | | 55 | 123 | 88/100 | 86/100 |
| Paul Island | 215 | 214 | 275 | | 104 | 164 | 105/100 | 84/100 |
| Petersfield | 271 | 236 | 295 | | 136 | 305 | 95/100 | 97/100 |
| Retreat | 345 | 376 | 365 | | 51 | 303 | 80/100 | 80/100 |
| Shrewsbury | 210 | 280 | 258 | | 42 | 329 | 88/100 | 70/100 |
| Spring Garden | 519 | 474 | 437 | | 100 | 324 | 78/100 | 79/100 |
| Springfield etc. | 320 | 247 | 131 | | 31 | 134 | 130/100 | 131/100 |
| Three Mile River | 279 | 238 | 218 | | 47 | 170 | 91/100 | 79/100 |
| **Totals** | **8,534** | **8,462** | **8,577** | **33** | **1,560** | **7,069** | **89/100** | **83/100** |

*Sources:* Westmoreland poll tax returns, 1807, 1814 (Barham b 34); Westmoreland slave registration returns, 1817, 1820, 1823, 1826, 1829, 1832, T 71/178–189, NA-UK; Westmoreland Slave Compensation Returns, 1834, T 71/723–724, NA-UK.

*Appendix 10* The Workforces on Twenty-Seven Westmoreland Sugar Estates, August 1, 1834

| | | HP | | Trade | Field | | Dom. | | Child | | Old | | | |
|---|---|---|---|---|---|---|---|---|---|---|---|---|---|---|
| Estate name | Slaves | M | F | M | M | F | M | F | M | F | M | F | Male | Female |
| Bath | 210 | 8 | 3 | 12 | 53 | 72 | 1 | 3 | 11 | 24 | 13 | 10 | 98 | 112 |
| Black Heath | 353 | 11 | 3 | 24 | 91 | 144 | 0 | 7 | 21 | 18 | 7 | 27 | 154 | 199 |
| Blackness | 263 | 6 | 2 | 14 | 66 | 96 | 4 | 10 | 21 | 16 | 0 | 28 | 111 | 152 |
| Blue Castle | 204 | 8 | 3 | 18 | 45 | 63 | 2 | 13 | 12 | 13 | 5 | 22 | 90 | 114 |
| Bog | 260 | 8 | 1 | 9 | 92 | 86 | 0 | 9 | 18 | 19 | 2 | 16 | 129 | 131 |
| Canaan | 200 | 6 | 2 | 7 | 55 | 71 | 0 | 4 | 8 | 16 | 12 | 19 | 88 | 112 |
| Carawina | 318 | 9 | 1 | 10 | 91 | 118 | 9 | 20 | 22 | 24 | 5 | 9 | 146 | 172 |
| Cornwall | 243 | 8 | 2 | 19 | 55 | 85 | 1 | 6 | 15 | 12 | 8 | 32 | 106 | 137 |
| Delve | 184 | 5 | 1 | 13 | 41 | 60 | 2 | 4 | 9 | 12 | 12 | 25 | 82 | 102 |
| Fontabelle etc. | 246 | 9 | 1 | 13 | 57 | 88 | 1 | 12 | 16 | 14 | 13 | 22 | 109 | 137 |
| Fort William etc. | 344 | 11 | 4 | 21 | 73 | 138 | 7 | 19 | 16 | 22 | 9 | 24 | 137 | 207 |
| Friendship | 262 | 10 | 1 | 21 | 51 | 90 | 4 | 12 | 19 | 18 | 10 | 26 | 115 | 147 |
| Georges Plain | 255 | 8 | 2 | 11 | 77 | 73 | 2 | 5 | 8 | 11 | 18 | 40 | 124 | 131 |
| Kings Valley | 324 | 10 | 3 | 16 | 86 | 114 | 3 | 13 | 16 | 11 | 18 | 34 | 149 | 175 |
| Lenox etc. | 309 | 9 | 3 | 13 | 98 | 94 | 4 | 19 | 26 | 29 | 2 | 12 | 152 | 157 |
| **Mesopotamia** | **316** | **11** | **3** | **20** | **81** | **90** | **1** | **13** | **20** | **24** | **15** | **38** | **148** | **168** |
| Meylersfield | 439 | 12 | 2 | 25 | 116 | 183 | 3 | 10 | 18 | 21 | 12 | 37 | 186 | 253 |
| Midgham | 232 | 10 | 1 | 11 | 65 | 80 | 1 | 6 | 16 | 14 | 12 | 16 | 115 | 117 |
| Mount Eagle | 255 | 7 | 4 | 16 | 71 | 102 | 0 | 13 | 15 | 13 | 5 | 9 | 114 | 141 |
| New Hope | 123 | 3 | 2 | 9 | 29 | 50 | 0 | 1 | 5 | 4 | 11 | 9 | 57 | 66 |
| Paul Island | 164 | 8 | 2 | 10 | 44 | 58 | 0 | 6 | 7 | 9 | 6 | 14 | 75 | 89 |
| Petersfield | 305 | 10 | 2 | 19 | 87 | 113 | 4 | 8 | 15 | 15 | 15 | 17 | 150 | 155 |
| Retreat | 303 | 11 | 3 | 18 | 72 | 114 | 0 | 4 | 22 | 21 | 12 | 26 | 135 | 168 |
| Shrewsbury | 329 | 10 | 2 | 20 | 70 | 112 | 4 | 13 | 21 | 27 | 10 | 40 | 135 | 194 |
| Spring Garden | 324 | 11 | 3 | 16 | 80 | 126 | 1 | 6 | 19 | 12 | 16 | 34 | 143 | 181 |
| Springfield | 134 | 8 | 0 | 8 | 49 | 39 | 1 | 3 | 7 | 5 | 3 | 11 | 76 | 58 |
| Three Mile River | 170 | 9 | 3 | 13 | 38 | 57 | 2 | 10 | 9 | 9 | 4 | 16 | 75 | 95 |
| **Totals** | **7,069** | **236** | **59** | **406** | **1,833** | **2,516** | **57** | **249** | **412** | **433** | **255** | **613** | **3,199** | **3,870** |
| **Percentage** | | **3.3** | **0.8** | **5.8** | **25.9** | **35.6** | **0.8** | **3.5** | **5.8** | **6.1** | **3.6** | **8.7** | **45.3** | **54.7** |

*Source:* Westmoreland Slave Compensation Returns, 1834, T 71/723–724, NA-UK.

*Note:* HP = head people, or supervisors; Trade = tradesmen or craft workers; Field = field laborers; Dom. = domestics; Child = nonworking boys and girls under age 6; Old = the nonworking sick and elderly.

*Appendix 11* Slaveholding in Six Virginia Counties, 1810

| County | Total white households (WH) | 0 slaves | % WH | 1–5 slaves | % WH | 6–24 slaves | % WH | 25–99 slaves | % WH | 100+ slaves | % WH |
|---|---|---|---|---|---|---|---|---|---|---|---|
| York in Tidewater | 390 | 89 | 23 | 139 | 36 | 141 | 36 | 21 | 5 | 0 | 0 |
| **Richmond in Tidewater** | **580** | **258** | **45** | **180** | **31** | **114** | **20** | **26** | **4** | **2** | **0** |
| Amelia in Piedmont | 595 | 107 | 18 | 163 | 27 | 252 | 42 | 69 | 12 | 4 | 1 |
| Amherst in Piedmont | 894 | 332 | 37 | 244 | 27 | 279 | 31 | 39 | 5 | 0 | 0 |
| Fauquier in Piedmont | 2,108 | 931 | 44 | 611 | 29 | 485 | 23 | 78 | 4 | 3 | 0 |
| Frederick in Shenandoah valley | 2,559 | 1,733 | 68 | 517 | 20 | 263 | 10 | 42 | 2 | 4 | 0 |
| Totals | 7,126 | 3,450 | 48 | 1,854 | 26 | 1,534 | 22 | 275 | 4 | 13 | 0 |

| County | Total whites | Total slaves | 1–5 slaves | % slaves | 6–24 slaves | % slaves | 25–99 slaves | % slaves | 100+ slaves | % slaves |
|---|---|---|---|---|---|---|---|---|---|---|
| York in Tidewater | 2,293 | 2,894 | 377 | 13 | 1,592 | 55 | 925 | 32 | 0 | 0 |
| **Richmond in Tidewater** | **2,775** | **3,178** | **474** | **15** | **1,388** | **44** | **939** | **29** | **377** | **12** |
| Amelia in Piedmont | 3,365 | 7,229 | 409 | 6 | 3,131 | 43 | 3,140 | 43 | 549 | 8 |
| Amherst in Piedmont | 5,371 | 5,177 | 620 | 12 | 3,352 | 65 | 1,205 | 23 | 0 | 0 |
| Fauquier in Piedmont | 12,333 | 10,356 | 1,497 | 14 | 5,569 | 54 | 2,810 | 27 | 480 | 5 |
| Frederick in Shenandoah valley | 16,075 | 6,499 | 1,175 | 18 | 2,899 | 45 | 1,708 | 26 | 717 | 11 |
| Totals | 42,212 | 35,333 | 4,552 | 13 | 17,931 | 51 | 10,727 | 30 | 2,123 | 6 |

*Note:* All data is compiled from the manuscript schedules for the 1810 Virginia census, NA-US.

# *Appendix 12* Sarah Affir's Mesopotamia Family Tree

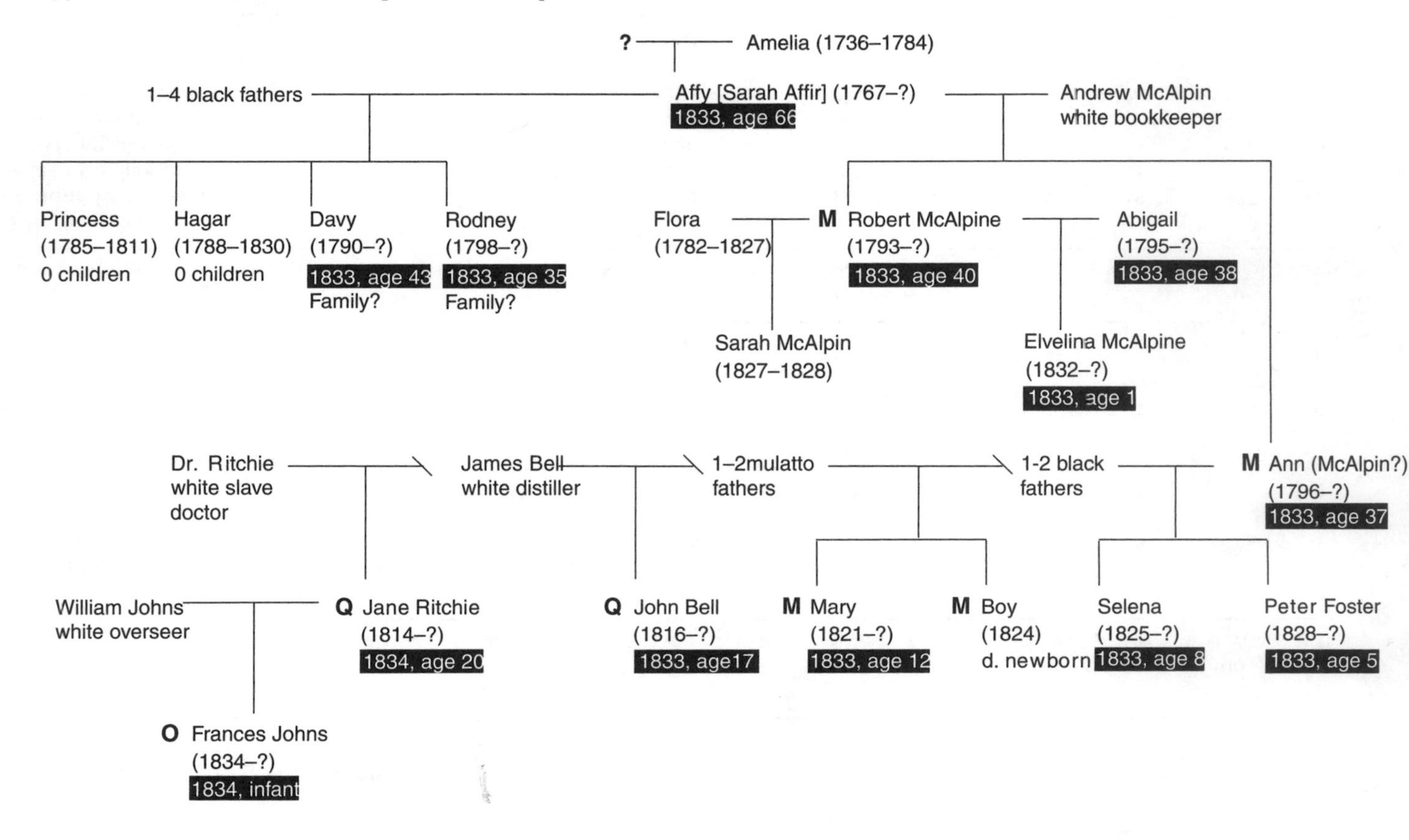

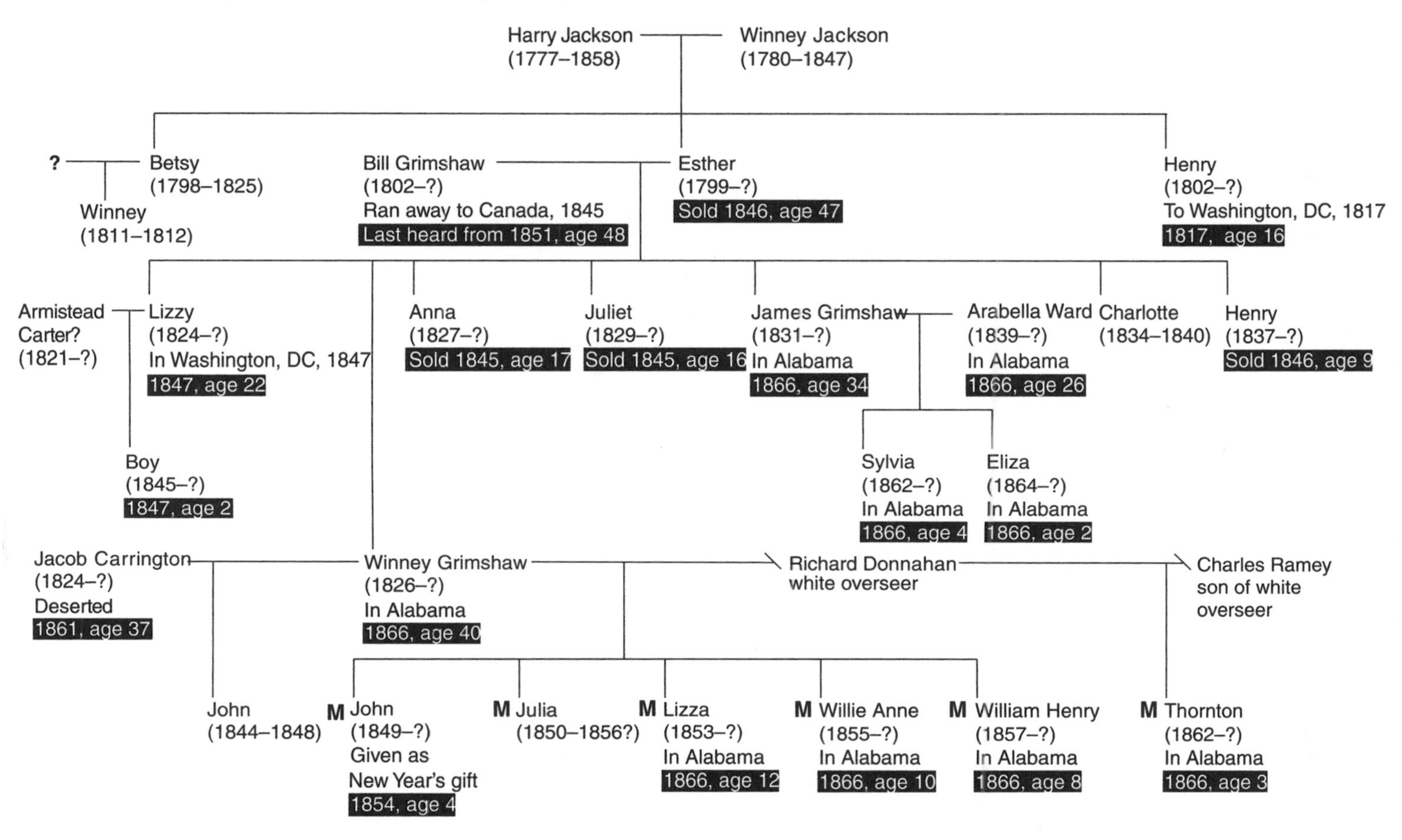
Harry Jackson
(1777–1858)
Winney Jackson
(1780–1847)
?
Betsy
(1798–1825)
Winney
(1811–1812)
Bill Grimshaw
(1802–?)
Ran away to Canada, 1845
Last heard from 1851, age 48
Esther
(1799–?)
Sold 1846, age 47
Henry
(1802–?)
To Washington, DC, 1817
1817, age 16
Armistead
Carter?
(1821–?)
Lizzy
(1824–?)
In Washington, DC, 1847
1847, age 22
Anna
(1827–?)
Sold 1845, age 17
Juliet
(1829–?)
Sold 1845, age 16
James Grimshaw
(1831–?)
In Alabama
1866, age 34
Arabella Ward
(1839–?)
In Alabama
1866, age 26
Charlotte
(1834–1840)
Henry
(1837–?)
Sold 1846, age 9
Boy
(1845–?)
1847, age 2
Sylvia
(1862–?)
In Alabama
1866, age 4
Eliza
(1864–?)
In Alabama
1866, age 2
Jacob Carrington
(1824–?)
Deserted
1861, age 37
Winney Grimshaw
(1826–?)
In Alabama
1866, age 40
Richard Donnahan
white overseer
Charles Ramey
son of white
overseer
John
(1844–1848)
M John
(1849–?)
Given as
New Year's gift
1854, age 4
M Julia
(1850–1856?)
M Lizza
(1853–?)
In Alabama
1866, age 12
M Willie Anne
(1855–?)
In Alabama
1866, age 10
M William Henry
(1857–?)
In Alabama
1866, age 8
M Thornton
(1862–?)
In Alabama
1866, age 3

*Appendix 14* Slave Population, Sugar Production, and Crop Value at Mesopotamia, 1751–1832

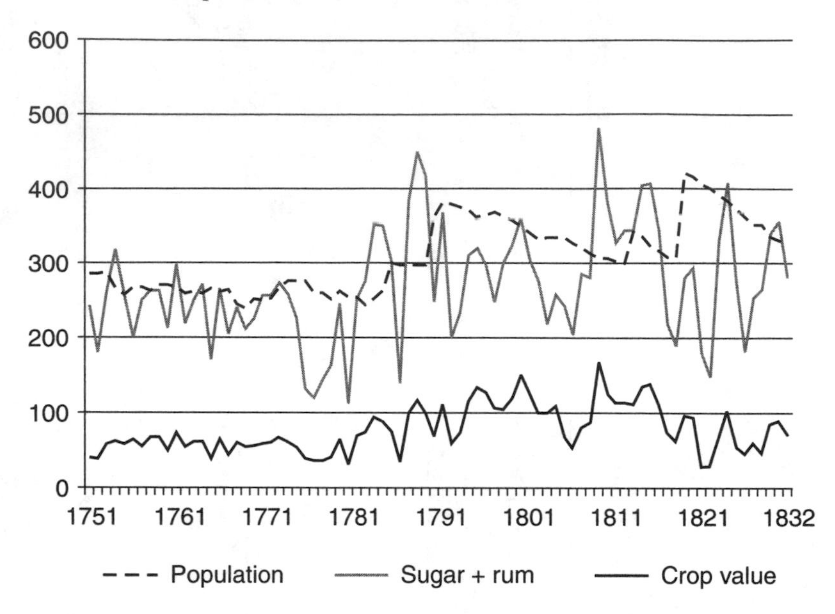


*Note:* The population line tracks changes in the Mesopotamia slave population from 1751 to 1832. The productivity line combines the annual totals for sugar and rum exports, reckoning each puncheon of rum as equivalent to three-fifths of a hogshead of sugar, which was the usual case at Mesopotamia. The crop value line gives the market value in hundreds of £ Jamaican currency. Crop valuations are estimated for 1777, 1824, and 1828–1832, because there are no valuations for those years in the Mesopotamia crop accounts.

*Appendix 15* The Primary Occupations of 877 Adult Mesopotamia Slaves, 1762–1833

| | Males | | Females | | Total workers | |
|---|---|---|---|---|---|---|
| Occupation | No. | % | No. | % | No. | % |
| Driver | 17 | 89 | 2 | 11 | 19 | 2.1 |
| Sugar worker | 23 | 100 | 0 | 0 | 23 | 2.6 |
| Craft worker | 71 | 100 | 0 | 0 | 71 | 8.1 |
| Stock keeper | 29 | 100 | 0 | 0 | 29 | 3.3 |
| Carter | 21 | 100 | 0 | 0 | 21 | 2.4 |
| Field worker | 256 | 46 | 305 | 54 | 561 | 64.0 |
| Domestic | 5 | 23 | 17 | 4 | 22 | 2.5 |
| Marginal worker | 46 | 57 | 35 | 43 | 81 | 9.3 |
| Nonworker | 17 | 33 | 33 | 66 | 50 | 5.7 |
| Totals | 485 | 55 | 392 | 45 | 877 | 100.0 |

*Appendix 16* The Occupation, Health, and Longevity of Mesopotamia Adult Slaves, 1762–1833

| | No. | Years prime job | Years marginal job(s) | Years at work | Years able | Years weak | Years invalid | No. died | Age at death |
|---|---|---|---|---|---|---|---|---|---|
| Male occupations | | | | | | | | | |
| Driver | 17 | 12.6 | 11.5 | 24.1 | 15.5 | 8.6 | 2.7 | 11 | 60.6 |
| Sugar worker | 23 | 9.6 | 13.8 | 23.4 | 11.7 | 11.7 | 1.1 | 23 | 53.4 |
| Craft worker | 71 | 17.5 | 2.3 | 19.8 | 13.4 | 6.4 | 1.3 | 50 | 43.9 |
| Stock keeper | 29 | 17.7 | 5.4 | 23.1 | 12.4 | 10.7 | 1.2 | 23 | 45.2 |
| Carter | 21 | 16.2 | 3.3 | 19.5 | 11.6 | 7.9 | 0.3 | 16 | 40.4 |
| Field worker | 256 | 11.9 | 5.8 | 17.7 | 9.0 | 8.7 | 0.9 | 178 | 42.0 |
| Domestic | 5 | 6.4 | 1.2 | 7.6 | 6.4 | 1.2 | 0.2 | 2 | 30.5 |
| Marginal worker | 46 | 0.0 | 12.0 | 12.0 | 0.5 | 11.5 | 1.9 | 39 | 55.1 |
| Nonworker | 17 | 0.0 | 0.0 | 0.0 | 0.0 | 0.0 | 3.9 | 17 | 41.2 |
| Totals | 485 | 11.9 | 5.8 | 17.7 | 9.2 | 8.5 | 1.2 | 359 | 45.0 |
| Female occupations | | | | | | | | | |
| Driver | 2 | 24.0 | 12.5 | 36.5 | 11.0 | 25.5 | 5.0 | 1 | 68.0 |
| Field worker | 305 | 14.5 | 5.1 | 19.6 | 10.2 | 9.4 | 2.5 | 191 | 44.9 |
| Domestic | 17 | 16.4 | 4.9 | 21.3 | 13.5 | 7.8 | 1.3 | 7 | 45.1 |
| Marginal worker | 35 | 0.0 | 11.0 | 11.0 | 0.6 | 10.4 | 5.6 | 28 | 60.6 |
| Nonworker | 33 | 0.0 | 0.0 | 0.0 | 0.0 | 0.0 | 8.7 | 31 | 61.6 |
| Totals | 392 | 12.1 | 5.2 | 17.3 | 8.6 | 8.7 | 3.3 | 258 | 48.5 |

*Appendix 17* Mesopotamia Adult Field Workers by Gender, 1762–1832

| | 1762 | 1772 | 1782 | 1792 | 1802 | 1812 | 1822 | 1832 |
|---|---|---|---|---|---|---|---|---|
| Slave population | 270 | 250 | 254 | 361 | 352 | 306 | 417 | 331 |
| % Women | 44% | 42% | 43% | 44% | 49% | 52% | 51% | 52% |
| Total workers | 176 | 197 | 218 | 278 | 275 | 246 | 316 | 257 |
| % Women | 39% | 35% | 41% | 41% | 47% | 50% | 49% | 50% |
| Adult field workers | 84 | 89 | 106 | 134 | 94 | 106 | 161 | 113 |
| % Women | 52% | 49% | 45% | 49% | 74% | 70% | 62% | 57% |
| First gang hands | 47* | 60* | 49* | 85* | 51 | 43 | 105 | 73 |
| Men | 25* | 33* | 30* | 44* | 21 | 15 | 48 | 36 |
| Women | 22* | 27* | 19* | 41* | 30 | 28 | 57 | 37 |
| % Female | 47%* | 45%* | 39%* | 48%* | 59% | 65% | 54% | 51% |

*Estimated.

*Appendix 18* Stated Causes of Slave Deaths at Mesopotamia (1762–1832) and at Worthy Park (1792–1838)

| | Mesopotamia | | | | | | Worthy Park | |
|---|---|---|---|---|---|---|---|---|
| | Male | | Female | | Total | | | |
| Cause | No. | % | No. | % | No. | % | No. | % |
| Debility, invalid | 44 | 10.6 | 62 | 20.2 | 106 | 14.7 | 43 | 10.7 |
| Old age | 50 | 12.1 | 55 | 17.9 | 105 | 14.6 | 89 | 22.2 |
| Tuberculosis | 30 | 07.3 | 21 | 06.8 | 51 | 07.0 | 15 | 03.8 |
| Dropsy | 35 | 08.5 | 14 | 04.5 | 49 | 06.8 | 38 | 09.5 |
| Flux, dysentery, diarrhea | 22 | 05.3 | 19 | 06.2 | 41 | 05.7 | 35 | 08.7 |
| Yaws, ulcers | 31 | 07.5 | 9 | 02.9 | 40 | 05.5 | 24 | 06.0 |
| Fever | 11 | 02.7 | 23 | 07.5 | 34 | 04.7 | 26 | 06.5 |
| Pleurisy | 25 | 06.0 | 5 | 01.6 | 30 | 04.2 | 10 | 02.5 |
| Accidents | 18 | 04.3 | 6 | 02.0 | 24 | 03.3 | 14 | 03.5 |
| Bowel complaint | 10 | 02.4 | 13 | 04.2 | 23 | 03.2 | 1 | 00.2 |
| Apoplexy, stroke, palsy | 19 | 04.6 | 2 | 00.7 | 21 | 02.9 | 5 | 01.3 |
| Lung, chest disease | 15 | 03.6 | 3 | 00.9 | 18 | 02.5 | 5 | 01.3 |
| Fits, convulsions, epilepsy | 13 | 03.1 | 4 | 01.3 | 17 | 02.4 | 7 | 01.8 |
| Leprosy | 11 | 02.7 | 6 | 02.0 | 17 | 02.4 | 2 | 00.5 |
| Geophagy | 11 | 02.7 | 6 | 02.0 | 17 | 02.4 | 9 | 02.2 |
| Smallpox | 9 | 02.2 | 6 | 02.0 | 15 | 02.1 | 2 | 00.5 |
| Venereal disease | 7 | 01.7 | 6 | 02.0 | 13 | 01.8 | 0 | 00.0 |
| Suddenly, act of God | 5 | 01.2 | 5 | 01.6 | 10 | 01.4 | 7 | 01.8 |
| Maternity | 0 | 00.0 | 10 | 03.3 | 10 | 01.4 | 9 | 02.2 |
| Cold, influenza, pneumonia | 5 | 01.2 | 4 | 01.3 | 9 | 01.2 | 14 | 03.5 |
| Whooping cough, measles | 4 | 01.0 | 5 | 01.6 | 9 | 01.2 | 9 | 02.2 |
| Encephalitis | 4 | 01.0 | 4 | 01.3 | 8 | 01.1 | 2 | 00.5 |
| Tetanus | 4 | 01.0 | 3 | 00.9 | 7 | 01.0 | 2 | 00.5 |
| Worms, worm fever | 3 | 00.7 | 4 | 01.3 | 7 | 01.0 | 14 | 03.5 |
| Bloated | 5 | 01.2 | 2 | 00.7 | 7 | 01.0 | 4 | 01.0 |
| Bladder, liver complaint | 4 | 01.0 | 1 | 00.3 | 5 | 00.7 | 0 | 00.0 |
| Rheumatism | 2 | 00.5 | 2 | 00.7 | 4 | 00.6 | 0 | 00.0 |
| Insane | 3 | 00.7 | 1 | 00.3 | 4 | 00.6 | 0 | 00.0 |
| Lethargy, sullen | 3 | 00.7 | 0 | 00.0 | 3 | 00.4 | 0 | 00.0 |
| Suicide, poisoned | 1 | 00.2 | 0 | 00.0 | 1 | 00.1 | 3 | 00.7 |
| Ruptured | 1 | 00.2 | 0 | 00.0 | 1 | 00.1 | 3 | 00.7 |
| Killed | 1 | 00.2 | 0 | 00.0 | 1 | 00.1 | 1 | 00.2 |
| Other | 8 | 01.9 | 6 | 02.0 | 14 | 01.9 | 8 | 02.0 |
| Total | 414 | 100.0 | 307 | 100.0 | 721 | 100.0 | 401 | 100.0 |

*Appendix 19* Recorded Births at Mesopotamia (1774–1833) and at Mount Airy (1809–1863)

| | Mesopotamia | | Mount Airy | |
|---|---|---|---|---|
| Number of children | Women | Total children | Women | Total children |
| 0 | 147 | 0 | 44 | 0 |
| 1 | 51 | 51 | 23 | 23 |
| 2 | 21 | 42 | 27 | 54 |
| 3 | 21 | 63 | 22 | 66 |
| 4 | 12 | 48 | 11 | 44 |
| 5 | 5 | 25 | 18 | 90 |
| 6 | 8 | 48 | 17 | 102 |
| 7 | 5 | 35 | 8 | 56 |
| 8 | 1 | 8 | 7 | 56 |
| 9 | 1 | 9 | 6 | 54 |
| 10 | 1 | 10 | 2 | 20 |
| 11 | 1 | 11 | 2 | 22 |
| 12 | 1 | 12 | 3 | 36 |
| 13 | 0 | 0 | 1 | 13 |
| 14 | 1 | 14 | 0 | 0 |
| Totals | 276 | 376 | 191 | 636 |

*Note:* This table includes births in Alabama by Mount Airy women.

*Appendix 20* Motherhood at Mesopotamia in 1802

| Occupation | Number of women | Mothers | | Children | | Children per mother |
|---|---|---|---|---|---|---|
| | | No. | % | No. | % | |
| First field gang | 30 | 24 | 37 | 107 | 48 | 4.5 |
| Second field gang | 30 | 14 | 22 | 45 | 20 | 3.2 |
| Domestics | 4 | 3 | 5 | 21 | 9 | 7.0 |
| Marginal workers | 37 | 16 | 25 | 28 | 13 | 1.8 |
| Invalids | 18 | 7 | 11 | 23 | 10 | 3.3 |
| Totals | 119 | 64 | 100 | 224 | 100 | 3.5 |

*Note:* Occupation is as of Jan. 1, 1802, when there were 119 women aged sixteen or older at Mesopotamia. All 224 of their recorded births are included in the tabulation.

*Appendix 21* Comparison of 678 Mesopotamia Adults by Gender, Color, and Origin, 1763–1832

| | No. | Entry age | Yrs. prime job | Yrs. sec. job(s) | Total yrs. worked | Yrs. able | Yrs. weak | Yrs. invalid | Died | Age at death |
|---|---|---|---|---|---|---|---|---|---|---|
| Males | | | | | | | | | | |
| Mulattoes and quadroons | 18 | 17.3 | 11.0 | 1.4 | 12.4 | 9.5 | 2.9 | 0.9 | 10 | 31.8 |
| Blacks born at Mesopotamia | 127 | 16.0 | 15.4 | 6.0 | 21.4 | 13.9 | 7.5 | 1.0 | 72 | 40.4 |
| Blacks from Jamaica | 136 | 28.6 | 8.9 | 5.3 | 14.2 | 7.1 | 7.1 | 1.1 | 79 | 45.0 |
| Blacks from Africa | 94 | 19.0 | 13.4 | 7.8 | 21.2 | 10.0 | 11.2 | 0.9 | 88 | 40.3 |
| Totals | 375 | 21.4 | 12.4 | 5.9 | 18.3 | 10.2 | 8.1 | 1.0 | 249 | 41.5 |
| Females | | | | | | | | | | |
| Mulattoes and quadroons | 9 | 20.8 | 11.5 | 0.2 | 11.7 | 9.7 | 2.0 | 1.6 | 3 | 45.7 |
| Blacks born at Mesopotamia | 124 | 16.0 | 16.3 | 4.6 | 20.9 | 13.7 | 7.2 | 2.1 | 59 | 38.1 |
| Blacks from Jamaica | 129 | 32.3 | 9.3 | 4.8 | 14.1 | 5.6 | 8.5 | 2.2 | 75 | 47.9 |
| Blacks from Africa | 41 | 17.6 | 18.5 | 6.3 | 24.8 | 13.4 | 11.4 | 2.5 | 33 | 41.1 |
| Totals | 303 | 23.3 | 13.5 | 4.8 | 18.3 | 10.1 | 8.2 | 2.2 | 170 | 43.2 |

*Appendix 22* The Mount Airy Slave Force in 1809

| | Male | | | | Female | | | | Total | |
|---|---|---|---|---|---|---|---|---|---|---|
| | No. men | No. boys | Total | % | No. women | No. girls | Total | % | No. | % |
| Home plantation | | | | | | | | | | |
| Domestic servants | 7 | 0 | 7 | 03.2 | 9 | 1 | 10 | 06.1 | 17 | 04.4 |
| Stable hands | 7 | 0 | 7 | 03.2 | 0 | 0 | 0 | 00.0 | 7 | 01.8 |
| Gardeners | 4 | 1 | 5 | 02.3 | 0 | 0 | 0 | 00.0 | 5 | 01.3 |
| Textile workers | 3 | 0 | 3 | 01.4 | 12 | 2 | 14 | 08.5 | 17 | 04.4 |
| Shoemakers | 3 | 1 | 4 | 01.8 | 0 | 0 | 0 | 00.0 | 4 | 01.1 |
| Carpenters | 8 | 3 | 11 | 05.0 | 0 | 0 | 0 | 00.0 | 11 | 02.9 |
| Joiners | 3 | 1 | 4 | 01.8 | 0 | 0 | 0 | 00.0 | 4 | 01.1 |
| Blacksmiths | 5 | 0 | 5 | 02.3 | 0 | 0 | 0 | 00.0 | 5 | 01.3 |
| Masons | 2 | 1 | 3 | 01.4 | 0 | 0 | 0 | 00.0 | 3 | 00.8 |
| Sailors | 4 | 0 | 4 | 01.8 | 0 | 0 | 0 | 00.0 | 4 | 01.1 |
| Jobbers | 6 | 1 | 7 | 03.2 | 0 | 0 | 0 | 00.0 | 7 | 01.8 |
| Miller | 1 | 0 | 1 | 00.5 | 0 | 0 | 0 | 00.0 | 1 | 00.2 |
| Nonworking: too young | 0 | 8 | 8 | 03.7 | 0 | 13 | 13 | 07.9 | 21 | 05.5 |
| | 53 | 16 | 69 | 31.6 | 21 | 16 | 37 | 22.5 | 106 | 27.7 |
| Farm quarters | | | | | | | | | | |
| Domestic servants | 1 | 0 | 1 | 00.5 | 6 | 0 | 6 | 03.7 | 7 | 01.8 |
| Blacksmiths | 2 | 0 | 2 | 00.9 | 0 | 0 | 0 | 00.0 | 2 | 00.5 |
| Field hands | 62 | 27 | 89 | 40.8 | 46 | 18 | 64 | 39.0 | 153 | 40.0 |
| Miller | 1 | 0 | 1 | 00.5 | 0 | 0 | 0 | 00.0 | 1 | 00.3 |
| Nonworking: too young | 0 | 52 | 52 | 23.9 | 0 | 50 | 50 | 30.5 | 102 | 26.7 |
| Nonworking: too old | 4 | 0 | 4 | 01.8 | 7 | 0 | 7 | 04.3 | 11 | 02.9 |
| | 70 | 79 | 149 | 68.4 | 59 | 68 | 127 | 77.5 | 276 | 72.3 |
| Totals | 123 | 95 | 218 | 100.0 | 80 | 84 | 164 | 100.0 | 382 | 100.0 |

*Appendix 23* Primary Occupations of the 542 Adult Slaves at Mount Airy and the 877 Adult Slaves at Mesopotamia

| | Males | | Females | | Total | |
|---|---|---|---|---|---|---|
| Occupation | No. | % | No. | % | No. | % |
| Mount Airy | | | | | | |
| Domestic servant | 28 | 44 | 35 | 56 | 63 | 11.6 |
| Craft worker | 72 | 100 | 0 | 0 | 72 | 13.3 |
| Cloth worker | 3 | 8 | 35 | 92 | 38 | 7.0 |
| Sailor | 11 | 100 | 0 | 0 | 11 | 2.0 |
| Carter | 1 | 100 | 0 | 0 | 1 | 0.2 |
| Field worker | 182 | 54 | 152 | 46 | 334 | 61.6 |
| Marginal worker | 14 | 93 | 1 | 7 | 15 | 2.8 |
| Nonworker | 5 | 63 | 3 | 37 | 8 | 1.5 |
| Totals | 316 | 58 | 226 | 42 | 542 | 100.0 |
| Mesopotamia | | | | | | |
| Domestic servant | 5 | 23 | 17 | 4 | 22 | 2.5 |
| Driver | 17 | 89 | 2 | 11 | 19 | 2.1 |
| Sugar worker | 23 | 100 | 0 | 0 | 23 | 2.6 |
| Craft worker | 71 | 100 | 0 | 0 | 71 | 8.1 |
| Stock keeper | 29 | 100 | 0 | 0 | 29 | 3.3 |
| Carter | 21 | 100 | 0 | 0 | 21 | 2.4 |
| Field worker | 256 | 46 | 305 | 54 | 561 | 64.0 |
| Marginal worker | 46 | 57 | 35 | 43 | 81 | 9.3 |
| Nonworker | 17 | 33 | 33 | 66 | 50 | 5.7 |
| Totals | 485 | 55 | 392 | 45 | 877 | 100.0 |

*Appendix 24* Harvest Teams at Doctor's Hall and Forkland, Mount Airy, 1816

The Doctor's Hall Harvest Team, 1816

| Cradler | Occupation | Age | Raker | Occupation | Age |
|---|---|---|---|---|---|
| 1 | Carpenter Charles | 49 | 1 | Spinner Lizzie | 35 |
| 2 | Carpenter Andrew | 42 | 2 | Spinner Winney | 20 |
| 3 | Carpenter Billy | 29 | 3 | Spinner Esther | 17 |
| 4 | Carpenter Moses | 20 | 4 | Spinner Jane | 27 |
| 5 | Carpenter Tom Spence | 32 | 5 | Free Judy | — |
| | | | 6 | Free Betsy | — |
| | | | 7 | Shoemaker Joe | 49 |

The Forkland Harvest Team, 1816

| Cradler | Occupation | Age | Raker | Occupation | Age |
|---|---|---|---|---|---|
| 1 | Joiner James | 31 | 1 | Marske George | 49 |
| 2 | House John | 31 | 2 | Weaver Jinney | 16 |
| 3 | Fork Tom | 33 | 3 | Spinner Nanny | 24 |
| 4 | Shoemaker Little Joe | 18 | 4 | Spinner Elsy | 32 |
| 5 | Blacksmith Jarrett | 21 | 5 | Spinner Eve | 29 |
| 6 | Carpenter Henry | 19 | 6 | Joiner Jemmy | 16 |

| | Age | Quarter | Occupation | Value 1810 | Comment | At Cloverdale, 1817 | Age | Value 1817 |
|---|---|---|---|---|---|---|---|---|
| Males | | | | | | | | |
| Cambridge | 22 | Doctor's Hall | Field | £70 | | Furnace filler | 26 | $500 |
| Charles | 39 | Oaken Brow | Field | £80 | Husband of Milly | Keeper | 50 | $400 |
| Dean | 18 | Old House | Field | £70 | | | | |
| Aleck | 23 | Forkland | Field | £80 | Family by 1817 | Stoker | 27 | $500 |
| Farmer | 17 | Doctor's Hall | Field | £70 | | Gutterman | 25 | $500 |
| George | 31 | Oaken Brow | Field | £80 | Family by 1817 | Miller | 45 | $450 |
| Harrington | 16 | Doctor's Hall | Field | £60 | | Furnace filler | 22 | $500 |
| Harry | 15 | Doctor's Hall | Field | £60 | | Bank hand | 19 | $500 |
| Jacob | 24 | Mount Airy | Jobber | £80 | | Wagoner | 27 | $500 |
| James | 21 | Hopyard | Field | £80 | | Collier | 27 | $500 |
| Joe | 24 | Hopyard | Field | £80 | | Wagoner | 38 | $500 |
| Joe | 21 | Doctor's Hall | Field | £70 | | | | |
| Joe | 8 | Oaken Brow | Not working | £50 | Son of Milly | Wheelwright | 14 | $600 |
| John | 26 | Hopyard | Field | £80 | | Collier | 32 | $500 |
| Lewis | 18 | Old House | Field | £70 | | Blacksmith | 26 | $600 |
| Michael | 19 | Gwinfield | Field | £80 | | | | |
| Nelson | 17 | Old House | Field | £70 | | Wood hauler | 23 | $500 |
| Paul | 18 | Gwinfield | Field | £80 | | | | |
| Peter | 16 | Old House | Field | £60 | Family by 1817 | House, now Woodcutter | 21 | $500 |
| Phil | 21 | Doctor's Hall | Field | £70 | Family by 1817 | Wood hauler | 30 | $500 |
| Phil | 18 | Gwinfield | Field | £80 | | | | |
| Prince | 30 | Old House | Field | £80 | Marries DH Rose | Collier | 32 | $500 |
| Ralph | 28 | Gwinfield | Field | £80 | | Collier | 30 | $500 |
| Tom | 41 | Mount Airy | Blacksmith | £80 | | Blacksmith | 55 | $400 |
| Females | | | | | | | | |
| Bridget | 22 | Doctor's Hall | Field | £60 | Son died, 1810 | Field | 27 | $400 |
| Delphy | 14 | Gwinfield | Field | £60 | | | | |
| Fanny | 2 | Oaken Brow | Not working | £30 | Daughter of Milly | Not working | 8 | $250 |
| Franky | 14 | Doctor's Hall | Field | £50 | | | | |
| Kate | 13 | Old House | Field | £50 | | | | |
| Milly | 38 | Oaken Brow | Field | £70 | Wife of Charles | Spinner | 40 | $350 |
| Milly | 6 | Oaken Brow | Not working | £40 | Daughter of Milly | Field | 18 | $400 |
| Nancy | 35 | Doctor's Hall | Field | £60 | Child at Oaken Brow | | | |
| Rose | 17 | Doctor's Hall | Field | £60 | 3 children by 1817 | Field | 28 | $400 |
| Rose | 14 | Gwinfield | Field | £60 | | Field | 17 | $400 |
| Sucky | 14 | Gwinfield | Field | £60 | | | | |
| Sucky | 13 | Old House | Field | £50 | | | | |

*Appendix 26* Mount Airy Slaves Sold in 1816

| Name | Age | Stated age | Quarter | Occupation | Comment | Price | Inventory value |
|---|---|---|---|---|---|---|---|
| Kesiah | 39 | 35 | Hopyard | Field | "Diseased" with 5 children | $150 | £60 |
| Milly | 16 | 12 | Hopyard | Field | Kesiah's daughter (called Patty's), "diseased in mind and body" | $300 | £60 |
| Delsy | 12 | 8 | Hopyard | Not working | Kesiah's daughter | $250 | £50 |
| Betty | 9 | 6 | Hopyard | Not working | Kesiah's daughter | $200 | £50 |
| Davy | 8 | 5 | Hopyard | Not working | Kesiah's son | $250 | £40 |
| Sam | 2 | 2 | Hopyard | Not working | Kesiah's son | $150 | £30 |
| Patty | 44 | 40 | Hopyard | Nurse | 2 of 6 children stayed at Hopyard | $150 | £40 |
| Winney | 20 | 20 | Hopyard | Field | Patty's daughter, "diseased in mind and body" | $300 | £60 |
| Charlotte | 18 | 16 | Oaken Brow | Field | Patty's daughter | $300 | £60 |
| Judy | 7 | 6 | Hopyard | Not working | Patty's daughter | $200 | £40 |
| Rose | 5 | 3 | Hopyard | Not working | Patty's daughter | $175 | £40 |
| Nancy | 34 | 40 | Oaken Brow | Field | Mother of Jenny | $250 | £60 |
| Jenny | 4 | Infant | Oaken Brow | Not working | Nancy's daughter | $100 | £30 |
| Celia | 17 | 18 | Gwinfield | Field | Misnamed Lucy; mother unknown | $350 | £60 |
| Reuben | 28 | 28 | Old House | Field | Husband of Eve | $500 | £80 |
| Eve | 23 | — | Old House | Domestic | Wife of Reuben | $410 | £60 |
| Lucy | 1 | Infant | Old House | Not working | Eve's daughter | 0 | £30 |
| Kate | 22 | — | Forkland | Field | Sold without son Alfred | $400 | £60 |
| Agga | 8 | — | Forkland | Not working | Daughter of Forkland Peggy | $200 | £30 |
| Peter | 40 | — | Mount Airy | Domestic | Wife dead; 6 children stay at Mount Airy | $500 | £100 |
| Tom | 14 | "boy" | Mount Airy | Miller | Mother unknown | $500 | £60 |
| Rachel | 40 | 45 | Gwinfield | Field | Misnamed Sarah; sold without son Bailor | $250 | £60 |
| Agga | 13 | 12 | Gwinfield | Field | Rachel's daughter | $350 | £60 |
| Jesse | 9 | 1 | Gwinfield | Not working | Rachel's daughter, misnamed Cornelius | $475 | £50 |
| Cornelius | 3 | 9 | Gwinfield | Not working | Rachel's son, misnamed Robert | $100 | £20 |
| Fanny | 19 | 20 | Gwinfield | Field | Gwinfield Kate's daughter | $400 | £60 |
| **Total** | | | | | | **$7,210** | £1,350 [$4,495] |

*Appendix 27* Moravian Missionaries and Diarists at Mesopotamia

| Year | Names of missionaries | Diaries (* = in English) |
|---|---|---|
| 1755 | Br. Zachariah George Caries | Caries, July 1755* |
| 1758 | Br. Caries | Caries, Oct.–Dec. 1758 |
| 1759 | Br. Caries | Caries, Jan.–Nov. 1759 |
| 1759 | Br. Christian Rauch & Sr. Anna Rauch, Br. Nicolaus Gandrup | Rauch, Nov.–Dec. 1759 |
| 1760 | Br. & Sr. Rauch, Br. Gandrup, Br. John Levering & Sr. Susanna Levering | Rauch, Jan.–Dec. 1760 |
| 1761 | Br. & Sr. Rauch, Br. Gandrup, Br. & Sr. Levering | Rauch, Jan.–Dec. 1761 |
| 1762 | Br. & Sr. Rauch, Br. Gandrup | |
| 1763 | Br. & Sr. Levering | Levering, Feb.–June 1763 |
| 1764 | Br. & Sr. Levering, Br. Peter Bader & Sr. Mary Bader | Levering/Bader, Jan.–Dec. 1764 |
| 1765 | Br. & Sr. Bader | Bader, Jan.–Dec. 1765 |
| 1766 | Br. Bader, Br. Friederich Schlegel | Schlegel, Jan.–Dec. 1766 |
| 1767 | Br. Joachim Senseman & Sr. Christina Senseman | Schlegel, Jan.–Dec. 1767 |
| 1768 | Br. & Sr. Senseman | Schlegel, Jan.–Dec. 1768 |
| 1769 | Br. & Sr. Senseman | Senseman, Dec. 1768–Aug. 1769 |
| 1770 | Br. John George Schnell & Sr. Helena Schnell | Schnell, Aug. 1769–Feb. 1770 |
| 1771 | Br. John Metcalf & Sr. Metcalf | Schlegel, Jan.–Dec. 1771 |
| 1772 | Br. & Sr. Metcalf | Metcalf, Jan.–Dec. 1772 |
| 1773 | Br. & Sr. Metcalf | Metcalf, Jan.–Dec. 1773* |
| 1774 | Br. & Sr. Metcalf | Metcalf, Jan.–Dec. 1774 |
| 1775 | Br. & Sr. Metcalf | Metcalf, Jan.–Dec. 1775 |
| 1776 | Br. & Sr. Metcalf, Br. Christian Franz | Metcalf/Franz, July–Dec. 1776 |
| 1777 | Br. Metcalf | Metcalf, Jan.–June, Dec. 1777 |
| 1778 | Br. Christian Franz & Sr. Johanna Franz | Franz, March–April, July–Dec. 1778* |
| 1779 | Br. & Sr. Franz | Anon., April–Dec. 1779 |
| 1780 | Br. & Sr. Franz | Anon./Franz, Jan.–Dec., 1780 |
| 1781 | Br. & Sr. Franz, Br. Philip Diemer & Sr. Rebecca Diemer | Franz, Jan.–April 1781 |
| 1782 | Br. & Sr. Diemer, Br. David Taylor & Sr. Johanna Taylor | Taylor, June–Dec. 1782* |
| 1783 | Br. & Sr. Taylor | Taylor, Jan.–Dec. 1783* |
| 1784 | Br. & Sr. Wagner, Br. & Sr. Taylor | Taylor, Jan.–Dec. 1784* |
| 1785 | Br. & Sr. Taylor, Br. Christian Zander & Sr. Sarah Zander | Taylor, Jan.–May 1785* |
| 1786 | Br. & Sr. Zander | Zander, Jan.–Dec. 1786 |
| 1787 | Br. & Sr. Zander | Zander, Jan.–Dec. 1787 |
| 1788 | Br. & Sr. Zander | Zander, Jan.–Dec. 1788 |
| 1789 | Br. & Sr. Zander | Zander, Jan.–Dec. 1789 |
| 1790 | Br. & Sr. Zander, Br. Jonas Herbst & Sr. Anna Herbst | Zander, Jan.–Feb. 1790 |
| 1791 | Br. & Sr. Herbst | |
| 1792 | Br. & Sr. Herbst | |
| 1793 | Br. & Sr. Herbst | |
| 1794 | Br. & Sr. Herbst | |

(*continued*)

*Appendix 27* (continued)

| Year | Names of missionaries | Diaries (* = in English) |
|---|---|---|
| 1795 | Br. & Sr. Herbst | |
| 1796 | Br. & Sr. Herbst | |
| 1797 | | |
| 1798 | Br. Nathaniel Brown & Sr. Elizabeth Brown | |
| 1799 | Br. & Sr. Brown | Brown, Jan.–Dec. 1799* |
| 1800 | | |
| 1801 | Br. Joseph Jackson & Sr. Rachel Jackson | |
| 1802 | Br. & Sr. Jackson | |
| 1803 | Br. & Sr. Jackson | |
| 1804 | Br. & Sr. Jackson | |
| 1805 | Br. & Sr. Jackson | |
| 1806 | Br. & Sr. Jackson | |
| 1807 | Br. & Sr. Jackson | |
| 1808 | Br. & Sr. Jackson, Br. John Lang | Lang, Dec. 1808* |
| 1809 | Br. John Lang | Lang, Nov. 1809* |
| 1810 | Br. John Lang | Lang, Nov. 1810* |
| 1811 | Br. John Samuel Gründer & Sr. Sarah Gründer | |
| 1812 | Br. & Sr. Gründer | Gründer, Jan.–Dec. 1812* |
| 1813 | Br. & Sr. Gründer | Gründer, Jan.–Dec. 1813 |
| 1814 | Br. & Sr. Gründer | |
| 1815 | Br. & Sr. Gründer | Gründer, Jan.–Dec. 1815* |
| 1816 | Br. Gründer | |
| 1817 | Br. & Sr. Gründer | Gründer, Jan.–Dec. 1817 |
| 1818 | Br. & Sr. Gründer, Br. Thomas Ward & Sr. Ward | |
| 1819 | Br. & Sr. Ward | |
| 1820 | | |
| 1821 | | |
| 1822 | | |
| 1823 | | |
| 1824 | Br. Lewis Stobwasser | Stobwasser, May 1824* |
| 1825 | | |
| 1826 | Br. Ellis | Ellis, July 1826* |
| 1827 | | |
| 1828 | | |
| 1829 | | |
| 1830 | Br. Jacob Zorn & Sr. Caroline Zorn, Br. Scholefield, Br. Peter Ricksecker & Sr. Sarah Ricksecker | Zorn, March–Sept. 1830* |
| 1831 | Br. & Sr. Ricksecker | Ricksecker, Jan.–Dec. 1831* |
| 1832 | Br. & Sr. Ricksecker | Ricksecker, Jan.–Dec. 1832* |
| 1833 | Br. & Sr. Ricksecker, Br. Zorn, Br. Pemsel & Sr. Pemsel | Zorn, April 1833* |
| 1834 | Br. & Sr. Pemsel | |
| 1835 | Br. Jacob Zorn | |

| | Baptized slaves | | | All slaves, Dec. 31, 1769 | | |
|---|---|---|---|---|---|---|
| | Males | Females | % | Males | Females | % |
| Occupation | | | | | | |
| Driver, field gang | 3 | 1 | 3.9 | 3 | 1 | 1.6 |
| House servant | 0 | 1 | 0.9 | 2 | 3 | 2.0 |
| Sugar maker | 6 | 0 | 5.8 | 3 | 0 | 1.2 |
| Craft worker | 5 | 0 | 4.8 | 17 | 0 | 6.8 |
| Carter | 3 | 0 | 2.9 | 2 | 0 | 0.8 |
| Livestock | 2 | 0 | 1.9 | 13 | 1 | 5.6 |
| Field hand | 7 | 22 | 28.0 | 63 | 60 | 49.4 |
| Watchman | 13 | 0 | 12.5 | 24 | 0 | 9.7 |
| Chapel attendant | 0 | 1 | 0.9 | 0 | 1 | 0.4 |
| Washerwoman | 0 | 3 | 2.9 | 0 | 3 | 1.2 |
| Midwife | 0 | 1 | 0.9 | 0 | 1 | 0.4 |
| Marginal job | 6 | 2 | 7.7 | 4 | 8 | 4.8 |
| Nonworking invalid | 2 | 23 | 24.0 | 4 | 14 | 7.2 |
| Nonworking child | 0 | 3 | 2.9 | 13 | 9 | 8.9 |
| Total | 47 | 57 | 100 | 148 | 101 | 100 |
| Age | | | | | | |
| 0–9 | 0 | 2 | 1.9 | 16 | 13 | 11.6 |
| 10–19 | 0 | 1 | 0.9 | 39 | 15 | 21.7 |
| 20–29 | 4 | 2 | 5.8 | 30 | 19 | 19.7 |
| 30–39 | 11 | 9 | 19.3 | 21 | 17 | 15.3 |
| 40–49 | 17 | 21 | 36.6 | 29 | 20 | 19.7 |
| 50–59 | 11 | 15 | 25.0 | 12 | 11 | 9.2 |
| 60–69 | 3 | 7 | 9.6 | 1 | 6 | 2.8 |
| 70+ | 1 | 0 | 0.9 | 0 | 0 | 0 |
| Total | 47 | 57 | 100 | 148 | 101 | 100 |
| Median age | 44.3 | 44.8 | | 27.2 | 31.2 | |
| Origin | | | | | | |
| Creole | 12 | 13 | 24.0 | 68 | 49 | 47.0 |
| Probably Creole | 6 | 8 | 13.5 | 10 | 12 | 8.8 |
| African | 10 | 12 | 21.2 | 56 | 22 | 31.3 |
| Probably African | 19 | 24 | 41.3 | 14 | 18 | 12.9 |
| Total | 47 | 57 | 100 | 148 | 101 | 100 |

*Note:* Hertford, baptized in 1767, and Phoebe, baptized in 1769, are excluded from this table because they have not been identified in the Mesopotamia records.

*Appendix 29* Mount Airy Slaves Sent to Alabama, 1833–1837

| Name | Party | Date | Age | Occupation | Family connections | Died | In 1863 |
|---|---|---|---|---|---|---|---|
| Billy Page | A | 1833 | 22 | Field | Brother of Marcus Page | 1863 | Larkin |
| David Moore | A | 1833 | 36 | Field | Brother of James and Tom Moore | | Larkin |
| James Moore | A | 1833 | 33 | Field | Brother of David and Tom Moore | 1847 | |
| James Thurston | A | 1833 | 27 | Field | | | Larkin |
| Ralph Elms | A | 1833 | 22 | Field | | | Larkin |
| Tom Moore | A | 1833 | 34 | Field | Brother of David and James Moore | 1862 | |
| John Moore | A | 1833 | 38 | Domestic | Probably brother to David, James, and Tom Moore | | Oakland |
| Emanuel | A | 1833 | 16 | Domestic | **Returned to VA, 1840** | 1852 | |
| Barneby | B | 1835 | 7 | | Son of Joe; **sold, 1838** | ? | |
| Jacob Smith | B | 1835 | 14 | Field | | | Oakland |
| Joe | B | 1835 | 35 | Field | Father of Barneby; **sold, 1836** | ? | |
| Peter Page | B | 1835 | 21 | Field | | 1848? | |
| Reuben Winters | B | 1835 | 17 | Field | Brother of George Bray | | Larkin |
| William | B | 1835 | 17 | Field | | 1838 | |
| Anna Marcus | B | 1835 | 12 | Field | Sister of Jane, Fanny, and Joe Moore | | Oakland |
| Becky Carrington | B | 1835 | 16 | Field | | | Oakland |
| Betsy Dixon | B | 1835 | 16 | Field | Sister of Rachel Dixon | | Oakland |
| Esther Barnes | B | 1835 | 31 | Field | Daughter of Nanny Glascow; sister of Judy and Mary Glascow; mother of Esther Barnes Jr. | | Oakland |
| Esther Barnes Jr. | B | 1835 | 14 | Field | Daughter of Esther Barnes | | Oakland |
| Grace | B | 1835 | 15 | Field | **Sold, 1835** | ? | |
| Jane Moore | B | 1835 | 13 | Field | Sister of Anna Marcus, Fanny and Joe Moore; **sold, 1844** | ? | |
| Judy Glascow | B | 1835 | 27 | Field | Daughter of Nanny Glascow; sister of Mary Glascow; mother of Mary Glascow Jr. | | Oakland |
| Mary Glascow Jr. | B | 1835 | 1 | | Daughter of Judy Glascow | | Oakland |
| Nanny Glascow | B | 1835 | 53 | Nurse | Mother of Esther Barnes, Judy and Mary | | Oakland |

| | | | | | | | |
|---|---|---|---|---|---|---|---|
| Mary | C | 1836 | 38 | Field | Mother of Cornelius and Catherine; **sold 1837** | ? | |
| Cornelius | C | 1836 | 15 | Field | Son of Mary; **sold, 1843** | ? | |
| Catherine | C | 1836 | 13 | Field | Daughter of Mary; **sold, 1838** | ? | |
| Rachel Dixon | C | 1836 | 15 | Field | Mother of Leana Dixon, sister of Betsy Dixon; **sold, 1837** | ? | |
| Leana Dixon | C | 1836 | 2 | | Daughter of Rachel Dixon | 1836 | |
| Mary Glascow | C | 1836 | 23 | Field | Daughter of Nanny Glascow; sister of Judy Glascow and Esther Barnes; **sold, 1836** | ? | |
| George Bray | C | 1836 | 23 | Field | Brother of Reuben Winters; **sold, 1838** | ? | |
| Marcus Page | C | 1836 | 16 | Field | Brother of Billy Page; **sold, 1838** | ? | |
| Fanny Moore | C | 1836 | 10 | Field | Sister of Anna Marcus, Jane and Joe Moore; **sold, 1837** | ? | |
| Frank | C | 1836 | 37 | Jobber | **Sold, 1838** | ? | |
| Judy | C | 1836 | 14 | Field | **Sold, 1837** | ? | |
| Sarah | C | 1836 | 12 | Field | **Sold, 1837** | ? | |
| Frank | C | 1836 | 39 | Field | | 1837 | |
| James Glascow | C | 1836 | 11 | Miller | Son of Mary Glascow, grandson of Nanny | | Oakland |
| Joe Moore | | 1837 | 17 | Ostler | Brother of Anna Marcus, Jane and Fanny Moore | | Oakland |

*Appendix 30* Mount Airy Slaves Sent to Alabama, 1845

| Name | Age | Occupation | Family connections | Died | In 1863 |
|---|---|---|---|---|---|
| Alfred Lewis | 38 | Jobber | Husband of Sinah Lewis; father of Alfred Jr., General, Anne, Matilda, Sally, and Eliza Lewis | | Oakland |
| Sinah Lewis | 30 | Field | Wife of Alfred Lewis; mother of Alfred Jr., General, Anne, Matilda, Sally, and Eliza Lewis | 1852 | |
| Alfred Lewis Jr. | 14 | Miller | Son of Sinah and Alfred Lewis | 1847 | |
| General Lewis | 12 | Field | Son of Sinah and Alfred Lewis | | Oakland |
| Anne Lewis | 10 | Field | Daughter of Sinah and Alfred Lewis | | Oakland |
| Matilda Lewis | 8 | | Daughter of Sinah and Alfred Lewis | | Larkin |
| Sally Lewis | 7 | | Daughter of Sinah and Alfred Lewis | | Oakland |
| Eliza Lewis | 3 | | Daughter of Sinah and Alfred Lewis | | Oakland |
| Lucy Dudley | 34 | Field | Mother of Carter, Gabriel, John, Thadeus, and Ibby Dudley | 1862 | |
| John Dudley | 16 | Field | Son of Lucy Dudley | | Oakland |
| Thadeus Dudley | 14 | Field | Son of Lucy Dudley | | Oakland |
| Carter Dudley | 10 | Field | Son of Lucy Dudley | | Oakland |
| Gabriel Dudley | 2 | | Son of Lucy Dudley | | Oakland |
| Ibby Dudley | 0 | | Daughter of Lucy Dudley | 1846 | |
| Letty Dixon | 21 | Field | Mother of Jesse Dixon; sister of James, Henry, Sophia, Mary Ann, Sophy, and Nancy Dixon | 1847 | |
| Jesse Dixon | 2 | | Son of Letty Dixon | 1847 | |
| James Dixon | 29 | Shoemaker | Brother of Letty Dixon | 1848 | |
| Henry Dixon | 19 | Shoemaker | Brother of Letty Dixon | | Oakland |
| Sophia Dixon | 15 | Field | Sister of Letty Dixon | | Oakland |

| | | | | | |
|---|---|---|---|---|---|
| Mary Ann Dixon | 14 | Field | Sister of Letty Dixon | | Larkin |
| Sophy Dixon | 10 | Field | Sister of Letty Dixon | | Larkin |
| Nancy Dixon | 7 | | Sister of Letty Dixon | | Oakland |
| Winney Grimshaw | 19 | Spinner | Mother of John Carrington; sister of James Grimshaw | | Larkin |
| John Carrington | 1 | | Son of Winney Grimshaw | 1847 | |
| James Grimshaw | 14 | Carpenter | Brother of Winney Grimshaw | | Oakland |
| William Page | 22 | Field | Brother of Grace and Chapman Page | | Larkin |
| Grace Page | 17 | Field | Sister of William Page | | Oakland |
| Chapman Page | 16 | Field | Brother of William Page | | Larkin |
| Lucy Thomas | 19 | Field | Sister of Carolina Thomas | | Oakland |
| Carolina Thomas | 17 | Field | Sister of Lucy Thomas | | Oakland |
| Austin Carrington | 20 | Scullion | Brother of David Carrington | | Oakland |
| David Carrington | 14 | Field | Brother of Austin Carrington | | Oakland |
| Travers Hilliard | 27 | Blacksmith | | | Oakland |
| Phill Carter | 21 | Field | | 1862 | |
| Tom Dudley | 19 | Field | | | Larkin |
| Scippio Bray | 16 | Field | | 1847 | |
| Tom Harwood | 16 | Field | | | Oakland |
| Albert Ward | 16 | Stable | | | Oakland |
| Dangerfield Perkins | 11 | Domestic | | | Oakland |
| Mary Lewis | 17 | Field | | 1860 | |
| Lucinda | 17 | Field | | 1847 | |
| Ann Richardson | 16 | Field | | | Oakland |
| Georgiana | 13 | Field | | | Larkin |
| Anthony | 32 | Jobber | **Sent to Walnut Grove; probably sold** | ? | |
| Rose | 17 | Domestic | **Sent to Walnut Grove; probably sold** | ? | |

*Appendix 31* Mount Airy Slaves Sent to Alabama, 1854

| Name | Age | Occupation | Family connections | Died | In 1863 |
|---|---|---|---|---|---|
| Marcus Bray | 32 | Field | Husband of Franky Bray; father of Reuben, Rose, and Rosetta Bray | 1857 | |
| Franky Bray | 28 | Field | Wife of Marcus Bray; mother of Reuben, Rose, and Rosetta Bray; sister of Becky Lewis | 1862 | |
| Reuben Bray | 9 | Field | Son of Franky and Marcus Bray | | Larkin |
| Rose Bray | 2 | | Daughter of Franky and Marcus Bray | 1854 | |
| Rosetta Bray | 0 | | Daughter of Franky and Marcus Bray | 1858 | |
| Becky Lewis | 31 | Field | Sister of Franky Bray; **sold, 1855** | ? | |
| Bridget Thurston | 42 | Field | Mother of Sarah Thurston; sister of Becky Hilliard; **sold, 1854** | ? | |
| Sarah Thurston | 20 | Field | Daughter of Bridget Thurston | 1858 | |
| Becky Hilliard | 34 | Field | Sister of Bridget Thurston | | Larkin |
| Abram Brown | 21 | Field | Brother of Archy, Eve, and Hannah Brown | | Larkin |
| Archy Brown | 17 | Field | Brother of Abram, Eve, and Hannah Brown | | Larkin |
| Eve Brown | 15 | Field | Sister of Abram, Archy, and Hannah Brown | | Larkin |
| Hannah Brown | 12 | Field | Sister of Abram, Archy, and Eve Brown | | Larkin |
| Lemuel Yeatman | 18 | Field | Brother of Lewis and Bill Yeatman | | Larkin |
| Lewis Yeatman | 17 | Field | Brother of Lemuel and Bill Yeatman | | Larkin |

| | | | | | |
|---|---|---|---|---|---|
| Bill Yeatman | 12 | Ditcher | Brother of Lemuel and Lewis Yeatman | 1858 | |
| Felicia Wormley | 15 | Field | Sister of Isaiah and Bill Wormley | | Oakland |
| Isaiah Wormley | 13 | Field | Brother of Felicia and Bill Wormley; **Sold, 1855** | ? | |
| Bill Wormley | 11 | Field | Brother of Felicia and Isaiah Wormley | | Oakland |
| Abner Dudley | 15 | Field | Brother of Gabriel Dudley | | Oakland |
| Gabriel Dudley | 11 | Field | Brother of Abner Dudley | | Oakland |
| Michael Thomas | 16 | Field | Half brother of Mary Thomas; **sold, 1855** | ? | |
| Mary Thomas | 11 | Field | Half sister of Michael Thomas | | Larkin |
| Sydney Ward | 21 | Jobber | Brother of Arabella Ward | | Oakland |
| Arabella Ward | 15 | Spinner | Sister of Sydney Ward | | Oakland |
| Marcus Denny | 32 | Field | **Sold, 1855** | ? | |
| Tom Moore | 26 | Field | | | Larkin |
| Paul Myers | 25 | Field | | | Larkin |
| Edmund Harwood | 20 | Field | | | Oakland |
| Amphy Wheeler | 17 | Field | | | Larkin |
| Simuel Page | 16 | Domestic | | | Oakland |
| Cornelius Elms | 15 | Field | | | Larkin |
| David Saunders | 14 | Blacksmith | | | Oakland |
| Elizabeth Carter | 25 | Field | | | Oakland |
| Leannah Yeatman | 11 | Field | Cousin of Anne Yeatman | | Oakland |
| Anne Yeatman | 9 | Domestic | Cousin of Leannah Yeatman; **sold, 1855** | ? | |

*Appendix 32* Slave Households at Oakland Plantation, 1859

| House | Male workers | Female workers | Young children |
|---|---|---|---|
| 2 | ♦ **Jim Grimshaw** 27 | ◘ **Arabella Ward Grimshaw** 19 | Felicia Grimshaw 1 |
| 3 | □ **Michael Smith** 48<br>Billy Smith 15<br>Harrington Smith 13 | □ **Patty Smith** 47 | Malinda Smith 3 |
| 5 | ♦ **Albert Ward** 30<br>◘ **Sydney Ward** 25<br>◘ **David Saunders** 19<br>◘ **Bill Wormley** 15 | | |
| 7 | ♦ **Travers Hilliard** 40 | □ **Anna Marcus Hilliard** 36 | Sam Hilliard 9<br>Sarah Hilliard 5<br>Brooke Hilliard 3<br>Emanuel Hilliard 1 |
| 9 | ♦ **David Carrington** 27 | ♦ **Anne Lewis Carrington** 24 | Emily Carrington 7<br>Washington Carrington 5<br>Selina Carrington 2 |
| 10 | □ **Jimmy Glascow** 34 | ♦ **Anne Richardson Glascow** 29 | Anne Glascow 2<br>Solomon Glascow 4 |
| 18 | □ Tom Flood 40<br>William Flood 18<br>Tom Flood 15 | □ **Becky Carrington Flood** 40 | John Flood 9<br>Weldon Flood 7<br>Godfrey Flood 5<br>Virginia Flood 2 |
| 20 | ♦ **Thadeus Dudley** 28 | ♦ **Sally Lewis Dudley** 20 | Henry Clay Dudley 3<br>Celia Dudley 1 |
| 22 | ♦ **General Lewis** 26 | ♦ **Nancy Dixon Lewis** 20 | Bladen Lewis 4<br>Louisa Lewis 2 |
| 24 | □ Peter Smith 41<br>♦ **John Dudley** 29<br>♦ **Carter Dudley** 24<br>◘ **Abner Dudley** 19<br>◘ **Gabriel Dudley** 16 | ♦ **Lucy Dudley Smith** 47 | |
| 26 | ♦ **Alfred Lewis** 52<br>Daniel Lewis 11 | ♦ **Sophia Dixon Lewis** 29 | Wise Dixon Lewis 3<br>Letty Dixon Lewis 1 |

*Key:* **Boldface** = came from **Mount Airy** in □ 1830s ♦ 1845 ◘ 1854

*Appendix 33* Motherhood among the Mount Airy Women in Virginia and Alabama

| | No. of women | No. of mothers | No. of children | Children per mother |
|---|---|---|---|---|
| Mount Airy women aged 16–40 | 147* | 116* | 502 | 4.3 |
| *Occupation at Mount Airy* | | | | |
| Field worker | 105 | 85 | 350 | 4.1 |
| Domestic | 18 | 13 | 63 | 4.8 |
| Textile worker | 24 | 18 | 89 | 4.9 |
| Alabama women aged 16–40 | 55* | 42* | 134 | 3.2 |
| *Occupation in Alabama* | | | | |
| Field worker | 53 | 40 | 126 | 3.2 |
| Domestic | 2 | 2 | 8 | 4.0 |
| Textile worker | 0 | 0 | 0 | |
| Totals | 191 | 147 | 636 | 4.3 |

*Includes 11 women who had children at Mount Airy and also in Alabama. These women had 47 children at Mount Airy and 21 in Alabama.

*Appendix 34* Mount Airy Slaves Sent to Alabama, December 1861

| Name | Age | Occupation | Family connections | Died | In 1863 |
|---|---|---|---|---|---|
| Anderson Hilliard | 38 | Field | Husband of Rose Hilliard; father of Travers, Anna, Martha, Jane, and Sallie Hilliard | | Larkin |
| Rose Hilliard | 37 | Field | Wife of Anderson Hilliard; mother of Martha, Jane, and Sallie Hilliard, and Henry Thomas | | Larkin |
| Anna Hilliard | 15 | Field | Daughter of Lizza (dec.) and Anderson Hilliard | | Oakland |
| Travers Hilliard | 13 | Field | Son of Lizza (dec.) and Anderson Hilliard | | Larkin |
| Martha Hilliard | 12 | | Daughter of Rose and Anderson Hilliard | | Larkin |
| Jane Hilliard | 8 | | Daughter of Rose and Anderson Hilliard | | Larkin |
| Sallie Hilliard | 1 | | Daughter of Rose and Anderson Hilliard | | Larkin |
| Henry Thomas | 16 | Miller | Son of Rose Hilliard | | Larkin |
| Harrington Smith | 43 | Field | Husband of Fanny Smith; father of Celia, John William, Alice, Sophie, Olivia, Lavinia, and Barnaby Smith | | Larkin |
| Fanny Smith | 29 | Field | Wife of Harrington Smith; mother of Celia, John William, Alice, Sophie, Olivia, Lavinia, and Barnaby Smith; brother of Ben Wright | | Larkin |
| Celia Smith | 12 | | Daughter of Fanny and Harrington Smith | | Larkin |
| John William Smith | 9 | | Son of Fanny and Harrington Smith | | Larkin |
| Alice Smith | 8 | | Daughter of Fanny and Harrington Smith | | Larkin |
| Sophie Smith | 6 | | Daughter of Fanny and Harrington Smith | | Larkin |
| Olivia Smith | 5 | | Daughter of Fanny and Harrington Smith | 1862 | |
| Lavinia Smith | 2 | | Daughter of Fanny and Harrington Smith | | Larkin |
| Barnaby Smith | 0 | | Daughter of Fanny and Harrington Smith | | Larkin |
| Ben Wright | 18 | Field | Brother of Fanny Smith | | Larkin |
| Israel Carrington | 39 | Field | Husband of Isadora Carrington | | Larkin |
| Isadora Carrington | 27 | Field | Wife of Israel Carrington | | Larkin |

| | | | | | |
|---|---|---|---|---|---|
| Sarah Yeatman | 17 | Field | Sister of Horace Yeatman | | Larkin |
| Horace Yeatman | 16 | Field | Brother of Sarah Yeatman | | Larkin |
| Philip Carrington | 18 | Field | Brother of Charles and James Carrington | | Larkin |
| Charles Carrington | 15 | Miller | Brother of James and Philip Carrington | | Larkin |
| James Carrington | 13 | Miller | Brother of Philip and Charles Carrington | | Larkin |
| Peter Yeatman | 19 | Field | Husband of Catherine Yeatman; brother of Thornton and Daniel Yeatman | | Larkin |
| Catherine Yeatman | 19 | Field | Wife of Peter Yeatman | | Larkin |
| Thornton Yeatman | 21 | Field | Brother of Peter and Daniel Yeatman | | Larkin |
| Daniel Yeatman | 15 | Field | Brother of Peter and Thornton Yeatman | | Larkin |
| Betsy Moore Lewis | 24 | Domestic | Mother of Middleton and Isabella Lewis; sister of Louisa Moore | | Larkin |
| Isabella Lewis | 5 | | Daughter of Betsy Lewis | | Larkin |
| Middleton Lewis | 0 | | Son of Betsy Moore Lewis | | Larkin |
| Louisa Moore | 15 | Domestic | Sister of Betsy Moore Lewis | | Larkin |
| Albert Wormley | 16 | Field | Brother of Mary Frances Wormley | | Larkin |
| Mary Frances Wormley | 13 | Field | Sister of Albert Wormley | | Larkin |
| Thomas Lewis | 22 | Field | Husband of Betsy Moore Lewis; half brother of Betsy and Airy Yeatman | | Larkin |
| Betsy Yeatman | 15 | Field | Sister of Thomas Lewis and Airy Yeatman | | Larkin |
| Airy Yeatman | 12 | Field | Sister of Thomas Lewis and Betsy Yeatman | | Larkin |
| Ailse Carter | 43 | Field | Mother of Amanda and Frances Carter | | Larkin |
| Frances Carter | 16 | Field | Daughter of Ailse Carter; sister of Amanda Carter | | Larkin |
| Amanda Carter | 15 | Field | Daughter of Ailse Carter; sister of Frances Carter | | Larkin |
| Martha Elms | 27 | Field | Mother of Georgiana, Malinda, and Maria Elms | | Oakland |
| Malinda Elms | 10 | | Daughter of Martha Elms | | Oakland |
| Georgiana Elms | 8 | | Daughter of Martha Elms | | Oakland |
| Maria Elms | 2 | | Daughter of Martha Elms | 1862 | |
| Ralph Wheeler | 19 | Field | | | Larkin |
| Beverley Harwood | 18 | Field | | | Larkin |
| Julia Dickinson Ward | 19 | Field | **Returned to Mount Airy, 1863** | | |

*Appendix 35* Mount Airy Slaves Sent to Alabama, March 1862

| Name | Age | Occupation | Family connections | Died | In 1863 |
|---|---|---|---|---|---|
| Gowen Page | 52 | Field | Husband of Georgina Page; father of John Page and Susan Carter | | Larkin |
| Georgina Page | 53 | Field | Wife of Gowen Page; mother of John Page and Susan Carter | | Larkin |
| John Page | 16 | Field | Son of Georgina and Gowen Page; brother of Susan Carter | | Larkin |
| William Carter | 25 | Carpenter | Husband of Susan Carter; father of Aggy and Lucy Carter | | Larkin |
| Susan Carter | 27 | Field | Wife of William Carter; mother of Aggy and Lucy Carter; daughter of Georgina and Gowen Page; sister of John Page | | Larkin |
| Lucy Carter | 5 | | Daughter of Susan and William Carter | | Larkin |
| Aggy Carter | 0 | | Daughter of Susan and William Carter | | Larkin |
| John Wormley | 49 | Field | Husband of Mildred Wormley; father of Patsy, Grace, Ralph, and Caroline Wormley | | Larkin |
| Mildred Wormley | 40 | Field | Wife of John Wormley; mother of Patsy, Grace, Ralph, and Caroline Wormley | | Larkin |
| Patsy Wormley | 13 | Field | Daughter of Mildred and John Wormley | | Larkin |
| Grace Wormley | 10 | | Daughter of Mildred and John Wormley | | Larkin |
| Ralph Wormley | 7 | | Son of Mildred and John Wormley | | Larkin |
| Caroline Wormley | 5 | | Daughter of Mildred and John Wormley | | Larkin |
| Godfrey Carrington | 44 | Field | Husband of Amelia Carrington; father of Milly, Brooke, Chris, Jacob, Mary, and Arthur Carrington | | Larkin |
| Amelia Carrington | 37 | Seamstress | Wife of Godfrey Carrington; mother of Milly, Brooke, Chris, Jacob, Mary, and Arthur Carrington | | Larkin |
| Milly Carrington | 12 | Field | Daughter of Amelia and Godfrey Carrington | | Larkin |
| Brooke Carrington | 10 | | Son of Amelia and Godfrey Carrington | | Larkin |
| Chris Carrington | 8 | | Son of Amelia and Godfrey Carrington | | Larkin |
| Jacob Carrington | 6 | | Son of Amelia and Godfrey Carrington | | Larkin |
| Mary Carrington | 4 | | Daughter of Amelia and Godfrey Carrington | | Larkin |
| Arthur Carrington | 1 | | Son of Amelia and Godfrey Carrington | | Larkin |

| | | | | | |
|---|---|---|---|---|---|
| Richard Yeatman | 43 | Gardener | Husband of Sally Yeatman; father of Nancy, Emma, Etta, Poinsetta, Ella, Richard Jr., and Susan Yeatman | | Larkin |
| Sally Yeatman | 39 | Field | Wife of Richard Yeatman; mother of Nancy, Emma, Etta, Poinsetta, Ella, Richard Jr., and Susan Yeatman | | Larkin |
| Nancy Yeatman | 14 | Field | Daughter of Sally and Richard Yeatman | | Larkin |
| Emma Yeatman | 12 | Field | Daughter of Sally and Richard Yeatman | | Larkin |
| Etta Yeatman | 9 | | Daughter of Sally and Richard Yeatman | | Larkin |
| Poinsetta Yeatman | 8 | | Daughter of Sally and Richard Yeatman | | Larkin |
| Ella Yeatman | 6 | | Daughter of Sally and Richard Yeatman | | Larkin |
| Richard Yeatman Jr. | 3 | | Son of Sally and Richard Yeatman | | Larkin |
| Susan Yeatman | 0 | | Daughter of Sally and Richard Yeatman | 1862 | |
| Phillis Owens | 32 | Field | Mother of Charlotte, Dorothea, Levicia, Clara, and Warren Owens | | Larkin |
| Charlotte Owens | 12 | Scullion | Daughter of Phillis Owens | | Larkin |
| Dorothea Owens | 9 | | Daughter of Phillis Owens | | Larkin |
| Levicia Owens | 8 | | Daughter of Phillis Owens | | Larkin |
| Clara Owens | 6 | | Daughter of Phillis Owens | | Larkin |
| Warren Owens | 4 | | Son of Phillis Owens | | Larkin |
| Joe Yeatman | 46 | Foreman | Husband of Letty Yeatman; father of Jesse and Nettie Yeatman | | Larkin |
| Letty Yeatman | 41 | Field | Wife of Joe Yeatman; mother of Jesse and Nettie Yeatman | | Larkin |
| Jesse Yeatman | 9 | | Son of Letty and Joe Yeatman | | Larkin |
| Nettie Yeatman | 1 | | Daughter of Letty and Joe Yeatman | | Larkin |
| William Richardson | 35 | Blacksmith | | | Larkin |
| John Moore | 37 | Gardener | Brother of James Moore | | Larkin |
| James Moore | 29 | Field | Brother of John Moore | | Larkin |
| Cornelius Ward | 30 | Field | | | Larkin |
| Austin Wormley | 16 | Domestic | Brother of Nannie Key Wormley | | Larkin |
| Nannie Key Wormley | 12 | Field | Sister of Austin Wormley | | Larkin |
| Jonathan Wheeler | 16 | Field | | | Larkin |
| Levinia Hilliard | 10 | | | | Larkin |
| William Carrington | 7 | | Brother of Josephine Carrington | | Larkin |
| Josephine Carrington | 4 | | Sister of William Carrington | | Larkin |

*Appendix 36* Profile of the Mesopotamia Congregation in 1761–1769, 1798–1818, and 1831

| | 1769 | | | 1818 | | | 1831 | | |
|---|---|---|---|---|---|---|---|---|---|
| | Males | Females | % | Males | Females | % | Males | Females | % |
| Occupation | | | | | | | | | |
| Driver | 3 | 1 | 4 | 1 | 2 | 4 | 2 | 1 | 4 |
| Sugar maker | 6 | 0 | 6 | 1 | 0 | 1 | 0 | 0 | 0 |
| Craft worker | 5 | 0 | 5 | 7 | 0 | 9 | 11 | 0 | 17 |
| Carter | 3 | 0 | 3 | 2 | 0 | 3 | 0 | 0 | 0 |
| Livestock | 2 | 0 | 2 | 1 | 0 | 1 | 1 | 0 | 2 |
| Field hand | 7 | 22 | 27 | 3 | 17 | 26 | 5 | 9 | 21 |
| Watchman | 13 | 0 | 12 | 3 | 0 | 4 | 1 | 0 | 2 |
| Domestic | 0 | 1 | 1 | 0 | 5 | 6 | 1 | 5 | 9 |
| Chapel attendant | 0 | 1 | 1 | 0 | 3 | 4 | 0 | 2 | 3 |
| Midwife | 0 | 1 | 1 | 0 | 1 | 1 | 0 | 2 | 3 |
| Washerwoman | 0 | 3 | 3 | 0 | 2 | 3 | 0 | 3 | 4 |
| Marginal job | 6 | 2 | 8 | 0 | 10 | 13 | 2 | 7 | 14 |
| Invalid | 2 | 23 | 24 | 8 | 11 | 25 | 1 | 10 | 17 |
| Child | 0 | 3 | 3 | 0 | 0 | 0 | 2 | 1 | 4 |
| Total | 47 | 57 | 100 | 26 | 51 | 100 | 26 | 40 | 100 |
| Age | | | | | | | | | |
| 0–9 | 0 | 2 | 2 | 0 | 0 | 0 | 2 | 1 | 4 |
| 10–19 | 0 | 1 | 1 | 0 | 3 | 4 | 1 | 4 | 7.5 |
| 20–29 | 4 | 2 | 6 | 2 | 4 | 8 | 2 | 3 | 7.5 |
| 30–39 | 11 | 9 | 19 | 11 | 11 | 28 | 10 | 7 | 26 |
| 40–49 | 17 | 21 | 37 | 7 | 15 | 28 | 5 | 5 | 15 |
| 50–59 | 11 | 15 | 25 | 2 | 7 | 12 | 5 | 10 | 23 |
| 60–69 | 3 | 7 | 9 | 3 | 9 | 16 | 1 | 6 | 11 |
| 70+ | 1 | 0 | 1 | 1 | 2 | 4 | 0 | 4 | 6 |
| Total | 47 | 57 | 100 | 26 | 51 | 100 | 26 | 40 | 100 |
| Median age | 44.2 | 45.1 | | 43.5 | 44.5 | | 37.2 | 45.9 | |
| Origin | | | | | | | | | |
| Creole | 12 | 13 | 24 | 16 | 34 | 65 | 20 | 28 | 73 |
| Probably Creole | 6 | 8 | 14 | 2 | 0 | 3 | 6 | 6 | 18 |
| African | 10 | 12 | 21 | 5 | 16 | 27 | 0 | 6 | 9 |
| Probably African | 19 | 24 | 41 | 3 | 1 | 5 | 0 | 0 | 0 |
| Total | 47 | 57 | 100 | 26 | 51 | 100 | 26 | 40 | 100 |

*Note:* The people in the 1769 group were baptized in 1761–1769; the people in the 1818 group were candidates for baptism in 1798–1818; the people in the 1831 group were inventoried as "Christians."

*Appendix 37* Slaves Inventoried at Mount Airy, February 1863

| Name | Age | Occupation | Comments by Tayloe | Family connections |
|---|---|---|---|---|
| Armistead Carter | 42 | House servant | | **Brother Edward Carter deserted, 1862** |
| "Crack" Fanny Carter | 70 | Retired cook | "chargable" | Widow; daughter in Alabama |
| Tom Carter | 69 | Carpenter | "worthless" | Husband of Franky Carter; 3 children in Alabama |
| Franky Carter | 64 | Retired cook | "chargable" | Wife of Tom Carter; 3 children in Alabama |
| Annie Flood | 75 | Retired spinner | Blind | Widow; 2 sons exiled to Deep Hole; **2 daughters, Mary and Lizzie Flood, were sold** |
| Ruffin Moore | 45 | Coachman | | Widower; father of Charles Moore; 2 daughters in Alabama; **daughter Caroline deserted**, 1862 |
| Charles Moore | 14 | Stable boy | | Son of Ruffin Moore; 2 sisters in Alabama; **sister Caroline deserted,** 1862 |
| Dick Yeatman | 76 | Retired gardener | "chargable" | Widower; father of George Yeatman; 2 sons in Alabama |
| George Yeatman | 53 | Field | | Son of Dick Yeatman; husband of Winney Yeatman; father of John, Marcus, Ruffin, William, and Judy Yeatman; 2 children in Alabama |
| Winney Yeatman | 41 | Field | "worthless" | Wife of George Yeatman; mother of John, Marcus, Ruffin, William, and Judy Yeatman; 2 children in Alabama |
| Ruffin Yeatman | 12 | Field | | Son of Winney and George Yeatman |
| Judy Yeatman | 10 | | | Daughter of Winney and George Yeatman |
| William Yeatman | 8 | | Died Sept. 1863 | Son of Winney and George Yeatman |
| Marcus Yeatman | 6 | | | Son of Winney and George Yeatman |
| John Yeatman | 0 | | | Son of Winney and George Yeatman |
| Dickerson Yeatman | 8 | | | Parents and 6 siblings in Alabama |
| George Wormley | 42 | Sailor | | Husband of Kitty Wormley; father of Ellick, Lloyd, and Rebecca Wormley; 2 children in Alabama |
| Kitty Wormley | 34 | Chambermaid | "crazy" | Wife of George Wormley; mother of Ellick, Lloyd, and Rebecca Wormley; 2 children in Alabama |
| Rebecca Wormley | 15 | Nurse | | Daughter of Kitty and George Wormley |
| Ellick Wormley | 11 | Mill boy | | Son of Kitty and George Wormley |
| Lloyd Wormley | 6 | | | Son of Kitty and George Wormley |
| Eliza Ward | 55 | Retired spinner | "worthless" | Mother of Eliza Jr., Ellen, and Julia Dickinson Ward; 4 children in Alabama; **husband Ralph Ward deserted, 1861; daughter Betsy Ward deserted, 1862** |
| Eliza Ward Jr. | 27 | Spinner | | Mother of Harry, Arabella, and Kitty Ward; 4 siblings in Alabama; **father Ralph and sister Betsy deserted, 1861–2** |
| Harry Ward | 9 | With Mrs. Sanford | | Son of Eliza Ward Jr.; **grandfather and aunt deserted, 1861–62** |
| Arabella Ward | 6 | | | Daughter of Eliza Ward Jr.; **grandfather and aunt deserted, 1861–62** |
| Kitty Ward | 3 | | | Daughter of Eliza Ward Jr.; **grandfather and aunt deserted, 1861** |

(*continued*)

| Name | Age | Occupation | Comments by Tayloe | Family connections |
|---|---|---|---|---|
| Ellen Ward | 14 | Field | | Daughter of Eliza Ward; sister of Eliza Jr. and Julia Dickinson Ward; 4 siblings in Alabama; **father Ralph and sister Betsy deserted, 1861–62** |
| Julia Dickinson Ward | 21 | Nurse | Returned from Alabama | Daughter of Eliza Ward; sister of Eliza Jr. and Ellen Ward; 4 siblings in Alabama; **father Ralph Ward and sister Betsy Ward deserted, 1861–62** |
| China Myers | 31 | Field | | Mother of Joseph, Lindsay, Nereus, and Fanny Myers; daughter of Winney Myers; brother in Alabama; **brother Urias Myers deserted, 1862** |
| Fanny Myers | 12 | Field | "deformed hand" | Daughter of China Myers; **uncle Urias deserted, 1862** |
| Nereus Myers | 9 | | | Son of China Myers; **uncle Urias deserted, 1862** |
| Lindsay Myers | 6 | | | Son of China Myers; **uncle Urias deserted, 1862** |
| Joseph Myers | 2 | | | Son of China Myers; **uncle Urias deserted, 1862** |
| Nancy Carter | 71 | Retired spinner | "worthless" | Mother of Martha Harrod and 12 other children: 8 have died, 3 were sold, 1 was transferred, and none are in Alabama |
| Martha Harrod | 47 | House cook | | Mother of Nancy Laws and Julius and Meta Harrod; 2 sons in Alabama; **son Elias Harrod deserted, 1861; daughter Betsy Harrod deserted, 1862** |
| Nancy Laws | 24 | House cook | | Mother of Sedgwick and Mary Jane Laws; daughter of Martha Harrod; sister of Julius and Meta Harrod; **brother Elias Harrod deserted, 1861; sister Betsy Harrod deserted, 1862** |
| Mary Jane Laws | 6 | | | Daughter of Nancy Laws; **uncle and aunt deserted, 1861–62** |
| Sedgwick Laws | 2 | | | Son of Nancy Laws; **uncle and aunt deserted, 1861–62** |
| Julius Harrod | 11 | | | Son of Martha Harrod; brother of Meta Harrod and Nancy Laws; 2 brothers in Alabama; **brother Elias and sister Betsy deserted, 1861–62** |
| Meta Harrod | 9 | | | Daughter of Martha Harrod; sister of Julius Harrod; 2 brothers in Alabama; **brother Elias and sister Betsy deserted, 1861–62** |
| Nancy Richardson | 63 | Retired spinner | "worthless" | Mother of Peter Richardson and 7 other children, 5 of whom are in Alabama; **son Peter deserted, 1863** |
| Peter Richardson | 29 | Sailor | "went to Yanks" | Husband of Alice Richardson; father of Ella, Harriet, and Richardann Richardson; **he deserted, May 1863** |
| Alice Richardson | 24 | Weaver | | Wife of Peter Richardson; mother of Becky, Ella, and Harriet Richardson; **brothers Richard Thomas and Jerry Glascow deserted, 1861; husband Peter deserted, 1863** |
| Harriet Richardson | 6 | | | Daughter of Alice and Peter Richardson; **2 uncles deserted, 1861; father deserted, 1863** |
| Ella Richardson | 3 | | | Daughter of Alice and Peter Richardson; **2 uncles deserted, 1861; father deserted, 1863** |
| Becky Richardson | 0 | | Born July 1862 | Daughter of Alice and Peter Richardson; **2 uncles deserted, 1861; father deserted,1863** |

| Name | Age | Occupation | Comments by Tayloe | Family connections |
|---|---|---|---|---|
| Jenny Glascow | 63 | Retired weaver | "worthless" | Mother of Alice Richardson; 1 son in Alabama; **son Richard Thomas deserted, 1861; son Jerry Glascow deserted, 1861; son-in-law Peter Richardson deserted, 1863** |
| Pressley Owens | 3 | | | Mother and 5 siblings in Alabama |
| Robert | 78 | Ferryman | "chargable" | Widower; 2 sons in Alabama |
| Seignor Wheeler | 68 | Field | "chargable" | Widower; father of Isabella Wheeler; 3 children in Alabama |
| Isabella Wheeler | 10 | | | Daughter of Seignor Wheeler; 3 siblings in Alabama |
| Tom Lawson | 49 | Retired fisherman | "chargable" | No wife or children identified |
| Tom Thomas | 49 | Carpenter | | Husband of Nancy Thomas; no children identified |
| Nancy Thomas | 42 | Spinner | | Wife of Tom Thomas; no children identified |
| William More | 48 | Blacksmith | | No wife or children identified |
| Delphy Wright | 55 | Field | | 4 surviving children all in Alabama |
| Sucky Page | 86 | Retired spinner | "worthless" | Mother of Tom Page, Winney Myers, and 9 other children: 2 are in Alabama and 2 were sold |
| Tom Page | 63 | Field | "chargable" | Son of Sucky Page; husband of Winney Page; 4 children in Alabama |
| Winney Page | 61 | Retired spinner | "chargable" | Wife of Tom Page; 4 children in Alabama |
| Winney Myers | 55 | Retired cook | "worthless" | Daughter of Sucky Page; mother of China Myers; 1 son in Alabama; **son Urias Myers deserted, 1862** |
| Marilla | 78 | Retired domestic | "worthless" | Mother of 6 children: 3 are in Alabama and **3 were sold** |
| Sally Jordan | ? | Retired | "worthless" | No husband or children identified |
| Evelina | 33 | Domestic | "worthless" | Hired to J. Clark; no husband or children identified |
| Fanny Bray | 71 | Retired cook | "worthless" | 2 sons in Alabama, **1 of them sold** |
| Ibby | 53 | Field | | Childless |
| Jemima D. | 14 | Chambermaid | | Parents unknown |
| Kesiah Brown | 52 | Overseer's cook | | Widow; 4 children in Alabama |

*Appendix 38* Field Laborers on August 1, 1834, and on October 29, 1838, at Eighteen Westmoreland Sugar Estates

| | 1834 | | | 1838 | | | | |
|---|---|---|---|---|---|---|---|---|
| Estate | Male | Female | Total | Male | Female | Total | Top wage | Job work |
| Bath | 53 | 72 | 125 | 30 | 10 | 40 | 1s. 8d. | 13s. per acre |
| Black Heath | 91 | 144 | 235 | 30 | 30 | 60 | 1s. 8d. | 16s. per acre |
| Blackness | 66 | 96 | 162 | 12 | 12 | 24 | ? | 18s. per acre |
| Blue Castle | 45 | 63 | 108 | 30 | 20 | 50 | ? | 20s. per acre |
| Carawina | 91 | 118 | 209 | 15 | 25 | 40 | 1s. 8d. | 20s. per acre |
| Fontabelle | 57 | 88 | 145 | 18 | 4 | 22 | 1s. 8d. | None |
| Fort William | 73 | 138 | 211 | 40 | 14 | 54 | 2s. 6d. | 20–30s. per acre |
| Friendship | 51 | 90 | 141 | 40 | 15 | 55 | ? | 16s. per acre |
| **Mesopotamia** | **81** | **90** | **171** | **60** | **50** | **110** | **2s. 6d.** | **20–30s. per acre** |
| Meylersfield | 116 | 183 | 299 | 20 | 20 | 40 | 1s. 8d. | ? |
| Midgham | 65 | 80 | 145 | 20 | 15 | 35 | ? | 16s. per acre |
| Mount Eagle | 71 | 102 | 173 | 20 | 20 | 40 | 1s. 8d. | 20s. per acre |
| New Hope | 29 | 50 | 79 | 30 | 15 | 45 | ? | 16s. per acre |
| Paul Island | 44 | 58 | 102 | 10 | 10 | 20 | ? | 20s. per acre |
| Petersfield | 87 | 113 | 200 | 40 | 20 | 60 | ? | 18s. per acre |
| Shrewsbury | 70 | 112 | 182 | 45 | 20 | 65 | ? | 18s. per acre |
| Spring Garden | 80 | 126 | 206 | 26 | 26 | 52 | 1s. 8d. | 18–20s. per acre |
| Three Mile River | 38 | 57 | 95 | 14 | 16 | 30 | ? | 22s. per acre |
| **Totals** | **1,208** | **1,780** | **2,988** | **500** | **342** | **842** | | |
| **Averages** | **67** | **99** | **166** | **28** | **19** | **47** | | |

*Sources:* For 1834, T71/723–725, NA-UK; for 1838, C.O. 137/232, p. 170, NA-UK.

# *Notes*

## *Abbreviations*

| | |
|---|---|
| Barham | The Barham Papers, Clarendon Manuscript Deposit, Bodleian Library, University of Oxford |
| BL | British Library, London |
| BOT | Benjamin Ogle Tayloe |
| *DNB* | *Dictionary of National Biography* |
| ETT | Edward Thornton Tayloe |
| HAT | Henry Augustine Tayloe, brother of William Henry Tayloe |
| HAT Jr. | Henry Augustine Tayloe, son of William Henry Tayloe |
| JA | The Jamaica Archives, Spanish Town |
| JB | John Foster Barham |
| JFB1 | Joseph Foster Barham I |
| JFB2 | Joseph Foster Barham II |
| JT1 | John Tayloe I |
| JT2 | John Tayloe II |
| JT3 | John Tayloe III |
| LVA | Library of Virginia, Richmond |
| MCHL | Moravian Church House, London |
| NA-UK | [British] National Archives, Kew |
| NA-US | [U.S.] National Archives, Washington, DC, and branches |
| Tayloe | The Tayloe Family Papers, Virginia Historical Society, Richmond |
| Thistlewood | Diary of Thomas Thistlewood, Monson 31, Lincolnshire County Archives, Lincoln, UK |
| UA | Unitätsarchiv, Herrnhut, Germany |
| UVA | Alderman Library, University of Virginia, Charlottesville |

VHS Virginia Historical Society, Richmond
WHT William Henry Tayloe
*WMQ* *The William and Mary Quarterly*

## Prologue

1. For background on Jamaica in the slave era, see Richard S. Dunn, *Sugar and Slaves: The Rise of the Planter Class in the English West Indies, 1624–1713* (Chapel Hill: University of North Carolina Press, 1972), chap. 5; Orlando Patterson, *The Sociology of Slavery: An Analysis of the Origins, Development, and Structure of Negro Slave Society in Jamaica* (Rutherford, NJ: Farleigh Dickinson University Press, 1969); Richard B. Sheridan, *Sugar and Slavery: An Economic History of the British West Indies, 1623–1775* (Baltimore: Johns Hopkins University Press, 1973); Andrew J. O'Shaughnessy, *An Empire Divided: The American Revolution and the British Caribbean* (Philadelphia: University of Pennsylvania Press, 2000); Edward Braithwaite, *The Development of Creole Society in Jamaica, 1770–1820* (Oxford: Oxford University Press, 1971); Roderick A. McDonald, *The Economy and Material Culture of Slaves: Goods and Chattels on the Sugar Plantations of Jamaica and Louisiana* (Baton Rouge: Louisiana State University Press, 1993); B.W. Higman, *Slave Population and Economy in Jamaica, 1807–1834* (Cambridge: Cambridge University Press, 1976); Higman, *Jamaica Surveyed: Plantation Maps and Plans of the Eighteenth and Nineteenth Centuries* (Kingston, Jamaica: University of the West Indies Press, 1988); and Christer Petley, *Slaveholders in Jamaica: Colonial Society and Culture during the Era of Abolition* (London: Pickering & Chatto, 2009).
2. There is a detailed description of all the Mesopotamia buildings (except for the slave village) in "The Present State of the Buildings on Mesopotamia Estate," June 1, 1814, Barham, b 34.
3. Obviously I am a rotten oral historian. I didn't even find out the name of the old age pensioner!
4. Sarah Arcedeckne's sister was Mary Barham, wife of Dr. Henry Barham; for more on the Barham family, see Chapter 1. The Arcedecknes were prominent Jamaica planters in the 1730s. Andrew Arcedeckne owned five sugar plantations in 1739, including Black Morass, which was close to Mesopotamia; he was probably Sarah's son. See Edward Long Papers, "Sugar Plantations in Jamaica with the Quantity of Sugars Made Generally for Some Years Past. Xmas 1739," Additional Manuscripts 12,434/1–12, BL.
5. For two highly informative analyses of archaeological investigations of slave villages in Jamaica, see Douglas V. Armstrong, *The Old Village and the Great House: An Archaeological and Historical Examination of Drax Hall Plantation, St. Ann's Bay, Jamaica* (Urbana: University of Illinois Press, 1990), and B.W. Higman, *Montpelier, Jamaica: A Plantation Community in Slavery and*

*Freedom, 1739–1912* (Kingston, Jamaica: University of the West Indies Press, 1998).

6. The Barham Papers are stored in the Clarendon Manuscript Deposit at the Bodleian Library. I thank the Earl of Clarendon for permitting me to use and cite the Barham Papers.
7. Eighty-five of the Mesopotamia inventories are in five boxes of the Barham Papers, b 34–38. They are dated 1736, 1743–1744, 1751–1752, 1754–1820, 1822–1832, with two inventories for the years 1756, 1763, and 1802. Another two slave inventories—the first and last of the Mesopotamia series—were taken in 1727 and 1833, and are in Jamaica Inventories, IB/11/3, vol. 14, p. 93, and vol. 150, p. 25, JA.
8. The Barham Papers also contain fifty-five slave inventories spanning the years 1757–1832 for Island estate in St. Elizabeth parish, a sugar plantation that JFB1 and JFB2 also owned. But the Island series is less complete and less informative than the Mesopotamia series, so I have not attempted to correlate or analyze the Island inventories.
9. There are Mesopotamia crop reports for sixty-six of the eighty-two years between 1751 and 1832, in Accounts Produce, vols. 4–75, JA. See Chapter 4.
10. The Mesopotamia diaries at Herrnhut span the dates 1758–1761, 1763–1790, 1799, 1812–1813, 1815, 1817, and 1831–1832. They are stored in the "Jamaika" section of the Unitätsarchiv, Herrnhut, Germany. See Chapter 6 and Appendix 27.
11. The Jamaican diaries and conference minutes relevant to Mesopotamia are all stored in the Fairfield section of the Moravian Archives of Jamaica, JA. The Minutes of Mesopotamia Conference volume, listing prayer sessions with the Saviour in 1798–1818, is an especially informative document; it is catalogued as Fairfield R/1. See Chapter 6.
12. I am of course by no means the first historian to present a case study of a Jamaican slave community in action. See in particular Michael Craton's study of the slaves at Worthy Park in *Searching for the Invisible Man: Slaves and Plantation Life in Jamaica* (Cambridge, MA: Harvard University Press, 1978), which is similarly based on an extended series of slave inventories. See also Philip D. Morgan, "Slaves and Livestock in Eighteenth-Century Jamaica: Vineyard Pen, 1750–1751," *WMQ* 52 (1995): 47–76; Armstrong's *The Old Village and the Great House;* and Higman's *Montpelier, Jamaica.*
13. For a perceptive discussion of Mount Airy's layout and design, see Dell Upton, "White and Black Landscapes in Eighteenth-Century Virginia," in *Material Life in America, 1600–1860,* ed. Robert Blair St. George (Boston: Northeastern University Press, 1988), 362–368. For further commentary on the architecture of the Mount Airy house and the landscape of the surrounding grounds, see Mills Lane, *Architecture of the Old South: Virginia* (Savannah, GA: Beehive Press, 1987), 68–69; Thomas Tileston Waterman and John A. Barrows, *Domestic Colonial Architecture of Tidewater Virginia* (New York, 1932; reprinted 1968, Da Capo Press), 125–137; and Thomas

Tileston Waterman, *The Mansions of Virginia, 1706–1776* (Chapel Hill: University of North Carolina Press, 1945), 253–260.

14. When historians like me focus on antebellum U.S. slavery, they necessarily challenge the family pride of people whose forbears were big slave owners, and I wish to thank the late Colonel and Mrs. H. Gwynne Tayloe and their sons Gwynne and William Tayloe for their many kindnesses to me as I have been conducting my research project.
15. See Chapter 7.
16. The Tayloes today call Mount Airy a farm, similar to Barham Farm. In 1809 the Mount Airy complex was called the Rappahannock Farms, while Mesopotamia (like all sugar production units in Jamaica) was known as an estate. Since large slave-based agricultural units are generically referred to as plantations, I use that label for both places in my book.
17. The Tayloe Family Papers in the VHS have been microfilmed in the *Records of Antebellum Southern Plantations from the Revolution through the Civil War, Series M,* reels 1–37. I thank the VHS for permission to use these documents. In researching this book I have primarily consulted the Tayloe manuscripts, but have also frequently used the microfilm edition to look for new documents or to recheck my research notes.
18. Tayloe d 538.
19. The Mount Airy slave inventories are dated 1808–1847, 1849–1855, and 1861–1865, and are found (chronologically) in Tayloe d 538, a 13, d 13410, d 13424, and d 23707.
20. Laura Croghan Kamoie, *Irons in the Fire: The Business History of the Tayloe Family and Virginia's Gentry, 1700–1860* (Charlottesville: University of Virginia Press, 2007), uses the Tayloe Papers to analyze the business practices of John Tayloe I, II, and III. Kamoie provides excellent background for my study, and I see our two approaches as complementary, since her focus is on the Tayloes, and mine is on their slaves.
21. See Chapter 5.
22. See Chapter 7.
23. The Tayloe Papers at the University of Virginia mainly contain family letters to BOT, who was WHT's older brother. This UVA collection of Tayloe Papers has been microfilmed in *Records of Antebellum Southern Plantations, Series E,* reels 1–5.
24. I am of course by no means the first historian to present a case study of an antebellum slave community in action. See in particular Lorena S. Walsh, *From Calabar to Carter's Grove: A History of a Virginia Slave Community* (Charlottesville: University of Virginia Press, 1997); Charles Joyner, *Down by the Riverside: A South Carolina Slave Community* (Urbana: University of Illinois Press, 1984); Jon F. Sensbach, *Separate Canaan: The Making of an Afro-American World in North Carolina, 1763–1840* (Chapel Hill: University of North Carolina Press, 1998); Edward Ball, *Slaves in the Family* (New York: Farrar, Straus and Giroux, 1998); and Brenda E. Stevenson, *Life in Black*

*and White: Family and Community in the Slave South* (New York: Oxford University Press, 1996), chaps. 6–10.

25. To complete my account of Mesopotamia, I decided to incorporate data from the fourteen earliest inventories in the Barham Papers, which have no age statements but provide useful information about the slaves alive in 1762. So I added new pages to my two Mesopotamia ledgers and worked backward from 1760 to 1736. I didn't yet know about the earliest Mesopotamia inventory in the Jamaica Archives, taken in 1727.
26. Mothers of newborns were not identified at Mesopotamia until 1774, but thereafter the mothers and the birth dates were always listed in an annual register. At Mount Airy, mothers of young children were always identified, but the Tayloes never kept a birth register, so the ages of children are usually approximate.
27. See Richard S. Dunn, "A Tale of Two Plantations: Slave Life at Mesopotamia in Jamaica and Mount Airy in Virginia, 1799 to 1828," *WMQ*, 3rd ser., 34 (January 1977): 32–65. In this essay I compared Mesopotamia from 1799 to 1818 with Mount Airy from 1809 to 1828.
28. Dunn, "A Tale of Two Plantations," 57, 64–65.
29. The review appeared in the *Newport News Daily Press*, January 16, 1977.
30. *The Age of Religious Wars, 1559–1689* (New York: W. W. Norton) was published in 1970; my second edition, *The Age of Religious Wars, 1559–1715*, with a new end date and a new final chapter, was issued by W. W. Norton in 1979.
31. *The Papers of William Penn, Volume One, 1644–1679* (1981) and *Volume Two, 1680–1684* (1982), ed. Mary Maples Dunn, Richard S. Dunn, Richard Ryerson, Jean Soderlund, and Scott Wilds; *Volume Three, 1685–1700* (1986) and *Volume Four, 1701–1718* (1987), ed. Richard S. Dunn, Mary Maples Dunn, Craig Horle, Alison Hirsch, Marianne Wokeck, and Joy Wiltenburg (Philadelphia: University of Pennsylvania Press).
32. *The Journal of John Winthrop, 1630–1649*, edited by Richard S. Dunn, James Savage, and Laetitia Yeandle (Cambridge, MA: Harvard University Press and the Massachusetts Historical Society, 1996). Laetitia Yeandle and I also produced an abridged, modernized text of *The Journal of John Winthrop* (Cambridge, MA: Harvard University Press, 1996).
33. See Herbert S. Klein, *Slavery in the Americas: A Comparative Study of Virginia and Cuba* (Chicago: University of Chicago Press, 1967); Carl Degler, *Neither Black nor White: Slavery and Race Relations in Brazil and the United States* (Madison: University of Wisconsin Press, 1971).
34. Compare Robert William Fogel and Stanley L. Engerman, *Time on the Cross: The Economics of American Negro Slavery*, 2 vols. (Boston: Little, Brown, 1974), with Herbert G. Gutman's sharply critical review essay, "The World Two Cliometricians Made," *Journal of Negro History* 60 (1975): 54–227. And for anti-cliometric approaches, see Gutman, *The Black Family in Slavery and Freedom, 1750–1925* (New York: Pantheon, 1976), and Lawrence W. Levine,

*Black Culture and Black Consciousness: Afro-American Folk Thought from Slavery to Freedom* (New York: Oxford University Press, 1977).

35. For example, Rhys Isaac, *The Transformation of Virginia, 1740–1790* (Chapel Hill: University of North Carolina Press, 1982); Jeffrey Bolster, *Black Jacks: African American Seamen in the Age of Sail* (Cambridge, MA: Harvard University Press, 1997); Douglas Egerton, *Gabriel's Rebellion: The Virginia Slave Conspiracies of 1800 and 1802* (Chapel Hill: University of North Carolina Press, 1993); Billy G. Smith and Richard Wojtowicz, *Blacks Who Stole Themselves: Advertisements for Runaways in the Pennsylvania Gazette, 1728–1790* (Philadelphia: University of Pennsylvania Press, 1989).
36. Stephanie Smallwood, *Saltwater Slavery: A Middle Passage from Africa to American Diaspora* (Cambridge, MA: Harvard University Press, 2007); Alexander Byrd, *Captives and Voyagers: Black Migrants across the Eighteenth-Century British Atlantic World* (Baton Rouge: Louisiana State University Press, 2008); Trevor Burnard, *Mastery, Tyranny, and Desire: Thomas Thistlewood and His Slaves in the Anglo-Jamaican World* (Chapel Hill: University of North Carolina Press, 2004); and Vincent Brown, *The Reaper's Garden: Death and Power in the World of Atlantic Slavery* (Cambridge, MA: Harvard University Press, 2010).
37. See Paul Gilroy, *The Black Atlantic: Modernity and Double Consciousness* (Cambridge, MA: Harvard University Press, 1993).
38. See Lisa A. Lindsay and John Wood Sweet, eds., *Biography and the Black Atlantic* (Philadelphia: University of Pennsylvania Press, 2014).
39. Jon F. Sensbach, *Rebecca's Revival: Creating Black Christianity in the Atlantic World* (Cambridge, MA: Harvard University Press, 2005); Sydney Nathans, *To Free a Family: The Journey of Mary Walker* (Cambridge, MA: Harvard University Press, 2012).
40. One measure of the difference between Rebecca Protten and Mary Walker is that Rebecca sat for a portrait painted by a Moravian artist (illustrated on pp. 66, 198 in Sensbach, *Rebecca's Revival*), whereas Nathans's twenty-page set of illustrations (between pp. 150 and 151 in *To Free a Family*) conspicuously lacks a photograph of Mary Walker.

### *1. Mesopotamia versus Mount Airy: The Demographic Contrast*

1. See Philip D. Curtin, *The Atlantic Slave Trade: A Census* (Madison: University of Wisconsin Press, 1969), 268.
2. See David Eltis, "The Volume and Structure of the Transatlantic Slave Trade: A Reassessment," *WMQ*, 3rd ser., 58 (2001): 35–37, 45.
3. See the summary statistics table in the Trans-Atlantic Slave Trade Database (www.slavevoyages.org).
4. The Trans-Atlantic Slave Trade Database total for Jamaica as of 1810 is 1,017, 206, including 29,594 slaves landed between 1806 and 1810. I assume that almost all of these slaves were delivered between 1806 and 1807.

5. Writing in 2001, David Eltis estimated in "The Volume and Structure of the Transatlantic Slave Trade," pp. 36, 45, that 1,070,000 Africans were imported to Jamaica, of whom 210,000 were reexported to the Spanish colonies.
6. The Database total for Virginia is somewhat higher than Lorena S. Walsh's calculation in "The Chesapeake Slave Trade: Regional Patterns, African Origins, and Some Implications," in *WMQ*, 3rd ser., 58 (2001): 166–169. Walsh estimated that about 78,000 Africans were imported to Virginia, 1698–1774, while the Database projects 95,025 imports for these years.
7. Edmund Stephenson's land purchases are in Barham c 360. He is described as the founder of Mesopotamia on his son Ephraim's tombstone; see Philip Wright, *Monumental Inscriptions of Jamaica* (London: Society of Genealogists, 1966), 192.
8. "A New and Exact Mapp of the Island of Jamaica" by Charles Bochart and Humphrey Knollis, published in *The Laws of Jamaica* (London, 1684), identifies 247 sugar works on the island, almost all of them in the eastern and central parishes. Three of the seven Westmoreland sugar works are located on the Cabarita River, and two of these—situated on opposite banks of the river on the site of Mesopotamia—are labeled "Stephens" and "Collins." The "Stephens" works was operated by Edmund Stephenson, and the "Collins" works was operated by Richard Collins. In 1698 Ephraim Stephenson acquired Collins's 630 acres and annexed this land to Mesopotamia. See Barham c 360.
9. The land acquisitions of Ephraim Stephenson are reported in Barham c 360 and c 376. Ephraim Stephenson's will, dated August 4, 1726, is in Barham c 386. His probate inventory for Mesopotamia estate, dated August 16, 1727, is in 1B/11/3, vol. 14, p. 93, JA.
10. I cannot trace Kickery's baby. In 1745 she had another child, known as Young Kickery, who was trained by her mother as a midwife and continued at this job from 1778 to her death in 1810.
11. Mesopotamia attorney James Graham thought that "Old Love . . . from the best account, could not have been less than 100 to 105 years, of age" when she died in 1795. Graham also told JFB2 that Love had thirty-nine descendants (Barham c 357). She married driver Matt, the slave leader at Mesopotamia in the 1760s. See Chapter 6.
12. I have not found the date of Mary Stephenson Heir's marriage to Henry Barham, but it took place before May 23, 1728, when they drew up a joint indenture (Barham c 386).
13. See *DNB*, vol. 3, p. 186.
14. Henry Barham's surviving Westmoreland business papers for 1728–1736 are in Barham c 360 and c 376. They include letters from the merchant Isaac Lomego in 1737–1739 about paying interest on his bonds (Barham c 376).

15. Wright, *Monumental Inscriptions of Jamaica,* 192. Mary Barham died on May 26, 1735, just four days after the death of her sister Sarah Arcedeckne, whose tombstone is described in the Prologue. Mary Barham's tombstone lies close to Sarah's, hidden under tropical vegetation in the Mesopotamia graveyard.
16. Barham's Mesopotamia inventory, dated April 18, 1736, is in Barham b 37.
17. Barham was a friend of Elizabeth Foster and her first husband, John Foster, in Jamaica. He seems to have married Elizabeth shortly after he arrived in England in 1736, because Isaac Lomego, a Kingston merchant, wrote on January 10, 1737, to congratulate him on his safe arrival and his new marriage with "so a Happy Lady, suitable to your merritts and fortune" (Barham c 376).
18. Barham was an executor of John Foster's estate in 1732. Foster's probate inventory, dated March 24, 1731/32, is in Barham c 386 and also in JA 1B/11/3, vol. 16, p. 15. For background on the Foster family, see Alkman Henryson Foster-Barham, *Genealogy of the Descendants of Roger Foster of Edreston, Northumberland* (London, 1897).
19. Joseph Foster's Jamaican inheritance, Island sugar estate, was situated in St. Elizabeth parish, about fifty miles by tortuous road southeast of Mesopotamia. It was a new plantation with seventy-two slaves when his father died in 1731. By 1757 the slave population had climbed to 170, but Island always had a smaller labor force and a lower sugar output than Mesopotamia.
20. Barham's will is in PROB11/749, NA-UK.
21. For JFB1's visit to Jamaica in 1750–1751, see Barham c 375 and c 376.
22. This was the advice that Vassall later gave to JFB1 in a letter dated May 4, 1767 (Barham c 376). There are also several letters from Vassall to JFB1 in 1750–1751, when they were both in Jamaica, in Barham c 376.
23. George (age unknown) seems to have served JFB1 in England from 1752 to 1760, and then was either freed or died.
24. Pool was JFB1's attorney at Mesopotamia for most of the time between 1751 and 1777.
25. See Chapter 6. I am greatly indebted to Dr. John Mason for his help in explicating JFB1's early relations with the Moravian Church.
26. There are comprehensive annual Mesopotamia sugar production figures for the years 1747–1833 in the Mesopotamia crop accounts in Accounts Produce, 1B/11/4, vols. 1–75, JA, supplemented by lists in Barham b 34, Barham b 37, and Barham c 360. For fuller discussion of Mesopotamia sugar production, see Chapter 4.
27. Daniel Barnjum to JFB1, July 19, 1762, Barham c 360. The 1762 inventory is in Barham b 37.
28. I have found no Mesopotamia inventories for the years 1737–1742, 1745–1750, and 1753, no death registers until 1751, and no birth registers until 1762.

29. Daniel Barnjum observed to JFB1 on July 15, 1768, that many of his "once ablest Negroes are now become old and useless" (Barham c 357).
30. In 1762, twenty-five of the forty-three Mesopotamia slave workers in their twenties had been imported. Among these were nineteen of the twenty-two new Africans who had arrived in 1756 and 1759; the other three had died.
31. This is my calculation from the Mesopotamia crop accounts in Accounts Produce, 1B/11/4, vols. 1–75, JA.
32. Barnjum to JFB1, December 23, 1760, July 19, 1762 (Barham c 360); Barnjum to JFB1, July 20, 1764, April 22, 1765, May 20, 1768 (Barham c 357).
33. For a summary of the evidence on male versus female mortality in Jamaica, see my essay "Sugar Production and Slave Women in Jamaica," in *Cultivation and Culture: Labor and the Shaping of Slave Life in the Americas,* ed. Ira Berlin and Philip D. Morgan (Charlottesville: University Press of Virginia, 1993), 49–52, 55.
34. The disparity between recorded girl and boy births, 1736–1762, is hard to believe: I have found only fifty-five baby girls in the Mesopotamia records for these years, as against seventy-seven boys. There was no such disparity after 1762.
35. Were they dying of starvation? We will see from the reports of Moravian missionaries in Chapter 6 that semistarvation was a recurrent condition at Mesopotamia. The death register for 1777 (Barham b 37) reports that three newborn babies died of convulsions, seven adults died of old age, five had the flux (or diarrhea, which may well indicate severe malnutrition), two had the yaws, two died from an inflammatory fever, and one had ulcers.
36. John Van Heilen to JFB1, January 8, 1786; John Wedderburn to JFB1, July 14, 1786 (Barham c 357). The Three Mile River slaves were priced at £3,150, or £79 per person, and are listed in family groups in Barham b 33.
37. See letters to JFB2 in Germany ca. 1774 from his parents, in Barham c 388. Dr. John Mason has kindly supplied me with additional information about JFB2's years in Germany.
38. JFB2's Jamaican accounts indicate that he arrived on the island in June 1779 and departed in July 1781.
39. JFB2's repairs to the Great House at Mesopotamia cost more than £500. For details, see Barham b 33, b 34.
40. Van Heilen to JFB1, April 7, 1781, June 4, 20, 1781; and Van Heilen to JFB2, April 27, 1782 (Barham c 357).
41. December 5, 1781, MO 356, Moravian Archives, Bedfordshire County Record Office. Dr. John Mason kindly supplied me with this reference.
42. See JFB2's obituary in *Gentleman's Magazine and Historical Chronicle* (London) 102, pt. 2 (1832): 573.

43. JFB2, Memorandum on the Negroes at Mesopotamia, n.d. (ca. 1825), Barham c 375; JFB2 to John Wedderburn and James Graham, June 23, 1792 (Barham c 428).
44. For JFB2's interest in bringing Chinese laborers to the West Indies, see the collection of documents in Barham c 366.
45. Entries on JFB2 in *The History of Parliament: The House of Commons, 1790–1820,* ed. R. Thorne, 5 vols. (London: Secker and Warburg, 1986); *The House of Commons, 1820–1832,* ed. D.R. Fisher, 7 vols. (Cambridge: Cambridge University Press, 2009).
46. For a full-scale discussion of "amelioration" at work, see J. R. Ward, *British West Indian Slavery, 1750–1834: The Process of Amelioration* (Oxford: Oxford University Press, 1988).
47. JFB2 to John Wedderburn and James Graham, September 8, 1789, October 16, 1790 (Barham c 428).
48. Wedderburn and Graham to JFB2, October 2, 1790; Graham to JFB2, March, 14, 18, 1791, September 7, 1792, December 8, 1792 (Barham c 357); JFB2 to Wedderburn and Graham, June 23, 1792 (Barham c 428). There is a list, dated March 9, 1791, of the Southfield slaves in family groups and identified by age, occupation, condition, and value, in Barham b 36.
49. JFB2 to H. W. Plummer and William Rodgers, n.d., Barham c 428.
50. Graham to JFB2, September 6, 1790; Wedderburn to JFB2, February 19, 1791 (Barham c 357).
51. There is a list of the Cairncurran slaves, dated January 29, 1814, and parallel in style to the 1791 Southfield list, in Barham b 34.
52. JFB2 to J.C. Grant and John Blyth, December 9, 1818, August 4, 1819 (Barham c 428); Grant and Blyth to JFB2, January 25, 28, July 5, 1819 (Barham c 358).
53. JFB2 paid about £21,000—or £91 per person—for the 221 people he acquired from Southfield, Cairncurran, and Springfield.
54. The ratio between Africans and Creoles can be estimated pretty accurately because the home government in 1817 mandated a comprehensive registration of all the British Caribbean slaves. In Jamaica, 345,252 slaves were recorded by name, sex, age, color, and origin, and 37 percent of them were African. At Mesopotamia in 1817, only 20 percent were African (T71/178, NA-UK). At Springfield in 1817 (one year before JFB2 bought the estate), 55 percent were African (T71/190, NA-UK). The registration evidence indicates that in 1817 about 45 percent of the Three Mile River, Southfield, and Cairncurran slaves living at Mesopotamia were African. For background, see B. W. Higman, *Slave Population and Economy in Jamaica, 1807–1834* (Cambridge: Cambridge University Press, 1976), 45–46, 75.
55. J. F. Barham, *Considerations on the Abolition of Negro Slavery: And the Means of Practically Effecting It* (London: James Ridgeway, 1823), 2–10, 14–18, 23, 35, 54–56, 79.

56. The Jamaican probate inventory for JFB2, dated August 23, 1833, is in JA 1B/11/3, vol. 150, p. 25. His 329 Mesopotamia slaves were given a much higher per capita valuation than his 327 slaves at Island sugar estate and at Windsor pen. They were priced collectively at £10,771.
57. JB's compensation claim for Mesopotamia is in T71/723, and his award is in T71/1328, NA-UK.
58. Lady Katherine Grimston, the widow of John Barham, married George Villiers, the fourth Earl of Clarendon, in 1839, and her first husband's family papers are now in the Clarendon Manuscript Deposit at the Bodleian Library, Oxford.
59. Douglas Hall, *In Miserable Slavery: Thomas Thistlewood in Jamaica, 1750–86* (Mona, Jamaica: University of the West Indies Press, 1999), 135.
60. For African background, see John Thornton, *Africa and Africans in the Making of the Atlantic World, 1400–1800,* 2nd ed. (Cambridge: Cambridge University Press, 1998), chaps. 2–4.
61. By 1833 the great majority of the fifty African slaves at Mesopotamia had come from Springfield in 1819. There were also a few Africans from Three Mile River, Southfield, and Cairncurran. Only 11 of the 137 Africans who came to Mesopotamia directly from the slave ships were still alive.
62. Another nine slaves (seven males and two females) are excluded from Appendix 5, because four were hired hands not owned by the Tayloes, and five were born in 1864–1865.
63. The recorded birth figures are firm: 502 new babies appear in the Virginia inventories, and 134 in the Alabama inventories. The 142 deaths in Virginia, 1809–1828, could be too high or too low, because it is unclear whether some of the people who disappeared from the records in these years died or moved away.
64. For a full discussion of the rise of the Tayloes as leading Virginia entrepreneurs, see Laura Croghan Kamoie, *Irons in the Fire: The Business History of the Tayloe Family and Virginia's Gentry, 1700–1860* (Charlottesville: University of Virginia Press, 2007).
65. On this topic I have greatly benefited from reading in manuscript John C. Coombs's book-in-progress, entitled "The Rise of Virginia Slavery, 1630–1730."
66. William Tayloe's public career can be traced in the earliest volumes of the Richmond County records, 1692–1724, LVA. His probate inventory, dated March 7, 1710, is in the Richmond County Wills and Inventories, 1709–1717, LVA.
67. JT1's account of his sales from a Bristol slave ship with 230 Africans in 1717 is in the Loyde-Tayloe account book, 1708–1778, Tayloe b 1. An account of Tayloe's management of sales from two other Bristol ships with 404 slaves in 1723 is in Elizabeth Donnan, ed., *Documents Illustrative of the History of the Slave Trade to America* (Washington, DC: Carnegie Institution, 1931–1935), 4:100–102, 185, 187.

68. JT1's will, dated January 31, 1744, is in Tayloe d 149; the probate inventory of his estate, dated November 2, 1747, is in Richmond County Wills and Inventories, 1725–1753, LVA. See also Kamoie, *Irons in the Fire,* chap. 1. Her slave totals for JT1 differ very slightly from mine.
69. Philip D. Morgan estimates that 21 percent of the Virginia slaves in 1750 were African-born. See Morgan, *Slave Counterpoint: Black Culture in the Eighteenth-Century Chesapeake and Low Country* (Chapel Hill: University of North Carolina Press, 1998), 61.
70. The nine slaves specifically identified as "child," "boy," or "girl" are priced at £22 on average. Five babies are lumped with their mothers and given no value. Only three slaves are identified as "old"; they are valued at £17 on average.
71. Tom and Sucky both appear in the Mount Airy tax lists taken in the 1780s (see later discussion), which increases the possibility that they lived their entire lives at Mount Airy, though they have very common slave names. Several other elderly slaves in 1809—Robin (married to Sucky), Dick, John, and Moses among the men; Judy and Patty among the women—could also have been at Mount Airy in 1747.
72. Fithian spent a year tutoring Robert Carter's children at nearby Nomini Hall. See Hunter Dickinson Farish, ed., *Journal and Letters of Philip Vickers Fithian, 1773–1774* (Williamsburg, VA: Colonial Williamsburg, Inc., 1943), 126–127.
73. JT2 to Edward Lloyd, November 27, 1775, Lloyd Papers, Maryland Historical Society, Baltimore.
74. See the list of JT2's lands (ca. 1770) in Tayloe d 161–162; see also Kamoie, *Irons in the Fire,* chaps. 2–3.
75. Advertisements placed by JT2 for runaway convicts appeared in the *Virginia Gazette* on September 2, 1757, December 24, 1772, July 8, 1773, October 7, 1773, September 22, 1774, and November 10, 1774.
76. JT2's will, dated May 22, 1773, is in Tayloe d 166 as well as in the Richmond County Will Book, 1767–1787, LVA. The will was proved on July 5, 1779.
77. See *Alumni Cantabrigienses: A Biographical List of All Known Students, Graduates, and Holders of Office at the University of Cambridge, from the Earliest Times to 1900,* ed. J.A. Venn, 10 vols. (Cambridge: Cambridge University Press, 1922–1953), pt. 2, vol. 6, p. 119.
78. The Mount Airy tax lists for 1783–1787 are incomplete and have to be tracked in three counties. Most of the slaves on this estate lived in Richmond, but some were on farm quarters in neighboring King George and Essex. The returns for 1785 and 1787 state the total Mount Airy slave population, while the Richmond return for 1783 and the Essex return for 1786 are the most informative because they list all the slaves by name and distinguish the taxable workers from the nontaxable children and old people.

79. For a discussion of JT3's switch from tobacco to grain, see Chapter 5.
80. Tayloe advertised his sale in the *Virginia Gazette and General Advertiser* (Richmond), the *Virginia Herald and Fredericksburg Advertiser* (Fredericksburg), the *Virginia Chronicle and Norfolk and Portsmouth General Advertiser* (Norfolk), the *Virginia Gazette and Alexandria Advertiser* (Alexandria), and the *Maryland Gazette* (Annapolis). I owe these newspaper references to Michael L. Nicholls and Michael Mullin.
81. The 1809 inventory is entitled "A General Inventory Taken at John Tayloes Esq Mount Airy Department the Beginning of January 1809" (Tayloe d 538).
82. For discussion of slave labor at Mount Airy during JT3's ownership, see Chapter 5.
83. The only time when the Mesopotamia gang had a larger proportion of young workers than the Mount Airy gang was in 1784–1792, when the Barhams acquired nearly a hundred new young slaves to bolster their workforce (see Appendix 3). But even in these years the Mesopotamia population was far more crippled by ill health than the Mount Airy population.
84. Of the 382 slaves inventoried in 1809, 134 were sold or transferred to other family work sites, and 70 died between 1809 and 1828, so that only 178 were still living on the estate when JT3 died.
85. See Chapter 5.
86. For discussion of this 1816 sale, see Chapter 5.
87. JT3's will, dated December 1827, is in Tayloe d 539. For discussion of the settlement of JT3's estate in 1828, see Chapter 5.
88. For fuller details, see Chapter 7.
89. The manuscript slave schedules for WHT's three plantations in 1850—Oakland, Adventure, and Walnut Grove—are very carelessly compiled. They are in U.S. Census, 1850, Schedule 2, Marengo County, Alabama, NA-US. The slave schedule for Mount Airy in 1850, which is equally defective, is in U.S. Census, 1850, Schedule 2, Richmond County, VA, NA-US.
90. Lists of WHT's wheat and corn production at Mount Airy from the 1830s into the 1860s are in Tayloe d 8521–8527, d 8528–8530, d 13410.
91. WHT's cotton receipts for 1858–1860 are in Tayloe d 19688–19830, d 19831–20010, d 20011–20126.
92. The manuscript slave schedules for Oakland and Larkin in 1860, which have the same defects as the 1850 schedules, are in U.S. Census, 1860, Schedule 2, Marengo and Perry Counties, Alabama, NA-US. The slave schedule for Mount Airy is in U.S. Census, 1860, Schedule 2, Richmond County, Virginia, NA-US.
93. WHT's Alabama census for 1863 is in Tayloe d 13453; HAT Jr.'s Mount Airy census for 1863 is in Tayloe d 8539–8590.
94. WHT to Courtenay Tayloe, June 8, 1865, Tayloe d 5292–5327.

95. For an island-wide analysis of this documentation, see Higman, *Slave Population and Economy in Jamaica.* I am greatly indebted to Barry Higman for his comprehensive analysis of Jamaican slavery in its final phase.
96. For 1740, see Edward Long Papers, Additional Manuscripts, 12,434/14–15, BL; for 1788 see Long, Additional Manuscripts, 12,435/43, BL. For 1807, the poll tax count of 21,414 should be increased by 1,600 to include the slaves held by nontaxpayers.
97. The thirty-six sugar estates with 200-plus slaves in 1807 are listed in a printed poll tax return, *Of Land, Slaves, Stock, Wheels, and Persons Saving Deficiency, for the Parish of Westmoreland, March 28, 1807* (Barham, b 34).
98. The twenty-eight estates with 200-plus slaves in 1814 are listed in a printed poll tax return, *Of Land, Slaves, Stock, Wheels, and Persons Saving Deficiency, for the Parish of Westmoreland, March 28, 1814* (Barham b 34).
99. Higman, *Slave Population and Economy in Jamaica,* 61, 256.
100. A Mesopotamia bookkeeper named Robert Williamson owned eleven slaves plus three horses. And Patrick Knight, the resident white carpenter at Mesopotamia who fathered several mulatto children on the estate, owned 120 acres, twenty-seven slaves, three horses, and a chaise.
101. Two of Wedderburn's properties—Spring Garden and Retreat—are listed among the twenty-seven estates in Appendix 9 and Appendix 10.
102. Higman, *Slave Population and Economy in Jamaica,* 255–256.
103. The Westmoreland slave registration for 1817 is in T71/178, NA-UK. Subsequent Westmoreland registrations for 1820–1832 are in T71/179–189, NA-UK.
104. The Mesopotamia inventories never identify people as African-born except sometimes by name such as "Chamba Jack" or "Coromantee Adam."
105. On Springfield estate in Hanover parish, which JFB2 purchased in 1818, more than half of the slaves in 1817 were African-born (63 out of 114) and only 4 of them were colored.
106. The Mesopotamia and Springfield returns for 1817 are in T71/178 and T71/190; for 1820, in T71/179 and T71/191; for 1823, T 71/180; for 1826, T71/181; for 1829, T71/182; for 1832, T71/188; and for 1834, T71/1007, NA-UK.
107. I have found returns for fifty-three Westmoreland sugar estates with over one hundred slaves in 1817–1834, but have omitted from my analysis eleven estates with incomplete records.
108. The compensation claims filed in 1834 for Westmoreland parish are in T71/723–725, NA-UK.
109. Four estates listed as Fontabelle, Fort William, Friendship, and Springfield are actually Fontabelle + Bellisle, Fort William + Roaring River, Friendship + Greenwich, and Springfield + Silver Spring. Springfield estate in Westmoreland parish should not be confused with Springfield estate in Hanover parish, purchased by Barham in 1818.

110. For discussion of the compensation claim data, see Higman, *Slave Populations of the British Caribbean, 1807–1834* (Baltimore: Johns Hopkins University Press, 1984), 46–47. While this data provides the best evidence that we have for the structure of the slave labor force on the eve of emancipation, most claimants probably exaggerated the number of prime workers and minimized the number of marginal workers or nonworkers in order to obtain as much government reimbursement as possible.
111. Because the handwritten Virginia manuscript census schedules for 1810 are difficult to decipher and are occasionally illegible, the census totals presented here (and in Appendix 11) are only approximate.
112. Richard S. Dunn, "Black Society in the Chesapeake, 1776–1810," in *Slavery and Freedom in the Age of the American Revolution*, ed. Ira Berlin and Ronald Hoffman (Charlottesville: University Press of Virginia, 1983), 49–62.
113. Michael Tadman, *Speculators and Slaves: Masters, Traders, and Slaves in the Old South* (Madison: University of Wisconsin Press, 1989), 12; Allan Kulikoff, "Uprooted Peoples: Black Migrants in the Age of the American Revolution, 1790–1820," in Berlin and Hoffman, *Slavery and Freedom in the Age of the American Revolution*, 147–152.
114. Tadman, *Speculators and Slaves*, 12.
115. Dunn, "Black Society," 59–65.
116. Louis Morton, *Robert Carter of Nomini Hall: A Virginia Tobacco Planter of the Eighteenth Century* (1941; repr., Charlottesville: University Press of Virginia, 1964), chap. 11.

## *2. Sarah Affir and Her Mesopotamia Family*

1. The Mesopotamia slave inventory for 1767, dated December 31, is in Barham b 37.
2. Douglas Hall, *In Miserable Slavery: Thomas Thistlewood in Jamaica, 1750–86* (Mona, Jamaica: University of the West Indies Press, 1999), 186–187, 189, 195, 208, 212. See also Richard B. Sheridan, *Doctors and Slaves: A Medical and Demographic History of Slavery in the British West Indies, 1680–1834* (Cambridge: Cambridge University Press, 1985), 200–201, 236–239.
3. Vincent Brown, *The Reaper's Garden: Death and Power in the World of Atlantic Slavery* (Cambridge, MA: Harvard University Press, 2010), 4.
4. See M. G. Lewis, *Journal of a West Indian Proprietor, Kept during a Residence in the Island of Jamaica* (London: John Murray, 1834), 240.
5. By 1804 the Mesopotamia grass gang had been subdivided into the hogmeat or fourth gang (for the most junior field workers) and the third gang (for young teenagers).
6. When Peg died of measles in 1784, the Moravian missionary Brother Taylor noted that she had been baptized in 1766 but fell away from the church as a "strayed sheep" until 1782. "She was sickly for a long time before her

end: which made her sometimes wish for her departure before it came: she was a person of an upright turn, & counted an honest & faithful Negroe by the overseer." Jamaica diary, May 1783–May 1785, UA R.15. Cb.4.7, p.40.

7. Mesopotamia males had a somewhat higher incidence of yaws than females, with 100 out of 166 recorded cases between 1762 and 1831. This was the standard pattern; see Sheridan, *Doctors and Slaves*, 83–85.
8. The inventory for January 1, 1785, is in Barham b 37. The Mesopotamia yaws house is described in an unusually full and detailed account of the plantation, drawn up in 1802 (Barham b 36). At this date seven patients (one of them being Affy's son Robert) were sequestered in the yaws house, attended by a slave nurse.
9. John Graham was the Mesopotamia overseer from 1778 to 1788; James Graham, probably his father, was JFB2's attorney from 1790 to 1799.
10. Mesopotamia diary, June 1782–April 1783, UA R.15.Cb.6.4, pp.4–5. Brother and Sister Taylor responded to Francisco that they "could not intermeddle in the affair of the overseer, and that we were here only for his [Francisco's] souls good, & had nothing to do with the outward matters."
11. Cudjoe's death is reported in the inventory for January 1, 1798, Barham b 34. The case was considered by a local jury, which acquitted Hector of murder.
12. Canes to cut for 1778 crop, in Barham b 34. Each cane piece had a name: Cedar Tree, Ralphs Hole, Lookout, Bridge, Cow Pen, Fig Tree, Pear Tree, etc. Barnjum's projection of 151 hogsheads turned out to be greatly exaggerated; Mesopotamia shipped only 92 hogsheads of sugar in 1778 (Accounts Produce, 1B/11/4/9/95, JA).
13. Unlike Affy, Doll had no recorded children. But like Affy she became a member of the Moravian Church, receiving baptism in 1802 (which was twelve years before Affy's baptism). And again like Affy, she was a tough, long-lived woman. When the last Mesopotamia inventory was taken in August 1833, Doll was valued at £5, which means that she was still working—probably as a washerwoman—at age sixty-five.
14. The two kitchen-washhouses are described in "The Present State of the Buildings on Mesopotamia Estate," June 1, 1814, Barham b 34.
15. For a full discussion of the Moravian mission at Mesopotamia, see Chapter 6.
16. Joseph Jackson to JFB2, November 7, 1804, May 6, 1805, December 17, 1805, Barham c 378.
17. Minutes of Mesopotamia Conference, 1798–1818, Moravian Archives of Jamaica, Fairfield R/1, JA. I thank Dr. John W. Pulis, who found this minute book in the JA and kindly photocopied it for me. For discussion of the conference book, see Chapter 6.
18. While the standard Moravian practice was to use three papers, the Jacksons may have used only two, since 48 percent of the answers they ob-

tained between 1801 and 1808 were yes. For discussion of the Moravian use and manipulation of the lot, see Gillian Lindt Gollin, *Moravians in Two Worlds: A Study of Changing Communities* (New York: Columbia University Press, 1967), chap. 3; and Jon F. Sensbach, *A Separate Canaan: The Making of an Afro-Moravian World in North Carolina, 1763–1840* (Chapel Hill: University of North Carolina Press, 1998), 26, 65, 107–108.

19. Annotated description of the Mesopotamia slave population, June 1802, Barham b 36. According to the inventory of January 1, 1810 (Barham b 36), there were sixteen working mothers on the estate with twenty-two young children aged two to five who needed looking after. Two of these children were cared for by their seventy-one-year-old grandmother, and the other twenty by the three nurses.
20. Princess had been listed as "able" in December 1810 and was in her eighth year on the first field gang. The death register stresses that she died "very suddenly."
21. Minutes of Mesopotamia Conference, 35–36, 42–43, 45–46, 49–50.
22. There is no surviving record of any of these four baptisms. Robert and Jane seem to have joined the Moravian congregation between 1818 (when Brother Gründer died) and 1830 (when Brother Ricksecker resumed record keeping at Mesopotamia). They are both listed among the baptized Christians on January 1, 1832 (Barham b 35). Davy and John seem to have joined the congregation in 1832 or 1833; they are both listed among the baptized Christians on August 23, 1833 (Jamaica Inventories, 1B/11/3, vol. 150, p. 25).
23. The Mesopotamia return, dated June 28, 1817, is filed among the Westmoreland parish slave registration returns for 1817 in T71/178, NA-UK. Barham's attorneys made subsequent slave registration returns in 1820, 1823, 1826, 1829, and 1832, but since these list only triennial births and deaths they provide no further information about Affy. For background on the Jamaican slave registration, see B. W. Higman, *Slave Population and Economy in Jamaica, 1807–1834* (Cambridge: Cambridge University Press, 1976), 45–46, 256.
24. So reported in the death register accompanying the January 1, 1831, inventory, Barham b 35.
25. The final Mesopotamia slave inventory of August 23, 1833, was a probate inventory taken after the death of JFB2 and recorded in Jamaica Inventories, 1B/11/3, vol. 150, p. 25, JA. Unlike the annual inventories compiled by the estate bookkeepers, it supplies monetary valuations for each of the 332 Mesopotamia slaves belonging to Barham, but gives no information about occupation, health, or family connections.
26. Collectively, the 332 Mesopotamia slaves in August 1833 were appraised at £20,195, which averages out to £61 per person.
27. John Blyth and J. C. Grant complained to JFB2 on October 17, 1819 (Barham c 358).

28. The birth register is in the inventory for January 1, 1794, Barham b 36.
29. Andrew McAlpin is listed as a bookkeeper in the inventories of 1791–1797 (Barham b 36), and his initial salary is reported in the plantation accounts for 1790–1791 (Barham b 33).
30. Robert McAlpin's short stay is recorded in the inventory for January 1, 1794, but his employment (if he had any) is not stated.
31. A man by the name of Andrew McAlpin—very possibly the former Mesopotamia bookkeeper—died in Westmoreland parish in 1815. He seems to have been an overseer, and was almost penniless, owning only a horse, saddle, and bridle and £5 in wearing apparel, plus a claim to £65 in back salary. See Jamaica Inventories, 1B/11/3/127, p. 8, JA.
32. Robert and the other inmates of the yaws house are identified in a detailed description of the Mesopotamia slave population in June 1802 (Barham b 36).
33. Horsley was paid £13 6s. 8d. for these operations. He also looked after the whites on the estate and was paid about £125 per year for his medicines and attendance according to the estate accounts for 1802 in Barham b 34. See also William Rogers to JFB2, July 5, 1802, March 11, 1805, Barham c 357.
34. My interpretation of this episode is conjectural. The Mesopotamia death register for 1806 reports that Tamerlane was kicked and killed "by a mulatto boy." Robert seems to me the most plausible candidate, but there were four other mulatto boys at Mesopotamia in March 1806 who could have done the deed. The oldest of them, Alexander, a seventeen-year-old carpenter, seems an unlikely choice because he had been badly ruptured since the age of six. John, a twelve-year-old waiting boy, was likewise in poor physical shape and had recently been hospitalized with the yaws. Andrew, an eleven-year-old apprentice carpenter, was a lot younger than Tamerlane and hence an unequal combatant. And fourteen-year-old Thomas had been attached to the Moravian chapel for the past five years; if he imbibed any of the Moravians' pacifist teachings during his long and close association with Brother Joseph Jackson, he too seems an unlikely candidate. Thomas traded places with Robert in 1807, Thomas moving to Robert's job in the overseer's house and Robert moving to Thomas's job at the Moravian chapel.
35. For more on Thomas Thistlewood, see Chapter 4.
36. Gründer claimed that three of his mulatto pupils at Mesopotamia "can read now pretty well"; J. S. Gründer to Christian Ignatius Latrobe, March 20, 1818, MCHL.
37. "The Present State of the Buildings on Mesopotamia Estate," June 1, 1814, Barham b 34; annotated description of the Mesopotamia slave population, June 1802, Barham b 36.
38. Between 1806 and 1811, JFB2's attorneys in Jamaica paid £1,599 for oak staves imported from America. When JFB2 objected to paying so much for imported lumber, attorney J. R. Webb explained that local Jamaican

timber could not be coopered. It was too green when newly cut and too brittle when dried. Webb to JFB2, July 27, 1812, Barham c 358.

39. The Mesopotamia crop account for 1810 is in Barham b 34. The sugar and rum exported to England that year fetched £18,173 sterling in London, the highest sale price achieved in 1762–1833.
40. Mesopotamia diary, 1815, UA R.15.Cb.6.11, pp. 3–6.
41. The Mesopotamia slave registration return, June 28, 1817, is in T71/178, NA-UK.
42. William Ridgard to JFB2, February 17, May 2, 10, 12, June 9, July 7, 1826, Barham c 359. The death register for 1826 is in the January 1, 1827, inventory, Barham b 35.
43. Ridgard to JFB2, August 4, 1826, Barham c 359.
44. Flora's mother, Marina, could also have nursed baby Sarah, but she was very sick and seventy years old, and died herself in 1828.
45. Patrickson was not fully liberated in 1831. He had contracted to pay £140 for his freedom and still owed £40, so the estate attorney kept his manumission papers in escrow. William Ridgard and Duncan Robertson to JFB2, July 17, 1830, Barham c 360; Ridgard and Robertson to JB, November 7, 1832, Barham c 389.
46. Mesopotamia accounts, July 1, 1816–June 30, 1817, Barham b 33.
47. Mesopotamia accounts, July 1, 1816–June 30, 1817, Barham b 33. This is the only information I have found about James Bell.
48. In 1823 Jane's grandmother Affy did not work at the overseer's house; she was a washerwoman at the Great House until 1824, when she became an invalid.
49. In 1800 Knight manumitted fourteen-year-old mulatto Susannah, born in 1780, whose mother was housekeeper Minny, and three-year-old Helen, born in 1797, whose mother was housemaid Sally.
50. Knight was described in his probate inventory, dated February 11, 1817, as a Westmoreland carpenter. Jamaica Inventories, 1B/11/3/128, p. 194, JA.
51. Ridgard to JFB2, July 20, 1824, Barham c 358; Mesopotamia accounts, 1825, Barham b 35.
52. Brother Zorn's diary, 1830, Moravian Archives of Jamaica, Fairfield H/7, pp. 20, 24, JA. Zorn lived in St. Elizabeth parish, but he made eight visits to Mesopotamia during 1830, since Mesopotamia had no resident missionaries. See Chapter 6 for further details.
53. Mary Barham was listed as a runaway in 1831, so like Jane Ritchie she cannot have enjoyed working for the Ricksecker.
54. Minutes of Moravian Conference, January 1830–April 1831, Moravian Archives of Jamaica, Fairfield H/5, p. 11, JA. The Rickseckers were especially upset when Prudence (Mary Barham) had an epileptic fit in April 1831.
55. Mesopotamia diary, 1832, UA R.15Cb.6.14, pp. 4, 12.

56. Mesopotamia diary, 1832, UA R.15Cb.6.14, pp. 13–14. This passage in Ricksecker's diary is so faint that it is almost impossible to decipher.
57. Mesopotamia diary, 1832, UA R.15Cb.6.14, pp. 4, 6, 8, 12–14.
58. The apprenticeship period was later shortened to four years for all ex-slaves; they were fully freed on August 1, 1838. For an account of emancipation at Mesopotamia, see Chapter 9.
59. Amended compensation claim for Mesopotamia, October 14, 1834, T71/1007, NA-UK.
60. Johns is listed as the Mesopotamia overseer, 1832–1835, and as the Mesopotamia attorney and overseer, 1839–1853, employed by the Reverend Charles Barham, in the annual crop accounts for the estate found in Accounts Produce, vols. 73–95, JA. He may have served longer at Mesopotamia, since no crop accounts survive for the estate between 1854 and 1863. By 1864 Mesopotamia had a new owner and a new attorney.
61. Only two other Mesopotamia apprentices were recorded as buying their freedom during the four-year apprenticeship period. Jane's apprenticeship valuation of £18 is listed in the 1836 Mesopotamia crop account (Accounts Produce, 1B/11/4, vol. 78, pp. 157–58, JA).

### *3. Winney Grimshaw and Her Mount Airy Family*

1. The 1808 inventory is in Tayloe d 538; the first few pages have been torn out, so that returns from three farm quarters—Forkland, Oaken Brow, and Old House—are missing, but the domestic and craft slaves at the home plantation are all identified.
2. Slave work logs kept by JT3 between 1805 and 1814 show that the Mount Airy craft and farm workers all labored more than 300 days per year. See Chapter 5. No work logs have survived for the domestic workers.
3. See the Prologue.
4. William H. Dorsey was the builder, and he charged JT3 $28,477 for land, materials, and labor between May 1799 and December 1801 (Tayloe d 359). Orlando Ridout, *Building the Octagon* (Washington, DC: American Institute of Architects Press, 1989), describes the design and construction of this house, the most significant private residence in the new federal city.
5. Mount Airy is a private residence, not open to the public. But the Octagon, which is now owned by AIA Legacy and has been extensively renovated, is open to visitors.
6. Laura Kamoie, *Irons in the Fire: The Business History of the Tayloe Family and Virginia's Gentry, 1700–1860* (Charlottesville: University of Virginia Press, 2007), 148.
7. A generation after the Revolution, the Mount Airy records from the 1800s and 1810s sometimes state values in pounds, shillings, and pence and sometimes in dollars; £1 was equivalent to $3.33.

8. On March 10, 1794, JT3 instructed Philadelphia coach maker Robert Monteath to adorn the new carriage he was building for him with "a wrist and hand holding a dagar with a boars head on its point" (Tayloe uncatalogued papers, box 1). This carriage may have been the coach inventoried in 1808.
9. In February 1815 President Madison, while living at the Octagon, ratified the Treaty of Ghent, which ended the War of 1812, in the circular room above the Octagon entrance hall.
10. The Mount Airy ratio between spinners and weavers was standard. Laurel Thatcher Ulrich observes that "in textile-producing areas of Europe, eight to ten spinners kept one weaver supplied with thread." See Ulrich, "Wheels, Looms, and the Gender Division of Labor in Eighteenth-Century New England," *WMQ*, 3rd ser., 55 (1998): 9.
11. The Neabsco inventories of 1825 that list Letty and Jim are in Tayloe d 538, d 8402–8422, and d 8471–8520. Letty appears, but Jim does not, in a Neabsco inventory for 1828 in Tayloe d 992.
12. The Old House inventory for 1827 is in Tayloe d 8539–8590. Inventories for the other slave quarters at Mount Airy in 1827 are in Tayloe d 538, d 8402–8422, and d 8539–8590.
13. For the 1835 reference, see Tayloe d 13410; for the 1845 reference, see Tayloe d 16253–16485; for the 1862 reference, see Tayloe d 8632–8667.
14. Ann Ogle Tayloe lived at the Octagon until her death in 1855.
15. Charlotte's death is reported in the 1840 inventory for Landsdown quarter, in Tayloe a 13.
16. This inventory of carpenters' tools, dated February 10, 1839, is in Tayloe d 8539–8590.
17. The fullest surviving Mount Airy work logs are dated 1805, 1811–1812, and 1813–1814. In these years the Mount Airy carpenters were frequently sent to Washington to make repairs at the Octagon, and they also did carpentry in JT3's outlying plantations and ironworks in Virginia and Maryland.
18. WHT diary, 1824–1831, Tayloe d 7923.
19. See Tayloe d 13410.
20. WHT to his younger brother HAT in Washington, September 17, 1828, Tayloe d 5849–5959.
21. See Tayloe d 13410.
22. Dr. W. G. Smith's medical bills, 1831–1840, are in Tayloe d 14410–14520, 15332–15579, 15919–16082.
23. J. W. Collins to WHT, April 30, 1838, Tayloe d 2860–2870.
24. David Carrington was living in Richmond County in 1830 according to the U.S. Census for that year. He was described as a single free colored man, aged between thirty-six and fifty-five (U.S. Census, Richmond County, VA, M 19, roll 194, p. 81, NA-US). His wife, Criss, who was thirty-six

in 1830, died in 1836, and David also died or moved away at about this time, for he is not listed in the indexes to the Virginia censuses of 1840, 1850, or 1860.

25. WHT to Benjamin Boughton, March 16, 1841, Tayloe d 2371–2452.
26. WHT paid $3.75 to run this ad for a month. His bill from the *Alexandria Gazette* is in Tayloe d 16253–16485.
27. For WHT's comments on the Grimshaws, see Tayloe a 13.
28. I do not know who Anna's prospective husband was.
29. Juliet forgets to mention that her fourteen-year-old brother James had also been sent to Alabama in October 1845.
30. Armistead Carter was a twenty-five-year-old dining room servant at Mount Airy in 1846, and was probably in a position where he could travel to the Octagon to see Lizza. He may well have been the father of her little boy. Armistead continued working as a domestic at Mount Airy until he was emancipated in 1865 at age forty-four. Instead of marrying Lizza, he married a slave named Maria who was owned by a neighboring planter.
31. Charles Tayloe, WHT's youngest brother, was the master of Oaken Brow plantation in King George County, which had previously been a Mount Airy farm quarter.
32. There were two women of this name at Mount Airy in 1846: Nancy Carter, age fifty-four, and Nancy Richardson, age forty-six. Both were spinners like Esther Grimshaw, and neither of them appears to be related to the Grimshaws.
33. Juliet's Aunt Betsy, whose brother Henry worked with Lizza at the Octagon, had died in 1825 before Juliet was born. Probably Juliet is here referring to Betsy Yeatman, a sixty-nine-year-old retired spinner in 1846. She had at least three living sons in 1846–1847; I cannot trace the son who died or identify the brother who worked at the Octagon.
34. This letter is filed in Tayloe d 27453–27504. The envelope is sealed, with the following address, indicating that the letter was mailed three days after it was written: "Warsaw Va / March 30 / Miss Elizabeth Grimshaw / care of Mrs John Tayloe / Washington City."
35. Robert Wormley Carter II, great-grandson of the famous diarist Landon Carter, was the owner of Sabine Hall in the 1840s. He and WHT were the two largest slaveholders in Richmond County.
36. Esther evidently came to Sabine Hall on March 21. Sunday was by far the most likely travel day for a slave, which leads me to believe that this letter was written in 1847 because March 21, 1847, was a Sunday.
37. Winney Jackson was last reported in the Mount Airy records as living at the Octagon in 1844, when she was sixty-four years old.
38. Probably Armistead Carter; see Juliet's letter of March 27, 1846, reproduced earlier in this chapter.

39. This letter is filed in Tayloe d 27453–27504. The envelope is sealed, with the following address, indicating that the letter was mailed four days after it was written: "Warsaw Va / March 26 / Elizabeth Grimshaw / Washington / D.C."
40. William Francis to WHT, August 14, 1851, Tayloe d 3280. The envelope is stamped "St John N.B. Au14 1851." It is addressed to "Mr William Tayloe / Warsaw Court House / Montery / Richmond County / Virginia."
41. W.A. Spray, *The Blacks in New Brunswick* (Fredericton, NB: Brunswick Press, 1972), 34–35.
42. Winney's party is enumerated in several lists in Tayloe a 13, a 2119, d 8632–8667, and d 13425; the Oakland overseer R.H. Donnahan describes the route they took in a letter to WHT, June 20, 1847, Tayloe d 2969–2985.
43. The forty-five Mount Airy slaves who were sent to Alabama in 1845 are all listed in Appendix 30.
44. The slave census of 1860 identifies the slaves on each property only by age, sex, and race, not by name. So WHT's ten mulatto slaves cannot be definitely identified.
45. WHT's comment on Mrs. Donnahan is in a memorandum he wrote in 1868 about his pre–Civil War problems in Alabama. In this memorandum WHT never talked very explicitly about Winney's relations with Donnahan, simply noting that Donnahan "had children by marriage and others by my servant Winney" (Tayloe d 2123–2135).
46. R.H. Donnahan to WHT, October 16, 1852, Tayloe d 2969–2985.
47. Lucy Tayloe's letter of thanks to WHT, dated January 1, 1854, is in Tayloe d 6342–6343. She promised to take great pains to train John properly, and asked WHT to assure Winney that her son would be well taken care of.
48. WHT memorandum, 1868, Tayloe d 2123–2135.
49. WHT to HAT Jr., April 20, 1857, Tayloe d 6046–6170.
50. For Jim's syphilis treatment, see the 1857 medical bill to WHT from Drs. Clarke and Langhorn, Tayloe d 19551–19697.
51. WHT to HAT Jr., March 25, 1858, Tayloe d 5960–6045.
52. WHT to HAT Jr., April 27, 1857, Tayloe d 6046–6170.
53. Dr. Browder billed WHT $251 for making seventy-six visits to Woodlawn between January 1 and September 20, 1858. He spent much more time with Winney than with any other slave patient. See Tayloe d 19688–19830.
54. WHT memorandum, 1868, Tayloe d 2123–2135.
55. WHT's valuations reflected the rising prices for slaves just before and during the war; in December 1858 at a sale near Selma, young men who were "nothing more than field hands" fetched $1,600 to $1,960, and girls aged fifteen to seventeen "commanded as many hundred dollars" (Tayloe d 2123–2135).
56. WHT's data on his Alabama slaves in 1863 is in Tayloe d 13453, d 8597–8605, and d 8632–8667.

57. See WHT at Oakland to his daughter-in-law Courtenay Tayloe in Virginia, June 8, 1865, Tayloe d 5292–5327.
58. The Oakland lists for 1866 are in Tayloe d 8632–8667 and d 13450.
59. For fuller discussion, see Chapter 9.
60. For a fuller account of the breakup at Larkin, see Chapter 9. A list of the Larkin hands, showing who had left by January 1866 and who stayed on, is in Tayloe d 13453.
61. Tayloe d 13450. On this Larkin list of mothers with children, the children are not named.
62. I have searched for Winney Grimshaw and her children, and also for Jacob Carrington, as well as for James Grimshaw and his family, in the 1870 and 1880 censuses via Ancestry.com, with no success. Possibly I found Winney's elder son: a thirty-one-year-old mulatto carter named William Henry Grimshaw was living with his family in Washington, D.C., in 1880, but this man was born in Virginia, whereas Winney's William Henry was born in Alabama and would have been only twenty-three years old in 1880.
63. WHT's full statement reads: "Archy Williams, Nat Moreton and Joe Saunders, Edward Hall and Winney Grimshaw could fill a volume with interesting events, if they could write" (WHT memorandum, 1868, Tayloe d 2123–2135). Williams, Moreton, and Saunders had also been WHT's slaves; like Winney they lived at Larkin and probably witnessed WHT's quarrel with Ramey. Williams and Moreton left with Ramey, and Saunders stayed on with WHT in 1866. Edward Hall had not been a slave of WHT's; he worked at Larkin as a freedman in 1865 and left at the end of the year, though not with Ramey.

### *4. "Dreadful Idlers" in the Mesopotamia Cane Fields*

1. JFB2, *Considerations on the Abolition of Negro Slavery: And the Means of Practically Effecting It* (London: James Ridgeway, 1823), 8.
2. The Mesopotamia evidence buttresses the arguments advanced by Barry W. Higman in *Slave Population and Economy in Jamaica, 1807–1834* (Cambridge: Cambridge University Press, 1976), 1–17, 121–124, 212–226; and in *Slave Populations of the British Caribbean, 1807–1834* (Baltimore: Johns Hopkins University Press, 1984), 158–199, 324–329, 332–336; and by Michael Craton in *Searching for the Invisible Man: Slaves and Plantation Life in Jamaica* (Cambridge, MA: Harvard University Press, 1978).
3. Diary of Thomas Thistlewood, 1748–1786, Monson 31, vols. 1–37, Lincolnshire County Archives, Lincoln, UK. I thank Lord and Lady Monson for permission to quote from the diaries.
4. The crop accounts and shipping lists sent to JFB1 and JFB2 in 1751–1832 are in Barham b 34, b 36, and b 37. Barham b 38 contains eleven lists of net sugar sales in London between 1777 and 1797. For the years 1786,

1804, 1806, 1817–1818, and 1822–1832, the crop accounts are incorporated with the slave inventories (stored in the same four boxes). There are informative Mesopotamia crop reports for sixty-six of these eighty-two years, in Accounts Produce, vols. 4–75, JA. The missing years are 1751–1760, 1767, 1777, 1791–1794, and 1805. There are no crop valuations in either the Barham Papers or in Accounts Produce for seven years: 1777, 1824, and 1828–1832. I have estimated these seven missing valuations.

5. For the 1787 reference, see Accounts Produce, 14/52–53, JA; for the 1801 reference, see Accounts Produce, 29/36–37, JA.
6. The Mesopotamia balance sheet for 1751–1777 is in Barham b 37; hired labor payments for 1777–1788 are in Barham b 33; those for 1798–1808 are in Barham b 34; those for 1814–1819 are in Barham b 36, and the 1777–1797 Plummer & Co. Mesopotamia balance sheet is in Barham b 38.
7. The Jamaican expense account for 1814–1820 is in Barham b 36.
8. Plummer & Co. Mesopotamia balance sheet, 1777–1797, Barham b 38.
9. JFB2's bankbooks for 1805–1825 are in Barham c 389. Some of the Plummer deposits represent earnings from Island estate, but Island shipped much less sugar than Mesopotamia and lost money in some years. See Plummer & Co., Mesopotamia v Island: debits and credits, 1788–1797, Barham b 38.
10. Accounts Produce 1B/11/4, vol. 42, p. 38, JA; Grant and Blyth to JFB2, August 11, 1810, Barham c 358.
11. JFB2 to William Ridgard, June 1, 1825; same to same, September 2, 1829, Barham c 428.
12. "The Present State of the Buildings on Mesopotamia Estate," June 1, 1814, Barham b 34.
13. Roderick A. McDonald discusses Jamaican slave housing in *The Economy and Material Culture of Slaves: Goods and Chattels on the Sugar Plantations of Jamaica and Louisiana* (Baton Rouge: Louisiana State University Press, 1993), 92–110. See also B. W. Higman, *Montpelier, Jamaica: A Plantation Community in Slavery and Freedom, 1729–1912* (Mona, Jamaica: University of the West Indies Press, 1998), chaps. 5–7.
14. Higman, *Montpelier, Jamaica,* 115–125; Higman, *Slave Population and Economy in Jamaica,* 156–173.
15. Higman, *Slave Populations in the British Caribbean,* 280–292, 534–535, 542–546; Higman, "Growth in Afro-Caribbean Slave Populations," *American Journal of Physical Anthropology* 1 (1979): 377–382; Gerald C. Friedman, "The Heights of Slaves in Trinidad," *Social Science History* 6 (1982): 493–501. The data analyzed by Higman and Friedman are from the slave registration returns in Trinidad, St. Lucia, and Berbice. The Jamaican slave registration returns supply no information on stature.
16. A note on the dating of the seventy-one Mesopotamia slave inventories in our study: Except for the first (July 10, 1762) and the last (August 23, 1833), all of the inventories were dated either December 31 or January 1

(January 12 in 1778), with fifty-six of them dated January 1. To eliminate confusion and achieve consistency, I date *all* of the inventories January 1, except for the inventories taken in 1762 and 1833. The inventory for 1821 is missing.

17. The June 1802 listing shows that the first gang was supervised by two male drivers and supported by two male carters, a mule man, two female cooks, and a female field nurse who took care of the gang's brood of young children (Barham b 36).
18. Martin Forster and S. D. Smith, "Surviving Slavery: Mortality at Mesopotamia, a Jamaican Sugar Estate, 1762–1832," *Journal of the Royal Statistical Society,* A (2011) 174, pt. 4, 907–929. The Forster-Smith data set (1,099 slaves) is very similar to mine (1,103 slaves), and we have arrived at much the same totals for births, deaths, years in good and poor health, and years spent in cane field labor. I am very pleased that Forster and Smith were enticed by my description of the Mesopotamia inventories in previous essays to undertake their own analysis, and I wish to thank Simon Smith for a series of e-mail exchanges that have stimulated me to reinterpret my own findings on Mesopotamia.
19. Forster and Smith, "Surviving Slavery," 907, 924–926.
20. Orlando Patterson, *Slavery and Social Death: A Comparative Study* (Cambridge, MA: Harvard University Press, 1982), 4.
21. For the gag, Mesopotamia accounts, 1780, Barham b 33; for the mouthpiece, Mesopotamia accounts, 1828, Barham b 35; for the stocks, Moravian Mesopotamia diary, July 11, 1760, UA R.15.Cb.1.3; Mesopotamia diary, 1812, UA R.15.Cb.6.9/8.
22. Kenneth E. Ingram, *Sources of Jamaican History, 1655–1838,* 2 vols. (Zug, Switzerland: Inter Documentation Company, 1976).
23. Thistlewood, vols. 1, 2, 16, 38, Monson 31.
24. My earliest published comment on Thistlewood was in Richard S. Dunn, "Servants and Slaves: The Recruitment and Employment of Labor," in *Colonial British America: Essays in the New History of the Early Modern Era,* ed. Jack P. Greene and J. R. Pole (Baltimore: Johns Hopkins University Press, 1984), 173–174.
25. Douglas Hall's *In Miserable Slavery: Thomas Thistlewood in Jamaica, 1750–1786* (Kingston, Jamaica: University of the West Indies Press, 1989) is a chronological tour (spiced with apt quotations) through the entire diary. Trevor Burnard's *Mastery, Tyranny, and Desire: Thomas Thistlewood and His Slaves in the Anglo-Jamaican World* (Chapel Hill: University of North Carolina Press, 2004) is more analytical and gives a much harsher view, stressing the violence and depravity of Thistlewood's slave management.
26. See Hall, *In Miserable Slavery,* 4–9.
27. Thistlewood, January 6–8, August 18, 1761, Monson 31/12.
28. Thistlewood, May 15, 18, 1750, Monson 31/1.

29. JFB1's Island overseer, Samuel Coulson, sent 250 plantains to Thistlewood in July 1750 to help feed his hungry slaves (Thistlewood, July 9, 1750, Monson 31/1).
30. Thistlewood, June 29, July 1–2, 11, 16, 1750, Monson 31/1.
31. Thistlewood, July 20–21, August 4, 1750, Monson 31/1.
32. Thistlewood, May 12, July 6–7, 1751, Monson 31/2. For a much fuller account of Thistlewood's year at Vineyard, stressing his positive interactions with the slaves, see Philip D. Morgan, "Slaves and Livestock in Eighteenth-Century Jamaica: Vineyard Pen, 1750–1751," *WMQ*, 3rd ser., 52 (1995): 47–76.
33. In 1764 Thistlewood's nephew John Thistlewood listed seventy-eight working slaves belonging to the owner of Egypt and eleven belonging to his uncle (Monson 31/38). In addition, Thistlewood hired two dozen extra slaves in January–February during crop time.
34. The year 1765 was a very bad one for sugar production in Westmoreland because the Cabarita River flooded the cane fields. In that year Egypt produced only 35 hogsheads of sugar and 13 puncheons of rum (Monson 31/16). By comparison the crop at Mesopotamia in 1765 was 104 hogsheads of sugar and 106 puncheons of rum, which was way below average.
35. Hall, *In Miserable Slavery*, 70–73; Burnard, *Mastery, Tyranny, and Desire*, 104, 260–261.
36. Hall, *In Miserable Slavery*, 94, 123, 129, 134, 218–220, 230, 237, 249, 251, 257, 261, 267, 275.
37. Daniel Barnjum to JFB1, August 28, December 23, 1760, Barham c 360.
38. Barnjum's twenty slaves are listed in his probate inventory, July 28, 1778, in JA Inventories 1B/11/3/60, pp. 57–58. Thistlewood's twenty-eight slaves in 1767 are listed by Hall, *In Miserable Slavery*, 143.
39. Moravian Mesopotamia diary, December 28, 1759, March 24, 1761, UA R.15.Cb.1.3.
40. See the summary statistics table in the Trans-Atlantic Slave Trade Database (www.slavevoyages.org).
41. Higman, *Slave Population and Economy in Jamaica*, 71–72, 255–256.
42. See Appendix 9.
43. For example, Kenneth F. Kiple discusses the disease and diet factors in *The Caribbean Slave: A Biological History* (Cambridge: Cambridge University Press, 1984), while Barry Higman stresses the labor factor in *Slave Population and Economy in Jamaica* and in *Slave Populations of the British Caribbean*.
44. Mesopotamia Food Allotments, June 1802, Barham b 36.
45. Mesopotamia accounts, May 1814, Barham b 36.
46. Moravian Mesopotamia diary, July 4, 1760, UA R.15.Cb.1.3; Mesopotamia diary, June 1782–April 1783, UA R.15.Cb.6.4, pp. 3, 23.
47. Moravian Mesopotamia diary, 1812, UA R.15.Cb.6.9, pp. 8, 17, 30–31, 35–36.

48. Michael Craton, *Searching for the Invisible Man,* 132–133. In his tabulation, Craton does not separate the Worthy Park male deaths from the female deaths.
49. Higman, *Slave Populations of the British Caribbean,* 272. For a somewhat more positive view of the slave doctors, see Richard B. Sheridan, *Doctors and Slaves: A Medical and Demographic History of Slavery in the British West Indies, 1680–1834* (Cambridge: Cambridge University Press, 1985).
50. These diseases were also among the leading causes of slave deaths in Grenada, Dominica, Tobago, Demerara, and Berbice. See Higman, *Slave Populations of the British Caribbean,* 340.
51. Kiple, *The Caribbean Slave,* chap. 6; Higman, *Slave Populations of the British Caribbean,* 295–297.
52. Barham's attorneys reported that they tried hard to save John (age thirty-six)—an excellent cooper, and the only blacksmith on the estate—after he injured himself severely from his fall, but he immediately contracted tetanus. William Ridgard and Duncan Robertson to JFB2, July 17, 1830, Barham c 360.
53. Patterson, *The Sociology of Slavery,* 98.
54. I am also attracted to Orlando Patterson's conception of slavery as human parasitism, in which the parasitic slaveholder leeches onto the slave host to his benefit and the slave's detriment, until the slave, losing "all claim to autonomous power, [is] degraded and reduced to a state of liminality," and becomes socially dead (*Slavery and Social Death,* 337). Patterson does not make a gender-based argument, but the leeching process would seem to be especially devastating to male slaves.
55. It is impossible to establish the exact percentage of African males and females at Mesopotamia, because the origin of some of the older people in 1762 can only be guessed at, and also because the origin of most of the slaves imported from Three Mile River in 1786 and Southfield in 1791 is unknown. But the Jamaican slave registration in 1817 identifies the thirty-one Africans from Three Mile River and Southfield still living at that date, as well as almost all of the Africans at Cairncurran and Springfield who came to Mesopotamia in 1814 and 1819. I base my estimate on the slave registration figures.
56. My account of Jamaican motherhood may be fruitfully compared with Jennifer L. Morgan's searching analysis of slave motherhood in early Barbados, based on an examination of wills and inventories, in which the slaveholders view slave women as reproductive property, and the mothers cope with sexual violation and the commodification of their children. See Morgan, *Laboring Women: Reproduction and Gender in New World Slavery* (Philadelphia: University of Pennsylvania Press, 2004), esp. chaps. 3–4.
57. J. C. Grant and John Blyth to JFB2, January 3, 1809 (Barham c 358).
58. A. Meredith John, *The Plantation Slaves of Trinidad, 1783–1816: A Mathematical and Demographic Enquiry* (Cambridge: Cambridge University Press,

1989), 159, argues that the problem on this island was high mortality—the quick deaths of slave infants and children—rather than low slave fertility. But this argument does not work at Mesopotamia, where mortality among young recorded children was no greater than at Mount Airy. One would need to assume that over 500 unrecorded infants died at Mesopotamia between 1762 and 1833, and none at Mount Airy between 1809 and 1863, in order to achieve a Mesopotamia fertility rate equivalent to Mount Airy's.

59. Rose E. Frisch, "Demographic Implications of the Biological Determinants of Female Fertility," *Social Biology* 22 (1975): 17–22; Frisch, "Population, Food Intake, and Fertility," *Science* 199 (1978): 2–30.
60. See Jane Menken, James Trussell, and Susan Watkins, "The Nutrition Fertility Link: An Evaluation of the Evidence," *Journal of Interdisciplinary History* 11 (Winter 1981): 425–441; Andrew and Ann Prentice, "Reproduction against the Odds," *New Scientist,* April 14, 1988.
61. Sally White was the midwife in 1802. William Rodgers to JFB2, April 16, July 5, August 3, 1802, Barham c 357; Mesopotamia accounts, 1802, b 34.
62. John Blyth to JFB2, September 14, 1811, Barham c 358.
63. Ridgard to JFB2, August 30, 1825, Barham c 359.
64. Ridgard to JFB2, November 9, 1824, Barham c 358; same to same, August 30, December 11, 1825, Barham c 359.
65. Barbara Bush, *Slave Women in Caribbean Society, 1650–1838* (Bloomington: Indiana University Press, 1990), is a leading proponent of this view. See also J. Morgan, *Laboring Women,* 113–114; Patterson, *Slavery and Social Death,* 133.
66. Dunn, "Sugar Production and Slave Women in Jamaica," in *Cultivation and Culture: Labor and the Shaping of Slave Life in the Americas,* ed. Ira Berlin and Philip D. Morgan (Charlottesville: University Press of Virginia, 1993), 49–72; " 'Dreadful Idlers' in the Cane Fields: The Slave Labor Pattern on a Jamaican Sugar Estate, 1762–1831," *Journal of Interdisciplinary History* 17 (Spring 1987): 795–822.
67. Clarinda, Judy, Sally, and Matura were all members of the first gang in 1802. Cooba at age fifteen was a member of the second gang, but her twelve children are not included in Appendix 20 since she was under age sixteen in 1802. Cooba was moved up to the first gang in 1807.
68. Survey of the Mesopotamia workforce, June 1802, Barham b 36. Eve's baby was not listed in the birth register for 1802; I found out about it by checking the payments made to midwife Sally White in Barham b 34.
69. On April 27, 1765, Dr. Robert Pinckney wrote to JFB1: "We had the misfortune the other day of a fine young Negro Wench getting her arm into the mill. I was obliged to take it off above the elbow and is now in a fair way of recovery" (Barham c 357). Rose was then fourteen years old.
70. The forty-one slaves from Three Mile River are listed in family groups in Barham b 33.

71. Bessy's Elsie may have been a sambo because she was put into the grass gang, but since she died at age ten it is hard to tell.
72. Eliza was born in 1829 and William in 1831. In the Mesopotamia ledger for 1831, William Parrott was listed as a distiller and bookkeeper, with a salary of £75 (Barham b 35).
73. Burnard, *Mastery, Tyranny, and Desire,* 156.
74. Diary of John Thistlewood, February 3, 1765, Monson 31/38.
75. John Thistlewood never mentioned Little Mimber in his diary, which ends abruptly on March 28, 1765, two days before he died (Monson 31/38).
76. An argument for suicide is that after John's death the other white man working on the estate told his uncle that John had been having bad dreams. Thomas Thistlewood gives his version of John's denouement in entries from February 4 to April 5, 1765, Monson 31/16.
77. For Thistlewood's observation of the differences between pre-initiation and post-initiation African youngsters, see his account in Chapter 1 of the slaves he bought in 1765. The age statements in the Mesopotamia inventories for newly arrived young Africans seem valid to me. Girls during puberty can be tracked by the most casual observer with considerable accuracy via growth spurt and breast development, and the bookkeepers surely noticed that some of the youngest African girls and boys had no country marks.
78. See Herbert S. Klein and Stanley L. Engerman, "Fertility Differentials between Slaves in the United States and the West Indies," *WMQ,* 3rd ser., 35 (1978): 357–373.
79. The Mesopotamia accounts show that the estate paid a total of £8 4s. in rewards for capturing Harriet and paying her workhouse fees in 1817–1818, Barham b 36.
80. One of Lizzie's children (born in 1817) was a sambo named Susan Forrester, suggesting that Susan's father was a mulatto son of overseer David Forrester, but Forrester's only known son, mulatto Alexander, had died in 1813. Another three of Lizzie's children—Ellen Clark, Frances Clark, and Susan Clark—all evidently had the same black father, whom I cannot identify.
81. JFB2 to H. W. Plummer and W. Rogers, n.d. (ca. 1800), Barham c 428.
82. Appendix 21 excludes carpenter William and mason John, who were working adults in 1762, aged twenty-two and eighteen, respectively. If they were added, the average male mulatto years worked would climb from 12.4 to 13.9, and age at death from 31.8 to 34.8.
83. For a contrary view, see Forster and Smith, "Surviving Slavery," 907–929. Throughout their analysis Forster and Smith focus on mortality rates, whereas I focus on productivity, and this I believe explains our differing opinions on whether it was better to buy African slaves or local Jamaican slaves.
84. Higman, *Slave Population and Economy in Jamaica,* 208–209.

85. John Patrickson's probate inventory (June 17, 1826) shows that he had seven slaves but no cash (Jamaica Inventories, 1B/11/3/142, p. 83, JA).
86. John Blyth to JFB2, September 1, 1823, Barham c 358; Ridgard and Robertson to JFB2, November 7, December 12, 1832, Barham c 389. Blyth described Henry Patrickson as "a very good and trusty Boy of his Colour," while Ridgard and Robertson found him "a very honest and well disposed young man."
87. The exact proportion of Mesopotamian, African, and Jamaican mothers cannot be established, because (as explained earlier) 44 of the 420 recorded babies had no listed mothers.
88. Thirty-seven of the Jamaican women were over age forty when they came to Mesopotamia, and only one of these women—Bathsheba from Cairncurran—had a recorded birth at Mesopotamia; she was forty-four when Rosetta was conceived.
89. In 1762, nine of the thirteen slaves listed as boilers or distillers were African-born. The role played by the sugar makers at Mesopotamia will be discussed in Chapter 8.
90. John Graham to JFB2, March 18, 1792, Barham c 357.
91. Vincent Brown, *The Reaper's Garden: Death and Power in the World of Atlantic Slavery* (Cambridge, MA: Harvard University Press, 2010), 255.

## 5. "Doing Their Duty" at Mount Airy

1. Laura Kamoie has revised Jackson Turner Main's list of the one hundred wealthiest Virginians in the revolutionary era (*Irons in the Fire: The Business History of the Tayloe Family and Virginia's Gentry, 1700–1860* [Charlottesville: University of Virginia Press, 2007], 172–174); she credits JT2 with 500 slaves at his death in 1779, which puts him in fourth place among the revolutionary era Virginia slaveholders. By 1791 JT3 probably had 350 slaves at Mount Airy and more than 600 slaves altogether.
2. The Mount Airy slave population can be estimated via tax records from Richmond, Essex, and King George counties. In 1782 there were about 265 slaves at Mount Airy, and 316 by 1787. From 1788 onward the three counties all taxed Negroes over the age of twelve, which in JT3's case meant approximately 65 percent of the total number of slaves. Extrapolating from the annual tax levies, JT3 had about 330 slaves in 1789 and 350 in 1791. The Personal Property Tax Lists for Richmond, Essex, and King George counties, 1782–1792, are in LVA.
3. Assuming that the Mount Airy population grew at the same rate as the overall southern U.S. slave population, which was 657,000 in 1790 and 1,980,000 in 1830, JT3 would have nearly tripled his slave force at Mount Airy by 1828.
4. I supposed initially that JT3 sold slaves in 1792 in order to pay off debts, and perhaps this was a major factor. But since I have found no record of

debt payments, while land purchases are recorded, I now believe that JT3 was mainly interested in expanding and improving his farmland.

5. Kamoie argues persuasively that JT3 was following the business model of his grandfather JT1 (1687–1747) and his father, JT2 (1721–1779), who had used their expanding slave populations to diversify into grain and livestock production, iron making, and other commercial activities. See Kamoie, *Irons in the Fire,* chaps. 1–3.
6. Alternately listed as Marshfield or Maskfield.
7. Tayloe's 9,700 acres were valued at over £52,000 (just under $174,000), or more than twice the valuation of his slaves. His plantation house and outbuildings were valued at £14,000 ($46,620). See the inventories for 1808 and 1809 in Tayloe d 538.
8. *Virginia Herald and Fredericksburg Advertiser,* April 17, 1794, cited in Gerald W. Mullin, *Flight and Rebellion: Slave Resistance in Eighteenth-Century Virginia* (New York: Oxford University Press, 1972), 107.
9. JT3 to William Gordon, June 7, 1801, JT3 letter book, 1801, Tayloe d 170.
10. Holburne to George Gresham, March 15, 1809, Holburne letter book, 1809.
11. Minute book, 1813–1818, Tayloe a 11.
12. JT3 to WHT, November 29, 1824, Tayloe d 249–252.
13. Cited by Kamoie, *Irons in the Fire,* 118.
14. *The Diary of Colonel Landon Carter of Sabine Hall, 1752–1778,* ed. Jack P. Greene, 2 vols. (Richmond: Virginia Historical Society, 1987); *Journal and Letters of Philip Vickers Fithian, 1773–1774,* ed. Hunter Dickinson Farish (Williamsburg: Colonial Williamsburg, Inc., 1943). Robert Carter's voluminous manuscript day books and letter books are in the Duke University Library and the Library of Congress.
15. Tayloe a 8.
16. Tayloe uncatalogued papers, box 1. The surviving pages of the minute book for 1806 are in two fragments which are in unreadable condition.
17. Tayloe uncatalogued papers, box 1. The 1807 minute book records twenty-three work weeks from January into June.
18. Tayloe a 10. The 1811–1812 minute book is the second most valuable of the surviving work logs; it records thirty-two weeks from January into August 1811, and eleven weeks in 1812, including the month of December, which is missing in the 1805 work log.
19. Tayloe a 12. This segment of a minute book has been torn out of Tayloe a 11; it covers twelve weeks, and when joined to the fifteen weeks recorded for 1814 in Tayloe a 11 (see note 20) gives us a full work record from January into July 1814.
20. Tayloe a 11. This 360-page minute book records complete January–December work routines for 1813, 1816, 1817, and 1818, with less complete entries for 1814 and 1815—throughout in briefer form than the 1805 work log. Regrettably, the 1813–1818 minute book is in very poor

condition, with ink bleeding through most of its pages. The entries for 1813 are extremely difficult to make out, and the entries for 1816–1818 are mostly illegible. The entries for August 1814–December 1815 are legible, but very brief. The most useful entries are for fifteen weeks in January–March and June–July 1814, which, when combined with Tayloe a 12 (see note 19), provide a record of twenty-seven consecutive work weeks in 1814.

21. The blacksmith book for 1793–1794 is in Tayloe uncatalogued papers, box 2; the spinning book for 1806–1807 is Tayloe a 9; the spinning books for 1805–1806 and 1816–1819 and the mill book for 1810–1813 are in Tayloe uncatalogued papers, box 1; the shoemaking book for 1816–1817 is in uncatalogued papers, box 3. There are also more fragmentary day books and minute books for 1795–1796, 1797, 1805, 1806, 1807, 1811, 1813, 1812, 1814, and 1815–1827 in uncatalogued papers, boxes 1–3, many of which are unreadable.
22. Personal Property Tax Lists for Richmond, Essex, and King George counties, 1782–1809, LVA.
23. Kamoie, *Irons in the Fire,* 96.
24. In the three years 1771–1773, the six British tobacco ports (principally Glasgow and London) imported one hundred million pounds of Chesapeake tobacco annually; in the three years 1784–1786 these six ports imported only forty-four million pounds annually; and by 1802 the total had fallen to twenty-five million pounds. Jacob M. Price, *Tobacco in Atlantic Trade: The Chesapeake, London and Glasgow, 1675–1775* (Aldershot: Variorum, 1995), chap. 3, appendix A.
25. JT3 account book, 1789–1828, Tayloe d 357.
26. In June and July 1801 JT3 consigned two shipments of tobacco totaling 130 hogsheads of "prime good quality" and "all of my own make" to a London merchant (JT3 letter book, 1801, Tayloe d 170). According to Kamoie, *Irons in the Fire,* 46, these shipments were as sizable as JT2's tobacco shipments had been in 1750–1770. Kamoie also discusses JT3's tobacco crop in 1819 (p. 110), but this was not grown at the Rappahannock farms.
27. The first several pages of the January 1, 1808, inventory, containing the names of approximately 104 people on three farm quarters, have been torn out. But the 1808 inventory introduces us to all of the craft workers and domestics who lived on the home plantation. The inventories for 1810 and 1822–1823 are likewise somewhat incomplete, but otherwise this twenty-one-year series of inventories is remarkably full and informative; see Tayloe d 538.
28. The parallel Mesopotamia inventory, dated January 1, 1808, is in Barham b 36; it lists 163 males and 159 females, for a total population of 322.
29. The farm boys started earliest, at age nine or ten, the craft boys at eleven or twelve, the farm and domestic girls at ten or eleven, and the textile girls at twelve or thirteen.

30. By comparison, JFB2 paid an average of £110 for the fifty-five Cairncurran slaves he bought in 1814—in Jamaican currency, which amounted to £80 sterling, or roughly the same valuation as at Mount Airy.
31. Kamoie tabulates much lower figures in *Irons in the Fire,* 102–103: 6,548 bushels of wheat and 4,222 barrels of corn, 1792–1823. But there are many missing farm quarter returns in the source she uses (JT3 account book, 1789–1828, Tayloe d 357), and her average includes the years 1792–1800, when three of the eight farms were not yet in operation. Her average for 1808–1823 is close to mine: 8,877 bushels of wheat and 4,907 barrels of corn. I have found full returns for eight years in Tayloe a 10, a 11, d 369–403, and uncatalogued papers, box 3.
32. JT3 to Charles Hallet, June 26, 1809, Holburne letter book, 1809.
33. May 1814 memorandum on sheep shearing in JT3 minute book, 1814, Tayloe a 12.
34. For hog killing, see the "Blotter" day book for 1811, Tayloe uncatalogued papers, box 2.
35. Susan Dunn, *Dominion of Memories: Jefferson, Madison and the Decline of Virginia* (New York: Basic Books, 2007), chaps. 1–2. Kamoie's *Irons in the Fire* pushes hard in the opposite direction, finding the Tayloes and their fellow slaveholding planters as entrepreneurial as any northern capitalist.
36. In 1809 there was no overseer at Marske quarter, a small operation with only eight slave field hands; it was managed by a forty-two-year-old slave named George and watched over by Cornelius Beazley, at adjacent Menokin quarter.
37. In 1809 three men named Beazley, probably brothers, worked as overseers for JT3: Ephraim was the overseer at Gwinfield, Richard at Forkland, and Cornelius at Menokin.
38. The surviving Tayloe work logs describe weekly work routines in four Richmond County farm quarters—Old House, Doctor's Hall, Forkland, and Marske—with some reference to Hopyard and Oaken Brow in King George County, but little mention of operations at the large Gwinfield quarter in Essex County. Menokin (Richmond County) is included in the 1805–1807 work logs, but John Tayloe Lomax (a cousin of JT3) took possession of this farm quarter in 1809.
39. 1805 work log, Tayloe a 8, p. 4.
40. For the whiskey, see the entry for June 18 in the 1805 work log, Tayloe a 8; for the extra rations, see the 1807 day book in Tayloe uncatalogued papers, box 3.
41. In 1805 the harvest began on June 22 and ended on July 5; in 1811 the harvest period was almost exactly the same, from June 21 to July 5.
42. Minute book, 1813–1818, Tayloe a 11.
43. Notes on agricultural experiments, 1794–95, Tayloe d 27700.
44. The farm implements are listed in the 1809 inventory, Tayloe d 538.

45. There is a list of forty-one harvest hands, dated June 24, 1811, in Tayloe a 10; a list of fifty-four hands, dated June 24, 1812, also in Tayloe a 10; and a list of forty-four hands, dated June 25, 1813, in Tayloe a 11. Only the 1812 list includes the artisans deployed at Gwinfield, Hopyard, and Oaken Brow.
46. In 1806 JT3 paid £7 to hire two slave cradlers for eleven days, and 10 shillings to hire one slave raker for four days (Tayloe uncatalogued papers, box 3). Each cradler cost more than twice as much per diem as the raker, demonstrating that skilled wheat cutters were specially prized.
47. Minute book, 1813–1818, Tayloe a 11.
48. Tayloe a 10. By the 1830s Tom was the foreman of the Forkland field gang; he was still listed as foreman in 1854 at age seventy-one, and died at Forkland in 1858.
49. JT3 to John Rose, June 18, 1801, JT3 letter book, 1801, Tayloe d 170.
50. Tayloe a 8.
51. Tayloe d 538.
52. Three carpenters worked uninterruptedly on this shingling project from October 26 to December 4, 1805. The joiners collaborated with them for fourteen days during this period, the jobbers put in twelve days, and the masons put in five days. See Tayloe a 8. The white supervisor appears to have been Aaron Dyer Jr., who submitted a bid for another shingling job that JT3 thought was too high (W. Holburne letter book, 1809).
53. JT3 to William Gordon, July 18, 1801, JT3 letter book, 1801; Holburne to Baber, November 4, 1794, JT3 letter book, 1794, uncatalogued MS, box 1.
54. JT3 to Ephraim Beazley, September 3, 1813, JT3 minute book, 1813–1827.
55. JT3 account book, 1789–1828, Tayloe d 357. Three Beazleys, two Redmans, and two Babers—probably all brothers—served as Mount Airy overseers. Ephraim Beazley was the longest-serving overseer, from 1790 to 1815; he managed Old House (1800–1802), then moved to Gwinfield (1803–1813).
56. Between 1795 and 1812, Gwinfield had two overseers, Yowel Rust and Ephraim Beazley; JT3 acquired Oaken Brow in 1805, and his only listed overseer was William Greenlaw (1807–1814).
57. Holburne to J. Parkins, October 10, 1794, JT3 letter book, 1794, uncatalogued MS, box 1.
58. Holburne to Gresham, March 15, 22, 1809, Holburne letter book, 1809. A week later Gresham sent another Hopyard hand named Joe (age twenty-three) to Mount Airy for punishment; Holburne gave him "the necessary correction" and ordered Gresham not to whip him further. In 1810 Joe was sent to Cloverdale.
59. JT3 to John Cannady, June 17, 1801, JT3 letter book, 1801 (Tayloe d 170).
60. In 1805 the hands at Doctor's Hall quarter very similarly worked 130 days on corn, 65 days on wheat, 22 days on fodder, 21 days on fencing, 16 days on hay, 14 days on oats, 13 days on grubbing, 7 days on cotton, and 15 days on all other tasks. Tayloe a 8.

61. By comparison, Philip D. Morgan estimates that slave tobacco labor in eighteenth-century Virginia averaged 113 days per year, including every month except December, and slave rice labor in eighteenth-century low-country South Carolina averaged 188 days, in all twelve months; Morgan, *Slave Counterpoint: Black Culture in the Eighteenth-Century Chesapeake and Lowcountry* (Chapel Hill: University of North Carolina Press, 1998), 176.
62. Lorena S. Walsh, "Plantation Management in the Chesapeake, 1620–1820," *Journal of Economic History* 49 (1989): 398–99.
63. I have followed the procedure Walsh outlines in her article in calculating the crop outputs per laborer at Mount Airy. A productivity calculation can also be made for the Mount Airy corn crop in 1808, when Tayloe's slaves produced thirty-five barrels per laborer.
64. Lois Green Carr and Lorena S. Walsh, "Economic Diversification and Labor Organization in the Chesapeake, 1650–1820," in *Work and Labor in Early America,* ed. Stephen Innes (Chapel Hill: University of North Carolina Press, 1988), 175–188.
65. JT3 to William Gordon, July 18, 1801, JT3 letter book, 1801, Tayloe d 170; Holburne to William Greenlaw, March 22, 1809, Holburne letter book, 1809.
66. In 1833 Lame Sam became a miller and was last reported working as a miller in 1855, when he was seventy-one. Like many of the Mount Airy slaves, Sam has no death date in the inventory records.
67. In December 1805 JT3 paid Joseph Fox $50 "for a cotton machine," which was probably a gin; JT3 account book, 1805–1812, Tayloe a 3.
68. Tayloe a 9. Fragments of similar spinning books for 1805–1806 and 1816–1819 are in Tayloe uncatalogued papers, box 1.
69. Kamoie, *Irons in the Fire,* 113.
70. 1806 day book, Tayloe uncatalogued papers, box 3.
71. "Blotter" day book, 1807, Tayloe uncatalogued papers, box 3.
72. Holburne to George Gresham, February 11, 1809. On the same day Holburne wrote to William Greenlaw Jr., the overseer at Oaken Brow, sending him thirty-four and a half pounds of sole leather and eighteen and a half yards of cloth. Holburne letter book, 1809.
73. The fourth shoemaker in 1808 was Ruffin (age twenty-five). Big Joe disappeared in 1809 or 1810, leaving only three shoemakers, but the work log for 1811–1812 identifies a shoemaker named Sam, who is not listed in the Mount Airy slave inventory and may have been a hired free black. In 1819, eight-year-old John started to learn the craft, and by 1828 Joe was sixty-one, Little Joe was thirty, Ruffin was forty-five, and John was seventeen.
74. Tayloe a 8.
75. The 1809 inventory lists the tools kept in all of the craft shops at Mount Airy; Tayloe d 538.
76. Shoemakers' account in the work log for 1811–1812, Tayloe a 10.
77. The 1816–1817 shoemakers' shop book is in Tayloe uncatalogued papers, box 3.

78. "Blotter" day book, 1807, Tayloe uncatalogued papers, box 3.
79. The 1805 work log entry for August 19 reports, "Harry getting beam for fencing at Marske, & getting it in the cart hurt his knee," and then revises the last three words to "broke his leg." The entry for September 11 reports, "Harry at work at Joiners shop for the first time since his leg was broke." Tayloe a 8, 132, 145.
80. "Blotter" day book, 1807, Tayloe uncatalogued papers, box 3.
81. Minute book for 1811–1812, Tayloe a 10.
82. Not all of the horses shod were plow horses; the smiths also shod Tayloe's carriage, riding, and race horses.
83. Kamoie discusses the productivity of JT3's Mount Airy mill in 1810 in *Irons in the Fire,* 111–112.
84. In 1808 there were four masons, but John (age sixty) and Cyrus (age fifteen) both disappeared from the records in 1809.
85. The 1808 Mount Airy inventory is in Tayloe d 538. By 1809, as Appendix 22 shows, the number of domestics had increased from twenty-eight to twenty-nine.
86. JT3 letter, written in 1817, was exhibited at the Octagon, and discussed by Benjamin Forgey in the *Washington Post,* October 28, 1995. I thank Marion Nelson for sending me this reference.
87. Inventory of JT4's estate at Mount Airy, July 8, 1824, Richmond County Will Book no. 10, 1822–1846, LVA.
88. The Mount Airy inventories listing the domestic staff in January 1815, 1818, 1822, 1825, and 1828 are all in Tayloe d 538.
89. Top Gallant was appraised at £600 in both 1808 and 1809; Pavilian and P.M. General were appraised only in 1809.
90. Holburne to John Milton, February 13, 1809, Holburne letter book, 1809.
91. Day book, 1814, Tayloe uncatalogued papers, box 3.
92. By way of comparison, the Mount Airy mansion house and dependencies (office and kitchen) were valued in 1808 at £10,000.
93. Nancy worked as a domestic and spinner at Mount Airy from 1808 to 1861. She was living in retirement at Mount Airy when she was finally freed from slavery, and her daughter Martha took her to Washington, D.C., in December 1865 at age seventy-three.
94. Between 1811 and 1827, JT3 sent another thirteen workers to Cloverdale from Mount Airy. All of these later migrants were males: five field hands from the farm quarters, and eight craftsmen—four carpenters, three jobbers, and a smith—from the home plantation.
95. Holburne to William Greenlaw, April 28, 1809, Holburne letter book, 1809.
96. Holburne to William Greenlaw, March 31, 1809, Holburne letter book, 1809.
97. On January 1, 1810, Nancy's William was five and Fanny was one; the inventory was later amended to state that both children were dead, and their names do not appear in the 1811 Doctor's Hall inventory.

98. The slave inventories for 1808–1828 always identify the mothers of young children, but never the fathers, and in only about twenty-five cases are husbands and wives identified. The post-1828 inventories compiled by WHT supply much fuller family information.
99. William Gordon to JT3, August 3, 1817, Tayloe Family Papers, UVA.
100. Cloverdale inventory, 1 December 1817.
101. For a succinct description of iron making in the early Republic, see John Bezís-Selfa, *Forging America: Ironworkers, Adventurers, and the Industrious Revolution* (Ithaca, NY: Cornell University Press, 2004), chap. 1. See also Charles B. Dew, *Bond of Iron: Master and Slave at Buffalo Forge* (New York: W. W. Norton, 1994), and Kamoie, *Irons in the Fire*, 121–122.
102. The proportion of young children continued to grow at Cloverdale. By 1822 a full third of the slaves—51 out of 151—were nonworking children.
103. The bill of sale to Bevan and Snide is in Tayloe d 414–440. The first twenty slaves listed in Appendix 26 were sold to Bevan in September 1816 for $5,135, and the last six soon afterward to Snide for $2,075.
104. Peggy's two eldest children, Sarah and James, had been sent away or sold in 1809.
105. The correspondence between WHT and BOT in 1824 is in Tayloe d 1217–1227.
106. The Mansion Hotel had one hundred rooms and was on the site of the present Willard Hotel. For full discussion of John III's enterprises, including a stagecoach and mail service and bank directorships, see Kamoie, *Irons in the Fire*, 128–138.
107. I have not been able to determine the exact size of Tayloe's slave force in 1828 because I have not found slave inventories for most of the outlying properties. I estimate that the grand total was closer to 800 than 700. Two years later, according to the 1830 census, his six surviving sons, widow, and grandson collectively held 723 slaves on their principal properties in Virginia, Maryland, and the District of Columbia, but these listings are clearly incomplete.
108. JT3's will, dated December 1827, is in Tayloe d 539.
109. The division of craft workers is in JT3's estate book, Tayloe d 928.
110. In 1828 William's four farms were populated by 111 slaves. Edward's Hopyard and Doeg were populated by sixty slaves, Henry's Gwinfield by fifty-five slaves, and Charles's Oaken Brow by forty-eight slaves.
111. The 1829 Mount Airy inventory is in Tayloe d 13410.

### *6. The Moravian Christian Community at Mesopotamia*

1. For discussion of Moravian missionary efforts around the world from the eighteenth to the twentieth century, see J. Taylor Hamilton and Kenneth G. Hamilton, *History of the Moravian Church: The Renewed Unitas Fratrum,*

*1722–1957* (Bethlehem, PA: Moravian Church, 1967), chaps. 3–4, 13, 24–29, 41–48.

2. Aaron Spencer Fogleman, *Jesus Is Female: Moravians and Radical Religion in Early America* (Philadelphia; University of Pennsylvania Press, 2007). Professor Fogleman, while doing research for his book, kindly pointed out to me Mesopotamia documents at Herrnhut of crucial importance for this chapter.
3. J.H. Buchner, *The Moravians in Jamaica: History of the Mission of the United Brethren's Church to the Negroes in the Island of Jamaica, from the Year 1754 to 1854* (London: Longman, Brown & Co., 1854), 66–67.
4. The Mesopotamia diaries at Herrnhut span the dates 1758–1761, 1763–1790, 1799, 1812–1813, 1815, 1817, and 1831–1832. Thirty of these diaries are written in German, eight in English. They range greatly in length, several of the early German diaries being close to 200 pages, while most—including all of the English diaries—are much shorter. They are catalogued in the "Jamaika" section of the Unitätsarchiv at Herrnhut. I wish to thank Dr. Paul M. Peucker, the archivist at Herrnhut, for kindly providing me with microfilms of all these documents.
5. The Jamaica Moravian records include a minute book kept by three sets of Mesopotamia missionaries from 1798 to 1818 (Minutes of Mesopotamia Conference, Moravian Archives of Jamaica, Fairfield R/1, JA), a travel diary by Brother Lang, who lived at Carmel from 1805 to 1809 (Fairfield Q/7, JA), and four diaries by Brother Zorn, who lived at Spring Vale and New Fulneck in 1830–1833 (Fairfield H/7; H/6; J/17; J/13, JA).
6. My discussion of the Moravians' interaction with slaves at Mesopotamia may be usefully compared with Jon F. Sensbach's discussion of the Moravians' interaction with slaves at Salem, North Carolina, in *A Separate Canaan: The Making of an Afro-Moravian World in North Carolina, 1763–1840* (Chapel Hill: University of North Carolina Press, 1998). The two situations differ, since Sensbach is tracing the emergence of segregation and white racism among Moravians in the Old South. But he too has richly detailed sources to draw upon, especially the *Lebensläufe* or spiritual biographies of the Salem slaves—which were not recorded in Jamaica.
7. "Diary of Brother Caries Voyage to Jamaica 1754," MCHL. Caries's initial Jamaica diary, written in German, opens on October 4, 1754, and closes on September 30, 1756. The MCHL copy is an English translation that was presented to JFB1. For commentary on Brother Caries's diary, see Shirley C. Gordon, *God Almighty Make Me Free: Christianity in Preemancipation Jamaica* (Bloomington: Indiana University Press, 1996), 17–25.
8. JFB1 to Brother Cennick, August 12, 1753, MCHL.
9. JFB1's proposal to the Brethren, May 23, 1754, UA R.15.C.a.1.3, courtesy of Katharine Gerbner (see notes 13, 36).
10. Caries diary, December 26, 1754, April 15, 1755, MCHL. For background, John Thornton has an interesting discussion of interaction between African

religions and Christianity in *Africa and Africans in the Making of the Atlantic World, 1400–1800,* 2nd ed. (Cambridge: Cambridge University Press, 1998), chap. 9.

11. Benjamin LaTrobe, *A Succinct View of the Missions Established among the Heathen by the Church of the Brethren, or Unitas Fratrum; In a Letter to a Friend* (London: M. Lewis, 1771), 3–5. Latrobe was a Moravian bishop and a close friend of JFB1. His son Christian Ignatius Latrobe was also a Moravian bishop and a close friend of JFB2. Christian's brother was the architect-engineer Benjamin Henry LaTrobe who had a celebrated career in the United States.
12. Caries diary, March 6, 9, 14, 21, 23, April 28, July 8, 1755, MCHL.
13. Katharine Gerbner, "'They Call Me Obea': The Rise, Fall and Resurrection of the Moravian Mission on Jamaica, 1754–1770" (paper presented at the American Historical Association Annual Conference, January 2011), 6. I thank Dr. Gerbner for permitting me to cite this essay.
14. Gerbner, "'They Call me Obea,'" 6–8.
15. Gerbner, "'They Call me Obea,'" 22–28. By the 1770s, the Bogue was the most vigorous Moravian congregation in Jamaica, with many more members and communicants than at Mesopotamia.
16. Gerbner, "'They Call me Obea,'" 5.
17. Buchner, *The Moravians in Jamaica,* 36–38.
18. Memorandum by JFB1, ca. 1755–1758, MCHL. JFB1 and William Foster each paid half of £3,300, which included £450 for 600 acres given to the Moravians, £240 for four slaves, £707 for hired black labor, and £1,080 for building and furnishing the mission house at Carmel.
19. Caries diary, July 10, 1755, MCHL.
20. Caries's entries on Mesopotamia in 1758 are in UA R.15.Cb.1.1; his Mesopotamia entries for 1759 are in UA R.15.Cb.1.2.
21. Rauch's 180-page Mesopotamia diary for November 1759–December 1761 is in UA R15.Cb.1.3.
22. For Pool's salary as attorney and his slave purchases and slave hires for Mesopotamia, see Barham b 34 and b 36.
23. Caries diary, October 22, November 19, 1758, UA R.15.Cb.1.1. I thank Rainer Hatoum for translating this and other passages for me.
24. Caries diary, February 11, May 6, 1759, UA R.15.Cb.1.2. Translation by Rainer Hatoum.
25. Mesopotamia diary, December 2, 8–9, 1759, UA R.15.Cb.1.3. Translations by Rainer Hatoum.
26. Mesopotamia diary, December 8, 10–11, 15, 20, 1759, UA R.15.Cb.1.3. Translations by Rainer Hatoum.
27. Love was inventoried in 1727; Primus and Ralph were her brothers.
28. Mesopotamia diary, January 10, February 15, 25, May 18, June 10, July 4, August 24, 31, December 6, 1760, UA R.15.Cb.1.3. Translations by Rainer Hatoum.

29. There was no Mesopotamia birth register in 1760–1761, and I have found no confirmatory record of either of Hannah's babies.
30. Mesopotamia diary, March 24, April 1, 4, July 5, August 5, 1761, UA R.15.Cb.1.3. I thank Katharine Gerbner for translating these and other passages for me.
31. Mesopotamia diary, April 17, June 22, July 9, 17, August 20, 22, 24–25, 1761, UA R.15.Cb.1.3. Translations by Katharine Gerbner.
32. Mesopotamia diary, June 28, 1763, UA R.15.Cb.1.3. This turned out to be Brother Christian Heinrich's final act at Mesopotamia; he returned to Carmel and died there in November 1763.
33. Mesopotamia diary, 1765, UA R.15.Cb.3.1/218; Mesopotamia diary, 1766, UA R.15.Cb.3.2/43. The missionaries received the Saviour's choices for who should be baptized and who should be confirmed via the lot—as will be described later in this chapter.
34. Mesopotamia diary, 1764, UA R.15.Cb.3.1/209, 216.
35. Katharine Gerbner has identified ninety-seven Mesopotamia people who were baptized in 1761–1769; the December 1768 slave inventory, which identifies eighty-three baptized people, adds seven not on Gerbner's list; and I have double checked the Moravians' Mesopotamia diaries for 1761–1769 and found two further slaves baptized in these years.
36. Katharine Gerbner has let me use her Excel spreadsheet of slaves baptized at Mesopotamia, 1761–1770, together with comments she extracted from the Moravian diaries on individual baptismal candidates. She has also given me data on Moravian arrivals, deaths, and departures from Jamaica, 1754–1770, taken from the Moravian Archives in Bethlehem, Pennsylvania. I am exceedingly grateful to Dr. Gerbner for so generously sharing with me these very valuable research findings from the Herrnhut and Bethlehem archives.
37. It is impossible to determine whether many of the slaves who appear on the earliest Mesopotamia inventories taken in 1727, 1736, 1743, and 1744 were born in Jamaica or came from Africa. Thus, the Creole and African totals in Appendix 28 are conjectural for over half of these older slaves.
38. Mesopotamia diary, July 8, 15, August 25, 1761, UA R.15.C.b.1.3. Translation by Katharine Gerbner.
39. Statistics on Mesopotamia confirmations are found in the Jamaica diary for 1766 (UA R.15.Cb.3.2/43); the Jamaica diary for 1767 (UA R.15. Cb.3.3/60); the Jamaica diary for 1769 (UA R.15.Cb.3.5/66); and the Jamaica diary for 1772 (UA R.15.Cb.3.8/119).
40. Mesopotamia diary, December 1768–August 1769, UA R.15.Cb.3.5/65–66.
41. The Jamaican government imposed a "deficiency" tax on estates with too few white residents in proportion to slaves, and the Barhams saved £50 per annum in taxes by counting the missionaries as residents of Mesopotamia. See JFB2 to John Foster, n.d. (ca. 1800), Barham c 378; JFB2 to Christian Ignatius Latrobe, January 4, 1802, MCHL.

42. JFB2 to P. White and J. R. Webb, July 1, 1804, Barham c 428; JFB2 to (?), November 22, 1800, Barham c 378.
43. For JFB2's efforts to strengthen mission work in Jamaica, see JFB2 to J. Harriet, March 1800, Barham c 378; JFB2 to J. Graham, October 16, 1790, Barham c 428; JFB2 to H. W. Plummer, July 6, 1799, Barham c 428; and JFB2 to Lord Wyndham, November 1, 1800, MCHL.
44. JFB2 to C. I. Latrobe, December 5, 1795, September 28, November 1, 1800, MCHL; subscription list, 1800, Barham c 378.
45. Brother Taylor's diary, May 1783–May 1785, UA R.15.Cb.4.7/39–40, 50, 55.
46. Mesopotamia diary, January–June 1773, UA R.15.Cb.3.9/140, 150.
47. Jamaica diary, 1778, UA R.15.Cb.4.2/2–4.
48. Douglas Hall, *In Miserable Slavery: Thomas Thistlewood in Jamaica, 1750–86* (Mona, Jamaica: University of the West Indies Press, 1999), 277–278; Trevor Burnard, *Mastery, Tyranny, and Desire: Thomas Thistlewood and His Slaves in the Anglo-Jamaican World* (Chapel Hill: University of North Carolina Press, 2004), 10–12.
49. Tony's leg was shattered in the hurricane, and when it was amputated he contracted tetanus and died.
50. Brother Taylor's diary, May 1783–May 1785, UA R.15.Cb.4.7/56.
51. The chapel is described in the accounts for 1780–1781 in Barham b 33 and b 37.
52. Mesopotamia diary, 1787–1790, UA R.15.-Cb.6.6/5, 15; James Graham to JFB2, May 11, 1789, December 16, 1790, Barham c 357; Mesopotamia accounts, 1789, Barham b 33.
53. H. W. Plummer to JFB2, September 23, 1799, Barham c 357; JFB2 to C.I. Latrobe, November 7, 1800, MCHL; W. Rodgers to JFB2, August 30, 1801, Barham c 357; J. Jackson to C.I. Latrobe, October 11, 1802, MCHL; "The Present State of the Buildings on Mesopotamia Estate," June 1, 1814, Barham b 34.
54. Mesopotamia diary, June 1782–April 1783, UA R.15.Cb.6.4/11, 14–17, 21.
55. Mesopotamia diary, 1799, UA R.15.1.Cb.6.8/5–6.
56. Brother Taylor's diary, May 1783–May 1785, UA R.15.Cb.4.7/27–28.
57. Brother Taylor's diary, May 1783–May 1785, UA R.15.Cb.4.7/22, 87.
58. Mesopotamia diary, June 1782–April 1783, UA R.15.Cb.6.4/3, 23.
59. John Graham served as JFB1's overseer at Mesopotamia from 1778 to 1788. James Graham, probably his father, served as JFB2's attorney for the estate from 1790 to 1799.
60. Mesopotamia diary, June 1782–April 1783, UA R.15.Cb.6.4/4–5, 11, 14–15, 22; Brother Taylor's diary, May 1783–May 1785, UA R.15. Cb.4.7/28.
61. Mesopotamia diary, January–June 1773, UA R.15 Cb.3.9/139, 147–150.
62. Mesopotamia diary, June 1782–April 1783, UA R.15.Cb.6.4/10–11, 16–17; Brother Taylor's diary, May 1783–May 1785, UA R.15.Cb.4.7/29–30.

63. Mesopotamia Conference, May 12, July 9, 1798; Mesopotamia diary, 1799, UA R.15.1.Cb.6.8/13–14.
64. Mesopotamia diary, 1799, UA R.15.1.Cb.6.8/12–13.
65. Mesopotamia diary, 1799, UA R.15.1.Cb.6.8/22–23.
66. Mesopotamia diary, 1787–1790, UA R.15.Cb.6.6/17.
67. Sarah Affir (in Chapter 2) contracted scrofula in the 1820s.
68. Mesopotamia diary, 1799, UA R.15.1.Cb.6.8/9–10, 19.
69. Jamaica diary, 1774, UA R.15.Cb.3.10/55.
70. Mesopotamia accounts, 1789, Barham b 33; J. Graham to JFB2, June 16, 1790, Barham c 357.
71. Mesopotamia diary, 1787–1790, UA R.15.Cb.6.6/41–47.
72. Brother Jackson's letter to Fulneck, June 28, 1769, UA R.15.1.Cb.3.5/51–58.
73. Philip Wright, *Monumental Inscriptions of Jamaica* (London: Society of Genealogists, 1966), 180, 192; Moravian arrivals, deaths, and departures from Jamaica, 1754–1770, Moravian Archives, Bethlehem, Pennsylvania (courtesy Katharine Gerbner).
74. J. Jackson to C.I. Latrobe, August 2, 1802, MCHL; Jackson to JFB2, November 7, 1804, Barham c 378; attorneys Wedderburn, Grant, and Blyth to JFB2, November 12, 1805, Barham c 357.
75. Jackson to JFB2, November 7, 1804, Barham c 378; Jackson to C.I. Latrobe, October 11, 1802, Jamaican Mission Papers, MCHL.
76. Jackson to JFB2, May 6, 1805, Barham c 378.
77. M.G. Lewis, *Journal of a West India Proprietor, 1815–1817*, ed. Mona Wilson (Boston, 1929 [orig. pub. London, 1834]), 152–153.
78. Diary of Brother J. Lang, 1805–1819, Fairfield Q7/26, 46, JA.
79. C.I. Latrobe to JFB2, October 4, 1809, Barham c 378; J. Blyth to JFB2, September 14, 1811, Barham c 358; J. Becker to C.I. Latrobe, February 13, 1813, MCHL; J. Lang to C.I. Latrobe, April 15, 1813, MCHL.
80. Mesopotamia diary, 1812, UA R.15.Cb.6.9/9, 40.
81. Mesopotamia diary, 1812, UA R.15.Cb.6.9/8, 17, 30–31, 35–36; Brother J. Lang to C.I. Latrobe, April 15, 1813, MCHL.
82. Mesopotamia diary, 1815, UA R.15.Cb.6.11/16, 22–23, 31; J.S. Gründer to C.I. Latrobe, March 20, 1813, MCHL.
83. Mesopotamia diary, 1812, UA R.15.Cb.6.9/7, 12; Mesopotamia diary, 1815, UA R.15.Cb.6.11/3–8.
84. Gründer to JFB2, April 19, 1814, Barham c 378; Mesopotamia diary, 1815, UA R.15.1.Cb.6.11/13–14, 21, 24.
85. Gründer to C.I. Latrobe, March 26, 1818, MCHL; Gründer to JFB2, March 27, 1818, MCHL; Blyth and Grant to JFB2, June 15, 1818, Barham c 358.
86. J.R. Webb to JFB2, May 5, 1818, Barham c 358; Blyth and Grant to JFB2, August 3, 1818, March 20, 1819, Barham c 358; C.I. Latrobe to JFB2, April 26, 1819, March 15, 1821, Barham c 378.
87. Minutes of Mesopotamia Conference, 1798–1818, Moravian Archives of Jamaica, Fairfield R1, JA.

88. Minutes of Mesopotamia Conference, Fairfield R1/1–2.
89. For Cooper Thomas, see Mesopotamia Conference, Fairfield R1/2, 4, 6, 9, 16, 18, 21–22; Mesopotamia diary, 1799, UA R.15.Cb.6.8/16–20.
90. Mesopotamia diary, 1815, UA R.15.Cb.6.11/27–28.
91. "I will ask the Father, and He will give you another Helper, that He may be with you forever."
92. L. Stobwasser's report on Mesopotamia, December 1, 1825, Barham c 378.
93. J. Blyth to JFB2, December 31, 1816, July 9, 1823, Barham c 358.
94. C.I. Latrobe to JFB2, July 14, 1829, Barham c 378; visits to Mesopotamia, 1828–1829, Barham b 35; Minutes of Moravian Conference, 1830–1831, Fairfield H5/3, 5, 7–8, JA.
95. Brother Zorn's diary, 1830, Fairfield H7/5–7, 12–13, 23–24, JA.
96. Mesopotamia diary, 1831, UA R.15.1.Cb.6.13/3, 5, 7, 9; Minutes of Moravian Conference, 1830–1831, p. 10, Fairfield H5, JA; Minutes of Moravian Conference, 1831–1833, Fairfield Q11/3, JA.
97. The 1832 inventory is in Barham b 35. It is dated January 1, but must have been taken a few days earlier, because when the slave revolt of 1831–1832 broke out the entire white staff at Mesopotamia left for militia duty on December 30 and didn't return until February 1832.
98. Mesopotamia diary, 1831, UA R.15.Cb.6.13/13–15.
99. Moravian hymn, 1768, composed on the death of William Foster, in UA R.15.Cb.3.5/24–25.

## *7. The Exodus from Mount Airy to Alabama*

1. Henrietta, known as Etta, was the niece of WHT's mother, Anne Ogle Tayloe; her sister Mary married WHT's brother Edward.
2. Ira Berlin, *The Making of African America* (New York: Viking, 2010), 263, maintains that the total number of slaves who were moved west and south between 1790 and 1860 can only be roughly estimated, but Robert William Fogel and Stanley L. Engerman, in *Time on the Cross: The Economics of American Negro Slavery* (Boston: Little, Brown, 1974), 46–47, use statistical modeling to posit an exact total of 835,000 interregional migrants.
3. Michael Tadman, *Speculators and Slaves: Masters, Traders, and Slaves in the Old South* (Madison: University of Wisconsin Press, 1989), presents convincing evidence (pp. 21–31) to argue that 60–70 percent of the slaves were sold to professional traders. Fogel and Engerman, in *Time on the Cross*, p. 48, claim that only 16 percent were sold to traders.
4. See Walter Johnson, *River of Dark Dreams: Slavery and Empire in the Cotton Kingdom* (Cambridge, MA: Harvard University Press, 2013), and Edward Baptist, *The Half Has Never Been Told* (New York: Basic Books, 2014). I am greatly indebted to Professor Baptist for sending me the draft chapters of his powerful book.

5. For the New Orleans slave market, see Walter Johnson, *Soul by Soul: Life inside the Antebellum Slave Market* (Cambridge, MA: Harvard University Press, 1999).
6. Charles Ball, *Fifty Years in Chains; or, The Life of an American Slave* (New York: H. Dayton, 1859), 29–37.
7. See Ira Berlin, *The Making of African America,* especially chap. 3, and Edward E. Baptist, "'Cuffy,' 'Fancy Maids,' and 'One-Eyed Men': Rape, Commodification, and the Domestic Slave Trade in the United States," *American Historical Review* 106, no. 5 (2001): 1–55.
8. Among Johnson's and Baptist's leading sources are Charles Ball, *Fifty Years in Chains* (1859); Henry Bibb, *Narrative of the Life and Adventures of . . . an American Slave* (1849); Leonard Black, *The Life and Sufferings of . . . a Fugitive from Slavery* (1847); John Brown, *Slave Life in Georgia* (1855); William Wells Brown, *Narrative of . . . a Fugitive Slave* (1847); *Narrative of the Life of Moses Grandy, Late a Slave* (1843); William Green, *Narrative of Events* (1853); William Hayden, *Narrative . . . Whilst a Slave in the South* (1846); Louis Hughes, *Thirty Years a Slave* (1897); *The Rev. J. W. Loguen, as a Slave and as a Freeman* (1859); Isaac Mason, *Life . . . as a Slave* (1893); Solomon Northup, *Twelve Years a Slave* (1853); John Parker, *His Promised Land* (1996); Moses Roper, *A Narrative of the Adventures and Escape . . .* (1838); and Jacob Stroyer, *Sketches of My Life in the South* (1879).
9. For details, see Chapter 5. BOT and WHT, the two oldest surviving sons, became JT3's executors and had to cope with his large debts and generous legacies, amounting to some $120,000, which—as BOT put it—"will be plaguing us the whole of our lives." BOT to WHT, May 13, 1830, Tayloe d 1217–1286.
10. The enumerated total was actually 213, but I have added four young girls and four young boys who were present on earlier and later inventories and seem to have been omitted in 1830 by accident.
11. WHT's corn and wheat production is well documented for the 1830s, but the crop and sale records for the 1840s and 1850s are very incomplete, and conflicting. Helpful lists are in Tayloe d 8521–8527, d 8528–8530, and d 13410.
12. WHT noted their departure in his Mount Airy inventory for 1832 (Tayloe d 13410). Jeffries reported their arrival in letters to WHT of January 23 and February 8, 1832 (Tayloe d 3741–3742).
13. BOT to WHT, January 5, 23, February 20, March 5, 1832; WHT to BOT, February 2, 1832, Tayloe d 1287–1351.
14. BOT to WHT, February 20, 1832, Tayloe d 1287–1351. BOT sold this Florida land in 1836.
15. Deep Hole was also a fishery. In 1835, 276 barrels of Deep Hole shad and herring were sold for $1,167, and additional barrels were distributed as slave food to Mount Airy and other Tayloe properties (Tayloe d 7925).
16. BOT to WHT, November 28, 1833, Tayloe d 1287–1351.

17. HAT to BOT, December 26, 1833, Tayloe Papers, UVA.
18. HAT to WHT, December 16, 1833, Tayloe d 5960–6045.
19. WHT contributed a total of thirteen slaves and BOT eleven. All of the forty-four people who headed to Alabama in December 1833, except for the ten Deep Hole slaves, had probably lived at Mount Airy during JT3's ownership.
20. Two lists enumerate the slaves in this initial Alabama party: Tayloe d 13410, p. 22; Tayloe d 14521–14625.
21. Travers had worked as a joiner at Mount Airy from age seven to twenty; I do not know whether Charles Tayloe managed to catch him. HAT to BOT, December 26, 1833, January 7, 1834, Tayloe Papers, UVA; HAT to WHT, December 29, 1833, Tayloe d 5849–5959; Charles Tayloe to WHT, (ca. January 1834), Tayloe d 5259–5291.
22. HAT to BOT, July 29, 1839, Tayloe Papers, UVA; see also Tayloe d 13410, p. 33.
23. Michael Tadman, in *Speculators and Slaves,* p. 12, calculates that 35,500 slaves were moved into Alabama in the 1810s, 54,156 in the 1820s, and 96,520 in the 1830s. See also Steven F. Miller, "Plantation Labor Organization and Slave Life on the Cotton Frontier: The Alabama-Mississippi Black Belt, 1815–1840," in *Cultivation and Culture: Labor and the Shaping of Slave Life in the Americas,* ed. Ira Berlin and Philip D. Morgan (Charlottesville: University Press of Virginia, 1993),156–157.
24. BOT to WHT, March 6, 1834, Tayloe Papers, UVA; HAT to WHT, February 7, 183[4], Tayloe d 5849–5959. HAT explored the Canebrake with his brother George, who was thinking of moving all his Cloverdale slaves to Alabama.
25. HAT claimed that "Negroes in the Cane Brake make from $400 to $450 per hand—Ought we to retain one in Virginia?" HAT to WHT, February 7, 183[4], Tayloe d 5849–5959.
26. I have not found a record of HAT's purchase of Walnut Grove. His indenture with BOT and WHT, July 21, 1834, is in Tayloe Papers, UVA.
27. HAT to WHT, October 22, 1834, Tayloe d 5849–5959.
28. HAT to WHT, July 15, 1835, Tayloe d 5849–5959; BOT to WHT, January 26, April 22, 1835, Tayloe d 1287–1351.
29. The following lists of the Mount Airy slaves sent to Alabama in October 1835—Tayloe d 13424, pp. 3, 17; d 13425; d 13410, p. 48; and d 8632–8667—do not entirely agree. These lists generally add up to fourteen migrants, not sixteen. Most omit Joe and all omit Grace, both of whom were sold on arrival.
30. BOT to WHT, August 2, 1835, Tayloe d 1287–1351; WHT memo, August 1835, Tayloe d 5259–5291.
31. ETT to WHT, August 13, 1835, Tayloe d 5339–5482.
32. HAT to BOT, December 1, 1834, Tayloe Papers, UVA.
33. HAT added that if Mary "had not been deaf more could have been obtained." HAT to WHT, May 1834, Tayloe d 5849–5959.

34. Mobile was about 150 miles south of HAT's Adventure plantation.
35. HAT to WHT, January 2, February 27, 1835, January 2, 1836, Tayloe d 5849–5959; HAT to BOT, January 5, February 6, 16, 1836, Tayloe Papers, UVA. Steven Miller stresses the harsh conditions endured by the pioneering Alabama and Mississippi slaves in "Plantation Labor Organization and Slave Life on the Cotton Frontier," 158–165.
36. HAT to WHT, November 16, 1834, Tayloe d 5849–5959.
37. HAT to WHT, November 9, 1835, Tayloe d 5849–5959.
38. HAT to WHT, January 2, 1836, Tayloe d 5849–5959; HAT to BOT, January 5, 1836, Tayloe Papers, UVA; BOT to WHT, December 7, 1835, January 21, February 26, 1836, Tayloe d 1287–1351.
39. Baptist, *The Half Has Never Been Told*, 295–299.
40. BOT to WHT, February 1, 1836, Tayloe d 1287–1351.
41. HAT to BOT, January 5, 1836, Tayloe Papers, UVA; BOT to WHT, April 4, May 16, 1836, Tayloe d 1287–1351.
42. BOT to WHT, October 26–27, 1836, HAT to BOT, November 14, 1836, Tayloe d 1287–1351; HAT to WHT, October 30, December 12, 1836, Tayloe d 5849–5959.
43. HAT to WHT, December 12, 1836, Tayloe d 5849–5959; HAT to BOT, January 28, 1837, Tayloe Papers, UVA; HAT account with BOT, Tayloe d 5849–5959.
44. Baptist, *The Half Has Never Been Told*, 299–301.
45. HAT's sale of five slaves belonging to WHT is recorded in Tayloe d 15,332–15,579. BOT received $2,070 as an initial payment for his slaves (BOT to WHT, March 8, 1837, Tayloe d 1352–1407), but I have found no record of a similar payment to WHT.
46. HAT to WHT, August 25, 1835, Tayloe d 5849–5959. HAT did, however, marry an Alabama woman, Narcissa Jamieson, in 1838.
47. There are lists of WHT's thoroughbreds dated 1834, 1835, 1837, and 1838 in Tayloe d 13410, pp. 31, 47; and Tayloe d 13424, pp. 12, 15.
48. Tayloe d 13424, p. 3. HAT and WHT also jointly owned a slave named Richard, an expert horse trainer, and HAT moved him from Virginia to Alabama in 1839.
49. HAT to BOT, January 5, 1836, Tayloe Papers, UVA.
50. BOT to WHT, March 10, 1838, WHT to BOT, April 11, 1838, Tayloe d 1352–1407.
51. HAT to BOT, February 21, 1839, WHT to BOT, January 23, 1845, Tayloe Papers, UVA.
52. HAT to BOT, May 10, 1840, Tayloe Papers, UVA; WHT to BOT, April 29, 1841, Tayloe d 1352–1407; BOT to WHT, March 24, July 8, 1842, Tayloe d 1408–1468.
53. BOT to WHT, May 24, July 13, 1843, Tayloe d 1408–1468.
54. ETT to WHT, October 26, 1843, Tayloe d 5339–5482. In May 1843, ETT had made a preliminary calculation in which he figured that HAT's

assets amounted to $55,825 and his liabilities to $153,846 (Tayloe d 20,960–21,010).

55. Baptist, *The Half Has Never Been Told,* 306, 312.
56. HAT to Tayloe & Company, January 1847, Tayloe d 20,943–20,946.
57. There are cotton production and sales records for Adventure, Walnut Grove, and Oakland, 1843–1845, in Tayloe b 58, d 5849–5959, and d 16,253–16,485.
58. WHT to BOT, January 23, 1845, Tayloe Papers, UVA. Giles was a carpenter from BOT's Nanjemoy estate in Virginia who was sent to Alabama in 1835; he was fifteen when inventoried at Nanjemoy in 1825, and thirty-four in 1844.
59. WHT to BOT, May 26, 1844; BOT to WHT, July 13, 24, 1844, Tayloe d 1408–1468.
60. John Jeffries to WHT, September 14, 1844, Tayloe d 3743–3746.
61. ETT to WHT, December 30, 1844, Tayloe d 5339–5482. Kitty Wormley continued as a chambermaid at Mount Airy until she was freed in 1865, and Jenny Moore also continued in this job until she died in 1852 at the age of thirty-five.
62. WHT notation on his 1844 slave inventory, Tayloe a 13.
63. BOT to WHT, December 26, 1844, Tayloe d 1408–1468; WHT memorandum, January 1, 1845, Tayloe Papers, UVA.
64. See Chapter 3.
65. WHT account book, 1835–1860, Tayloe b 58. The carpenters' contract for $2,600 is in Tayloe d 7926–7978.
66. WHT's Alabama memorandum book (Tayloe d 13425) records his Oakland transactions in 1843–47; see pp. 33, 38–39, 52, 56. He paid BOT $3,227 for his one-third share of Oakland land and stock, and $8,538 to Tayloe & Company for the remaining one-third share of Oakland slaves, land, and stock.
67. BOT to WHT, September 26, 1845, Tayloe d 1408–1468.
68. A Virginian named George S. Saunders sent the wife and three children of WHT's shoemaker James Dixon to his Alabama relative G. B. Saunders in this party, and GBS then hired James from WHT at $150 per year. GBS to WHT, January 4, 1847, October 2, December 25, 1848, Tayloe d 4957–4968.
69. Tayloe a 2118, p. 13; R.H. Donnahan to WHT, June 20, 1847, Tayloe d 2969–2985.
70. Donnahan to WHT, July 9, October 3, 1846, June 6, 20, 1847, Tayloe d 2969–2985.
71. Tayloe a 2118, p. 13; d 13425, p. 56.
72. WHT's memorandum is in Tayloe d 15581–15722; his 1847 Oakland inventory is in Tayloe d 13425. Baptist notes that almost 45 percent of the slaves imported to the cotton states were fourteen to twenty-five years old (*The*

*Half Has Never Been Told,* 128). Johnson, in *River of Dark Dreams,* 159, finds the slave trade focused on "prime age" slaves aged fifteen to twenty-five.

73. R.H. Donnahan to WHT, October 7, December 6, 1848, Tayloe d 2969–2985.
74. Johnson, *River of Dark Dreams,* 176–180.
75. Donnahan to WHT, December 6, 1848, Tayloe d 2969–2985.
76. Donnahan to WHT, November 10, 1849, Tayloe d 2969–2985.
77. There were also many cases of chills and fever at Oakland in 1848. Donnahan to WHT, June 6, 20, 1847, October 7, 1848, Tayloe d 2969–2985.
78. Donnahan to WHT, December 14, 1850, December 20, 1851, Tayloe d 2969–2985. The Oakland slaves were also short of blankets and hats, but had shoes and winter outside clothing.
79. Tayloe b 58; Tayloe a 2118, p. 13.
80. Donnahan to WHT, May 13, December 20, 1851, Tayloe d 2969–2985. Sinah might have been a daughter of Joe Moore, who came to Alabama from Mount Airy in 1837 and lived at Oakland with his wife Sinah, a non–Mount Airy woman. Or she might have been a daughter of Alfred Lewis and his wife Sinah, who came to Oakland from Mount Airy in 1845. There is no surviving slave inventory for Oakland between 1847 and 1853, and Sinah is not among the children listed in 1853, so she must have died young.
81. 1853 Oakland inventory, Tayloe d 13425, p. 117.
82. Tayloe d 13425, pp. 47–48, 52, 55. WHT paid Tayloe & Company $11,552 for the half share of Adventure, and $20,297 for the half share of Walnut Grove.
83. Tayloe d 13453. WHT sold his share of Walnut Grove to his brother George Tayloe for $13,240, and his share of Adventure to BOT and HAT for $12,795.
84. Tayloe a 13.
85. I have found no sale record for these three males and three females.
86. The Oakland and Woodlawn inventories for 1855 are in Tayloe d 13425.
87. BOT to WHT, June 9, 1857, Tayloe d 1730–1770; the Larkin deeds are in Tayloe d 13,453.
88. WHT's railroad platform and depot at Larkin cost him $1,859 (Tayloe d 19831–20010). For background, Alabama and Mississippi Rivers Railroad to WHT, February 9, 1857, June 1, 1859, Tayloe d 4846–4847; WHT to HAT Jr., March 6, 1858, Tayloe d 5960–6045; Alabama and Mississippi Rivers Railroad receipts, Tayloe d 19551–19697. For Tayloe's Crossing on the Marion Railroad, see WHT to HAT Jr., September 1, 1862, Tayloe d 5960–6045; also Marion Railroad receipts, Tayloe d 19688–19870.
89. Directions to Richmond, n.d., Tayloe d 13430; WHT to HAT Jr., May 5, 1859, Tayloe d 5980–6045.
90. These figures represent WHT's net receipts, after deducting transportation charges and commissions, which averaged 8 percent of the sale price. WHT's cotton receipts for 1858–1860 are in Tayloe d 19688–19830, d 19831–20010, d 20011–20126.

91. For Mount Airy wheat and corn production in the 1850s, see Tayloe d 8521–8527, d 8528–8530, and d 13410.
92. BOT to WHT, January 6, 1859, June 3, 1860, Tayloe d 1771–1811.
93. Tayloe d 19831–20010. This house burned down in 1866.
94. R.J. Booth to WHT, December 21, 23, 1858, Tayloe d 2352–2354; WHT to HAT Jr., n.d., Tayloe d 6046–6170.
95. WHT to HAT Jr., March 3, 1858, Tayloe d 5960–6045; WHT to HAT Jr., n.d. [ca. 1858], Tayloe d 6046–6170.
96. WHT to HAT Jr., March 25, 1858, Tayloe d 5960–6045; WHT to HAT Jr., ca. 1857, ca. 1858, April 8, 1858, Tayloe d 6046–6170.
97. E.W. Holcroft to WHT, October 8, 1857, Tayloe d 3702–3706; WHT to HAT Jr., February 15, [1859?], Tayloe d 5960–6045.
98. Holcroft to WHT, October 8, 1857, Tayloe d 3702–3706; WHT to [HAT Jr.?], July 16, 1858, Tayloe d 2123–2135. Another of the 1854 migrants, Sarah Thurston Moore, also died of typhoid in 1858.
99. WHT to HAT Jr., May 23, 1857, n.d. [ca. 1857], Tayloe d 6046–6170; WHT to [HAT Jr.?], July 16, 1858, Tayloe d 2123–2135.
100. The Oakland list (in Tayloe d 8402–8422) is undated, but internal evidence places it in 1859. The author was barely literate. Jim Grimshaw is "Jimgrimsker"; Poinsetta Glascow is "Pineset"; and Celia Dudley is "Sealy."
101. U.S. Census, 1860, Schedule 2, Perry County, Alabama (M653, roll 34); Marengo County, Alabama (M653, roll 31), NA-US.
102. BOT had 149 slaves, George P. Tayloe had 67 slaves, HAT had 54 slaves, BOT's son Thornton had 104 slaves, and George Tayloe's son John had 125 slaves.
103. There are also six undocumented years at Mount Airy: 1848 and 1856–1860. But the gaps at Mount Airy are much easier to bridge than in Alabama.
104. Baptist, *The Half Has Never Been Told,* 138–139, 142–143, 151–155.
105. The figure for 1859 is approximate. In 1859 WHT produced 860 bales of cotton at Oakland and Larkin (Tayloe d 19831–20010). There were 123 slaves (and 68 cotton hands) at Oakland in 1859, but no inventory for Larkin, which was expanding very rapidly from 47 slaves in 1858 to 152 in 1860, as WHT moved all his slaves at Woodlawn and Adventure to Larkin. Estimating that Oakland and Larkin had workforces of about the same size in 1859, I have allocated 430 of the 860 bales to Oakland.
106. Herbert G. Gutman, *The Black Family in Slavery and Freedom, 1750–1925* (New York: Pantheon, 1976), chaps. 1–4.
107. I have compiled this list of the largest Richmond County slaveholders from the manuscript census returns for 1810, 1820, 1830, 1840, 1850, and 1860 in the NA-US.
108. These free black families lived in Lunenburg parish, where Mount Airy is located.

109. John was sold at age eighteen, Mary at age eighteen, and Jane at age twenty.
110. WHT to BOT, December 17, 1859, Tayloe Papers, UVA. For close analysis of the political and economic divisions within Alabama on the eve of the Civil War, see William L. Barney, *The Secessionist Impulse: Alabama and Mississippi in 1860* (Princeton, NJ: Princeton University Press, 1974).
111. WHT to HAT Jr., December 11, 1860, Tayloe d 5960–6045. This statement suggests that WHT rented some of his Mount Airy land to tenants, but I have found no rent payments in his business records.
112. Letter book of BOT, "On the Subject of the Disunion, 1859–1860," Tayloe c 1. See also BOT to WHT, November 27, 1860, March 16, 1861, Tayloe d 1771–1811.
113. WHT to HAT Jr., December 11, 1860, Tayloe d 5960–6045
114. HAT Jr. account book, 1861–1866, Tayloe d 23707. HAT Jr. led the party, but WHT paid J.P. Watkins $100 for "services rendered bringing lot negroes from Richmond Va to Selma Ala" (Tayloe d 24021–24175).
115. Tayloe d 23707. See also list of deserted slaves, ca. March 1862 (Tayloe d 8402–8422).
116. Urias Myers was thirty-four years old and Edward Carter was forty-seven years old when they fled Mount Airy.
117. Note by WHT: "April 1, 1862, Overseer Reynolds brought 50 Negroes to Alabama" (Tayloe d 13425); Conway Reynolds to WHT, July 12, 1862, Tayloe d 4875–4876.
118. Lists of the slaves sent to Alabama in 1861–1862 are in Tayloe d 8597–8605, d 13453, and d 23707.
119. A Mount Airy slave named John broke his thigh in December 1860 (Tayloe d 23871–23927), but there were three other Johns at Mount Airy at this date, so it seems unlikely that John Page had two breaks in two years. For John Page's December 1861 break, see HAT Jr. account book, d 23707; also WHT to HAT Jr., August 1, 1862, Tayloe d 6046–6170.
120. Slave impressment list, February 3, 1865, Marion, Alabama, Tayloe d 8234–8255.
121. J.W. Ramey to WHT, June 16, July 14, 1861, Tayloe d 4851–4863; WHT to HAT Jr., March 20, August 12, 1862, Tayloe d 6046–6170; WHT to B.C. Chinn, September 2, 1862, Tayloe d 2751–2781.
122. WHT to HAT Jr., March 15, 20, May 28, 1862, Tayloe d 6046–6170.
123. David Carrington was described in the census of 1830 as a free colored man aged between thirty-six and fifty-five; he owned no slaves and had no dependents (U.S. Census, 1830, Richmond County, Virginia [M19, roll 194], US-NA).
124. WHT to HAT Jr., July 4, 1862, Tayloe d 6046–6170.
125. Tayloe d 5960–6045, d 13453; WHT to HAT Jr., January 12, 1865, Tayloe d 6046–6170.
126. Tayloe d 20126–20257.

127. This was the Morrow and Burnett plantation in Perry County. WHT hired William J. Ramey (probably a son of Larkin overseer J. W. Ramey) to manage it (Tayloe d 8597–8605).
128. Tayloe d 20258–20359.
129. WHT to HAT Jr., May 28, August 1, 1862, Tayloe d 6046–6170; same to same, September 1, 1862, Tayloe d 5960–6045; WHT to Courtenay Tayloe, May 3, 1864, Tayloe d 5292–5327.
130. WHT's 1863 census is in Tayloe d 13453. Two other sets of 1863 lists (stating a price for each slave) are in Tayloe d 8597–8605 and d 8632–86670.
131. WHT to HAT Jr., August 1, 1862, May 29, 1863, Tayloe d 6046–6170; WHT to Courtenay Tayloe, July 22, 1863, Tayloe d 5292–5327.
132. Tayloe d 20258–20359. There were two other recorded Tayloe impressments in 1863. On July 30 a forty-five-year-old Oakland hand named Peter Smith (not from Mount Airy) was dispatched to work on the fortifications at Mobile. And on December 17 an unidentified pair of Tayloe "boys" who had been hauling coal for the Confederate government were given a pass to travel back to Larkin with their team and wagon.
133. WHT to HAT Jr., July 24, 1864, Tayloe d 6046–6170.
134. WHT to Courtenay Tayloe, May 3, 1864, Tayloe d 5292–5327; for the Canebrake Relief Association, Tayloe d 20258–20359; for impressments in 1864, Tayloe d 8234–8255.
135. WHT to B. C. Chinn, n.d., January 3, 1865, Tayloe d 2751–2781; WHT to HAT Jr., January 5, 1865, Tayloe d 6046–6170; also Tayloe d 8234–8255.
136. WHT's valuation of his Alabama slaves, ca. 1863, is in Tayloe d 8597–8605 and d 8632–8667. His valuations ranged from $100 for a newborn infant to $2,500 for his most prized craft workers.
137. Donnahan to WHT, May 31, 1851, Tayloe d 2969–2985.
138. Baptist, *The Half Has Never Been Told,* 128.

## *8. Mesopotamia versus Mount Airy: The Social Contrast*

1. This is Trevor Burnard's interpretation in his *Mastery, Tyranny, and Desire: Thomas Thistlewood and His Slaves in the Anglo-Jamaican World* (Chapel Hill: University of North Carolina Press, 2004). See especially chapter 5, subtitled "Thistlewood's War with His Slaves."
2. Thistlewood, July 31, August 10, October 10, 24, 1750, Monson 31/1; March 24, May 30, 1751, Monson 31/2.
3. Thistlewood, January 2, February 6, March 9, April 30, July 23, 25, 29, August 27, October 16, 1765, Monson 31/16.
4. When John Barham's attorneys put in a claim for compensation for the 316 Mesopotamia slaves on August 1, 1834, they valued the 14 head people at £80 each, the 20 craft workers ("tradesmen") at £56 each, the 171 field workers at £54 each, the 14 domestics at £49 each, the 44 children under six at £15 each, and the 53 aged or sickly "non-effective"

people at £12 each. Westmoreland Slave Compensation, 1834, T71/723, #77, NA-UK.

5. Parry died after a long period of invalidism in 1783 at approximately age eighty-two. If Parry's age seems exaggerated, the Moravian missionaries judged him to be nearly seventy in 1765, when the slave inventory listed him as sixty-four.
6. Matt was sixty-five when he died, and Ralph after many years as an invalid died in 1802 at age eighty.
7. Francisco, born at Mesopotamia, was only twenty-eight when he became head driver. In 1781 James was demoted from driver to rat catcher by Graham, and he died a broken man in 1783.
8. B. W. Higman, *Slave Populations of the British Caribbean, 1807–1834* (Baltimore: Johns Hopkins University Press, 1984), 171–172.
9. Mesopotamia diary, January 2, 8, 9, 10, 11, 15, 17, 20, 1759, May 23, 1760, UA R.15.Cb.1.3.
10. Mesopotamia diary, January 5, 27, 1760, UA R.15.Cb.1.3.
11. Mesopotamia diary, December 28, 1759, UA R.15.Cb.1.3. I have not identified Augustus's wife.
12. For details on these three groups, see Chapter 6. The table excludes two people in the 1761–1769 group, because I cannot identify Hertford (baptized in 1767) or Phoebe (baptized in 1769). The slave inventory identifying "Christians" was dated January 1, 1832, but was probably taken in late 1831, since the entire white staff at Mesopotamia left for militia duty on December 30, 1831, to put down the slave rebellion of 1831–1832, and didn't return to Mesopotamia until late January or early February 1832. See below.
13. See Vincent Brown's *Slave Revolt in Jamaica, 1760–1761: A Cartographic Narrative,* at www.revolt.axismaps. Brown supplies both a place-name map and a terrain map, accompanied by contemporary accounts of what was happening, and invites the viewer to track the progress of the revolt day by day. I thank Professor Brown for presenting his array of documentary findings to the public and for giving me permission to cite this compelling Internet presentation.
14. Brown, *Slave Revolt in Jamaica,* May 26, 1760.
15. D. Barnjum reported to JFB1 on December 23, 1760, that he had been out in the woods during the rebellion, and that many of the ablest Mesopotamia slaves "have been obliged to attend the [scouting] Parties with the Baggage" during the rebellion (Barham c 360).
16. Higman, *Slave Populations in the British Caribbean,* 186–188.
17. Douglas Hall, *In Miserable Slavery: Thomas Thistlewood in Jamaica, 1750–1786* (Mona, Jamaica: University of the West Indies Press, 1999), 97–99; Brown, *Slave Revolt in Jamaica,* May 25–28, 1760.
18. Brown, *Slave Revolt in Jamaica,* May 29, June 2, 1760; Mesopotamia diary, June 1, 1760, UA R.15.Cb.1.3.

19. Brown, *Slave Revolt in Jamaica,* June 5, 10, 16, 1760; Mesopotamia diary, June 10, 1760, UA R.15.Cb.1.3.
20. Brown, *Slave Revolt in Jamaica,* June 22, 1760. This entry in the Mesopotamia diary was written in English instead of the missionaries' usual German, which may explain the odd locution.
21. Brown, *Slave Revolt in Jamaica,* June 25, 1760; Mesopotamia diary, June 26, 1760, UA R.15.Cb.1.3.
22. Mesopotamia diary, July 4, 11, 1760, UA R.15.Cb.1.3.
23. Barnjum to JFB1, December 23, 1760, Barham c 360.
24. Brown, *Slave Revolt in Jamaica,* November 2, 1760.
25. Mesopotamia accounts, 1764, Barham b 37.
26. Barnjum to JFB1, September 14, 1765, Barham c 357; Mesopotamia accounts, 1766, Barham b 37.
27. Barnjum to JFB1, May 8, August 30, 1767, Barham c 357. Dirt eating or geophagy is discussed in Chapter 4.
28. J. Graham to JFB2, June 27, 1791 (Barham c 357). Some of these incorrigible slaves were from Southfield. Graham wanted to transport them to the Spanish Main, but on September 15, 1791, JFB2 objected to this (Barham c 428).
29. J. Blyth and J. C. Grant to JFB2, July 3, August 27, 1820, Barham c 358; W. Ridgard to JFB2, April 16, June 6, 1824, Barham c 358.
30. Ridgard to JFB2, August 5, 1825, May 10, 12, June 9, 1826, Barham c 359.
31. *Jamaica Courant and Public Advertiser* (Kingston), January 1, 6, 10, 18, 24, February 26, April 19, June 14, 16, July 28, October 4, November 7, 22, December 29–30, 1831, University of the West Indies Library, Mona, Jamaica. This four-page newspaper was published Monday through Saturday; the first two pages were filled with ads and notices, and there were generally one to three slave runaway ads per issue.
32. *Jamaica Courant and Public Advertiser,* May 12, July 22, August 19, September 14–15, October 15, 25, November 7, December 5, 1831.
33. Higman, *Slave Population and Economy in Jamaica,* 227–230; B. W. Higman, *Montpelier, Jamaica: A Plantation Community in Slavery and Freedom, 1739–1912* (Kingston, Jamaica: University of the West Indies Press, 1998), chap. 8; Mary Turner, *Slaves and Missionaries: The Disintegration of Jamaican Slave Society, 1784–1834* (Urbana: University of Illinois Press, 1982), chap. 6.
34. Brother Zorn's diary, 1831, pp. 27–29, Fairfield H6, JA.
35. Brother Zorn's diary, 1832, Fairfield J17/1–3, JA.
36. *Jamaica Courant and Public Advertiser,* December 30, 1831; Mesopotamia diary, 1831, UA R.15.1.Cb.6.13/14; Mesopotamia diary, 1832, UA R.15.1.Cb.6.14/1.
37. Mesopotamia diary, 1832, UA R.15.1.Cb.6.14/1. Unfortunately, pages 2–4 of Ricksecker's 1832 diary are so faint that some passages are illegible.
38. Mesopotamia diary, 1832, UA R.15.1.Cb.6.14/1–2.
39. Mesopotamia diary, 1832, UA R.15.1.Cb.6.14/2–3.

40. Jamaica Slave Rebellion Courts Martial, January 20, 1832, CO 137/185/ Pt 3/594, NA-UK.
41. Jamaica Slave Rebellion Courts Martial, January 19, 1832, CO 137/185/Pt 3/589–90, NA-UK.
42. I have surveyed the January 1832 issues of the *St. Jago de la Vega Gazette* (University of West Indies Library, Jamaica) and the *Watchman and Jamaica Free Press* and *Royal Gazette* (NA-UK).
43. I have compiled these statistics from the Jamaica Slave Rebellion trial records in CO 137/185/pt. 1 and pt. 3, NA-UK. They differ somewhat from Higman's figures in *Slave Population and Economy in Jamaica,* 229.
44. Ridgard and D. Robertson to JFB2, February 10, 1832, Barham c 389.
45. Ridgard and Robertson to JFB2, February 10, March 16, December 12, 1832, Barham c 389; J.H. Buchner, *The Moravians in Jamaica,* 86. Mary Turner, in *Slaves and Missionaries,* 158–159, stresses that only a small minority of slaves acted energetically (as at Mesopotamia) to protect their estates during the rebellion.
46. Ridgard and Robertson to JB, January 31, June 4, July 5, 1833, Barham c 360. The attorneys refer to William Samuels as "Samuel Williams," but he is clearly identified as William Samuels both in the Mesopotamia inventories of 1831 and 1833 and in the Jamaica rebellion trial records.
47. Minutes of the Missions Conference, June 1831–December 1833, conference at New Eden, March 7, 1832, Fairfield Q 11/ 8–9, JA.
48. Mesopotamia diary, 1832, UA R.15.1.Cb.6.14/4, 6, 12–17. Ricksecker also complained on December 30 that "our servants were away 2 days without our leave."
49. The Mesopotamia inventory for August 1833 (JFB2's probate inventory) is in Jamaica Inventories, IB/11/3, vol. 150/25, JA.
50. See Appendix 25.
51. See Appendix 26.
52. See Appendix 29.
53. See Appendix 30.
54. See Appendix 31.
55. See Appendix 34 and Appendix 35.
56. As the war progressed, however, and planters in the Confederacy could no longer get clothing from the northern manufacturers, WHT fixed up looms to make cloth and put his slave shoemakers to work. WHT to HAT, August 1, 1862, Tayloe d 6046–6170.
57. See Appendix 37.
58. WHT to HAT Jr., March 15, 20, 1862, Tayloe d 6046–6170.
59. *Virginia Herald & Fredericksburg Advertiser,* September 25, 1794, May 12, 1795 (for George); *Virginia Argus,* May 21, 1799 (for postilion Jerry).
60. *Virginia Herald & Fredericksburg Advertiser,* June 6, 1793. No slave named Sambo lived at Mount Airy between 1808 and 1865.

61. *Virginia Herald & Fredericksburg Advertiser,* June 12, 1792, March 28, 1793, April 17, 1794, cited by Gerald W. Mullin, *Flight and Rebellion: Slave Resistance in Eighteenth-Century Virginia* (New York: Oxford University Press, 1972), 107. Bob and Duke do not appear in the 1808–1809 Mount Airy inventories.
62. Mount Airy day book, 1795–96, uncatalogued Tayloe papers (for Garrett, Isaac, and James); *Virginia Herald & Fredericksburg Advertiser,* July 27, 1798 (for Harry).
63. See Appendix 26.
64. John Arnold to JT3, March 3, 1819, Tayloe d 174.
65. Charles Tayloe to WHT, n.d. [January 1834], Tayloe d 5259–5291; HAT to WHT, December 29, 1833, Tayloe d 5849–5959.
66. WHT to HAT Jr., December 11, 1860, Tayloe d 5960–6045.
67. Elias's brother William Henry was William Tayloe's body servant.
68. BOT to WHT, May 8, 1862, Tayloe d 1771–1811. BOT himself lost three sixteen-year-old boys from Nanjemoy, who fled to the Yankees at Fort Monroe.
69. WHT to HAT Jr., August 12, 1862, Tayloe d 6046–6170.
70. In his inventories for 1861 and 1862, HAT Jr. had identified most of these people as "not worth moving." The Mount Airy census for 1863 is in Tayloe d 8539–8590.

## *9. Emancipation*

1. Entry on John Foster Barham in *The History of Parliament: The House of Commons, 1820–1832,* ed. D. R. Fisher (Cambridge: Cambridge University Press, 2009).
2. I have greatly profited from Thomas C. Holt's treatment of apprenticeship and emancipation in Jamaica, *The Problem of Freedom: Race, Labor, and Politics in Jamaica and Britain, 1832–1938* (Baltimore: Johns Hopkins University Press, 1992), chaps. 1–5. Also very helpful are Philip D. Curtin, *Two Jamaicas: The Role of Ideas in a Tropical Colony, 1830–1865* (Cambridge, MA: Harvard University Press, 1955); W. L. Burn, *Emancipation and Apprenticeship in the British West Indies* (London: Jonathan Cape, 1937), which focuses on the Jamaica special magistrates; Sidney W. Mintz, *Caribbean Transformations* (Chicago: Aldine, 1974), chaps. 5–7, which focuses on the rise of the Jamaican peasantry; Edward Bartlett Rugemer, *The Problem of Emancipation: The Caribbean Roots of the American Civil War* (Baton Rouge: Louisiana State University Press, 2008), which traces the impact of British Caribbean emancipation in the United States; Douglas Hall, *Free Jamaica, 1838–1865: An Economic History* (New Haven: Yale University Press, 1959); and William A. Green, *British Slave Emancipation: The Sugar Colonies and the Great Experiment, 1830–1865* (Oxford: Clarendon, 1976).
3. Attorney Duncan Robertson wrote JB on November 25, 1833, that it was impossible to convey the atmosphere of excitement in Jamaica, where de-

spite great opposition the assembly was almost certainly going to pass the parliamentary bill for the total abolition of slavery in order to claim their "pittance of compensation" from the home government (Barham c 360).

4. I have found no satisfactory account of why the Jamaica Assembly reduced required weekly labor from forty-five to forty and a half hours per week. W.L. Burn briefly discusses the issue in *Emancipation and Apprenticeship,* 171. The other colonies seem to have accepted the parliamentary formula. For example, the St. Vincent Act for the Abolition of Slavery, passed on April 2, 1834, specified a maximum of forty-five hours of unpaid labor for apprentices; see Roderick A. McDonald, ed., *Between Slavery and Freedom: Special Magistrate John Anderson's Journal of St. Vincent during the Apprenticeship* (Philadelphia: University of Pennsylvania Press, 2001), 218.
5. W. Ridgard and D. Robertson to JB, December 18, 1833, July 31, 1834, Barham c 360; Accounts Produce, vol. 73/152; vol. 75/12; vol. 76/39, JA.
6. Sligo to the Jamaica slaves, April 10, ca. May 1834; Sligo to the Colonial Office, August 12, 1834, C.O. 137/192/148, 197, 435, NA-UK.
7. Ridgard and Robertson to JB, September 13, December 29, 1834, Barham c 360.
8. Hankey, Plummer, and Wilson to JB, September 3, October 27, 1834, Barham c 364, bundle 4.
9. JB, "Copies [of] WB's Jamaica Papers," December 1834; memorandum on "Moravian Missionaries," n.d., memorandum on "Attornies," n.d.; Barham c 373. JB characterized attorney Ridgard at Mesopotamia as "a careful plodding man of business a good Planter but not over bright, and quite at sea when put out of his daily routine." He saw Robertson at Island as "a shrewd Scotchman" who didn't visit the estate often "and does his duty too much by deputy."
10. Ridgard and Robertson to JB, December 29, 1834, January 21, April 11, May 12, 1835, Barham c 360.
11. Robertson to JB, May 19, July 3, 1835, Barham c 360. Robertson reported that the Mesopotamia people also became more difficult to manage after William Barham departed for England.
12. JB to William Barham, December 26, 1835, Barham c 373.
13. Ridgard and Robertson to JB, December 29, 1835, Barham c 360.
14. S.M. Edward Philp to Gov. Sligo, March 17, 1835, C.O. 137/198/152–153.
15. Holt, *The Problem of Freedom,* 63–66.
16. McDonald, *Between Slavery and Freedom,* 102, 118–119, 128–130, 156–160, 191–193. For the special magistrates in Jamaica, see Burn, *Emancipation and Apprenticeship,* chap. 5.
17. On January 2, 1834, Oliver wrote to Secretary Stanley in the Colonial Office, very anxious about his application for a job because he needed to support two sons in school (C.O. 137/196/140, NA-UK). I do not know Oliver's date of dismissal, but his successor, George Gordon, was at work in January 1838.

18. C.O. 137/216/91–82, 139–140, 339–340; C.O. 137/222/140. Oliver's reports also show that in 1835 about 40 percent of the estates he visited had complaints, but by 1836 only about 15 percent had complaints.
19. Governor Sir Lionel Smith reported to the Colonial Office in an undated memorandum in 1838 that Oliver was a "Bad Magistrate—so involved he has been lately several Months in Prison and took the Benefit of the Insolvent [Debtors] Act" (C.O. 137/231/290–91).
20. C.O. 137/222/7 lists Jamaica sugar export figures in hogsheads for seven years during 1772–1827 compared with 1832–1836, and Douglas Hall lists Jamaica sugar export figures in hundredweight during 1831–1838 (in *Free Jamaica, 1838–1865,* 270). When these lists are set against each other, it appears that Jamaica annually shipped about 115,000 hogsheads during 1799–1817, 84,235 hogsheads during 1831–1834, and 63,750 hogsheads during 1835–1838.
21. Jamaica sugar exports in 1838 were almost identical to exports in 1836 (about 62,000 hogsheads). Since Mesopotamia produced 211 hogsheads in 1836 (Accounts Produce, 78/157–158, JA), the 1838 crop was likely very similar.
22. Gordon reported on January 2, 1838, "There are to my knowledge establishments upon Carawina, Mesopotamia, Petersfield and Friendship Estates for the instruction of the young," and he thought that children from neighboring estates might also be attending these schools (C.O. 137/231 58–59).
23. Burn, *Emancipation and Apprenticeship,* chaps. 7–9; Holt, *The Problem of Freedom,* chaps. 2–3.
24. C.O. 137/231/302; C.O. 137/232/74, 103–106, 151.
25. George Gordon and Thomas Abbott, "RETURN of Properties in the Parish of Westmorland—29 October 1838," C.O. 137/232/170.
26. C.O. 137/241/84; C.O. 137/242/173.
27. It is of course probable that some of the ejected people had adopted new surnames. For example, Rachel Howell in 1838 may have been Rachel Myrie in 1833, with a new married surname. But Thomas Morning and Thomas Allen must have been newcomers, because no adult slave was named Thomas in 1833. Four of the ejectees were named Marianne or Mary Ann, and there was only one adult named Mary Ann on the 1833 slave list, which indicates that at least three more of the ejectees were new to Mesopotamia.
28. At Frome estate in February 1839, only thirteen of the forty-eight crop workers were formerly apprentices at Frome (C.O. 137/242/171–172, 288).
29. C.O. 137/242/136, 292–293.
30. Accounts Produce, 82/206–207, JA.
31. Accounts Produce, 82/85 [1840]; 86/84–85 [1841]; 87/64–65 [1842]; 88/173–174 [1843]; 90/5–6 [1844]; 92/6–7 [1846]; 94/85–87 [1847–1850]; 95/193–194 [1851–1853], JA.
32. JB to Hankey, Plummer, and Wilson, November 12, 1835, Barham c 364, bundle 7.

33. Hall, *Free Jamaica,* 21–22.
34. Accounts Produce, 87/64–65 [1842]; 88/173–174 [1843]; 90/5–6 [1844], JA; B.W. Higman, ed., *The Jamaican Censuses of 1844 and 1861: A New Edition Derived from the Manuscript and Printed Schedules in the Jamaica Archives* (Mona, Jamaica: University of the West Indies, 1980), 4–6, 31–39.
35. Accounts Produce, 94/85 [1847], JA; Hall, *Free Jamaica,* 272. Mesopotamia may also have imported a few contract workers from West Africa; in 1849 William Johns charged J.W. Foster (very likely the Joseph Foster mentioned in 1843) a tax of £2 on his "African boy" (Accounts Produce, 94/86 [1849], JA).
36. JFB2 to F. Zincke, June 4, 1806, Barham c 428; JFB2's papers on Chinese labor in the West Indies are in Barham c 366.
37. Accounts Produce, 98/12–13 [1864], JA; Hall, *Free Jamaica,* 274; Holt, *The Problem of Freedom,* 197, 200, 289–290.
38. At Mesopotamia, after the Barhams sold the place, sugar production increased in the late nineteenth century. By 1883, when there were twenty-four sugar estates operating in Westmoreland (compared with sixty-five in 1800), Mesopotamia was shipping 298 hogsheads of sugar and 273 puncheons of rum—quantities that equaled the largest sugar shipment of the slave era in 1810; see *The Handbook for Jamaica for 1884–1885* (Kingston, 1884), 399. The Barhams' Island estate in St. Elizabeth parish abandoned sugar production in 1881.
39. I derive these estimates from Curtin, *Two Jamaicas,* 250, and Higman, *The Jamaican Censuses of 1844 and 1861,* 1.
40. See Mintz, *Caribbean Transformations,* chaps. 5, 7; and Roderick A. McDonald, *The Economy and Material Culture of Slaves: Goods and Chattels on the Sugar Plantations of Jamaica and Louisiana* (Baton Rouge: Louisiana State University Press, 1993), chap. 1.
41. J.C. Grant and J.R. Webb to JFB2, September 4, 1809, Barham c 358, cited by McDonald, *The Economy and Material Culture of Slaves,* 18.
42. Michael Craton, *Searching for the Invisible Man: Slaves and Plantation Life in Jamaica* (Cambridge, MA: Harvard University Press, 1978), 277–279, 283, 287, 290. Worthy Park collected £342 in rent from its workers in 1839, but only £50 by 1842. At Mesopotamia, rent and pasturage peaked in 1842–1844, but William Johns still collected £98 per annum in 1846–1849.
43. Thomas Carlyle, "Occasional Discourse on the Negro Question," *Fraser's Magazine for Town and Country* (1849): 528–529, 535–536.
44. Holt, *The Problem of Freedom,* 165, 173.
45. Higman, *The Jamaican Censuses of 1844 and 1861,* 1, 16.
46. WHT to Courtenay Tayloe, June 8, 1865, Tayloe d 5292–5327.
47. For background, see Stephanie McCurry, *Confederate Reckoning: Power and Politics in the Civil War South* (Cambridge, MA: Harvard University Press, 2010), prologue, chaps. 6–8; Eric Foner, *Reconstruction: America's Unfinished*

*Revolution, 1863–1877* (New York: Harper and Row, 1988), chaps. 2–5; and Leon F. Litwack, *Been in the Storm So Long: The Aftermath of Slavery* (New York: Vintage, 1980), chaps. 3–5. Two books look at Alabama from opposing perspectives: Peter Kolchin, *First Freedom: The Responses of Alabama's Blacks to Emancipation and Reconstruction* (1972; reissued Tuscaloosa: University of Alabama Press, 2008), is a pioneering interpretation from the Alabama ex-slaves' point of view, while Jonathan M. Wiener, *Social Origins of the New South: Alabama, 1860–1885* (Baton Rouge: Louisiana State University Press, 1978), focuses on the white planters.

48. WHT's accounts, November 9, December 13, 1865, Tayloe d 20360–20506.
49. WHT to HAT Jr., June 22, 1865, Tayloe d 6046–6170.
50. WHT to HAT Jr., October 21, November 13, December 4, 1865, Tayloe d 6046–6170.
51. WHT, "Classification of Hands Nov 1865," Tayloe d 13450.
52. WHT wrote accounts of his quarrel with Ramey in a letter to HAT Jr., December 20, 1865, Tayloe d 6046–6170; in a letter to W.W. Gwathmey, May 24, 1867, Tayloe d 3288–3366; and in an 1868 memorandum, Tayloe d 2123–2135.
53. WHT compiled several lists of the freed people who went with Ramey, who went to Booth, to Pool, and to other places, and who stayed at Larkin (Tayloe d 6171–6296; d 13453). In January 1866 the remaining Larkin hands were paid $421 for their work in 1865, while the Oakland hands received $862 (Tayloe d 20360–29506).
54. WHT's census of the Oakland hands in early 1866 is in Tayloe d 13450; his list of the people who left Oakland in December 1865 is in Tayloe d 13453.
55. WHT to Courtenay Tayloe, March 23, 1866, Tayloe d 5292–5327.
56. Foner, *Reconstruction,* chaps. 5–9.
57. Mary Gwathmey was a daughter of WHT's brother George Tayloe.
58. Emma Tayloe had married Thomas Munford in 1866; his first wife, who died in 1863, was Mary Gwathmey's sister and Emma's cousin.
59. WHT's will is in Tayloe d 8683. Daughter Sophia had been mainly provided for at her marriage in 1856; in 1869 WHT gave her a house on Pennsylvania Avenue in Washington, D.C., and land in Perry County, Alabama, near Larkin. WHT's income tax returns show that he had taxable income of $35,595 in 1867 and $33,214 in 1868, but only $13,610 in 1869 and $10,044 in 1870 (Tayloe d 13,453; d 20507–20660).
60. Kolchin, *First Freedom,* 39.
61. W.W. Gwathmey to WHT, August 13, 1866, Tayloe d 3288–3366; Mary Gwathmey to WHT, December 28, 1866, Tayloe d 3284–3287.
62. WHT notes the Yeatmans' departure from Larkin in Tayloe d 13453.
63. WHT notes the Dixons' departure from Oakland in Tayloe d 13453; Mary Gwathmey to WHT, December 28, 1866, Tayloe d 3284–3287; W.W. Gwathmey to WHT, February 5, 1868, labels Holcroft "an infernal scoundrel" (Tayloe d 3288–3366).

64. T.T. Munford to WHT, December 17, 1869, Tayloe d 4474–4548.
65. WHT to HAT, October 21, December 20, 1865, Tayloe d 6046–6170; WHT to W.W. Gwathmey, May 24, 1867, Tayloe d 3288–3366.
66. W.W. Gwathmey to WHT, September 10, 1866, Tayloe d 3288–3366; WHT's account book, 1855–1871, Tayloe d 13453.
67. "Paid Hands at Oakland—1866," Tayloe d 8632–8667.
68. Kolchin, *First Freedom*, 39.
69. W.W. Gwathmey to WHT, July 12, 1867, January 4, 1868, Tayloe d 3288–3366.
70. J. Taylor to WHT, January 11, 1869, Tayloe d 6424–6462; T.T. Munford to WHT, November 7, 1869, Tayloe d 4474–4548.
71. W.W. Gwathmey to WHT, October 28, December 5, 1866, Tayloe d 3288–3366.
72. W.W. Gwathmey to WHT, July 12, August 5, 1867, February 5, 1868, Tayloe d 3288–3366; John Chapman to WHT, February 11, 1868, Tayloe d 2742–2745.
73. WHT to HAT Jr., February 7, 1866, Tayloe d 6171–6296; WHT to Courtenay Tayloe, March 23, 1866, Tayloe d 5292–5327.
74. Mary Gwathmey to WHT, June 10, December 28, 1866, Tayloe d 3284–3287; W.W. Gwathmey to WHT, February 5, 1867, Tayloe d 3288–3366.
75. W.W. Gwathmey to WHT, July 12, August 10, 25, 1867, Tayloe d 3288–3366.
76. Arthur Thomas lived at Oakland in 1870 with his wife and two sons, and was unusual among the Tayloe freed people in having personal property worth $150, according to the census.
77. W.W. Gwathmey to WHT, December 5, 1866, August 5, November 9, 1867, October 25, 1868, Tayloe d 3288–3366.
78. Emma Munford to WHT, January 29, 1869, Tayloe d 4461–4469; T.T. Munford to WHT, January 30, December 17, 1869, Tayloe d 4474–4548.
79. This includes retired spinner Annie Flood, aged seventy-five in 1863, who was also chargeable because she was blind. See Appendix 37.
80. WHT to Courtenay Tayloe, June 8, 1865, February 20, 1866, Tayloe d 5292–5327.
81. HAT Jr. account book, 1862–1868, Tayloe d 23,706; HAT Jr. account book, 1861–1866, Tayloe d 23707; HAT Jr. account book, 1866, Tayloe d 23709; further contracts in Tayloe d 24270–24385.
82. See Appendix 22.
83. HAT Jr. to WHT ca. January 1867, Tayloe d 6046–6170; Indenture, January 1, 1867; "Allowance List," 1868, HAT Jr. account book, 1862–1868, Tayloe d 23706; HAT Jr. to WHT, January 22, February 14, 1871, Tayloe d 6171–6296.
84. WHT to HAT Jr., February 7, 1866, Tayloe d 6171–6296; WHT to Courtenay Tayloe, February 20, March 23, 1866, Tayloe d 5292–5327; HAT Jr. account book, 1862–1868, Tayloe d 23706; W.W. Gwathmey to WHT,

December 5, 1866, Tayloe d 3288–3366. I would love to see Glascow's letter to Hilliard, which I have not found in the Tayloe Papers.

85. WHT to HAT Jr., January 25, 1867, Tayloe d 5960–6045; HAT Jr. to WHT, March 25, 1870, February 14, 1871, Tayloe d 6171–6296. Tom Thomas, a carpenter at Mount Airy, was said to laugh at "the idea of anyone from Alabama returning here if they are doing anything there" (HAT Jr. to WHT, September 18, 1870, Tayloe d 6171–6296).
86. HAT Jr. to WHT, March 25, 1870, February 14, 1871, Tayloe d 6171–6296.
87. *Ninth Census, United States, 1870, Instructions to Assistant Marshalls* (Washington DC: Government Printing Office, 1870), 5–11.
88. The 243 Tayloe freed people I found are in National Archives M593, microfilm roll 18 (Hale, Alabama), roll 28 (Marengo, Alabama), roll 33 (Perry, Alabama), rolls 124, 126 (Washington, D.C.), and roll 1674 (Richmond, Virginia). Since Ancestry.com puts primacy on date and place of birth, faulty age statements are a big problem, and I often had better luck finding a family by tracking the children, whose stated ages are generally more accurate than the ages of their parents.
89. Before the Civil War, Oakland had been in Marengo County on the western border of Perry County. In 1867 Hale County was established, incorporating the northern section of Marengo, where Oakland was situated. Oakland was now on the Hale/Perry border.
90. Householder #542 was a white teacher; #555 was John Munford, who was managing Oakland for Thomas Munford; #556 was a white mechanic; #570 was Charles Weaver, a black laborer who was not a former Tayloe slave; and #572 was a white farm agent. All the intervening twenty-six householders, #543–554, 557–569, and 571, were ex-Tayloe slaves.
91. The black workers' cabins at Larkin were dwellings #123–158 in township 17, route 6, Perry County, Alabama. Tayloe freed people occupied all of these cabins except #126, #131, #134, #144–149, #153, and #156.
92. WHT to HAT Jr., May 23, 1857, Tayloe d 6046–6170.
93. WHT to Courtney Tayloe, March 23, 1866, Tayloe d 5292–5327; W.W. Gwathmey to WHT, July 9, 1866, Tayloe d 3288–3366.
94. WHT to [?], July 16, 1858, Tayloe d 2123–2135.
95. Joseph Taylor to WHT, April 30, 1867, Tayloe d 6424–6462; HAT Jr. to WHT, March 25, 1870, February 14, 1871, Tayloe d 6171–6296.
96. WHT to HAT Jr., July 4, 1862, Tayloe d 6046–6170.
97. WHT to HAT Jr., May 28, 1862, Tayloe d 6046–6170; WHT to Courtenay Tayloe, May 3, 1864, Tayloe d 5292–5327.
98. Edward Thornton Tayloe Jr. to WHT, January 11, 1870, Tayloe d 5483–5600.
99. Mintz, *Caribbean Transformations*, 155.

# *Acknowledgments*

In the forty years that I have been working on this book, I have received generous support from far more people than can be listed in this brief space. It has been a great privilege to talk and argue about the history of slavery in the Americas with many of the leading figures in the field, including Edward Baptist, Hilary Beckles, Ira Berlin, Timothy Breen, Vincent Brown, Trevor Burnard, John Coombs, Emilia da Costa, Michael Craton, Philip Curtin, Paul David, David Brion Davis, Seymour Drescher, Stanley Elkins, David Eltis, Stanley Engerman, Robert Engs, Drew Faust, Robert Fogel, John Hope Franklin, David Galenson, Barry Gaspar, Eugene Genovese, Jack Greene, Herbert Gutman, Sheldon Hackney, Jerome Handler, Barry Higman, Ronald Hoffman, Rhys Isaac, Winthrop Jordan, Alan Kulikoff, Lawrence Levine, Leon Litwack, Roderick McDonald, James McPherson, Joseph Miller, Walter Minchinton, Sidney Mintz, Kenneth Morgan, Philip Morgan, Gerald Mullin, Gary Nash, Sydney Nathans, Nell Painter, Orlando Patterson, Jacob Price, Arnold Rampersad, Marcus Rediker, Stuart Schwartz, Richard Sheridan, Lorena Walsh, James Walvin, Peter Wood, and Gavin Wright. Out of this maelstrom of opinion I have tried to forge my own approach.

I am enormously grateful for fellowship support during the formative years of this project to the National Endowment for the Humanities, the Institute for Advanced Study at Princeton, the American Council of Learned Societies, and the Center for Advanced Study in the Behavioral Sciences at Stanford. I am also deeply indebted to the staffs of the libraries and archives where I did most of my research. I worked on Mesopotamia principally in the Bodleian Library at the University of Oxford and in the Jamaica Archives at Spanish Town, but also in the British National Archives at Kew, the British Library, the Lincolnshire County Archives in Britain, and (via microfilm) the Unitätsarchiv in Herrnhut, Germany. I worked on Mount Airy principally in the Virginia Historical Society, but also in the Library of Virginia, the University of Virginia Library, and the U.S. National Archives. A year at Oxford as the Harmsworth Professor of American History enabled me to complete my research in the Barham Papers, and the microfilm edition of the Tayloe Papers has opened new angles of research that I had not thought of when working with the manuscripts at Richmond. I am also indebted to the staffs of the Newberry Library, the American Antiquarian Society, the John Carter Brown Library, the Library Company of Philadelphia, the Schlesinger Library at Radcliffe, and the libraries of the University of Pennsylvania, Princeton University, Stanford University, and Harvard University.

In my years at the Penn History Department, a host of students have helped me to shape this book, in particular Roseanne Adderley, Dee Andrews, Edward Baptist, Todd Barnett, Rose Beiler, John Bezis-Selfa, Nicholas Canny, Nicole Eustace, Alison Games, Joseph Illick, Ned Landsman, Barry Levy, Marion Nelson, Marcus Rediker, Liam Riordan, Beverly Smaby, Karim Tiro, and Joseph Torsella. Similarly, colleagues at the McNeil Center for Early American Studies who have taken active interest in my project include Richard Beeman, Wayne Bodle, Vincent Brown, Trevor Burnard, Elaine Crane, Aaron Fogleman, James Green, Douglas Hamilton, Dallett Hemphill, Ruth Herndon, Maurice Jackson, Susan Klepp, Michelle and Roderick McDonald, Kenneth Morgan, John Murrin, Simon Newman, Andrew O'Shaughnessy, William Pencak, Rosalind Remer, Daniel Richter, Billy Smith, Patrick Spero, John Wood Sweet, Camilla Townsend, David Waldstreicher, and Michael Zuckerman. The McNeil Center, with its annually changing community of young scholars, is the perfect venue for trying out new strategies, and I have presented several of my draft chapters at our Friday seminars. I take this opportunity to express my deep gratitude to the late Robert L. McNeil Jr. for his splendid support of the McNeil Center, and I also extend my thanks to Dan Richter for his outstanding leadership of the Center since my retirement in 2000.

As I began to pull my book together, I benefited greatly from appearances at Alan Kulikoff's Georgia Workshop, and at Early American seminars at the Massachusetts Historical Society and at Columbia University. I have presented my ideas to Caribbean scholars in conferences at Curaçao and Barbados. And I especially thank Edward Baptist for inviting me to give the Carl Becker lectures at Cornell University, which spurred me into redesigning my concluding chapters.

Portions of Chapter 2 were previously published in "The Story of Two Jamaican Slaves: Sarah Affir and Richard McAlpine of Mesopotamia Estate," in *West Indies Accounts: Essays on the History of the British Caribbean and the Atlantic Economy in Honour of Richard Sheridan*, ed. Roderick A. McDonald (Kingston, Jamaica: University of the West Indies Press, 1996), 188–210; portions of Chapter 3 were previously published in "Winney Grimshaw, a Virginia Slave, and Her Family," *Early American Studies* 9 (2011): 493–521; and portions of Chapter 5 were previously published in "After Tobacco: The Slave Labour Pattern on a Large Chesapeake Grain-and-Livestock Plantation in the Early Nineteenth Century," in *The Early Modern Atlantic Economy*, ed. John J. McCusker and Kenneth Morgan (Cambridge: Cambridge University Press, 2000), 344–363.

I have also published five other essays that explore various aspects of my project: "A Tale of Two Plantations: Slave Life at Mesopotamia in Jamaica and

Mount Airy in Virginia, 1799 to 1828," *WMQ*, 3rd ser., 34 (January 1977): 32–65; "'Dreadful Idlers' in the Cane Fields: The Slave Labor Pattern on a Jamaican Sugar Estate, 1762–1831," *Journal of Interdisciplinary History* 17 (Spring 1987): 795–822; "Sugar Production and Slave Women in Jamaica," in *Cultivation and Culture: Labor and the Shaping of Slave Life in the Americas*, ed. Ira Berlin and Philip D. Morgan (Charlottesville: University Press of Virginia, 1993), 49–72; "Moravian Missionaries at Work in a Jamaican Slave Community, 1754–1835," James Ford Bell Lecture no. 32, The Associates of the James Ford Bell Library, University of Minnesota, 1994; and "The Demographic Contrast between Slave Life in Jamaica and Virginia, 1760–1865," *Proceedings of the American Philosophical Society* 151 (2007): 43–60.

I wish to pay special thanks to Gwynne Tayloe for inviting me to visit Mount Airy and meet his family. I also thank John Pulis and Barry Higman for showing me key sources in the Jamaica Archives, Ian Foster for sharing valuable information about Mesopotamia, Aaron Fogleman for help with the Moravian Archives at Herrnhut, and Kenneth Morgan for help with the National Archives at Kew. Katharine Gerbner has been extraordinarily generous in sharing her research findings on the founding of the Moravian mission at Mesopotamia, to the great benefit of Chapter 6. Roderick McDonald has brought his deep knowledge of Jamaican slave life to bear on my Mesopotamia chapters, and while we disagree on some issues, his careful scrutiny has been hugely helpful. Gwendolyn Jensen, another dear friend, has read every draft chapter in turn with her poet's eye, offering valuable interpretive commentary and catching stylistic glitches. Joyce Seltzer, editor extraordinaire, has put up with my first-person mode of presentation and my barrage of charts and tables, and showed me how to strengthen the framework for my argument. Brian Distelberg has been extremely helpful in preparing the manuscript for publication. Barb Goodhouse has done a terrific job of copyediting my complex manuscript, and Melody Negron has expertly scrutinized all of the tricky additions and corrections to the text. The two reviewers for the Press, Edward Rugemer and Vincent Brown, have both been enormously encouraging and constructive, pointing out where I needed to revise and where I needed to expand.

The seven extensive slave family trees that I have produced for this book pose a problem that I did not know how to solve because they are far too large for the printed page. Fortunately for me, Vincent Brown has come to my rescue. As Director of the History Design Studio at Harvard University, he has engaged Sean Treacy, Alec Harrison, and Luke Peters of Grafton Studio to design and build a website, *www.twoplantations.com*, that presents the 430 members of these seven multigenerational slave families in an exciting digital format. JT White of the History Design Studio has also helped to push this project.

My two daughters have long supported me in this project. Rebecca has read the draft manuscript on her computer, and Ceci on her Kindle. Rebecca has helped me devise the population pyramids in Chapter 1 and has constructed the family trees in Appendix 12 and Appendix 13. And my beloved wife, Mary

Maples Dunn, has done everything within her power to get me across the finish line. Not wishing me to end up like Mr. Casaubon in *Middlemarch,* she has become my Dorothea, engaging in every angle of the story of Mesopotamia and Mount Airy, critiquing every line of every draft of every chapter—as always, my comrade and partner, and the mainstay of my life.

# Index